MIMO RADAR
SIGNAL PROCESSING

MIMO RADAR SIGNAL PROCESSING

Edited by

JIAN LI
PETRE STOICA

WILEY

A JOHN WILEY & SONS, INC., PUBLICATION

Published by John Wiley & Sons, Inc., Hoboken, New Jersey
Published simultaneously in Canada

For general information on our other products and services or for technical support, please contact our Customer Care Department within the United States at (800) 762-2974, outside the United States at (317) 572-3993 or fax (317) 572-4002.

Wiley also publishes its books in variety of electronic formats. Some content that appears in print may not be available in electronic format. For more information about Wiley products, visit our web site at www.wiley.com.

Library of Congress Cataloging-in-Publication Data:

Li, Jian
MIMO radar signal processing / Jian Li, Petre Stoica.
p. cm.
Includes bibliographical references and index.
ISBN 978-0-470-17898-0
1. Radar. 2. MIMO systems. I. Stoica, Petre. II. Title.
TK6575.L64 2008
621.3848- -dc22 2008022134

10 9 8 7 6 5 4 3 2 1

CONTENTS

PREFACE

Multiple-input multiple-output (MIMO) radar has been receiving increasing attention in recent years from researchers, practitioners, and funding agencies. MIMO radar is characterized by using multiple antennas to simultaneously transmit diverse (possibly linearly independent) waveforms and by utilizing multiple antennas to receive the reflected signals. Like MIMO communications, MIMO radar offers a new paradigm for signal processing research. MIMO radar possesses significant potentials for fading mitigation, resolution enhancement, and interference and jamming suppression. Fully exploiting these potentials can result in significantly improved target detection, parameter estimation, as well as target tracking and recognition performance.

The objective of this contributed book is to introduce more recent developments on MIMO radar, to stimulate new concepts, theories, and applications of the topic, and to foster further cross-fertilization of ideas with MIMO communications. This book, which is the first to present a coherent picture of the MIMO radar topic, includes an excellent list of contributions by distinguished authors from both academia and research laboratories.

The book is organized as follows. The first seven chapters focus on the merits of the waveform diversity, allowed by transmit and receive antenna arrays containing elements that are collocated, to improve the radar performance, while the last three chapters exploit the diversity, offered by widely separated transmit/receive antenna elements, to achieve performance gains.

Chapter 1, by J. Li (University of Florida) and P. Stoica (Uppsala University), shows that waveform diversity enables MIMO radar superiority in several fundamental aspects, including improved parameter identifiability, direct applicability of many adaptive as well as parametric techniques to the received data to improve target detection and parameter estimation performance, and better flexibility of transmit beampattern designs. Chapter 2, by K. W. Forsythe and D. W. Bliss (MIT Lincoln Laboratory), provides an interesting historical review of radar as well as the current state of the MIMO radar research. This chapter also covers a wide range of fundamental topics from MIMO virtual aperture, performance bounds, and waveform optimization, to minimum detectable velocity of MIMO ground moving-target indicator (GMTI) radars. Chapter 3, by G. San Antonio, D. R. Fuhrmann (Washington University), and F. C. Robey (MIT Lincoln Laboratory), addresses a basic radar issue: how to extend the conventional Woodward's ambiguity function to MIMO radars. The MIMO ambiguity functions provided in the chapter can simultaneously characterize the effects of array geometry, transmitted waveforms, and target scattering on resolution performance. Chapter 4, by J. Tabrikian (Ben-Gurion University of the Negev) presents performance bounds and techniques for target localization using MIMO radar. Insights into and properties of the target localization techniques are also given. Chapter 5, by B. Friedlander (University of California at Santa Cruz), considers waveform design, based on target and clutter statistics, to improve the radar target detection and parameter estimation performance. Chapter 6, by C.-Y. Chen and P. P. Vaidyanathan (California Institute of Technology), focuses on fast-time MIMO spacetime adaptive processing (STAP) and provides new algorithms to fully utilize the geometry and structure of the covariance matrix of the jammer and clutter to achieve reduced computational complexity while maintaining a good signal-to-interference-and-noise ratio (SINR). Chapter 7, by V. F. Mecca (Duke University) and D. Ramakrishnan (Qualcomm Inc.), F. C. Robey (MIT Lincoln Laboratory), and J. L. Krolik (Duke University), is concerned with slow-time MIMO spacetime adaptive processing and its application to over-the-horizon radar clutter mitigation. The waveform orthogonality is achieved by phase coding from pulse-to-pulse (and hence the term "slowtime"), which has the important advantage of hardware implementation simplicity.

Chapter 8, by H. D. Griffiths (Cranfield University), C. J. Baker and P. F. Sammartino (University College London), and M. Rangaswamy (Air Force Research Laboratory), studies the performance and utilities of distributed MIMO radar networks that exploit the target scintillation as an advantage, and provides insights into the MIMO framework as applied to radar. Chapter 9, by H. Godrich, and A. M. Haimovich (New Jersey Institute of Technology) and R. S. Blum (Lehigh University), contains a comprehensive overview of the concepts and applications of a MIMO radar system with widely separated antennas. This chapter also discusses ambiguity functions and performance bounds, as well as techniques for high-resolution target localization. Finally, Chapter 10, by A. De Maio (Università degli Studi di Napoli "Federico II") and M. Lops (Università degli Studi di Cassino), concentrates on developing statistical MIMO techniques via optimizing spacetime code matrices and on providing useful insights into the interplay between detection performance and code matrix choice.

We are grateful to the authors who have contributed the chapters of this book for their excellent work. We would also like to acknowledge the contributions of several other people and organizations to the completion of this book. Most of our work in the area of waveform diversity exploitation and its applications to MIMO radar and biomedical engineering has been an outgrowth of our research programs in array signal processing. We would like to thank those who have supported our research in this area: the National Science Foundation (NSF), the Office of Naval Research (ONR), the Army Research Office (ARO), the Defense Advanced Research Projects Agency (DARPA), and the Swedish Science Council (VR). We also wish to thank George Telecki (Associate Publisher) and Melissa Valentine as well as Rachel Witmer (Editorial Assistants) at Wiley for their efforts on the publication of this book. Finally, we gratefully acknowledge Mr. Xing Tan, who helped us put this book together.

JIAN LI AND PETRE STOICA

CONTRIBUTORS

C. J. Baker, Department of Electronic and Electrical Engineering, University College, London, WC1E TJE, UK

Daniel W. Bliss, MIT Lincoln Laboratory, Lexington, MA 02420

Rick S. Blum, Department of Electrical and Computer Engineering, Lehigh University, Bethlehem, PA 18015

Chun-Yang Chen, Department of Electrical Engineering, California Institute of Technology, Pasadena, CA 91125

Antonio De Maio, Università degli Studi di Napoli "Federico II," DIET Via Claudio 21, I-80125 Napoli, Italy

Keith W. Forsythe, MIT Lincoln Laboratory, Lexington, MA 02420

Benjamin Friedlander, Department of Electrical Engineering, University of California, Santa Cruz, CA 95064

Daniel R. Fuhrmann, Department of Electrical and System Engineering, Washington University, St. Louis, MO 63130

Hana Godrich, Department of Electrical and Computer Engineering, New Jersey Institute of Technology, Newark, NJ 07102

H. D. Griffiths, DCMT, Shrivenham, Cranfield University, Shrivenham, Swindon, SN6 8LA, UK

Alexander M. Haimovich, Department of Electrical and Computer Engineering, New Jersey Institute of Technology, Newark, NJ 07102

Jeffrey L. Krolik, Department of Electrical and Computer Engineering, Duke University, PO Box 90291, Durham, NC 27708

Jian Li, Department of Electrical and Computer Engineering, University of Florida, Gainesville, FL 32611

Marco Lops, Università degli Studi di Cassino, DAEIMI Via Di Biasio 43, I-03043 Cassino, Italy

Vito F. Mecca, Department of Electrical and Computer Engineering, Duke University, PO Box 90291, Durham, NC 27708

Dinesh Ramakrishnan, Audio Systems, Qualcomm Inc., 5775 Morehouse Dr, San Diego, CA, 92121

M. Rangaswamy, Air Force Research Laboratory (AFRL) Sensors Directorate, Hanscom Air Force Base, MA 01731

Frank C. Robey, MIT Lincoln Laboratory, Lexington, MA 02420

P. F. Sammartino, Department of Electronic and Electrical Engineering, University College, London, WC1E 7JE, UK

Geoffrey San Antonio, Department of Electrical and System Engineering, Washington University, St. Louis, MO 63130

Petre Stoica, Information Technology Department, Uppsala University, PO Box 337, SE-751 05 Uppsala, Sweden

Joseph Tabrikian, Department of Electrical and Computer Engineering, Ben-Gurion University of the Negev, Beer-Sheva, 84105, Israel

P. P. Vaidyanathan, Department of Electrical Engineering, California Institute of Technology, Pasadena, CA 91125

1

MIMO RADAR — DIVERSITY MEANS SUPERIORITY

JIAN LI

Department of Electrical and Computer Engineering, University of Florida, Gainesville

PETRE STOICA

Information Technology Department, Uppsala University, Uppsala, Sweden

1.1 INTRODUCTION

MIMO radar is an emerging technology that is attracting the attention of researchers and practitioners alike. Unlike a standard phased-array radar, which transmits scaled versions of a single waveform, a MIMO radar system can transmit via its antennas multiple probing signals that may be chosen quite freely (see Fig. 1.1). This waveform diversity enables superior capabilities compared with a standard phased-array radar. In Refs. 1–4, for example, the diversity offered by widely separated transmit/receive antenna elements was exploited. Many other papers, including, for instance, Refs. 5–29, have considered the merits of a MIMO radar system with collocated antennas. The advantages of a MIMO radar system with both collocated and widely separated antenna elements are investigated in Ref. 30.

For collocated transmit and receive antennas, the MIMO radar paradigm has been shown to offer higher resolution (see, e.g., Refs. 6 and 9), higher sensitivity to detecting slowly moving targets [11], better parameter identifiability [15,18], and direct applicability of adaptive array techniques [15,26,27]. Waveform optimization has also been shown to be a unique capability of a MIMO radar system. For example, it has been used to achieve flexible transmit beampattern designs

MIMO Radar Signal Processing, edited by Jian Li and Petre Stoica

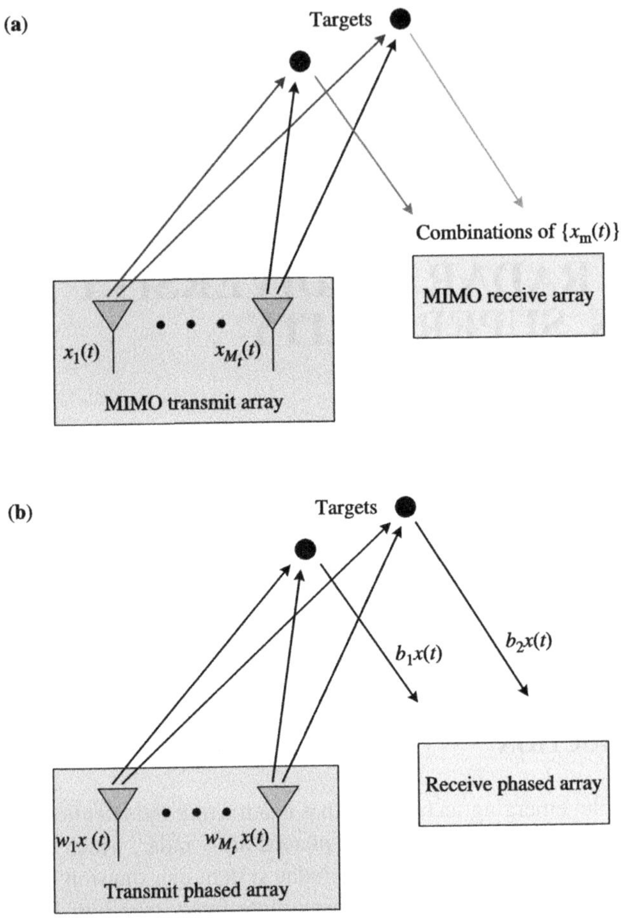

Figure 1.1 (a) MIMO radar versus (b) phased-array radar.

(see, e.g., Refs. 5, 12, 13, 17, and 23) as well as for MIMO radar imaging and parameter estimation [10,19,29].

In this chapter we present more recent results showing that this waveform diversity enables the MIMO radar superiority in several fundamental aspects. We focus on the case of collocated antennas. Without loss of generality, we consider targets associated with a particular range and Doppler bin. Targets in adjacent range bins can be viewed as interferences for the range bin of interest.

In Section 1.3 we address a basic aspect on MIMO radar — its parameter identifiability, which is the maximum number of targets that can be uniquely identified by the radar. We show that the waveform diversity afforded by MIMO radar enables a much improved parameter identifiability over its phased-array counterpart; specifically, the maximum number of targets that can be uniquely identified by the MIMO radar is up to M_t times that of its phased-array counterpart, where M_t is the number of transmit antennas. The parameter identifiability is further demonstrated in a numerical study using both the Cramér–Rao bound (CRB) and a least-squares method for target parameter estimation.

In Section 1.4 we consider nonparametric adaptive MIMO radar techniques that can be used to deal with multiple targets. Linearly independent waveforms can be transmitted simultaneously via the multiple transmit antennas of a MIMO radar. Because of the different phase shifts associated with the different propagation paths from the transmitting antennas to targets, these independent waveforms are linearly combined at the target locations with different phase factors. As a result, the signal waveforms reflected from different targets are linearly independent of each other, which allows for the direct application of Capon (after J. Capon) and of other adaptive array algorithms. In the absence of array steering vector errors, we discuss the application of several existing data-dependent beamforming algorithms, including Capon, APES (amplitude and phase estimation) and CAPES (combined Capon and APES), and then present an alternative estimation procedure, referred to as the *combined Capon–approximate maximum-likelihood* (CAML) method. In the presence of array steering vector errors, we apply the robust Capon beamformer (RCB) and doubly constrained robust Capon beamformer (DCRCB) approaches to the MIMO radar system to achieve accurate parameter estimation and superior interference and jamming suppression performance.

In Section 1.5 we discuss parametric methods for parameter estimation and number detection of MIMO radar targets. Cyclic optimization algorithms are presented to obtain the maximum-likelihood (ML) estimates of the target parameters, and a Bayesian information criterion (BIC) is used for target number detection. Specifically, an approximate cyclic optimization (ACO) approach is first presented, which maximizes the likelihood function approximately. Then an exact cyclic optimization (ECO) approach that maximizes the exact likelihood function is introduced for target parameter estimation. The ACO and ECO target parameter estimates are used with the BIC for target number determination.

In Section 1.6, we show that the probing signal vector transmitted by a MIMO radar system can be designed to approximate a desired transmit beampattern and also to minimize the cross-correlation of the signals bounced from various targets of interest — an operation that would hardly be possible for a phased-array radar. An efficient semidefinite quadratic programming (SQP) algorithm can be used to solve the signal design problem in polynomial time. Using this design, we can significantly improve the parameter estimation accuracy of the adaptive MIMO radar techniques. In addition, we consider a minimum sidelobe beampattern design. We demonstrate the advantages of these MIMO transmit beampattern designs over their phased-array counterparts. Because of the significantly larger number of degrees of freedom of a MIMO system, we can achieve much better transmit beampatterns with a MIMO radar, under the practical uniform elemental transmit power constraint, than with its phased-array counterpart. We also present an application of the MIMO transmit beampattern designs to the ultrasound hyperthermia treatment of breast cancer. By choosing a proper covariance matrix of the transmitted waveforms under the uniform elemental power constraint, the ultrasound system can provide a focal spot matched to the entire tumor region, and simultaneously minimize the impact on the surrounding healthy breast tissue.

1.2 PROBLEM FORMULATION

Consider a MIMO radar system with M_t transmit antennas and M_r receive antennas. Let $x_m(n)$ denote the discrete-time baseband signal transmitted by the mth transmit antenna. Also, let θ denote the location parameter(s) of a generic target, for example, its azimuth angle and its range. Then, under the assumption that the transmitted probing signals are narrowband and that the propagation is nondispersive, the baseband signal at the target location can be described by the expression (see, e.g., Refs. 12 and 17 and Chapter 6 in Ref. 31):

$$\sum_{m=1}^{M_t} e^{-j2\pi f_0 \tau_m(\theta)} x_m(n) \stackrel{\Delta}{=} \mathbf{a}^*(\theta)\mathbf{x}(n), \qquad n = 1, \ldots, N \tag{1.1}$$

where f_0 is the carrier frequency of the radar, $\tau_m(\theta)$ is the time needed by the signal emitted via the mth transmit antenna to arrive at the target, $(\cdot)^*$ denotes the conjugate transpose, N denotes the number of samples of each transmitted signal pulse

$$\mathbf{x}(n) = [x_1(n) \quad x_2(n) \quad \cdots \quad x_{M_t}(n)]^T \tag{1.2}$$

and

$$\mathbf{a}(\theta) = \left[e^{j2\pi f_0 \tau_1(\theta)} \quad e^{j2\pi f_0 \tau_2(\theta)} \quad \ldots \quad e^{j2\pi f_0 \tau_{M_t}(\theta)}\right]^T \tag{1.3}$$

where $(\cdot)^T$ denotes the transpose. By assuming that the transmit array of the radar is calibrated, $\mathbf{a}(\theta)$ is a known function of θ.

Let $y_m(n)$ denote the signal received by the mth receive antenna; let

$$\mathbf{y}(n) = [y_1(n) \quad y_2(n) \quad \cdots \quad y_{M_r}(n)]^T, \qquad n = 1, \ldots, N \tag{1.4}$$

and let

$$\mathbf{b}(\theta) = \left[e^{j2\pi f_0 \tilde{\tau}_1(\theta)} \quad e^{j2\pi f_0 \tilde{\tau}_2(\theta)} \quad \ldots \quad e^{j2\pi f_0 \tilde{\tau}_{M_r}(\theta)}\right]^T \tag{1.5}$$

where $\tilde{\tau}_m(\theta)$ is the time needed by the signal reflected by the target located at θ to arrive at the mth receive antenna. Then, under the simplifying assumption of point targets, the received data vector can be described by the equation (see, e.g., Refs. 3 and 28)

$$\mathbf{y}(n) = \sum_{k=1}^{K} \beta_k \mathbf{b}^c(\theta_k)\mathbf{a}^*(\theta_k)\mathbf{x}(n) + \boldsymbol{\epsilon}(n), \qquad n = 1, \ldots, N \tag{1.6}$$

where K is the number of targets that reflect the signals back to the radar receiver, $\{\beta_k\}$ are complex amplitudes proportional to the radar cross sections (RCSs) of those targets, $\{\theta_k\}$ are the target location parameters, $\epsilon(n)$ denotes the interference-plus-noise term, and $(\cdot)^c$ denotes the complex conjugate. The unknown parameters, to be estimated from $\{\mathbf{y}(n)\}_{n=1}^N$, are $\{\beta_k\}_{k=1}^K$ and $\{\theta_k\}_{k=1}^K$.

The problem of interest here is to determine the maximum number of targets that can be uniquely identified, to devise parametric and nonparametric approaches to estimate the target parameters, and to provide flexible transmit beampattern designs by optimizing the covariance of matrix of $\mathbf{x}(n)$ under practical constraints.

1.3 PARAMETER IDENTIFIABILITY

Parameter identifiability is basically a consistency aspect: we want to establish the uniqueness of the solution to the parameter estimation problem as either the signal-to-interference-plus-noise ratio (SINR) or the snapshot number N goes to infinity [18]. It is clear that in either case, assuming that the interference-plus-noise term $\epsilon(n)$ is uncorrelated with $\mathbf{x}(n)$, the identifiability property of the first term in (1.6) is not affected by the second term. In particular, it follows that asymptotically we can handle any number of interferences; of course, for a finite snapshot number N and a finite SINR, the accuracy will degrade as the number of interferences increases, but that is a different issue — the parameter identifiability is *not* affected.

1.3.1 Preliminary Analysis

The "identifiability equation" is as follows:

$$\sum_{k=1}^{K} \breve{\beta}_k \mathbf{b}^c(\breve{\theta}_k)\mathbf{a}^*(\breve{\theta}_k)\mathbf{x}(n) = \sum_{k=1}^{K} \beta_k \mathbf{b}^c(\theta_k)\mathbf{a}^*(\theta_k)\mathbf{x}(n), \quad n = 1, \ldots, N \tag{1.7}$$

For the identifiability of parameters to hold, (1.7) should have a unique solution: $\breve{\beta}_k = \beta_k$, $\breve{\theta}_k = \theta_k$, $k = 1, \ldots, K$. Assume that the M_t transmitted waveforms are linearly independent of each other, which implies that

$$\text{rank}\{[\mathbf{x}(1) \cdots \mathbf{x}(N)]\} = M_t \tag{1.8}$$

Then (1.7) is equivalent to

$$\sum_{k=1}^{K} \breve{\beta}_k \mathbf{b}^c(\breve{\theta}_k)\mathbf{a}^*(\breve{\theta}_k) = \sum_{k=1}^{K} \beta_k \mathbf{b}^c(\theta_k)\mathbf{a}^*(\theta_k) \tag{1.9}$$

or

$$\breve{\mathbf{A}}\breve{\boldsymbol{\beta}} = \mathbf{A}\boldsymbol{\beta} \tag{1.10}$$

where

$$\boldsymbol{\beta} = [\beta_1 \ \cdots \ \beta_K]^T \tag{1.11}$$

$$\breve{\boldsymbol{\beta}} = [\breve{\beta} \ \cdots \ \breve{\beta}_K]^T \tag{1.12}$$

$$\mathbf{A} = [\mathbf{a}^c(\theta_1) \otimes \mathbf{b}^c(\theta_1) \ \cdots \ \mathbf{a}^c(\theta_K) \otimes \mathbf{b}^c(\theta_K)] \tag{1.13}$$

and

$$\breve{\mathbf{A}} = [\mathbf{a}^c(\breve{\theta}_1) \otimes \mathbf{b}^c(\breve{\theta}_1) \ \cdots \ \mathbf{a}^c(\breve{\theta}_K) \otimes \mathbf{b}^c(\breve{\theta}_K)] \tag{1.14}$$

with the symbol $\otimes$ denoting the Krönecker product operator, and $\mathbf{a}^c \otimes \mathbf{b}^c$ denoting the virtual steering vector of the MIMO radar.

In the next subsection we present conditions for the uniqueness of the solution to (1.10). However, before doing so, we discuss some features of (1.10) based on several examples of special cases of MIMO radar. First, consider the case where the transmit array is also the receive array, which in particular implies that $M_t = M_r \triangleq M$. Then (1.10) may contain quite a few redundant equations. In such a case, we have

$$\mathbf{b}(\theta) = \mathbf{a}(\theta) \tag{1.15}$$

and hence the generic column of $\mathbf{A}$ is the $M^2 \times 1$ vector $\mathbf{a}^c(\theta) \otimes \mathbf{a}^c(\theta)$. For a uniform linear array (ULA), the vector $\mathbf{a}^c(\theta) \otimes \mathbf{a}^c(\theta)$ has only $2M - 1$ distinct elements: 1, $e^{-j\omega}, \ldots, e^{-j(2M-1)\omega}$, where $\omega = 2\pi f_0 \tau(\theta)$, where $\tau(\theta)$ is the inter-element delay difference. However, a nonuniform but still linear array may have up to $(M^2 + M)/2$ distinct elements. For example, this is the case for the minimum redundancy linear array [32] with $M = 4$ and $\mathbf{a}^c(\theta) = [1 \quad e^{-j\omega} \quad e^{-j4\omega} \quad e^{-j6\omega}]^T$.

Next, consider the more general case of $M_r \neq M_t$ and of possibly different receive and transmit arrays. When the transmit (receive) array is a ULA that is a contiguous subset of the receive (transmit) ULA, $\mathbf{b}^c(\theta) \otimes \mathbf{a}^c(\theta)$ has only $M_t + M_r - 1$ distinct elements; in fact, this appears to be the smallest possible number of distinct elements. When the transmit and receive arrays share few or no antennas, all equations of (1.10) may well be distinct. For example, let

$$\mathbf{b}^c(\theta) = [1 \quad e^{-j\omega} \quad \cdots \quad e^{-j(M_r-1)\omega}]^T \tag{1.16}$$

and

$$\mathbf{a}^c(\theta) = [1 \quad e^{-jM_r\omega} \quad \cdots \quad e^{-j(M_t-1)M_r\omega}]^T \tag{1.17}$$

Then

$$\mathbf{b}^c(\theta) \otimes \mathbf{a}^c(\theta) = [1 \quad e^{-j\omega} \quad \cdots \quad e^{-j(M_tM_r-1)\omega}]^T \tag{1.18}$$

which is a (virtual) ULA with M_tM_r elements.

1.3.2 Sufficient and Necessary Conditions

Let

$$\breve{\mathbf{B}}\boldsymbol{\beta} = \mathbf{B}\boldsymbol{\beta} \tag{1.19}$$

denote the system of equations in (1.10) from which the identical equations have been eliminated. Let $\mathbf{c}(\theta)$ denote a (generic) column of $\mathbf{B}$, and let L_c denote the dimension of $\mathbf{c}(\theta)$. Then, according to the discussion at the end of Section 1.3.1, we obtain

$$L_c \in [M_t + M_r - 1, \;\; M_t M_r] \tag{1.20}$$

Using results from Refs. 33 and 34, we can show that when the M_t transmitted waveforms are linearly indepenent of each other [as assumed before; see (1.8)], a *sufficient and generically (for almost every vector* β*) necessary condition for parameter identifiability* is

$$L_c + 1 > 2K, \quad \text{i.e.,} \quad K_{\max} = \left\lceil \frac{L_c - 1}{2} \right\rceil \tag{1.21}$$

where $\lceil \cdot \rceil$ denotes the smallest integer greater than or equal to a given number. In view of (1.20), we thus have

$$K_{\max} \in \left[\frac{M_t + M_r - 2}{2}, \; \frac{M_t M_r + 1}{2} \right) \tag{1.22}$$

depending on the array geometry and on how many antennas are shared between the transmit and receive arrays [18].

Furthermore, generically (i.e., for almost any vector β), the identifiability can be ensured under the following condition [18,34]

$$L_c > \frac{3}{2}K, \quad \text{i.e.,} \quad K_{\max} = \left\lceil \frac{2L_c - 3}{3} \right\rceil \tag{1.23}$$

and, similarly to (1.22), we have

$$K_{\max} \in \left[\frac{2(M_t + M_r) - 5}{3}, \; \frac{2M_t M_r}{3} \right) \tag{1.24}$$

[which, typically, yields a larger number $K_{\max}$ than the one given in (1.22)].

For a phased-array radar (which uses M_r receiving antennas, and for which we can basically assume that $M_t = 1$), the condition similar to (1.22) is

$$K_{\max} = \left\lceil \frac{M_r - 1}{2} \right\rceil \tag{1.25}$$

and that similar to (1.24) is

$$K_{\max} = \left\lceil \frac{2M_r - 3}{3} \right\rceil \tag{1.26}$$

Hence, the maximum number of targets that can be uniquely identified by a MIMO radar can be up to M_t times that of its phased-array counterpart. To illustrate the extreme cases, note that when a filled [i.e., half-wavelength (0.5λ) inter-element spacing] uniform linear array (ULA) is used for both transmitting and receiving, which appears to be the worst MIMO radar scenario from the parameter identifiability standpoint, the maximum number of targets that can be identified by the MIMO radar is about twice that of its phased-array counterpart. This is because the virtual aperture $\mathbf{b}^c(\theta) \otimes \mathbf{a}^c(\theta)$ of the MIMO radar system has only $M_t + M_r - 1$ distinct elements. On the other hand, when the receive array is a filled ULA with M_r elements and the transmit array is a sparse ULA comprising M_t elements with $M_r/2$-wavelength inter-element spacing, the virtual aperture of the MIMO radar system is a filled (M_tM_r)-element ULA; that is, the virtual aperture length is M_t times that of the receive array [9,18]. This increased virtual aperture size leads to the result that the maximum number of targets that can be uniquely identified by the MIMO radar is M_t times that of its phased-array counterpart.

1.3.3 Numerical Examples

We present several numerical examples to demonstrate the parameter identifiability of MIMO radar, as compared to its phased-array counterpart. First, consider a MIMO radar system where a ULA with $M_t = M_r = M = 10$ antennas and half-wavelength spacing between adjacent antennas is used both for transmitting and for receiving. The transmitted waveforms are orthogonal to each other. Consider a scenario in which K targets are located at $\theta_1 = 0°$, $\theta_2 = 10°$, $\theta_3 = -10°$, $\theta_4 = 20°$, $\theta_5 = -20°$, $\theta_6 = 30°$, $\theta_7 = -30°, \ldots,$ with identical complex amplitudes $\beta_1 = \cdots = \beta_K = 1$. The number of snapshots is $N = 256$. The received signal is corrupted by a spatially and temporally white circularly symmetric complex Gaussian noise with mean zero and variance 0.01 (i.e., SNR = 20 dB) and by a jammer located at 45° with an unknown waveform (uncorrelated with the waveforms transmitted by the radar) with a variance equal to 1 (i.e., INR = 20 dB).

Consider the Cramér–Rao bound (CRB) of $\{\theta_k\}$, which gives the best performance of an unbiased estimator. By assuming that $\{\epsilon(n)\}_{n=1}^N$ in (6) are independently and identically distributed (i.i.d.) circularly symmetric complex Gaussian random vectors with mean zero and unknown covariance $\mathbf{Q}$, the CRB for $\{\theta_k\}$ can be obtained using the Slepian–Bangs formula [31]. Figure 1.2a shows the CRB of θ_1 for the MIMO radar as a function of K. For comparison purposes, we also provide the CRB of its phased-array counterpart, for which all the parameters are the same as for the MIMO radar except that $M_t = 1$ and that the amplitude of the transmitted waveform is adjusted so that the total transmission power does not change. Note that the phased-array CRB

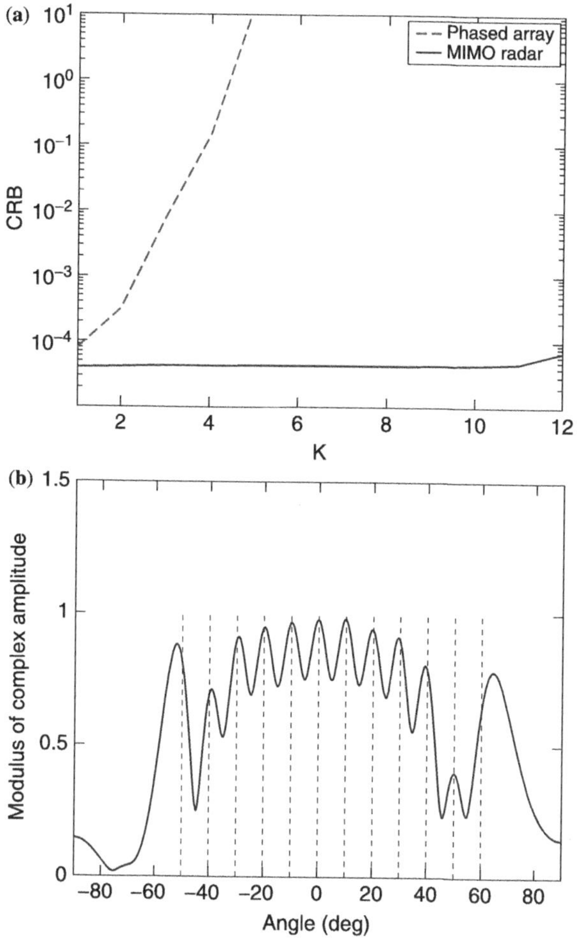

Figure 1.2 Performance of a MIMO radar system where a ULA with $M = 10$ antennas and 0.5-wavelength interelement spacing is used for both transmitting and receiving: (a) Cramér–Rao bound of θ_1 versus K; (b) LS spatial spectrum when $K = 12$.

increases rapidly as K increases from 1 to 6. The corresponding MIMO CRB, however, is almost constant when K is varied from 1 to 12 (but becomes unbounded for $K > 12$). Both results are consistent with the parameter identifiability analysis: $K_{\max} \leq 6$ for the phased-array radar and $K_{\max} \leq 12$ for the MIMO radar.

We next consider a simple nonparametric data-independent least-squares (LS) method [28] [see also Eq. (1.30)] for MIMO radar parameter estimation. Figure 1.2b shows the LS spatial spectrum as a function θ, when $K = 12$. Note that all 12 target locations can be approximately determined from the peak locations of the LS spatial spectrum.

Consider now a MIMO radar system with $M_t = M_r = 5$ antennas. The distance between adjacent antennas is 0.5λ for the receiving ULA and 2.5λ for the transmitting ULA. We retain all the simulation parameters corresponding to Fig. 1.2 except that the targets are located at $\theta_1 = 0°, \theta_2 = 8°, \theta_3 = -8°, \theta_4 = 16°, \theta_5 = -16°, \theta_6 = 24°, \theta_7 = -24°, \ldots$ in this example. Figure 1.3a shows the CRB of θ_1, for both the MIMO radar and the phased-array counterpart, as a function of K. Again, the MIMO CRB is much lower than the phased-array CRB. The behavior of both CRBs is consistent with the parameter identifiability analysis: $K_{\max} \leq 3$ for the phased-array radar and $K_{\max} \leq 16$ for the MIMO radar. Moreover, the

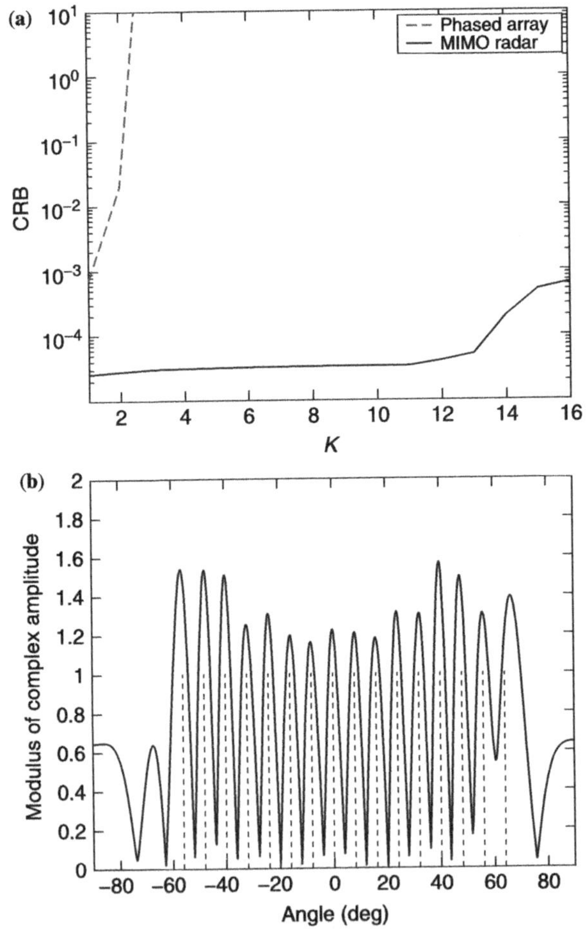

Figure 1.3 Performance of a MIMO radar system with $M_t = M_r = 5$ antennas, and with half-wavelength interelement spacing for the receive ULA and 2.5-wavelength interelement spacing for the transmit ULA: (a) Cramér–Rao bound of θ_1 versus K; (b) LS spatial spectrum when $K = 16$.

parameters of all $K = 16$ targets can be approximately determined with the simple LS method, as shown in Fig. 1.3b.

1.4 NONPARAMETRIC ADAPTIVE TECHNIQUES FOR PARAMETER ESTIMATION

As shown in Refs. 26–28, a MIMO radar makes it possible to use *adaptive* localization and detection techniques directly, unlike a phased-array radar. This is another significant advantage of a MIMO radar system since adaptive techniques are known to have much better resolution and much better interference rejection capability than their data-independent counterparts. For example, in spacetime adaptive processing applications, secondary range bins are needed to be able to use adaptive techniques [35–37]; however, selecting quality secondary range bins is in itself a major challenge [38]. Since the MIMO-probing signals reflected back by the targets are actually linearly independent of each other, the direct application of adaptive techniques is made possible for a MIMO radar system without the need for secondary range bins or even for range compression [19,29].

Let

$$\tilde{\mathbf{A}} = [\beta_1^* \mathbf{a}(\theta_1) \quad \beta_2^* \mathbf{a}(\theta_2) \quad \cdots \quad \beta_K^* \mathbf{a}(\theta_K)] \tag{1.27}$$

Then the sample covariance matrix of the target reflected waveforms is $\tilde{\mathbf{A}}^* \hat{\mathbf{R}}_{xx} \tilde{\mathbf{A}}$, where

$$\hat{\mathbf{R}}_{xx} = \frac{1}{N} \sum_{n=1}^{N} \mathbf{x}(n)\mathbf{x}^*(n) \tag{1.28}$$

is the sample covariance matrix of the transmitted waveforms. When orthogonal waveforms are used for MIMO probing, for example, and $N \geq M_t$, then $\hat{\mathbf{R}}_{xx}$ is a scaled identity matrix. Then $\tilde{\mathbf{A}}^* \tilde{\mathbf{R}}_{xx} \tilde{\mathbf{A}}$ has full rank; in other words, the target reflected waveforms are not completely correlated with each other (or coherent), if the columns of $\tilde{\mathbf{A}}$ are linearly independent of each other, which requires that $K \leq M_t$. The fact that the target reflected waveforms are noncoherent allows the direct application of many adaptive techniques for target localization [28].

Let θ be the location parameter of a generic target, for example, the direction of arrival (DOA) when the targets are in the far field of the arrays. We express the signal matrix at the output of the receiving array in the form

$$\mathbf{Y} = \mathbf{b}^c(\theta)\beta(\theta)\mathbf{a}^*(\theta)\mathbf{X} + \mathbf{Z} \tag{1.29}$$

where $\mathbf{X} = [\mathbf{x}(1) \quad \mathbf{x}(2) \quad \cdots \quad \mathbf{x}(N)]$, the columns of $\mathbf{Y} \in \mathcal{C}^{M_r \times N}$ are the received data samples $\{\mathbf{y}(n)\}_{n=1}^{N}$, and $\beta(\theta) \in \mathcal{C}$ denotes the complex amplitude of the reflected signal from θ, which is proportional to the radar cross section (RCS) of the focal point θ.

The matrix $\mathbf{Z} \in \mathcal{C}^{M_r \times N}$ denotes the residual term, which includes the unmodeled noise, interferences from targets at locations other than θ and at other range bins, and intentional or unintentional jamming. For notational simplicity, we do not explicitly show the dependence of $\mathbf{Z}$ on θ.

The problem is to estimate $\beta(\theta)$ for each θ of interest from the observed data matrix $\mathbf{Y}$. The estimates of $\{\beta(\theta)\}$ can be used to form a spatial spectrum. We can then estimate the locations of the targets and their complex amplitudes by searching for the peaks in the so-obtained spectrum.

Note that in (1.29) we consider a particular range bin of interest, whose time of arrival (TOA) is known a priori. This data model can be readily extended to the unknown TOA case by including an unknown TOA parameter τ in $\mathbf{X}$. By estimating β for each θ and τ using the presented methods, we can form a two-dimensional (2D) radar image, whose peaks will indicate existence of targets and their locations (DOA and TOA). Note also that, compared with the conventional techniques, where pulse compression is done separately, the range compression step is implicit in the following proposed methods.

1.4.1 Absence of Array Calibration Errors

A simple way to estimate $\beta(\theta)$ in (1.29) is via the least-squares (LS) method

$$\hat{\beta}_{\text{LS}}(\theta) = \frac{\mathbf{b}^T(\theta)\mathbf{Y}\mathbf{X}^*\,\mathbf{a}(\theta)}{N\,\|\mathbf{b}(\theta)\|^2[\mathbf{a}^*(\theta)\hat{\mathbf{R}}_{xx}\mathbf{a}(\theta)]} \tag{1.30}$$

where $\|\cdot\|$ denotes the Euclidean norm. However, as any other data-independent beamforming-type method, the LS method suffers from high sidelobes and low resolution. In the presence of strong interference and jamming, this method completely fails to work.

We present below several data-dependent beamforming-based methods, which, as shown via numerical examples in Section 1.4.3, have much higher resolution and much better interference rejection capability than does the LS method.

1.4.1.1 Capon The Capon estimator consists of two main steps: (1) the Capon beamforming step [31,39] and (2) a LS estimation step, which involves basically a matched filtering.

The Capon beamformer can be formulated as follows

$$\min_{\mathbf{w}} \mathbf{w}^*\hat{\mathbf{R}}_{yy}\mathbf{w} \quad \text{subject to} \quad \mathbf{w}^*\mathbf{b}^c(\theta) = 1 \tag{1.31}$$

where $\mathbf{w} \in \mathcal{C}^{M_r \times 1}$ is the weight vector used to achieve noise, interference, and jamming suppression while keeping the desired signal undistorted; and $\hat{\mathbf{R}}_{yy}$ is the sample covariance matrix of the observed data samples:

$$\hat{\mathbf{R}}_{yy} = \frac{1}{N}\mathbf{Y}\mathbf{Y}^* \tag{1.32}$$

Following Refs. 31 and 39, we can readily obtain the solution to (1.31) as follows:

$$\hat{\mathbf{w}}_{\text{Capon}} = \frac{\hat{\mathbf{R}}_{yy}^{-1}\mathbf{b}^c(\theta)}{\mathbf{b}^T(\theta)\hat{\mathbf{R}}_{yy}^{-1}\mathbf{b}^c(\theta)} \tag{1.33}$$

The output of the Capon beamformer is thus given by

$$\frac{\mathbf{b}^T(\theta)\hat{\mathbf{R}}_{yy}^{-1}\mathbf{Y}}{\mathbf{b}^T(\theta)\hat{\mathbf{R}}_{yy}^{-1}\mathbf{b}^c(\theta)} \tag{1.34}$$

Substituting (1.29) into (1.34) yields

$$\frac{\mathbf{b}^T(\theta)\hat{\mathbf{R}}_{yy}^{-1}\mathbf{Y}}{\mathbf{b}^T(\theta)\hat{\mathbf{R}}_{yy}^{-1}\mathbf{b}^c(\theta)} = \beta(\theta)\mathbf{a}^*(\theta)\mathbf{X} + \frac{\mathbf{b}^T(\theta)\hat{\mathbf{R}}_{yy}^{-1}\mathbf{Z}}{\mathbf{b}^T(\theta)\hat{\mathbf{R}}_{yy}^{-1}\mathbf{b}^c(\theta)} \tag{1.35}$$

Applying the LS method to (1.35), we get the Capon estimate of $\beta(\theta)$ as follows:

$$\hat{\beta}_{\text{Capon}}(\theta) = \frac{\mathbf{b}^T(\theta)\,\hat{\mathbf{R}}_{yy}^{-1}\mathbf{Y}\mathbf{X}^*\,\mathbf{a}(\theta)}{N\,[\mathbf{b}^T(\theta)\hat{\mathbf{R}}_{yy}^{-1}\mathbf{b}^c(\theta)][\mathbf{a}^*(\theta)\hat{\mathbf{R}}_{xx}\mathbf{a}(\theta)]} \tag{1.36}$$

where $\hat{\mathbf{R}}_{xx}$ is defined in (1.28). Note that (1.36) is a function of θ. In practice, we need to compute (1.36) for each θ of interest to form a spatial spectrum.

1.4.1.2 APES

The amplitude and phase estimation (APES) approach [40–42] is a nonparametric spectral analysis method with superior estimation accuracy [43,44]. We apply this method to the proposed MIMO radar system to achieve better amplitude estimation accuracy.

By following Ref. 41, we can formulate the APES method as

$$\min_{\mathbf{w},\beta} \|\, \mathbf{w}^*\mathbf{Y} - \beta(\theta)\mathbf{a}^*(\theta)\mathbf{X}\,\|^2 \quad \text{subject to} \quad \mathbf{w}^*\mathbf{b}^c(\theta) = 1 \tag{1.37}$$

where $\mathbf{w} \in C^{M_r \times 1}$ is the weight vector. Intuitively, the goal of (1.37) is to find a beamformer whose output is as close as possible to a signal with the waveform given by $\mathbf{a}^*(\theta)\,\mathbf{X}$.

Minimizing the cost function in (1.37) with respect to $\beta(\theta)$ yields

$$\hat{\beta}_{\text{APES}}(\theta) = \frac{\mathbf{w}^*\mathbf{Y}\mathbf{X}^*\mathbf{a}(\theta)}{N\,\mathbf{a}^*(\theta)\hat{\mathbf{R}}_{xx}\mathbf{a}(\theta)} \tag{1.38}$$

Then the optimization problem in (1.37) reduces to

$$\min_{\mathbf{w}} \mathbf{w}^*\hat{\mathbf{Q}}\mathbf{w} \quad \text{subject to} \quad \mathbf{w}^*\mathbf{b}^c(\theta) = 1 \tag{1.39}$$

with

$$\hat{\mathbf{Q}} = \hat{\mathbf{R}}_{yy} - \frac{\mathbf{Y}\mathbf{X}^*\,\mathbf{a}(\theta)\mathbf{a}^*(\theta)\mathbf{X}\mathbf{Y}^*}{N^2\,\mathbf{a}^*(\theta)\hat{\mathbf{R}}_{xx}\mathbf{a}(\theta)} \tag{1.40}$$

Solving the optimization problem in (1.39) gives the APES beamformer weight vector [41]:

$$\hat{\mathbf{w}}_{\text{APES}} = \frac{\hat{\mathbf{Q}}^{-1}\mathbf{b}^c(\theta)}{\mathbf{b}^T(\theta)\hat{\mathbf{Q}}^{-1}\mathbf{b}^c(\theta)} \tag{1.41}$$

By inserting (1.41) into (1.38), the APES estimate of $\beta(\theta)$ is readily obtained as

$$\hat{\beta}_{\text{APES}}(\theta) = \frac{\mathbf{b}^T(\theta)\hat{\mathbf{Q}}^{-1}\mathbf{Y}\mathbf{X}^*\,\mathbf{a}(\theta)}{N\,[\mathbf{b}^T(\theta)\hat{\mathbf{Q}}^{-1}\mathbf{b}^c(\theta)]\,[\mathbf{a}^*(\theta)\hat{\mathbf{R}}_{xx}\mathbf{a}(\theta)]} \tag{1.42}$$

Note that the only difference between the Capon estimator and the APES estimator is that the sample covariance matrix $\hat{\mathbf{R}}_{yy}$ in (1.36) is replaced in (1.42) by the residual covariance estimate $\hat{\mathbf{Q}}$. However, this seemingly minor difference makes these two methods behave quite differently (see, e.g., Section 1.4.3).

1.4.1.3 CAPES and CAML As discussed in Ref. 40 (see also Section 1.4.3), the two data-dependent methods described above behave differently. Capon has high resolution, and hence can provide accurate estimates of the target locations. However, the Capon amplitude estimates are significantly biased downward. The APES estimator gives much more accurate amplitude estimates at the target locations, but at the cost of lower resolution (i.e., greater minimum distance needed for two targets to be resolved). To reap the benefits of both Capon and APES, an alternative estimation procedure, referred to as *CAPES*, has been proposed [45]. CAPES first estimates the peak locations using the Capon estimator and then refines the amplitude estimates at these locations using the APES estimator. CAPES can be directly applied to the MIMO radar target localization and amplitude estimation problem considered here.

Inspired by the CAPES method, we propose below a new approach, referred to as CAML, which combines Capon and the more recently proposed approximate maximum likelihood (AML) estimator based on a diagonal growth curve (DGC) model [46]. In CAML, AML, instead of APES, is used to estimate the target amplitudes at the target locations estimated by Capon. Since, unlike APES, AML estimates the amplitudes of all targets at the given locations jointly rather than one at a time, the latter can provide better estimation accuracy than can the former.

Let $\hat{\theta}_k (k = 1, 2, \ldots, K)$ denote the estimated target locations, that is, the peak locations of the spatial spectrum estimated by Capon. Here K is assumed known. If K is unknown, it can be determined accurately using a generalized likelihood ratio test (GLRT), the details of which are given in the Appendix 1A. Let

$$\mathbf{A}_r = [\mathbf{b}(\hat{\theta}_1)\,\mathbf{b}(\hat{\theta}_2)\,\cdots\,\mathbf{b}(\hat{\theta}_K)] \tag{1.43}$$

$$\mathbf{A}_t = [\mathbf{a}(\hat{\theta}_1)\,\mathbf{a}(\hat{\theta}_2)\,\cdots\,\mathbf{a}(\hat{\theta}_K)] \tag{1.44}$$

and

$$\beta = [\beta(\hat{\theta}_1)\,\beta(\hat{\theta}_2)\,\cdots\,\beta(\hat{\theta}_K)]^T \tag{1.45}$$

Then the data model in (1.29) can be reformulated as a DGC model [46]:

$$\mathbf{Y} = \mathbf{A}_r^c\mathbf{B}(\mathbf{A}_t^*\mathbf{X}) + \tilde{\mathbf{Z}} \quad \text{with} \quad \mathbf{B} = \text{diag}(\beta) \tag{1.46}$$

In (1.46), diag(β) denotes a diagonal matrix whose diagonal elements are equal to the elements of β, and $\tilde{\mathbf{Z}}$ is the residual term, whose columns are assumed to be independently and identically distributed (i.i.d.) circularly symmetric complex Gaussian random vectors with mean zero and unknown covariance matrix. Since the maximum-likelihood (ML) estimate of β in (1.46) cannot be produced in closed form, we use the AML estimator [46] instead, which is asymptotically equivalent to the ML for a large data sample number N.

By using the AML method in Eq. (18) of Ref. 46, the estimate of β can be readily obtained as

$$\hat{\beta}_{\text{AML}} = \frac{1}{N}\left[(\mathbf{A}_r^T\mathbf{T}^{-1}\mathbf{A}_r^c) \odot (\mathbf{A}_t^T\hat{\mathbf{R}}_{xx}^c\mathbf{A}_t^c)\right]^{-1} \text{vecd}\left(\mathbf{A}_r^T\mathbf{T}^{-1}\mathbf{Y}\mathbf{X}^*\mathbf{A}_t\right) \tag{1.47}$$

where $\odot$ denotes the Hadamard product [47], vecd($\cdot$) denotes a column vector formed by the diagonal elements of a matrix, and

$$\mathbf{T} = N\hat{\mathbf{R}}_{yy} - \frac{1}{N}\mathbf{Y}\mathbf{X}^*\mathbf{A}_t(\mathbf{A}_t^*\hat{\mathbf{R}}_{xx}\mathbf{A}_t)^{-1}\mathbf{A}_t^*\mathbf{X}\mathbf{Y}^* \tag{1.48}$$

1.4.2 Presence of Array Calibration Errors

The previous data-dependent methods assume that the transmitting and receiving arrays are perfectly calibrated, that is, that $\mathbf{a}(\theta)$ and $\mathbf{b}(\theta)$ are accurately known as functions of θ. However, in practice, array calibration errors are often inevitable. The presence of array calibration errors and the related small data sample number problem [48] can significantly degrade the performance of the data-dependent beamforming methods discussed so far.

1.4.2.1 RCB We consider the application of the robust Capon beamformer (RCB) (see Refs. 49–51 and references cited therein) to a MIMO radar system that suffers from calibration errors. RCB allows $\mathbf{b}(\theta)$ to lie in an uncertainty set. Without loss of generality, we assume that $\mathbf{b}(\theta)$ belongs to an uncertainty sphere

$$\| \mathbf{b}(\theta) - \bar{\mathbf{b}}(\theta) \|^2 \leq \epsilon_r \tag{1.49}$$

where both $\bar{\mathbf{b}}(\theta)$, the nominal receiving array steering vector, and ϵ_r are given. (For the more general case of ellipsoidal uncertainty sets, see Refs. 49–51 and references cited therein.) Note that the calibration errors in $\mathbf{a}(\theta)$ will also degrade the accuracy of the estimate of $\beta(\theta)$ [see (1.36)]. However, the LS (or matched filtering) approach of (1.36) is quite robust against calibration errors in $\mathbf{a}(\theta)$.

The RCB method is based on the following covariance fitting formulation [49,50,52]

$$\begin{aligned} \max_{\sigma^2(\theta),\mathbf{b}(\theta)} \sigma^2(\theta) \quad \text{subject to} \quad & \hat{\mathbf{R}}_{yy} - \sigma^2(\theta)\mathbf{b}^c(\theta)\mathbf{b}^T(\theta) \geq 0 \\ & \| \mathbf{b}(\theta) - \bar{\mathbf{b}}(\theta) \|^2 \leq \epsilon_r \end{aligned} \tag{1.50}$$

where $\sigma^2(\theta)$ denotes the power of the signal of interest and $\mathbf{P} \geq 0$ means that the matrix $\mathbf{P}$ is positive semidefinite. The optimization problem in (1.50) can be simplified as follows [49]:

$$\min_{\mathbf{b}(\theta)} \mathbf{b}^T(\theta)\hat{\mathbf{R}}_{yy}^{-1}\mathbf{b}^c(\theta) \quad \text{subject to} \quad \| \mathbf{b}(\theta) - \bar{\mathbf{b}}(\theta) \|^2 \leq \epsilon_r \tag{1.51}$$

By using the Lagrange multiplier methodology [49], the solution to (1.51) is found to be

$$\hat{\mathbf{b}}(\theta) = \bar{\mathbf{b}}(\theta) - [\mathbf{I} + \lambda(\theta)\hat{\mathbf{R}}_{yy}^c]^{-1}\bar{\mathbf{b}}(\theta) \tag{1.52}$$

where $\mathbf{I}$ denotes the identity matrix. The Lagrange multiplier $\lambda(\theta) \geq 0$ in (1.52) is obtained as the solution to the constraint equation

$$\| [\mathbf{I} + \lambda(\theta)\hat{\mathbf{R}}_{yy}^c]^{-1}\, \bar{\mathbf{b}}(\theta) \|^2 = \epsilon_r \tag{1.53}$$

which can be solved efficiently by using the Newton method since the left side of (1.53) is a monotonically decreasing function of $\lambda(\theta)$ (see Ref. 49 for more details). Once the Lagrange multiplier $\lambda(\theta)$ has been determined, $\hat{\mathbf{b}}(\theta)$ can be obtained from (1.52). To eliminate a scaling ambiguity [49], we scale $\hat{\mathbf{b}}(\theta)$ such that $\| \hat{\mathbf{b}}(\theta) \|^2 = M_r$. Replacing $\mathbf{b}(\theta)$ in (1.36) by the scaled steering vector $\hat{\mathbf{b}}(\theta)$ yields the RCB estimate of $\beta(\theta)$.

1.4.2.2 DCRCB A doubly constrained robust Capon beamformer (DCRCB) has been proposed [53]. In the DCRCB, in addition to the spherical constraint in (1.49), the steering vector $\mathbf{b}(\theta)$ is constrained to have a constant norm:

$$\| \mathbf{b}(\theta) \|^2 = M_r \tag{1.54}$$

Then the covariance fitting formulation [see (1.50)] becomes

$$\begin{aligned} \max_{\sigma^2(\theta),\mathbf{b}(\theta)} \sigma^2(\theta) \quad \text{subject to} \quad & \hat{\mathbf{R}}_{yy} - \sigma^2(\theta)\mathbf{b}^c(\theta)\mathbf{b}^T(\theta) \geq 0 \\ & \| \mathbf{b}(\theta) - \bar{\mathbf{b}}(\theta) \|^2 \leq \epsilon_r \\ & \| \mathbf{b}(\theta) \|^2 = M_r \end{aligned} \tag{1.55}$$

The optimization problem in (1.55) can be simplified to [53]

$$\begin{aligned} \min_{\mathbf{b}(\theta)} \mathbf{b}^T(\theta)\hat{\mathbf{R}}_{yy}^{-1}\mathbf{b}^c(\theta) \quad \text{subject to} \quad & \text{Re}[\bar{\mathbf{b}}^T(\theta)\mathbf{b}^c(\theta)] \geq M_r - \frac{\epsilon_r}{2} \\ & \| \mathbf{b}(\theta) \|^2 = M_r \end{aligned} \tag{1.56}$$

where Re(·) denotes the real part of a complex number. Let $\mathbf{u}_1$ be the principal eigenvector of $\hat{\mathbf{R}}_{yy}$, and let

$$\tilde{\mathbf{b}}^c(\theta) = M_r^{1/2}\, \mathbf{u}_1 \, \exp\{j \arg [\, \mathbf{u}_1^* \bar{\mathbf{b}}^c(\theta)]\} \tag{1.57}$$

with arg(·) denoting the phase of a complex number. If $\tilde{\mathbf{b}}(\theta)$ in (1.57) satisfies

$$\text{Re}[\bar{\mathbf{b}}^T(\theta)\tilde{\mathbf{b}}^c(\theta)] \geq M_r - \frac{\epsilon_r}{2} \tag{1.58}$$

then, obviously, $\hat{\hat{\mathbf{b}}}(\theta) = \tilde{\mathbf{b}}(\theta)$ is the optimal solution of (1.56). Otherwise, by using the Lagrange multiplier methodology [53], the optimal solution to (1.56) is

$$\hat{\hat{\mathbf{b}}}(\theta) = \left(M_r - \frac{\epsilon_r}{2}\right) \frac{[\hat{\mathbf{R}}_{yy}^{-c} + \lambda(\theta)\mathbf{I}]^{-1}\,\bar{\mathbf{b}}(\theta)}{\bar{\mathbf{b}}^*(\theta)\,[\hat{\mathbf{R}}_{yy}^{-c} + \lambda(\theta)\mathbf{I}]^{-1}\,\bar{\mathbf{b}}(\theta)} \tag{1.59}$$

where $(\cdot)^{-c} = [(\cdot)^c]^{-1}$. The Lagrange multiplier $\lambda(\theta)$ in (1.59), which can be negative, is obtained as the solution to the constraint equation

$$\frac{\bar{\mathbf{b}}^T(\theta)\,[\hat{\mathbf{R}}_{yy}^{-1} + \lambda(\theta)\mathbf{I}]^{-2}\,\bar{\mathbf{b}}^c(\theta)}{\left\{\bar{\mathbf{b}}^T(\theta)\,[\hat{\mathbf{R}}_{yy}^{-1} + \lambda(\theta)\mathbf{I}]^{-1}\,\bar{\mathbf{b}}^c(\theta)\right\}^2} = \frac{M_r}{(M_r - (\epsilon_r/2))^2} \tag{1.60}$$

As in the case of RCB, the constraint equation (1.60) can be solved efficiently by using the Newton method since the left side of (1.60) is a monotonically decreasing function of $\lambda(\theta)$ (see Ref. 53 for more details).

We summarize the DCRCB estimator of $\beta(\theta)$ as follows:

1. Compute $\tilde{\mathbf{b}}(\theta)$ by (1.57).
2. Assuming that $\tilde{\mathbf{b}}(\theta)$ satisfies (1.58), estimate $\beta(\theta)$ by replacing $\mathbf{b}(\theta)$ in (1.36) by $\hat{\hat{\mathbf{b}}}(\theta) = \tilde{\mathbf{b}}(\theta)$, and stop.
3. Assuming that (1.58) is not satisfied, solve (1.60) for $\lambda(\theta)$. Calculate $\hat{\hat{\mathbf{b}}}(\theta)$ by (1.59) using the so-obtained $\lambda(\theta)$.
4. Compute the DCRCB estimate of $\beta(\theta)$ by replacing $\mathbf{b}(\theta)$ in (1.36) by $\hat{\hat{\mathbf{b}}}(\theta)$, and stop.

1.4.3 Numerical Examples

Consider a MIMO radar system where a uniform linear array with $M_t = M_r = M = 10$ antennas and half-wavelength spacing between adjacent antennas is used for both transmitting and receiving. The transmitted waveforms (rows of $\mathbf{X}$) are orthogonal quadrature phase-shift-keyed (QPSK) sequences, and hence we have $\hat{\mathbf{R}}_{xx} = \mathbf{I}$ (this choice of $\hat{\mathbf{R}}_{xx}$ is optimal in a maximum-SNR sense [54]). Consider a scenario in which $K = 3$ targets are located at $\theta_1 = -40^\circ$, $\theta_2 = -25^\circ$, and $\theta_3 = -10^\circ$ with complex amplitudes $\beta_1 = 4$, $\beta_2 = 4$, and $\beta_3 = 1$, respectively. There is a strong jammer at 0° with an unknown waveform and with amplitude 1000 (i.e., 60 dB above β_3). The received signal has $N = 256$ data samples and is corrupted by a zero-mean spatially colored Gaussian noise with an unknown covariance matrix. The (p,q)th element of the unknown noise covariance matrix is $(1/\mathrm{SNR})0.9^{|p-q|}e^{j(p-q)\pi/2}$.

We first consider the case of no array calibration errors. Let SNR = 10 dB. The moduli of the spatial spectral estimates of $\beta(\theta)$, versus θ, obtained by using LS, Capon, APES, CAPES, and CAML are given in Figs. 1.4a,b,c,e,f, respectively. For comparison purpose, we also show the true spatial spectrum via dashed lines in these figures. As seen in Fig. 1.4a, the LS method suffers from high sidelobes and low resolution. In the presence of the strong jamming signal, the LS estimator fails to work properly. Capon and APES possess great interference and jamming suppression capabilities. The Capon method gives very narrow peaks around the target locations. However, the Capon amplitude estimates are biased downward. The APES method gives more accurate amplitude estimates around the target locations, but its resolution is slightly worse than that of Capon. Note that in Figs. 1.4a–c a false peak occurs at $\theta = 0^\circ$ owing to the presence of the strong jammer. Although the jammer waveform is statistically independent of the waveforms transmitted by the MIMO radar, a false peak still exists since the jammer is 60 dB stronger than the weakest target and the data sample number is finite.

To reject the false peak, we use a generalized likelihood ratio test (GLRT) for each angle of interest; the generalized likelihood ratio (GLR) is given by (see Appendix 1A for details)

$$\rho(\theta) = 1 - \frac{\mathbf{b}^T(\theta)\hat{\mathbf{R}}_{yy}^{-1}\mathbf{b}^c(\theta)}{\mathbf{b}^T(\theta)\hat{\mathbf{Q}}^{-1}\mathbf{b}^c(\theta)} \tag{1.61}$$

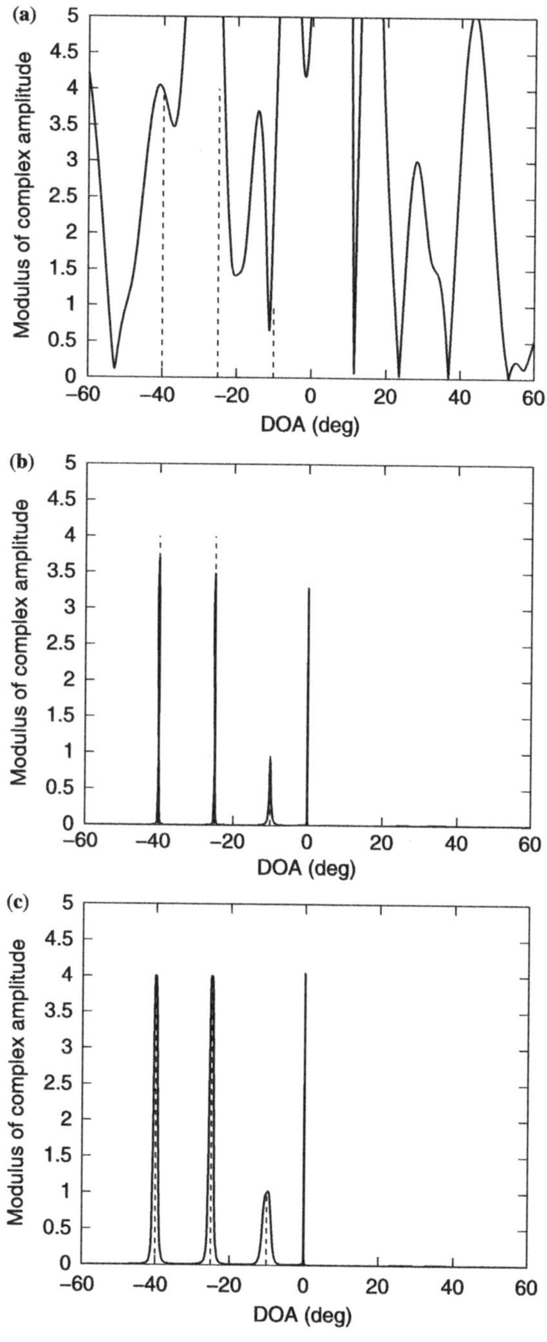

Figure 1.4 Spatial spectral estimates and GLR in the absence of array calibration errors, when $\theta_1 = -40°$, $\theta_2 = -25°$, and $\theta_3 = -10°$: (a) LS; (b) Capon; (c) APES; (d) GLR; (e) CAPES; (f) CAML.

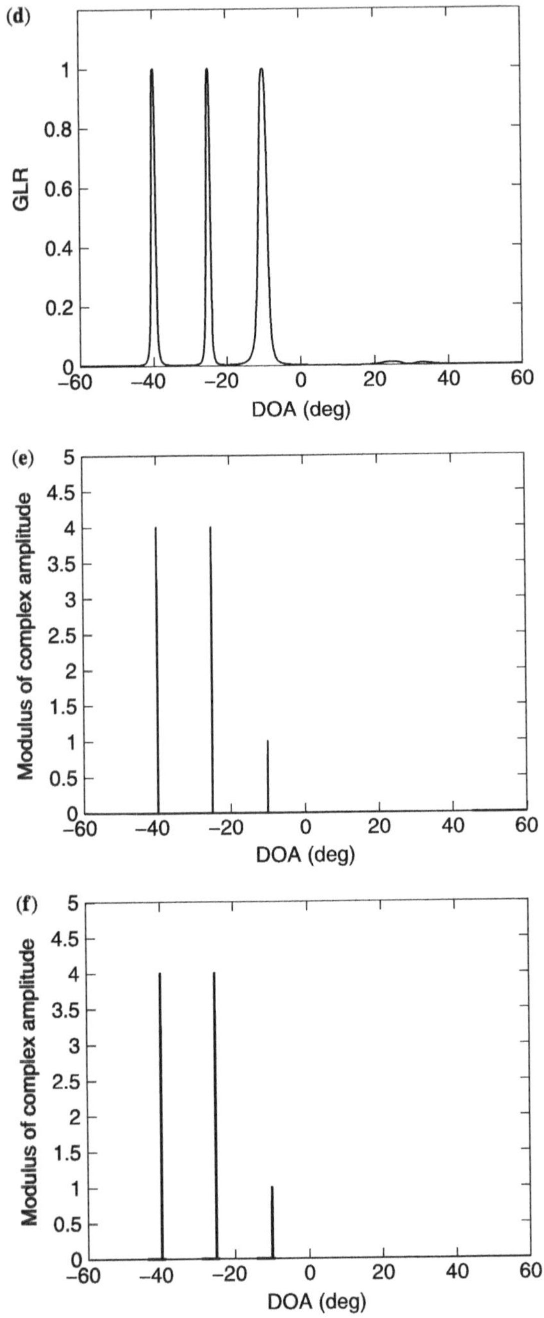

Figure 1.4 *Continued.*

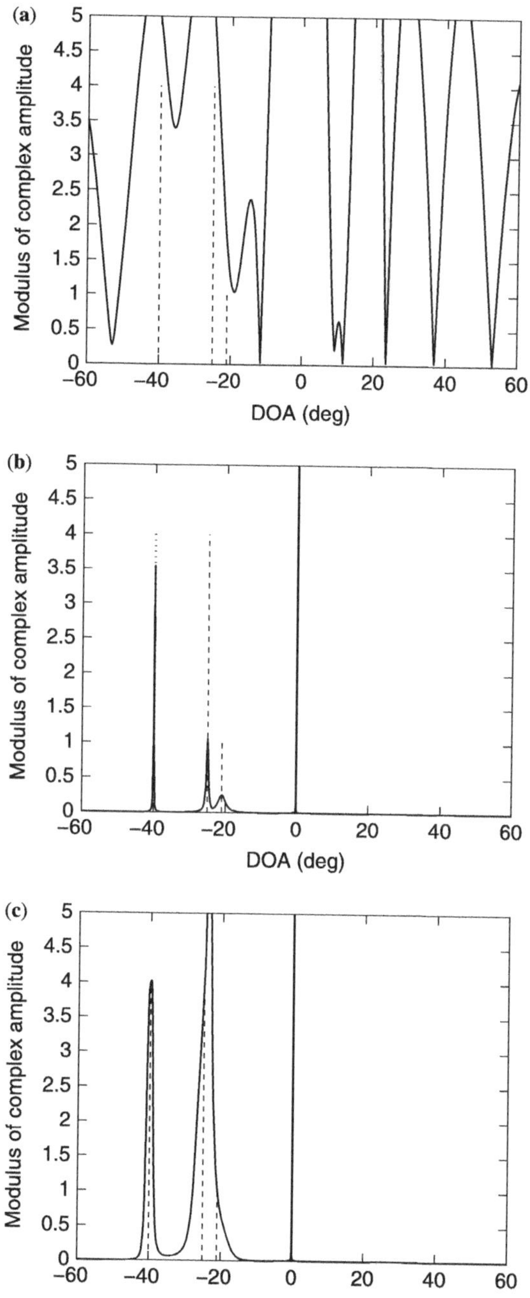

Figure 1.5 Spatial spectral estimates and GLR in the absence of array calibration errors, when $\theta_1 = -40°$, $\theta_2 = -25°$, and $\theta_3 = -21°$: (a) LS; (b) Capon; (c) APES; (d) GLR; (e) CAPES; (f) CAML.

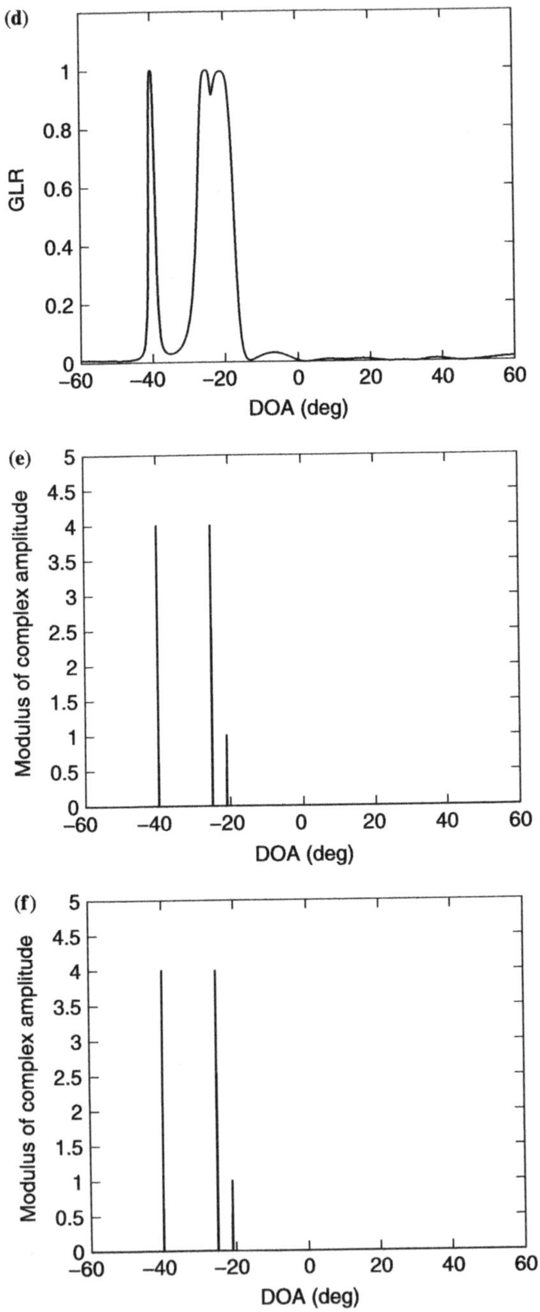

Figure 1.5 *Continued.*

where $\hat{\mathbf{R}}_{yy}$ and $\hat{\mathbf{Q}}$ are as given in (1.32) and (1.40), respectively. Figure 1.4d gives the GLR as a function of the target location parameter θ. As expected, we get high GLRs at the target locations and low GLRs at other locations, including the jammer location. By comparing the GLR with a threshold, the false peak due to the strong jammer can be readily detected and rejected, and a correct estimate of the number of the targets can be obtained. This information can be passed to CAPES and CAML to obtain Figs. 1.4e,f. As seen from these figures, both CAPES and CAML give accurate estimates of the target locations and amplitudes.

Next we consider a more challenging example where θ_3 is $-21°$ while all the other simulation parameters are the same as before. As shown in Fig. 1.5c, in this example

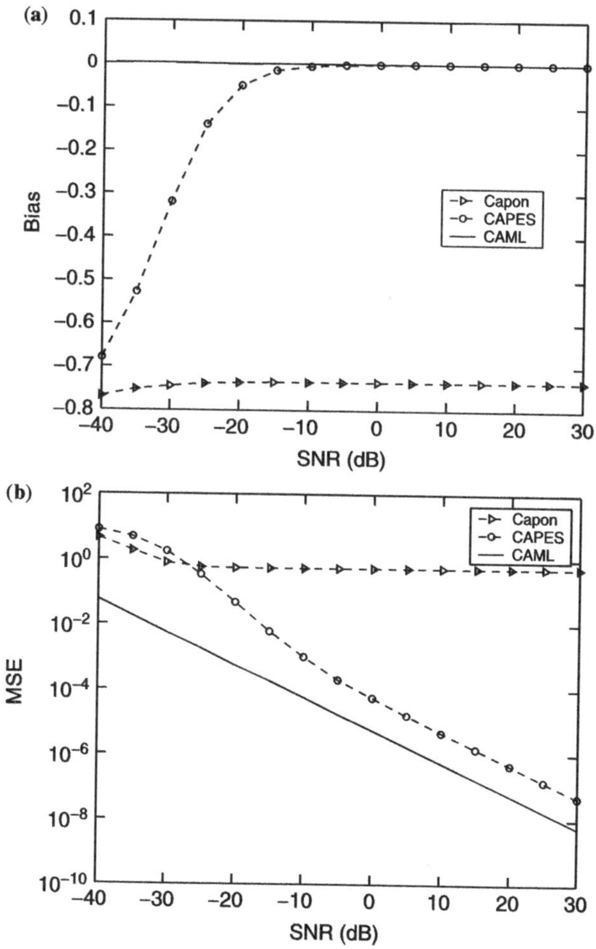

Figure 1.6 Empirical bias and MSE of the amplitude estimate for the target at θ_3 versus SNR, in the absence of array calibration errors, when $\theta_1 = -40°$, $\theta_2 = -25°$, and $\theta_3 = -21°$: (a) bias; (b) MSE.

the APES method fails to resolve the two closely spaced targets at $\theta_2 = -25^\circ$ and $\theta_3 = -21^\circ$. On the other hand, Capon gives well-resolved peaks around the target locations, but its amplitude estimates are significantly biased downward. On the basis of the GLR in Fig. 1.5d, the false peak due to the strong jammer can be readily rejected. Once again CAPES and CAML outperform the other methods and provide quite accurate target parameter estimates. To illustrate the improvement of CAML over CAPES for target amplitude estimation, we consider the amplitude estimation performance of Capon, CAPES, and CAML as a function of SNR. All the other simulation parameters are the same as in Fig. 1.5. Figures 1.6a,b show, respectively, the bias and mean-squared error (MSE) of the amplitude estimate for the target

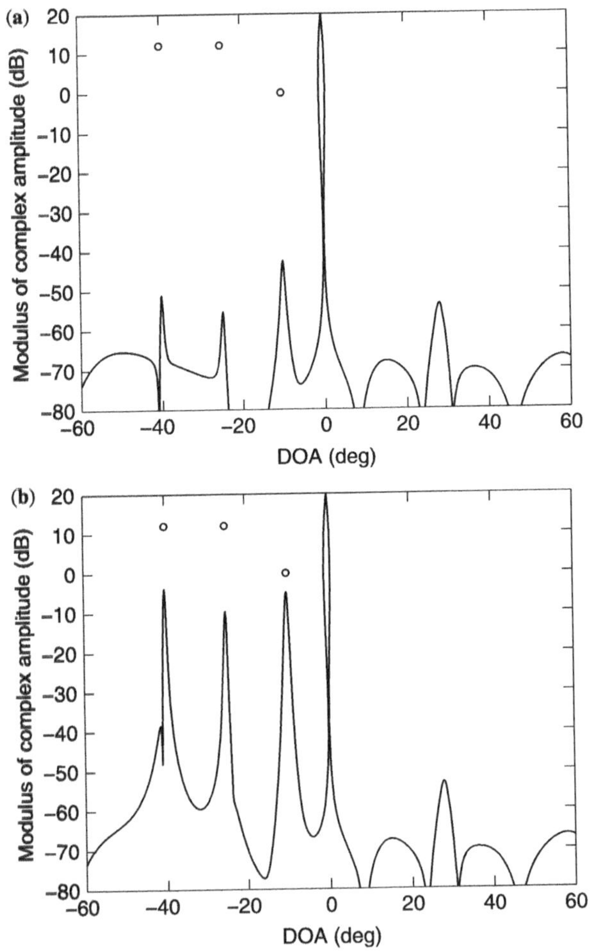

Figure 1.7 Spatial spectral estimates in the presence of array calibration errors, when $\theta_1 = -40^\circ$, $\theta_2 = -25^\circ$, and $\theta_3 = -10^\circ$: (a) Capon; (b) APES; (c) RCB with $\epsilon_r = 0.1$; (d) DCRCB with $\epsilon_r = 0.1$.

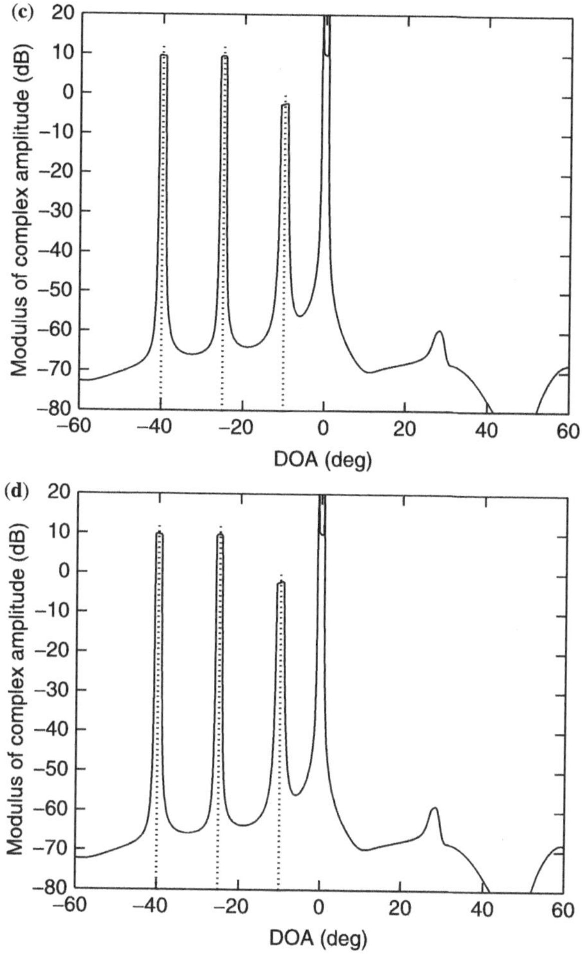

Figure 1.7 *Continued.*

at θ_3. The results are obtained using 1000 Monte Carlo simulations. As seen from these figures, the Capon estimate is biased downward significantly. Because of this bias, an error floor occurs in the MSE of the Capon estimate. CAPES is also biased downward at very low SNR. The CAML estimator is unbiased and outperforms CAPES for the entire range of the SNR considered, especially at low SNR.

We next consider the case where array calibration errors are present. To simulate the array calibration error, each element of the steering vector $\mathbf{a}(\theta) = \mathbf{b}(\theta)$, for each incident signal, is perturbed with a zero-mean circularly symmetric complex Gaussian random variable with variance 0.005 and then scaled to have norm $\sqrt{M_r}$, so that the squared Euclidean norm of the difference between the true array steering vector and the presumed one is about 0.05. We let SNR $= 30$ dB. The other simulation parameters are the same as those in Fig. 1.4. For the sake of description convenience, the moduli of the amplitude estimates, that is, the y axes of Figs. 1.7a–d, are on a dB scale. As shown in Fig. 1.7a, the Capon method fails to work properly in the presence of array calibration errors, as expected: its amplitude estimates at the target locations are severely biased downward (by >60 dB for some of them). Although APES gives much better performance than Capon, its amplitude estimates at the target locations are ~10 dB lower than the true amplitudes. On the other hand, both RCB and DCRCB provide accurate estimates of the target amplitudes as well as target locations, but their peaks are wider (and hence their resolutions

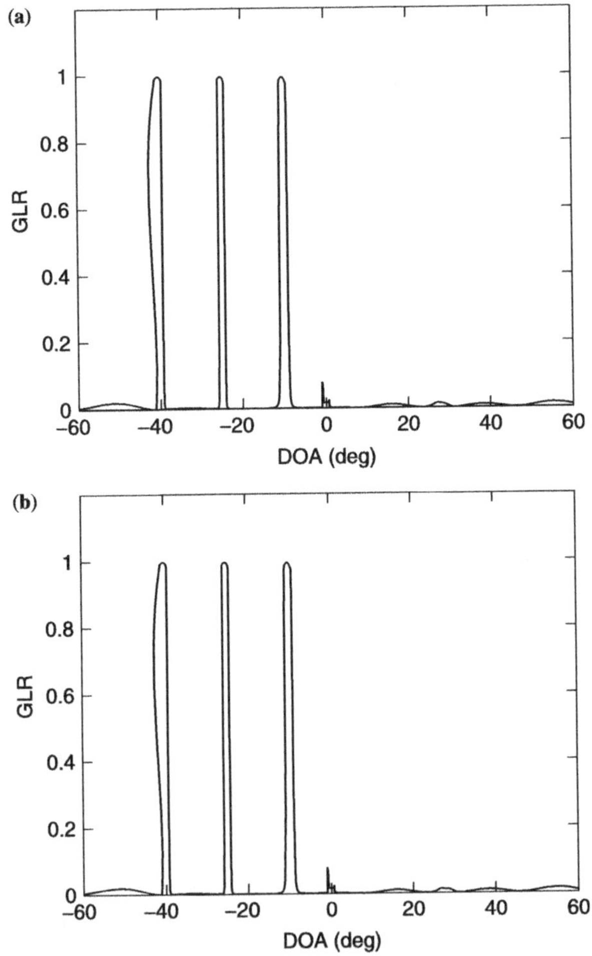

Figure 1.8 GLRs and refined spatial spectral estimates in the presence of array calibration errors, when $\theta_1 = -40°$, $\theta_2 = -25°$, and $\theta_3 = -10°$: (a) GLR of RCB; (b) GLR of DCRCB; (c) refined spatial spectral estimate of RCB; (d) refined spatial spectral estimate of DCRCB.

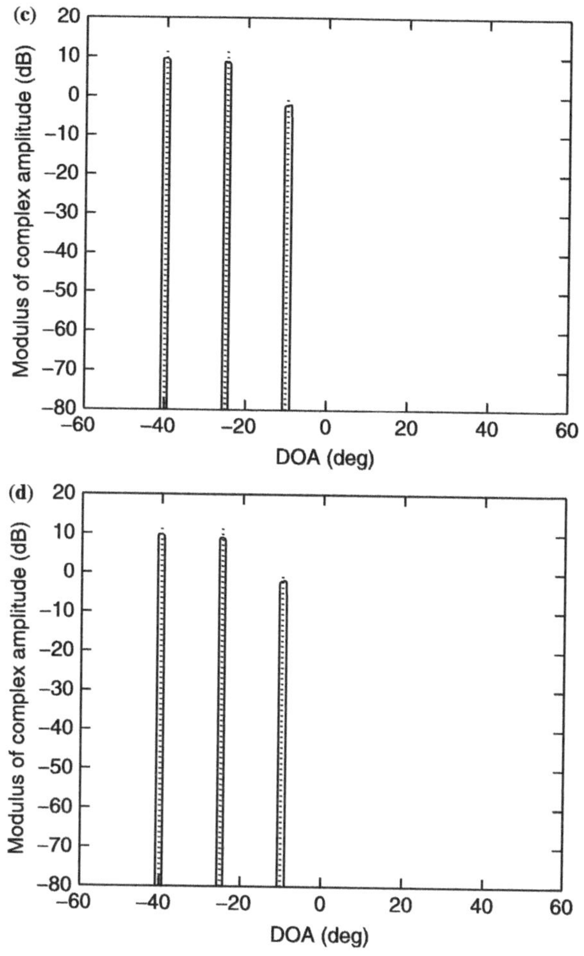

Figure 1.8 *Continued.*

are poorer) compared to what is shown in Fig. 1.4b, as expected (robustness to array calibration errors inherently reduces the resolution).

Figures 1.8a,b show the GLRs corresponding to RCB and DCRCB, as functions of θ, which are obtained by replacing $\mathbf{b}(\theta)$ in (1.61) by $\hat{\mathbf{b}}(\theta)$ and $\hat{\hat{\mathbf{b}}}(\theta)$ from Sections 1.4.2.1 and 1.4.2.2, respectively. As we can see, both RCB and DCRCB give high GLRs at the target locations and a low GLR at the jammer location. From the GLRs, we can again readily and correctly estimate the number of targets to be 3. Plotting the spatial spectral estimates in Figs. 1.7c,d only for the angles at which the corresponding GLRs are above a given threshold (say, 0.8), we obtain the refined spatial spectral estimates in Figs. 1.8c,d. These refined spatial spectral estimates obtained by RCB and DCRCB provide an accurate description of the target scenario.

1.5 PARAMETRIC TECHNIQUES FOR PARAMETER ESTIMATION

Several nonparametric data adaptive target parameter estimation methods for MIMO radar with collocated antennas have been presented in Section 1.4. We now consider parametric techniques for target parameter estimation as well as target number detection for MIMO radar with collocated antennas. Cyclic optimization algorithms are presented to obtain the maximum-likelihood (ML) estimates of the target parameters, and a Bayesian information criterion (BIC) is used for target number detection. Specifically, an approximate cyclic optimization (ACO) approach is first presented, which maximizes the likelihood function approximately. Then an exact cyclic optimization (ECO) approach that maximizes the exact likelihood function is introduced for target parameter estimation. The ACO and ECO target parameter estimates are used with the BIC for target number determination.

To remind the reader with some notation, we assume that the number of targets in a particular range and Doppler bin is K. The received data matrix can be written as

$$\mathbf{Y} = \sum_{k=1}^{K} \mathbf{b}^c(\theta_k)\beta_k \mathbf{s}^T(\theta_k) + \tilde{\mathbf{Z}}, \quad \text{with} \quad \mathbf{s}^T(\theta_k) = \mathbf{a}^*(\theta_k)\mathbf{X} \tag{1.62}$$

where the parameters are defined similarly as in (1.6) and (1.46), and the column of $\tilde{\mathbf{Z}} \in \mathbb{C}^{M_r \times N}$ comprise independently and identically distributed (i.i.d.) circularly symmetric complex Gaussian random vectors with mean zero and unknown covariance matrix $\mathbf{Q}$. Our goal here is to detect the unknown target number K and estimate the unknown target location parameters $\{\theta_k\}_{k=1}^K$ as well as complex amplitudes $\{\beta_k\}_{k=1}^K$.

1.5.1 ML and BIC

Consider first the ML estimation of the target parameters $\{\beta_k\}_{k=1}^K$ and $\{\theta_k\}_{k=1}^K$ by assuming that the target number K is known *a priori*. The BIC estimate of the unknown K is described later.

The negative loglikelihood function (LF) of $\{\theta_k\}_{k=1}^K$, $\{\beta_k\}_{k=1}^K$, and the unknown elements in $\mathbf{Q}$ for the data model in (1.62) can be written, to within a constant, as

$$\begin{aligned} f_1(\mathbf{Q}, \{\theta_k\}, \{\beta_k\}) &= N \ln(|\mathbf{Q}|) \\ &\quad + \operatorname{tr}\left\{ \left[\mathbf{Y} - \sum_{k=1}^{K} \mathbf{b}^c(\theta_k)\beta_k \mathbf{s}^T(\theta_k) \right] \right. \\ &\quad \times \left. \left[\mathbf{Y} - \sum_{k=1}^{K} \mathbf{b}^c(\theta_k)\beta_k \mathbf{s}^T(\theta_k) \right]^* \mathbf{Q}^{-1} \right\} \end{aligned} \tag{1.63}$$

where $|\cdot|$ and $\operatorname{tr}(\cdot)$ denote the determinant and trace of a matrix, respectively.

The minimization of the negative loglikelihood function in (1.63) with respect to $\mathbf{Q}$ yields the following concentrated negative log-LF of the unknown parameters $\{\theta_k, \beta_k\}$:

$$f_2(\{\theta_k\}, \{\beta_k\}) = N \ln \left| \left[\mathbf{Y} - \sum_{k=1}^{K} \mathbf{b}^c(\theta_k)\beta_k \mathbf{s}^T(\theta_k) \right] \times \left[\mathbf{Y} - \sum_{k=1}^{K} \mathbf{b}^c(\theta_k)\beta_k \mathbf{s}^T(\theta_k) \right]^* \right| \tag{1.64}$$

The minimization of this cost function must be conducted with respect to $3K$ unknown real-valued variables $\{\theta_k\}_{k=1}^K$ and the real and imaginary parts of $\{\beta_k\}_{k=1}^K$. Because this optimization problem does not appear to admit a closed-form solution, we present two cyclic optimization algorithms for target parameter estimation in the following.

The estimates of the target parameters $\{\beta_k\}_{k=1}^K$ and $\{\theta_k\}_{k=1}^K$ are associated with the assumed target number K. Once we got the estimates of $\{\beta_k\}_{k=1}^K$ and $\{\theta_k\}_{k=1}^K$ for various values of K, the target number K can be detected by minimizing the following BIC cost function (see, e.g., Ref. 55)

$$\text{BIC}(K) = 2f_2(\{\hat{\theta}_k\}_{k=1}^K, \{\hat{\beta}_k\}_{k=1}^K) + 3K \ln N \tag{1.65}$$

where $\hat{\theta}_k$ and $\hat{\beta}_k$ are the estimates of θ_k and β_k, respectively.

1.5.1.1 ACO ACO can be used to obtain initial target parameter estimates, which can be refined via ECO. The received data matrix in (1.62) can be rewritten in the form of the diagonal growth curve (DGC) model [46]. Using the AML estimator in [83], we can then obtain an estimate of $\{\beta_k\}_{k=1}^K$ that can be used to further concentrate the negative log-LF in (1.64), at least approximately.

For convenience, let

$$\mathbf{B}_r(\boldsymbol{\theta}) = [\mathbf{b}^c(\theta_1) \ \cdots \ \mathbf{b}^c(\theta_K)] \tag{1.66}$$

$$\mathbf{S}(\boldsymbol{\theta}) = [\mathbf{s}(\theta_1) \ \cdots \ \mathbf{s}(\theta_K)]^T \tag{1.67}$$

and

$$\boldsymbol{\theta} = [\theta_1 \ \cdots \ \theta_K]^T \tag{1.68}$$

Then, (1.64) can be rewritten as

$$f(\boldsymbol{\beta}, \boldsymbol{\theta}) = N \ln \left| [\mathbf{Y} - \mathbf{B}_r \operatorname{diag}(\boldsymbol{\beta})\mathbf{S}][\mathbf{Y} - \mathbf{B}_r \operatorname{diag}(\boldsymbol{\beta})\mathbf{S})]^* \right| \tag{1.69}$$

where $\boldsymbol{\beta}$ is as defined in (1.11). By using the AML estimator [46], we can estimate $\boldsymbol{\beta}$ approximately, for any fixed vector θ, as follows

$$\hat{\boldsymbol{\beta}}_{\text{AML}} = [(\mathbf{B}_r^* \tilde{\mathbf{Q}}^{-1} \mathbf{B}_r) \odot (\mathbf{S}^c \mathbf{S}^T)]^{-1} \text{vecd}(\mathbf{B}_r^* \tilde{\mathbf{Q}}^{-1} \mathbf{Y}\mathbf{S}^*) \tag{1.70}$$

where

$$\tilde{\mathbf{Q}} = \hat{\mathbf{R}}_{yy} - \frac{1}{N}\mathbf{Y}\mathbf{S}^*(\mathbf{S}\mathbf{S}^*)^{-1}\mathbf{S}\mathbf{Y}^* \tag{1.71}$$

and $\hat{\mathbf{R}}_{yy}$ is given as in (1.31). Note that (1.70) is similar to (1.47). Note also that for notational convenience we have omitted the argument $\boldsymbol{\theta}$ of $\mathbf{B}_r$, $\mathbf{S}$, and $\tilde{\mathbf{Q}}$ in (1.70). Substituting (1.70) in (1.69), followed by some straightforward manipulations, yields the concentrated approximate negative log-LF of $\boldsymbol{\theta}$ as follows:

$$\begin{aligned} f_3(\boldsymbol{\theta}) = {} & N \ln |\tilde{\mathbf{Q}}| + N \ln \{1 - M_r + \text{tr}(\tilde{\mathbf{Q}}^{-1}\hat{\mathbf{R}}_{yy}) \\ & - \frac{1}{N}\text{vecd}^*(\mathbf{B}_r^*\tilde{\mathbf{Q}}^{-1}\mathbf{Y}\mathbf{S}^*) \\ & \times [(\mathbf{B}_r^*\tilde{\mathbf{Q}}^{-1}\mathbf{B}_r) \odot (\mathbf{S}^c\mathbf{S}^T)]^{-1}\text{vecd}(\mathbf{B}_r^*\tilde{\mathbf{Q}}^{-1}\mathbf{Y}\mathbf{S}^*)\} \end{aligned} \tag{1.72}$$

Note that by using AML, the number of unknowns in (1.72) has been reduced to K. Cyclic optimization (CO) techniques can be used to minimize (1.72) with respect to $\{\theta_k\}_{k=1}^K$. Within each substep of the CO algorithm, we estimate one target location parameter, say, θ_k, by using the most recent estimates of the other parameters. The iteration of the CO algorithm is terminated when "practical convergence" is reached, which may be determined by checking whether the relative change of the log-LF between two consecutive iterations is less than some predetermined threshold ϵ_t. The ACO algorithm can be described as follows:

Step 0: Initialization. Set $i = 1$.

Step 1: Cyclic Optimization:

(a) Do the following until "practical convergence" is reached. For $k = \text{i}, 1, 2, \ldots, i-1$ estimate θ_k by searching the minimum of the cost function (1.72) using the most recent estimates of $\{\theta_p\}^i_{p \neq k,\, p=1}$.

(b) Set $i = i + 1$ and go to step 2(a) until i reaches a prescribed maximum number.

Step 2: BIC Order Selection. Find the minimum of $\text{BIC}(i) = 2f_2(\hat{\theta}^i) + 3i \ln N$ with respect to i.

The so-obtained i corresponding to the minimum of BIC(i) is the estimate of target number K, and the corresponding $\{\hat{\theta}_k\}$ are the estimates of the target location parameters with $\hat{\theta}^i = \{\hat{\theta}_k\}_{k=1}^i$. Once $\hat{\theta}^i$ has been determined, the corresponding $\hat{\beta}$ can be obtained by using (1.70).

1.5.1.2 ECO Inspired by the RELAX method [56], ECO determines the parameters of each target iteratively. At each step, ECO estimates the parameters of one target by assuming that the other target parameters are given. Let

$$\tilde{\mathbf{Y}}_k = \mathbf{Y} - \sum_{p \neq k} \mathbf{b}^c(\hat{\theta}_p)\hat{\beta}_p \mathbf{s}^T(\hat{\theta}_p) \tag{1.73}$$

Then, the cost function in (1.64) reduces to

$$f(\theta_k, \beta_k) = N \ln \left| [\tilde{\mathbf{Y}}_k - \beta_k \mathbf{b}^c(\theta_k)\mathbf{s}^T(\theta_k)][\tilde{\mathbf{Y}}_k - \beta_k \mathbf{b}^c(\theta_k)\mathbf{s}^T(\theta_k)]^* \right| \tag{1.74}$$

Minimizing (1.74) with respect to β_k yields

$$\hat{\beta}_k = \frac{\mathbf{b}^T(\theta_k)\tilde{\mathbf{Q}}_k^{-1}\tilde{\mathbf{Y}}_k \mathbf{s}^c(\theta_k)}{\mathbf{b}^T(\theta_k)\tilde{\mathbf{Q}}_k^{-1}\mathbf{b}^c(\theta_k)\ \|\mathbf{s}(\theta_k)\|^2} \tag{1.75}$$

where $\|\cdot\|$ denotes the Euclidean norm, and

$$\tilde{\mathbf{Q}}_k(\theta_k) = \hat{\mathbf{R}}_k - \frac{\tilde{\mathbf{Y}}_k \mathbf{s}^c(\theta_k)\mathbf{s}^T(\theta_k)\tilde{\mathbf{Y}}_k^*}{N\ \|\mathbf{s}(\theta_k)\|^2} \tag{1.76}$$

and where

$$\hat{\mathbf{R}}_k = \frac{\tilde{\mathbf{Y}}_k \tilde{\mathbf{Y}}_k^*}{N}. \tag{1.77}$$

Inserting (1.75) into (1.74) and after some matrix manipulations (see, e.g., Ref. 30), we obtain the following concentrated negative log-LF function of θ_k:

$$f(\theta_k) = N \ln |\hat{\mathbf{R}}_k| + N \ln [\mathbf{b}^T(\theta_k)\hat{\mathbf{R}}_k^{-1}\mathbf{b}^c(\theta_k)] - N \ln [\mathbf{b}^T(\theta_k)\tilde{\mathbf{Q}}_k^{-1}\mathbf{b}^c(\theta_k)] \tag{1.78}$$

Then, θ_k can be estimated by searching for the minimum of (1.78), and β_k can be estimated by substituting the so-obtained $\hat{\theta}_k$ into (1.75).

Combining BIC and ACO with a RELAX-type procedure [56], whose main step is as described above, leads to ECO, which can be summarized as follows:

Step 0: Initialization. Set $i = 1$.

Step 1: Cyclic Optimization:

(a) Obtain initial parameter estimates of $\{\theta_k\}_{k=1}^i$ and $\{\beta_k\}_{k=1}^i$ by running step 1(a) of ACO and Eq. (1.70).

(b) Do the following until "practical convergence" is reached. For $k = i, 1, 2, \ldots, i-1$.

Compute $\tilde{\mathbf{Y}}_k$ in (1.73) by using the most recent estimates of $\{\beta_p\}_{p\neq k, p=1}^i$ and $\{\theta_p\}_{p\neq k, p=1}^i$.

Estimate θ_k by searching for the minimum of the cost function in (1.78);

Estimate β_k by inserting the so-obtained $\hat{\theta}_k$ in (1.75).

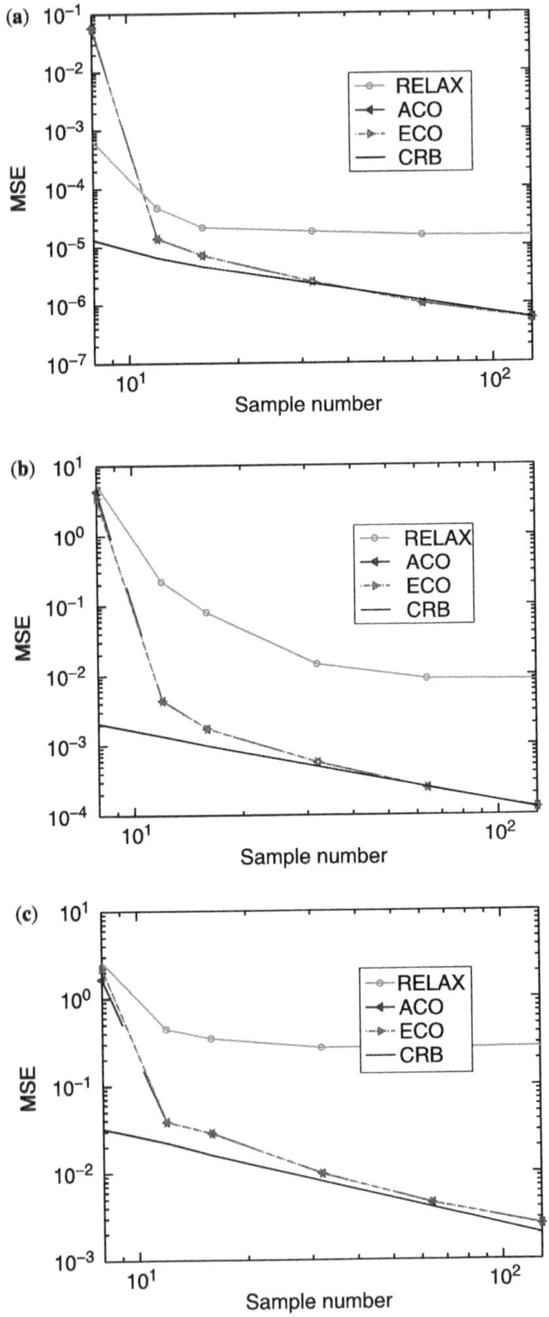

Figure 1.9 Empirical MSEs of the estimates of target location and amplitude parameters versus the number of data samples when SNR = 20 dB: (a) θ_1; (b) θ_2; (c) θ_3; (d) β_1; (e) β_2; (f) β_3.

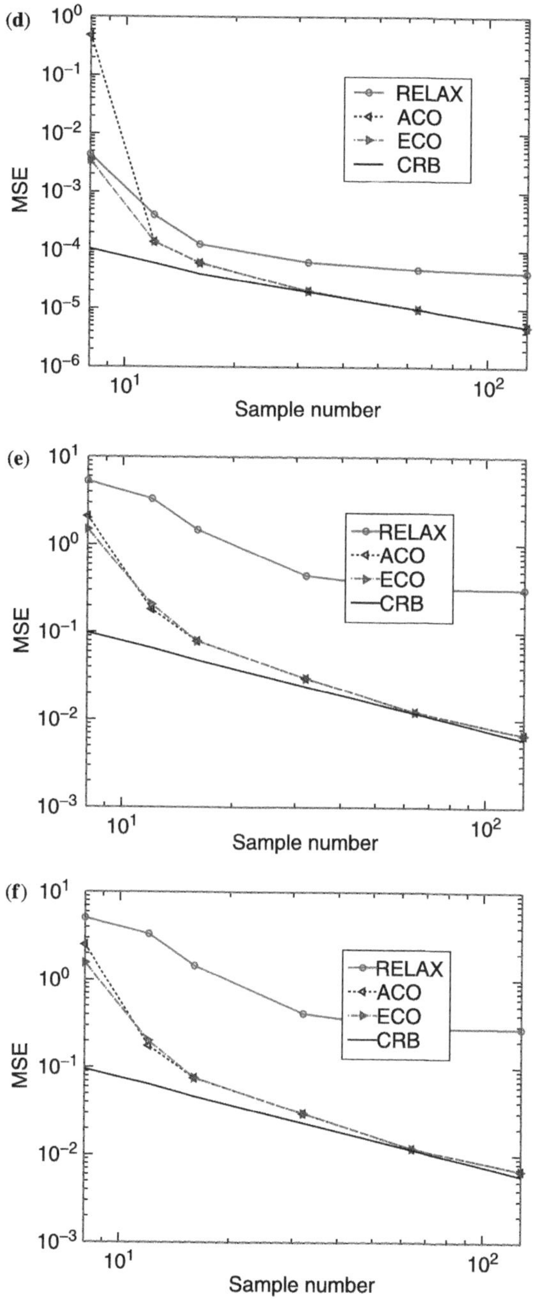

Figure 1.9 *Continued.*

(c) Compute BIC(i) using (1.65), and the so-obtained $\{\hat{\beta}_k\}_{k=1}^{i}$ and $\{\hat{\theta}_k\}_{k=1}^{i}$.
(d) Set $i = i + 1$ and go to step 1(a) until i reaches a prescribed maximum number.

Step 2: BIC Order Selection. Find the minimum of the so-obtained BIC(i) with respect to i.

1.5.2 Numerical Examples

Consider a MIMO radar system where uniform linear arrays with $M_t = M_r = 5$ antennas are used for both transmitting and receiving. The interelement spacing is 0.5λ for the receive array and 2.5λ for the transmit array. The transmitted waveforms are orthogonal QPSK sequences that satisfy $\hat{\mathbf{R}}_{xx} \triangleq (1/N)\mathbf{X}\mathbf{X}^* = \mathbf{I}$ [see also (1.28)].

Consider a scenario in which $K = 3$ targets are at $\theta_1 = -20°$, $\theta_2 = -10°$, and $\theta_3 = -9°$ with the corresponding complex amplitudes $\beta_1 = 4$, $\beta_2 = 4$, and $\beta_3 = 1$, respectively. There is a strong jammer at 0° with an unknown waveform and with amplitude 100 (i.e., 40 dB stronger than β_3). The received signal is corrupted by a zero-mean spatially colored Gaussian noise with an unknown covariance matrix. The (p,q)th element of the noise covariance matrix is $(1/\text{SNR})\, 0.9^{|p-q|} e^{j[(p-q)\pi]/2}$. We choose the "practical convergence" threshold as $\epsilon_t = 10^{-6}$.

We first determine the empirical mean-squared errors (MSEs) of the target location and amplitude parameters estimated by ACO and ECO, as functions of the number of data samples, when the number of targets is assumed to be known *a priori*. For comparison purposes, we also give the performance of a RELAX algorithm, which is essentially a simplified version of ECO without the initialization step, namely, step 1(a). We also provide the corresponding Cramér–Rao bounds (CRBs) (see, e.g., Ref. 30), which represent the best possible performance of any unbiased estimators. The empirical MSEs are all obtained by 100 Monte Carlo simulations. From Figs. 1.9a–f, we can see that RELAX suffers from an error floor problem. ACO and

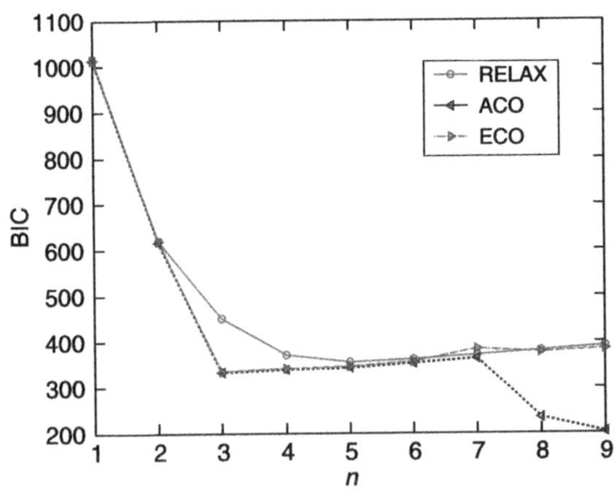

Figure 1.10 BIC cost functions obtained by RELAX, ACO, and ECO when $N = 32$ and SNR = 20 dB.

ECO outperform RELAX significantly. Their MSEs approach the CRB as the number of data samples increases. At a small number of data samples, ECO can provide slightly better performance than does ACO.

Figure 1.10 shows the BIC cost function obtained by RELAX, ACO, and ECO in one of the Monte Carlo simulations when $N = 32$ and SNR = 20 dB. As shown, ECO can be used with BIC to correctly detect the target number, while RELAX and ACO fail to work properly. In fact, ECO detected the target number accurately in all 100 Monte Carlo simulations.

1.6 TRANSMIT BEAMPATTERN DESIGNS

The probing signal vector transmitted by a MIMO radar system can be designed to approximate a desired transmit beampattern and also to minimize the cross-correlation of the signals bounced from various targets of interest — an operation that, like the direct application of adaptive techniques, would be hardly possible for a phased-array radar [12,13,17,23].

The power of the probing signal at a generic focal point with location θ is given by [see (1.1)]

$$P(\theta) = \mathbf{a}^*(\theta)\mathbf{R}\mathbf{a}(\theta) \tag{1.79}$$

where $\mathbf{R}$ is the covariance matrix of $\mathbf{x}(n)$:

$$\mathbf{R} = E\{\mathbf{x}(n)\mathbf{x}^*(n)\} \tag{1.80}$$

The "spatial spectrum" in (1.79), as a function of θ, will be called the *transmit beampattern*.

We assume in this section that the transmit and receive antennas are identically located $[\mathbf{b}(\theta) = \mathbf{a}(\theta)]$. The received data vector in (1.6) can then be expressed as

$$\mathbf{y}(n) = \sum_{k=1}^{K} \beta_k \mathbf{a}^c(\theta_k)\mathbf{a}^*(\theta_k)\mathbf{x}(n) + \boldsymbol{\epsilon}(n), \quad n = 1, \ldots, N \tag{1.81}$$

1.6.1 Beampattern Matching Design

The first problem we will consider in this subsection consists of choosing $\mathbf{R}$, under a uniform elemental power constraint

$$R_{mm} = \frac{c_p}{M_t}, \quad m = 1, \ldots, M_t; \quad \text{with } c_p \text{ given} \tag{1.82}$$

where R_{mm} denotes the (m, m)th element of $\mathbf{R}$, to achieve the following goals [17,23]:

1. Control the spatial power at a number of given target locations by matching (or approximating) a desired transmit beampattern.

2. Minimize the cross-correlation between the probing signals at a number of given target locations; note from (1.1) that the cross-correlation between the probing signals at locations θ and $\bar{\theta}$ is given by $\mathbf{a}^*(\theta)\mathbf{R}\mathbf{a}(\bar{\theta})$.

Let $\phi(\theta)$ denote a desired transmit beampattern, and let $\{\mu_l\}_{l=1}^{L}$ be a fine grid of points that cover the location sectors of interest. We assume that some of these grid-points are good approximations of the locations $\{\theta_k\}_{k=1}^{\tilde{K}}$ of the targets of interest, and that we dispose of (initial) estimates $\{\hat{\theta}_k\}_{k=1}^{\tilde{K}}$ of $\{\theta_k\}_{k=1}^{\tilde{K}}$, where $\tilde{K}$ denotes the number of targets of interest that we wish to probe further. We can obtain $\phi(\theta)$ and $\{\hat{\theta}_k\}_{k=1}^{\tilde{K}}$, for instance, using the Capon and GLRT approaches presented in Section 1.4.

As stated above, our goal is to choose $\mathbf{R}$ such that the transmit beampattern, $\mathbf{a}^*(\theta)\mathbf{R}\mathbf{a}(\theta)$, matches or rather approximates [in a least-squares (LS) sense] the desired transmit beampattern, $\phi(\theta)$, over the sectors of interest, and also such that the cross-correlation (beam)pattern, $\mathbf{a}^*(\theta)\mathbf{R}\mathbf{a}(\bar{\theta})$ (for $\theta \neq \bar{\theta}$), is minimized (once again, in a LS sense) over the set $\{\hat{\theta}_k\}_{k=1}^{\tilde{K}}$. Mathematically, we want to solve the following problem

$$\min_{\alpha,\mathbf{R}} \left\{ \frac{1}{L}\sum_{l=1}^{L} w_l[\alpha\phi(\mu_l) - \mathbf{a}^*(\mu_l)\mathbf{R}\mathbf{a}(\mu_l)]^2 \right.$$
$$\left. + \frac{2w_c}{\tilde{K}^2 - \tilde{K}} \sum_{k=1}^{\tilde{K}-1} \sum_{p=k+1}^{\tilde{K}} \left|\mathbf{a}^*(\hat{\theta}_k)\mathbf{R}\mathbf{a}(\hat{\theta}_p)\right|^2 \right\}$$

subject to

$$R_{mm} = \frac{c_p}{M_t}, \quad m = 1, \ldots, M_t$$
$$\mathbf{R} \geq 0 \tag{1.83}$$

where α is a scaling factor, $w_l \geq 0$, $l = 1, \ldots, L$, is the weight for the lth gridpoint, and $w_c \geq 0$ is the weight for the cross-correlation term. The reason for introducing α in the design problem is that typically $\phi(\theta)$ is given in a "normalized form" [e.g., satisfying $\phi(\theta) \leq 1$, $\forall\theta$], and our interest lies in approximating an appropriately scaled version of $\phi(\theta)$, not $\phi(\theta)$ itself. The value of w_l should be larger than that of w_k if the beampattern matching at μ_l is considered to be more important than the matching at μ_k. Note that by choosing $\max_l w_l > w_c$ we can give more weight to the first term in the design criterion above, and vice versa for $\max_l w_l < w_c$. We have shown [17,23] (see also below) that this design problem can be efficiently solved in polynomial time as a semidefinite quadratic program (SQP).

Let vec($\mathbf{R}$) denote the $M_t^2 \times 1$ vector obtained by stacking the columns of $\mathbf{R}$ on top of each other. Let $\mathbf{r}_v$ denote the $M_t^2 \times 1$ real-valued vector made from

R_{mm} $(m = 1, \ldots, M_t)$ and the real and imaginary parts of R_{mp}, $(m, p = 1, \ldots, M_t; p > m)$. Then, given the Hermitian symmetry of $\mathbf{R}$, we can write

$$\operatorname{vec}(\mathbf{R}) = \mathbf{J}\mathbf{r}_v \tag{1.84}$$

for a suitable $M_t^2 \times M_t^2$ matrix $\mathbf{J}$ whose elements are easily derived constants $(0, \pm j, \pm 1)$. Making use of (1.84) and of some simple properties of the vec operator, the reader can verify that (the symbol $\otimes$ denotes the Krönecker product operator):

$$\begin{aligned}\mathbf{a}^*(\mu_l)\mathbf{R}\mathbf{a}(\mu_l) &= \operatorname{vec}\,[\mathbf{a}^*(\mu_l)\mathbf{R}\mathbf{a}(\mu_l)]\\ &= [\mathbf{a}^T(\mu_l) \otimes \mathbf{a}^*(\mu_l)]\mathbf{J}\mathbf{r}_v\\ &\triangleq -\mathbf{g}_l^T\mathbf{r}_v\end{aligned} \tag{1.85}$$

$$\begin{aligned}\mathbf{a}^*(\hat{\theta}_k)\mathbf{R}\mathbf{a}(\hat{\theta}_p) &= [\mathbf{a}^T(\hat{\theta}_p) \otimes \mathbf{a}^*(\hat{\theta}_k)]\mathbf{J}\mathbf{r}_v\\ &\triangleq \mathbf{d}_{k,p}^*\mathbf{r}_v\end{aligned} \tag{1.86}$$

Inserting (1.85) and (1.86) into (1.83) yields the following more compact form of the design criterion (which clearly shows the quadratic dependence on $\mathbf{r}_v$ and α)

$$\begin{aligned}&\frac{1}{L}\sum_{l=1}^{L} w_l\left[\alpha\phi(\mu_l) + \mathbf{g}_l^T\mathbf{r}_v\right]^2 + \frac{2w_c}{\tilde{K}^2 - \tilde{K}}\sum_{k=1}^{\tilde{K}-1}\sum_{p=k+1}^{\tilde{K}}\left|\mathbf{d}_{k,p}^*\mathbf{r}_v\right|^2\\ &= \frac{1}{L}\sum_{l=1}^{L} w_l\left\{\begin{bmatrix}\phi(\mu_l) & \mathbf{g}_l^T\end{bmatrix}\begin{bmatrix}\alpha\\ \mathbf{r}_v\end{bmatrix}\right\}^2\\ &\quad + \frac{2w_c}{\tilde{K}^2 - \tilde{K}}\sum_{k=1}^{\tilde{K}-1}\sum_{p=k+1}^{\tilde{K}}\left|\begin{bmatrix}0 & \mathbf{d}_{k,p}^*\end{bmatrix}\begin{bmatrix}\alpha\\ \mathbf{r}_v\end{bmatrix}\right|^2\\ &\triangleq \boldsymbol{\rho}^T\boldsymbol{\Gamma}\boldsymbol{\rho}\end{aligned} \tag{1.87}$$

where

$$\boldsymbol{\rho} = \begin{bmatrix}\alpha\\ \mathbf{r}_v\end{bmatrix} \tag{1.88}$$

and

$$\begin{aligned}\boldsymbol{\Gamma} &= \frac{1}{L}\sum_{l=1}^{L} w_l\begin{bmatrix}\phi(\mu_l)\\ \mathbf{g}_l\end{bmatrix}\begin{bmatrix}\phi(\mu_l) & \mathbf{g}_l^T\end{bmatrix}\\ &\quad + \operatorname{Re}\left\{\frac{2w_c}{\tilde{K}^2 - \tilde{K}}\sum_{k=1}^{\tilde{K}-1}\sum_{p=k+1}^{\tilde{K}}\begin{bmatrix}0\\ \mathbf{d}_{k,p}\end{bmatrix}\begin{bmatrix}0 & \mathbf{d}_{k,p}^*\end{bmatrix}\right\}\end{aligned} \tag{1.89}$$

where Re(·) denotes the real part. The matrix $\mathbf{\Gamma}$ above is usually rank-deficient. For example, in the case of an M_t sensor uniform linear array with half-wavelength (or smaller) interelement spacing and for $w_c = 0$, one can show that the rank of $\mathbf{\Gamma}$ is $2M_t$. The rank deficiency of $\mathbf{\Gamma}$, however, does not pose any serious problem for the SQP solver outlined below.

Making use of the form in (1.87) of the design criterion, we can rewrite (1.83) as the following SQP (see, e.g., Refs. 57 and 58)

$$\begin{aligned} \min_{\delta, \varrho} \quad & \delta \\ \text{subject to} \quad & \|\varrho\| \leq \delta \\ & R_{mm}(\varrho) = \frac{c_p}{M_t}, \qquad m = 1, \ldots, M_t \\ & \mathbf{R}(\varrho) \geq 0 \end{aligned} \tag{1.90}$$

where ($\mathbf{\Gamma}^{1/2}$ denotes a Hermitian square root of $\mathbf{\Gamma}$)

$$\varrho = \mathbf{\Gamma}^{1/2}\boldsymbol{\rho} \tag{1.91}$$

and where we have indicated explicitly the (linear) dependence of $\mathbf{R}$ on ϱ.

In some applications, we would like the synthesized beampattern at some given locations to be very close to the desired values. As already mentioned, to a certain extent, this design goal can be achieved by the selection of the weights $\{w_l\}$ of the design criterion in (1.83). However, if we want the beampattern to match the desired values *exactly*, then selecting the weights $\{w_l\}$ is not enough, and we have to modify the design problem, as we now explain.

Consider, for instance, that we want the transmit beampattern at a number of points to be equal to certain desired levels. Then the optimization problem we need to solve is (1.83) with the following additional constraints

$$\mathbf{a}^*(\breve{\mu}_l)\mathbf{R}\mathbf{a}(\breve{\mu}_l) = \zeta_l, \quad l = 1, \ldots, \breve{L} \tag{1.92}$$

where $\{\zeta_l\}$ are predetermined levels. A similar modification of (1.83) takes place when the transmit beampattern at a number of points $\{\breve{\mu}_l\}_{l=1}^{\breve{L}}$ is restricted to be less than or equal to certain desired levels. The extended problems (with additional either equality or inequality constraints) are also SQPs, and therefore, similarly to (1.83), they can be solved efficiently using readily available software [57,58].

1.6.2 Minimum Sidelobe Beampattern Design

Another beampattern design problem we consider consists of choosing $\mathbf{R}$, under the uniform elemental power constraint in (1.82), to achieve the following goals [17,23]:

1. Minimize the sidelobe level in a prescribed region.
2. Achieve a predetermined 3 dB width of the mainbeam.

This problem can be formulated as follows (where s.t. denotes "subject to")

$$\begin{aligned}
\min_{t,\mathbf{R}} \quad & -t \\
\text{s.t.} \quad & \mathbf{a}^*(\theta_0)\mathbf{R}\mathbf{a}(\theta_0) - \mathbf{a}^*(\mu_l)\mathbf{R}\mathbf{a}(\mu_l) \geq t, \quad \forall \mu_l \in \Omega \\
& \mathbf{a}^*(\theta_1)\mathbf{R}\mathbf{a}(\theta_1) = 0.5\mathbf{a}^*(\theta_0)\mathbf{R}\mathbf{a}(\theta_0) \\
& \mathbf{a}^*(\theta_2)\mathbf{R}\mathbf{a}(\theta_2) = 0.5\mathbf{a}^*(\theta_0)\mathbf{R}\mathbf{a}(\theta_0) \\
& \mathbf{R} \geq 0 \\
& R_{mm} = \frac{c_p}{M_t}, \quad m = 1, \ldots, M_t
\end{aligned} \tag{1.93}$$

where $\theta_2 - \theta_1$ (with $\theta_2 > \theta_0$ and $\theta_1 < \theta_0$) determines the 3 dB width of the mainbeam and Ω denotes the sidelobe region of interest. As shown in Refs. 17 and 23, this minimum sidelobe beampattern design problem can be efficiently solved in polynomial time as a semidefinite program (SDP). Note that we can relax somewhat the constraints in (1.93) defining the 3 dB width of the mainbeam; for instance, we can replace them by $(0.5 - \delta)\mathbf{a}^*(\theta_0)\mathbf{R}\mathbf{a}(\theta_0) \leq \mathbf{a}^*(\theta_i)\mathbf{R}\mathbf{a}(\theta_i) \leq (0.5 + \delta)\mathbf{a}^*(\theta_0)\mathbf{R}\mathbf{a}(\theta_0)$, $i = 1,2$, for some small δ. Such a relaxation leads to a design with lower sidelobes, and to an optimization problem that is feasible more often than (1.93).

We can also introduce some flexibility in the elemental power constraint by allowing the elemental power to be within a certain range around c_p/M_t, while still maintaining the same total transmit power of c_p. Such a relaxation of the design problem allows lower sidelobe levels and smoother beampatterns, as we will show later via some numerical examples.

1.6.3 Phased-Array Beampattern Design

Finally, consider the conventional phased-array beampattern design problem in which only the array weight vector can be adjusted and therefore all antennas transmit the same differently scaled waveform. We can readily modify the previously described beampattern matching or minimum sidelobe beampattern designs for the case of phased arrays by adding the constraint

$$\text{rank}(\mathbf{R}) = 1 \tag{1.94}$$

However, because of the rank 1 constraint, both these originally convex optimization problems become nonconvex. The lack of convexity makes the rank 1–constrained problems much harder to solve than the original convex optimization problems [59]. Semidefinite relaxation (SDR) is often used to obtain approximate solutions to such rank-constrained optimization problems [58]. Typically, the SDR is obtained by omitting the rank constraint. Hence, interestingly, *the MIMO beampattern design problems are SDRs of the corresponding phased-array beampattern design problems*.

In the numerical examples below, we have used the Newton-like algorithm presented in Ref. 59 to solve the rank 1–constrained design problems for phased arrays. This algorithm uses SDR to obtain an initial solution. Although the convergence of the said Newton-like algorithm is not guaranteed [59], we did not encounter any apparent problem in our numerical simulations. An interesting detail here is that the approach in Ref. 59 is for real-valued vectors and matrices; therefore we had to rewrite the rank 1 constraint in (1.94) in terms of real-valued quantities

$$\text{rank}(\tilde{\mathbf{R}}) = 2 \tag{1.95}$$

where

$$\tilde{\mathbf{R}} = \begin{bmatrix} \text{Re}\{\mathbf{R}\} & -\text{Im}\{\mathbf{R}\} \\ \text{Im}\{\mathbf{R}\} & \text{Re}\{\mathbf{R}\} \end{bmatrix} \tag{1.96}$$

where Re$\{\mathbf{x}\}$ and Im$\{\mathbf{x}\}$ denote the real and imaginary parts of $\mathbf{x}$, respectively. The equivalence between (1.94) and (1.95) is proved in Appendix 1B.

1.6.4 Numerical Examples

We present several numerical examples to demonstrate the merits of the probing signal designs for MIMO radar systems. We consider a MIMO radar with a uniform linear array (ULA) constituting $M_t = M_r = M = 10$ antennas with half-wavelength spacing between adjacent antennas. This array is used both for transmitting and for receiving. Without loss of generality, the total transmit power is set to $c_p = 1$.

1.6.4.1 Beampattern Matching Design Consider first a scenario where $K = 3$ targets are located at $\theta_1 = -40°$, $\theta_2 = 0°$, and $\theta_3 = 40°$ with complex amplitudes equal to $\beta_1 = \beta_2 = \beta_3 = 1$. There is a strong jammer at 25° with an unknown waveform (uncorrelated with the transmitted MIMO radar waveforms) with a power equal to 10^6 (60 dB). Each transmitted signal pulse has $N = 256$ samples. The received signal is corrupted by zero-mean circularly symmetric spatio temporally white Gaussian noise with variance σ^2. We assume that only the targets reflect the transmitted signals. In practice, the background can also reflect the signals. In the latter case, transmitting most of the power toward the targets should generate much less clutter returns than when transmitting power omnidirectionally. Therefore, a MIMO radar system with a proper transmit beampattern design might provide even greater performance gains than those demonstrated here.

Since we do not assume any prior knowledge about the target locations, orthogonal waveforms are used for MIMO probing. (We refer to this as "initial probing," since after we get the target location estimates with this probing, we can optimize the transmitted beampattern to improve the estimation accuracy.) Using the data collected as a result of this initial probing, we can obtain the Capon spatial spectrum and the GLRT function (see Ref. 28 and Section 1.4). An example of the Capon spectrum for $\sigma^2 = -10$ dB is shown in Fig. 1.11a, where

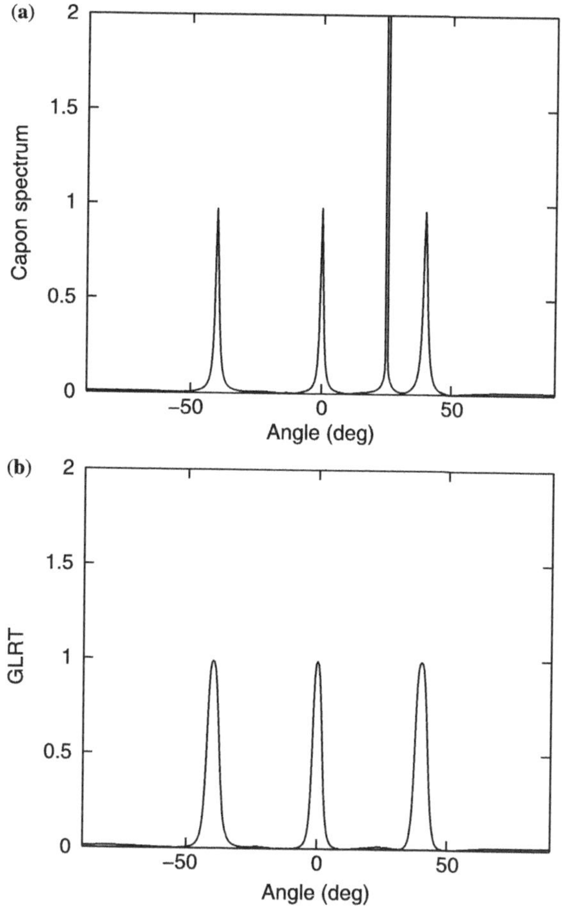

Figure 1.11 The Capon spatial spectrum and the GLRT pseudospectrum as functions of θ, for the initial omnidirectional probing: (a) Capon; (b) GLRT.

very narrow peaks occur around the target locations. Note that in Fig. 1.11a, a false peak occurs around $\theta = 25^\circ$, due to the presence of the very strong jammer. The corresponding GLRT pseudospectrum as a function of θ is shown in Fig. 1.11b. Note that the GLRT is close to one at the target locations and close to zero at any other locations, including the jammer location. Therefore, the GLRT can be used to reject the jammer peak in the Capon spectrum. The remaining peak locations in the Capon spectrum can be taken as the estimated target locations.

To illustrate the beampattern matching design, consider the example considered in Fig. 1.11. The initial target location estimates obtained using Capon or GLRT can be used to derive a desired beampattern. In the following numerical examples, we form the desired beampattern by using the dominant peak locations of the GLRT pseudospectrum, denoted as $\hat{\theta}_1, \ldots, \hat{\theta}_{\hat{K}}$, as follows (where $\hat{K}$ is the resulting estimate of K, and $\tilde{K} = \hat{K}$)

$$\phi(\theta) = \begin{cases} 1, & \theta \in [\hat{\theta}_k - \Delta, \hat{\theta}_k + \Delta],\ k = 1, \ldots, \hat{K} \\ 0, & \text{otherwise} \end{cases} \tag{1.97}$$

where 2Δ is the chosen beamwidth for each target (Δ should be greater than the expected error in $\{\hat{\theta}_k\}$). Figure 1.12b is obtained using (1.97) with $\Delta = 5°$ in the beampattern matching design in (1.83) along with a mesh grid size of $0.1°$, $w_l = 1$, $l = 1, \ldots, L$, and either $w_c = 0$ or $w_c = 1$. Note that the designs obtained with $w_c = 1$ and with $w_c = 0$ are similar to one another. However, the cross-correlation

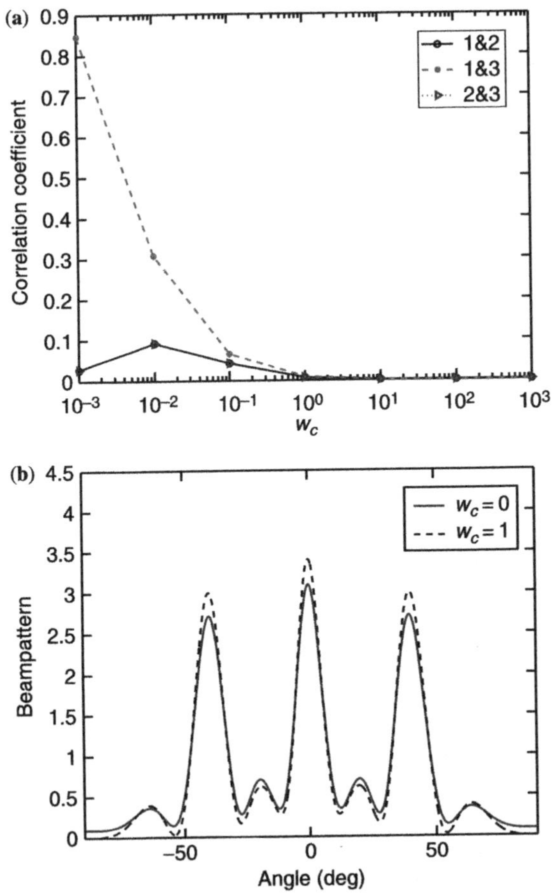

Figure 1.12 MIMO beampattern matching designs with $\Delta = 5°$ and $c_p = 1$, under the uniform elemental power constraint: (a) cross-correlation coefficients of the three target reflected signals as functions of w_c; (b) comparison of the beampatterns obtained with $w_c = 0$ and $w_c = 1$.

behavior of the former is much better than that of the latter in that the reflected signal waveforms corresponding to using $w_c = 1$ are almost uncorrelated with each other, as shown in Fig. 1.12a.

In practice, the theoretical covariance matrix $\mathbf{R}$ of the transmitted signals is realized via the sample covariance matrix $\hat{\mathbf{R}}_{xx} = (1/N)\sum_{n=1}^{N} \mathbf{x}(n)\mathbf{x}^*(n)$, which may cause the synthesized transmit beampattern to be slightly different from the designed beampattern, unless $\hat{\mathbf{R}}_{xx} = \mathbf{R}$, which holds, for instance, if $\mathbf{x}(n) = \mathbf{R}^{1/2}\mathbf{w}(n)$ and $(1/N)\sum_{n=1}^{N} \mathbf{w}(n)\mathbf{w}^*(n) = \mathbf{I}$ exactly; in what follows, however, we assume that $\{\mathbf{w}(n)\}$ is a temporally and spatially white signal for which the last equality holds only approximately in finite samples. Let $\epsilon(\theta)$ denote the relative difference of the beampatterns obtained by using $\hat{\mathbf{R}}_{xx}$ and $\mathbf{R}$:

$$\epsilon(\theta) = \frac{\mathbf{a}^*(\theta)(\hat{\mathbf{R}}_{xx} - \mathbf{R})\mathbf{a}(\theta)}{\mathbf{a}^*(\theta)\mathbf{R}\mathbf{a}(\theta)}, \quad \theta \in [-90^\circ, 90^\circ] \tag{1.98}$$

Figure 1.13a shows an example of $\epsilon(\theta)$, as a function of θ, for the beampattern design in Fig. 1.12b with $w_c = 1$ and for $N = 256$. Note that the difference is quite small. We define the mean-squared error (MSE) between the beampatterns obtained by using $\hat{\mathbf{R}}_{xx}$ and $\mathbf{R}$ as the average of the square of (1.98) over all mesh gridpoints and over the set of Monte Carlo trials. The MSE as a function of N, obtained from 1000 Monte Carlo trials, is shown in Fig. 1.13b. As expected, the larger the sample number N, the smaller the MSE.

Next, we examine the MSEs of the location estimates $\{\hat{\theta}_k\}$ obtained by Capon and of the complex amplitude estimates $\{\hat{\beta}_k\}$ obtained by AML (see Ref. 46 or Section 1.4). In particular, we compare the MSEs obtained using the initial omnidirectional probing with those obtained using the optimal beampattern matching design shown in Fig. 1.12 with $\Delta = 5^\circ$ and $w_c = 1$. Figures 1.14a,b show the MSE curves of the location and complex amplitude estimates obtained for the target at -40° from 1000 Monte Carlo trials (results for the other targets are similar). The estimates obtained using the optimal beampattern matching design are much better; the SNR gain over the omnidirectional design is larger than 10 dB.

Now consider an example where two of the targets are closely spaced. We assume that there are $K = 3$ targets, located at $\theta_1 = -40^\circ$, $\theta_2 = 0^\circ$, and $\theta_3 = 3^\circ$ with complex amplitudes equal to $\beta_1 = \beta_2 = \beta_3 = 1$. There is a strong jammer at 25° with an unknown waveform, which is uncorrelated with the transmitted MIMO radar waveforms, and with a power equal to 10^6 (60 dB). Each transmitted signal pulse has $N = 256$ samples. The received signal is corrupted by zero-mean circularly symmetric spatially and temporally white Gaussian noise with variance $\sigma^2 = -10$ dB. Figures 1.15a,b show the Capon spectrum and the GLRT pseudospectrum, respectively, for the initial omnidirectional probing; as can be seen from these figures, the two closely spaced targets cannot be resolved. Using this initial probing result, we derive an optimal beampattern matching design using (1.83) with a mesh grid size of 0.1°, $w_l = 1$, $l = 1, \ldots, L$, and $w_c = 1$. Since the initial probing indicated only two dominant peaks, only the locations of these two peaks

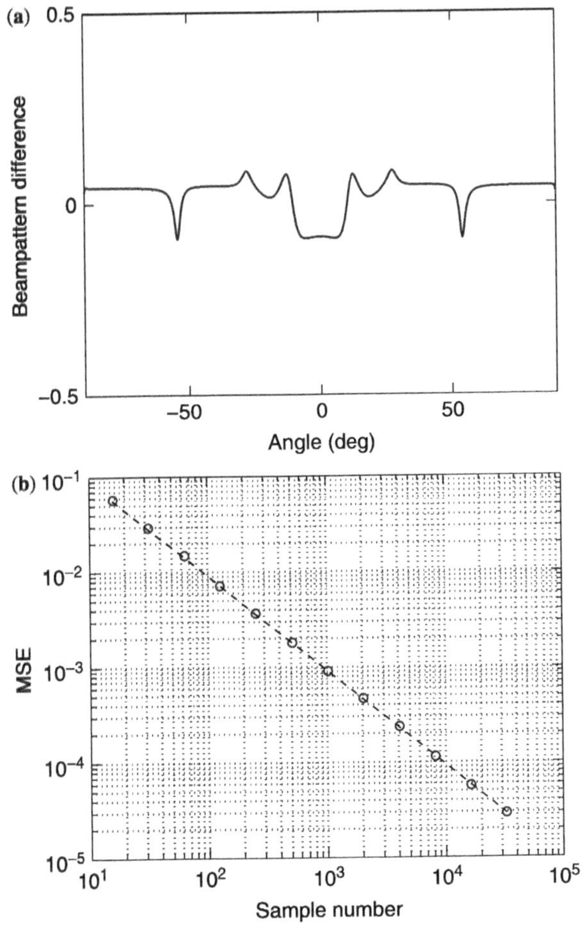

Figure 1.13 Analysis of the beampattern difference resulting from using $\hat{\mathbf{R}}_{xx}$ in lieu of $\mathbf{R}$: (a) beampattern difference versus θ when $N = 256$; (b) average MSE of the beampattern difference as a function of the sample number N.

are used in (1.83). The desired beampattern is given by (1.97) with $\Delta = 10^\circ$ and $\hat{K} = 2$. Figures 1.15c,d, respectively, show the Capon spectrum and the GLRT pseudo-spectrum for the optimal probing. In principle, the two closely spaced targets are now resolved.

To conclude this section, we consider an example where the desired beampattern has only one wide beam centered at 0° with a width of 60°. Figure 1.16a shows the result for the beampattern matching design in (1.83) with a mesh grid size of 0.1°, $w_l = 1$, $l = 1, \ldots, L$, and $w_c = 0$. Figure 1.16b shows the corresponding phased-array beampattern obtained by using the additional constraint of $\text{rank}(\mathbf{R}) = 1$ in (1.83). We note that, under the elemental power constraint, the number of degrees of freedom (DOF) of the phased array that can be used for beampattern design is

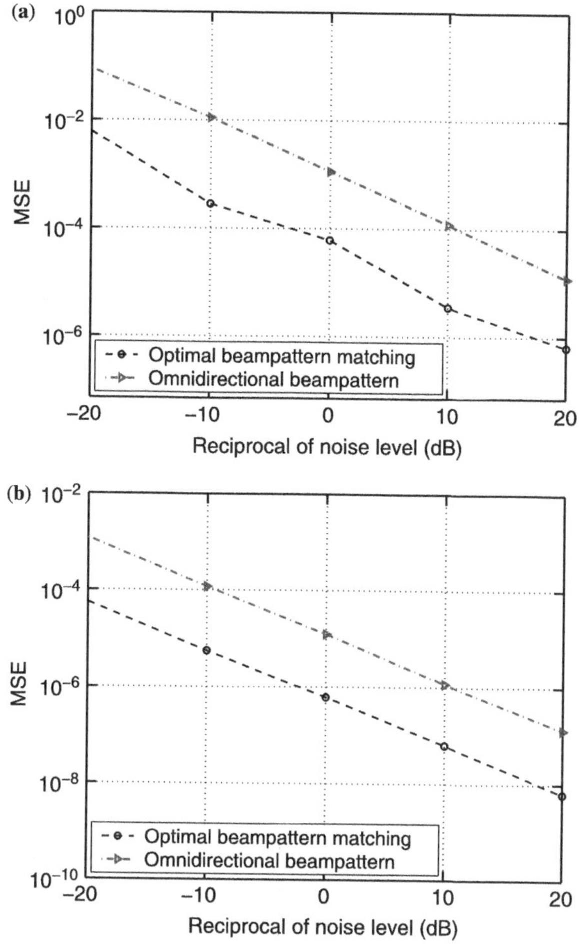

Figure 1.14 MSEs of (a) the location estimates and of (b) the complex amplitude estimates for the first target, as functions of $-10\log_{10}\sigma^2$, obtained with initial omnidirectional probing and with probing using the beampattern matching design for $\Delta = 5^\circ$, $w_c = 1$, and $c_p = 1$.

equal to only $M_t - 1$ (real-valued parameters); consequently, it is difficult for the phased array to synthesize a proper wide beam. The MIMO design, however, can be used to achieve a beampattern significantly closer to the desired beampattern because of its much higher DOF value: $M_t^2 - M_t$.

1.6.4.2 Minimum-Sidelobe Beampattern Design Consider the minimum sidelobe beampattern design problem in (1.93), with the mainbeam centered at $\theta_0 = 0^\circ$, with a 3 dB width equal to 20° ($\theta_2 = -\theta_1 = 10^\circ$), and with $c_p = 1$, for the same MIMO radar scenario as the one considered in Fig. 1.11. The sidelobe region is $\Omega = [-90°, -20°] \cup [20°, 90°]$. The MIMO minimum-sidelobe beampattern design is shown in Fig. 1.17a. Note that the peak sidelobe level achieved by the MIMO design is approximately 18 dB below the mainlobe peak level. Figure 1.17b shows the corresponding phased-array beampattern obtained by using the additional constraint rank($\mathbf{R}$) = 1. The phased-array design fails to provide a proper mainlobe (it suffers from peak splitting), and its peak sidelobe level is much higher than that of its MIMO counterpart.

Figure 1.18 is similar to Fig. 1.17, except that now we allow the elemental powers to be between 80% and 120% of $c_p/M_t = 1/10$, while the total power is still constrained to be $c_p = 1$. Observe that by allowing such a flexibility in setting the elemental powers, we can bring down the peak sidelobe level of the MIMO beampattern by $\sim$3 dB. The phased-array design, on the other hand, does not appear to improve in any significant way.

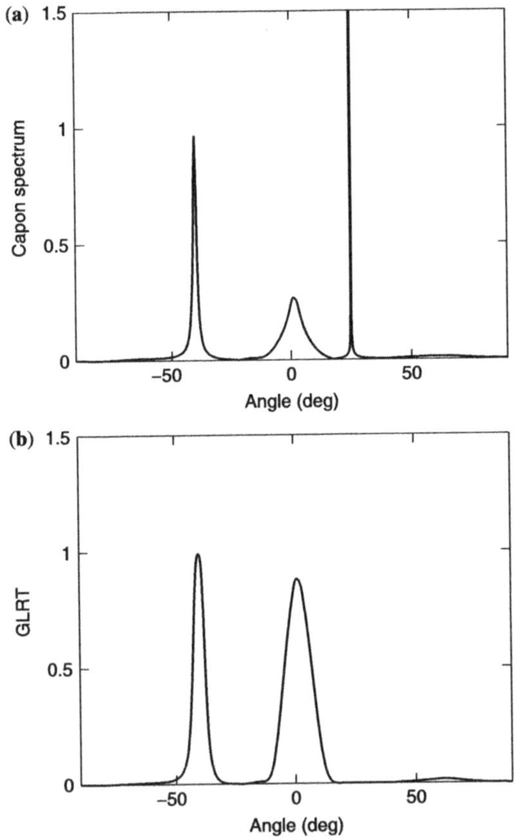

Figure 1.15 The Capon spatial spectra and the GLRT pseudospectra as functions of θ: (a) Capon for initial omnidirectional probing; (b) GLRT for initial omnidirectional probing; (c) Capon for optimal probing; (d) GLRT for optimal probing.

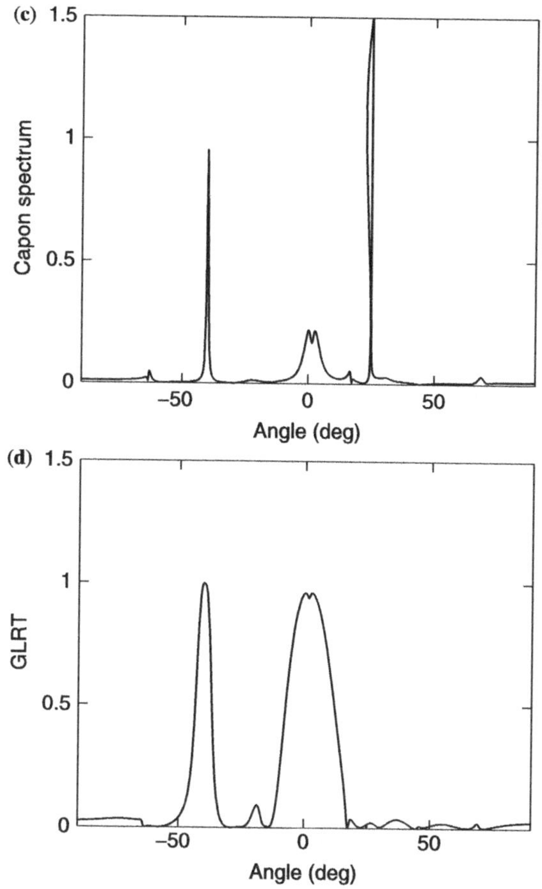

Figure 1.15 *Continued.*

1.6.5 Application to Ultrasound Hyperthermia Treatment of Breast Cancer

Breast cancer is the most common nonskin malignancy in women and the second leading cause of female cancer mortality [60]. There are over 200,000 new cases of invasive breast cancer diagnosed each year in the United States, where one out of every seven women will be diagnosed with breast cancer in her lifetime (the American Cancer Society, 2006, URL: http://www.cancer.org).

The development of breast cancer imaging techniques, such as microwave imaging [61,62], ultrasound imaging [63,64], thermal acoustic imaging [65], and magnetic resonance imaging (MRI), has improved the ability to visualize and accurately locate the breast tumor without the need for surgery [66]. The possibility of noninvasive local hyperthermia treatment of breast cancer is also under investigation. Many studies have been performed to demonstrate the effectiveness of local

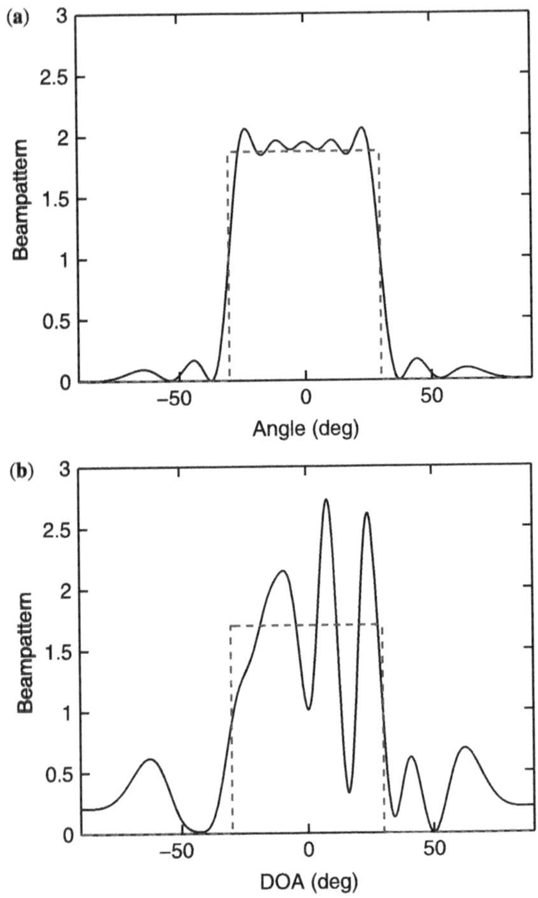

Figure 1.16 Beampattern matching designs under the uniform elemental power constraint: (a) MIMO; (b) phased array.

hyperthermia on the treatment of breast cancer [67,68]. A challenge in the local hyperthermia treatment of breast cancer is to heat the malignant tumors to a temperature above 43°C for ~30–60 min, and at the same time maintain a low temperature level in the surrounding healthy breast tissue region.

There are two major classes of local hyperthermia techniques: microwave hyperthermia [69] and ultrasound hyperthermia [70]. The penetration of microwave in biological tissues is poor. Moreover, the focal spot generated by microwave is undesirable at the normal–cancerous tissues interface because of the long wavelength of the microwave. Ultrasound can achieve much better penetration depths than can microwave. However, because the acoustic wavelength is very short, the focal spot generated by ultrasound is very small (<1 mm in diameter) compared to the large tumor region (centimeters in diameter, on the average). Thus, many focal spots are required for complete tumor coverage, and this results in a long treatment time and missed cancer cells.

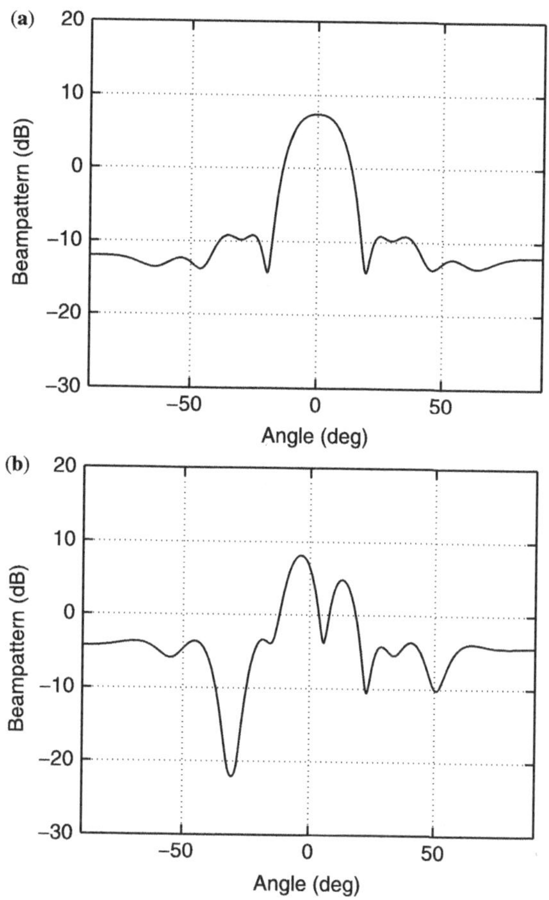

Figure 1.17 Minimum sidelobe beampattern designs, under the uniform elemental power constraint, when the 3 dB mainbeam width is 20°: (a) MIMO; (b) phased array.

We apply below the MIMO (or waveform diversity based) transmit beampattern designs to the ultrasound hyperthermia treatment of breast cancer.

1.6.5.1 Waveform Design We consider an ultrasound hyperthermia system as shown in Fig. 1.19. Let $\mathbf{r}_0$ denote the center location of the tumor, which is assumed to be previously estimated accurately using breast cancer imaging techniques. There are M_t acoustic transducers deployed around the breast at locations $\mathbf{r}_m(m = 1, 2, \ldots, M_t)$. We assume that the transmitted acoustic signals $\{x_m(n)\}$ are narrowband and that each acoustic transducer is omnidirectional. The baseband signal at a location $\mathbf{r}$ inside the breast can be described as

$$y(\mathbf{r}, n) = \sum_{m=1}^{M_t} \frac{e^{-j2\pi f_0 \tau_m(\mathbf{r})}}{\| \mathbf{r}_m - \mathbf{r} \|^{1/2}} x_m(n), \qquad n = 1, 2, \ldots, N \tag{1.99}$$

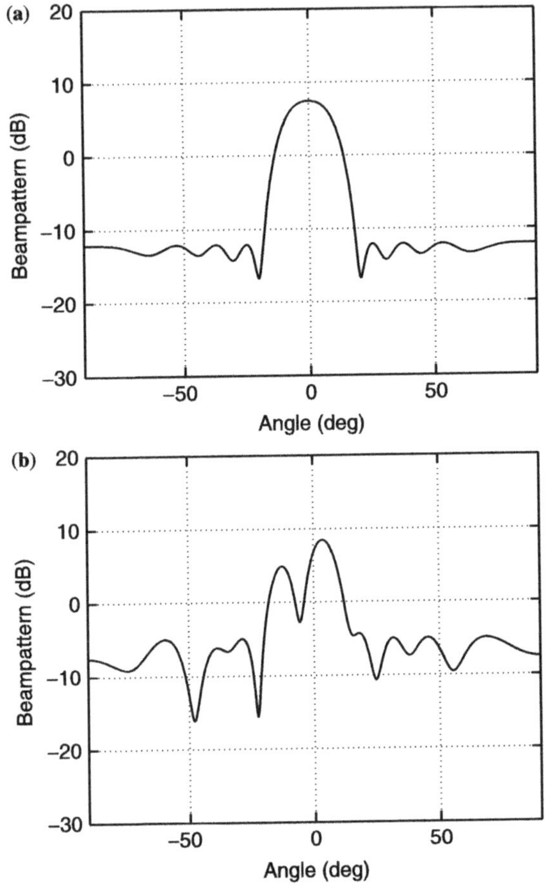

Figure 1.18 Minimum sidelobe beampattern designs, under a relaxed ($\pm 20\%$) elemental power constraint, when the 3 dB mainbeam width is 20°: (a) MIMO; (b) phased array.

where f_0 is the carrier frequency

$$\tau_m(\mathbf{r}) = \frac{\|\mathbf{r}_m - \mathbf{r}\|}{c} \tag{1.100}$$

is the time needed by the signal emitted via the mth transducer to arrive at the location $\mathbf{r}$, where c is the sound speed inside the breast tissues, and $1/(\|\mathbf{r}_m - \mathbf{r}\|^{1/2})$ is the propagation attenuation of the acoustic wave. Therefore the steering vector is

$$\mathbf{a}(\mathbf{r}) = \left[\frac{e^{j2\pi f_0 \tau_1(\mathbf{r})}}{\| \mathbf{r}_1 - \mathbf{r} \|^{1/2}} \quad \frac{e^{j2\pi f_0 \tau_2(\mathbf{r})}}{\| \mathbf{r}_2 - \mathbf{r} \|^{1/2}} \quad \cdots \quad \frac{e^{j2\pi f_0 \tau_{M_t}(\mathbf{r})}}{\| \mathbf{r}_{M_t} - \mathbf{r} \|^{1/2}}\right]^T \tag{1.101}$$

Equation (1.99) can be rewritten as

$$y(\mathbf{r}, n) = \mathbf{a}^*(\mathbf{r})\mathbf{x}(n), \qquad n = 1, 2, \ldots, N \tag{1.102}$$

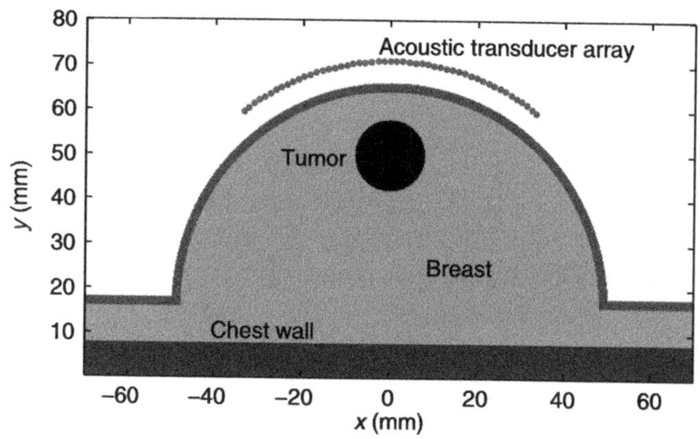

Figure 1.19 Acoustic array and breast model.

The power of the transmitted signals at location **r** (i.e., the transmit beampattern) is given by

$$P(\mathbf{r}) = \mathrm{E}\{y(\mathbf{r}, n)y^*(\mathbf{r}, n)\} = \mathbf{a}^*(\mathbf{r})\mathbf{R}\mathbf{a}(\mathbf{r}) \tag{1.103}$$

The transmit beampattern is a function of the location **r**.

The goal here is to focus the acoustic power onto the entire tumor region while minimizing the peak power level in the surrounding healthy breast tissue region. The corresponding MIMO or waveform diversity-based beampattern design problem is to choose the covariance matrix **R**, under the uniform elemental power constraint [see (1.82)], to achieve the following goals (see Ref. 71 and also (1.93) for a related design problem):

1. Achieve a predetermind main-beam width that is matched to the entire tumor region (be within 10% of the power deposited at the tumor center).
2. Minimize the peak sidelobe level in a prescribed region (the surrounding healthy breast tissue region).

This problem can be formulated as

$$\begin{aligned}
\min_{t,\mathbf{R}} \quad & -t \\
\text{s.t.} \quad & \mathbf{a}^*(\mathbf{r}_0)\mathbf{R}\mathbf{a}(\mathbf{r}_0) - \mathbf{a}^*(\mu)\mathbf{R}\mathbf{a}(\mu) \geq t, && \forall\, \mu \in \Omega_B \\
& \mathbf{a}^*(\nu)\mathbf{R}\mathbf{a}(\nu) \geq 0.9\mathbf{a}^*(\mathbf{r}_0)\mathbf{R}\mathbf{a}(\mathbf{r}_0), && \forall\, \nu \in \Omega_T \\
& \mathbf{a}^*(\nu)\mathbf{R}\mathbf{a}(\nu) \leq 1.1\mathbf{a}^*(\mathbf{r}_0)\mathbf{R}\mathbf{a}(\mathbf{r}_0), && \forall\, \nu \in \Omega_T \\
& \mathbf{R} \geq 0 \\
& R_{mm} = \frac{c_p}{M_t}, \quad m = 1,2, \ldots, M_t,
\end{aligned} \tag{1.104}$$

where Ω_T and Ω_B, which are assumed to be given, denote the tumor region and the surrounding healthy breast tissue region (sidelobe region), respectively.

As shown in Section 1.6.2, this beampattern design problem is a semidefinite program (SDP) that can be efficiently solved in polynomial time using public-domain software. Once $\mathbf{R}$ is determined, a signal sequence $\{\mathbf{x}(n)\}$ that has $\mathbf{R}$ as its covariance matrix can be synthesized as

$$\mathbf{x}(n) = \mathbf{R}^{1/2}\mathbf{w}(n), \qquad n = 1, 2, \ldots, N \tag{1.105}$$

where $\{\mathbf{w}(n)\}$ is a sequence of i.i.d. random vectors with mean zero and covariance matrix $\mathbf{I}$, and $\mathbf{R}^{1/2}$ denotes a Hermitian square root of $\mathbf{R}$.

By transmitting $\mathbf{x}(n)$ given in (1.105) using the acoustic transducer array, we can approximately get a desired high acoustic power deposition in the entire tumor region, and at the same time minimize the power deposition in the surrounding healthy breast tissue region.

1.6.5.2 Simulation Examples For simulation purposes, the 2D breast model shown in Fig. 1.19 is considered. This breast model is a semicircle with a diameter equal to 10 cm, which includes breast tissues, skin, and chest wall. The acoustic properties of the breast tissues are assumed to be random with a variation of $\pm 5\%$ around the nominal values. A tumor with a diameter of 16 mm is embedded below the skin, with the tumor center location at $x = 0$ mm, $y = 50$ mm. There are 51 acoustic transducers deployed uniformly around the breast model. The distance between two adjacent acoustic transducers is 1.5 mm (this is 0.5λ of the carrier frequency, which is equal to 500 kHz).

The two basic linear acoustic wave generation equations are [72,73]

$$\rho\frac{\partial}{\partial t}\mathbf{u}(\mathbf{r}, t) = -\nabla p(\mathbf{r}, t) \tag{1.106}$$

and

$$\nabla \cdot \mathbf{u}(\mathbf{r}, t) = -\frac{1}{\rho c^2}\frac{\partial}{\partial t}p(\mathbf{r}, t) + \alpha p(\mathbf{r}, t) \tag{1.107}$$

where $\mathbf{u}(\mathbf{r}, t)$ is the acoustic velocity vector, $p(\mathbf{r}, t)$ is the acoustic pressure field, ρ is the mass density, $\nabla \cdot \mathbf{u}$ is the divergence of $\mathbf{u}$, and α is the attenuation coefficient. The nominal values of ρ, c, and α for different breast tissues are listed in Table 1.1 [64,74,75]. The values for the tumor are considered to be the same as those for the chest wall, as tumor-specific values are not available in the literature. The finite-difference time-domain (FDTD) approach is used to compute the acoustic field distribution based on Eqs. (1.106) and (1.107). More details about FDTD for acoustic simulations can be found in the literature [76,77].

TABLE 1.1 Typical Acoustic Parameters of Breast Tissue

Tissue	ρ (kg/m^3)	c (m/s)	α (dB/cm)
Breast tissue	1020	1510	0.26
Skin	1100	1537	1.0
Chest wall	1041	1580	0.28
Tumor	1041	1580	0.28

TABLE 1.2 Typical Thermal Parameters of Breast Tissue

Tissue	K [W/(m · °C)]	A (W/m^3)	B [W/(°C · m^3)]	C [J/(kg · °C)]
Breast tissue	0.499	480	2700	3550
Skin	0.376	1620	9100	3500
Chest wall	0.564	480	2700	3510
Tumor	0.564	480	2700	3510

Once the acoustic pressure has been calculated, the acoustic power deposition at location **r**, denoted as $Q(\mathbf{r})$, is given by [73]

$$Q(\mathbf{r}) = \frac{\alpha}{\rho c} |p(\mathbf{r})|^2 \tag{1.108}$$

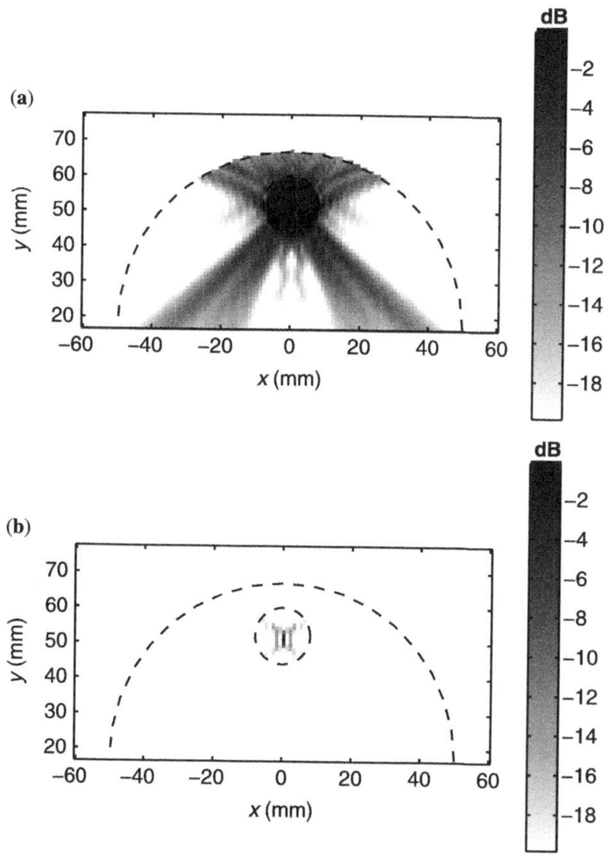

Figure 1.20 Transmit beampatterns $P(\mathbf{r})$ in (1.103) for (a) and in (1.111) for (b): (a) waveform diversity; (b) phased array.

After obtaining the acoustic power deposition, a 2D thermal model is used to calculate the temperature distribution in the breast tissues. The thermal model is based on the bioheat equation [5]

$$\begin{aligned} \nabla \cdot (K(\mathbf{r})\, \nabla T(\mathbf{r}, t)) + A(\mathbf{r}) + Q(\mathbf{r}) - B(\mathbf{r})(T(\mathbf{r}, t) - T_B) \\ = C(\mathbf{r})\rho(\mathbf{r}) \frac{\partial T(\mathbf{r}, t)}{\partial t} \end{aligned} \tag{1.109}$$

where $K(\mathbf{r})$ is the thermal conductivity, $A(\mathbf{r})$ is metabolic heat production, $B(\mathbf{r})$ represents the heat exchange mechanism due to capillary blood perfusion, $C(\mathbf{r})$ is the specific heat, and T_B is the blood temperature, which can be assumed as the body temperature. The thermal parameters used for our breast model are listed in Table 1.2 (more detailed discussions can be found in Ref. 78).

The thermal model is also simulated using the FDTD method [79]. The body temperature and the environmental temperature are set to 36.8°C and 20°C,

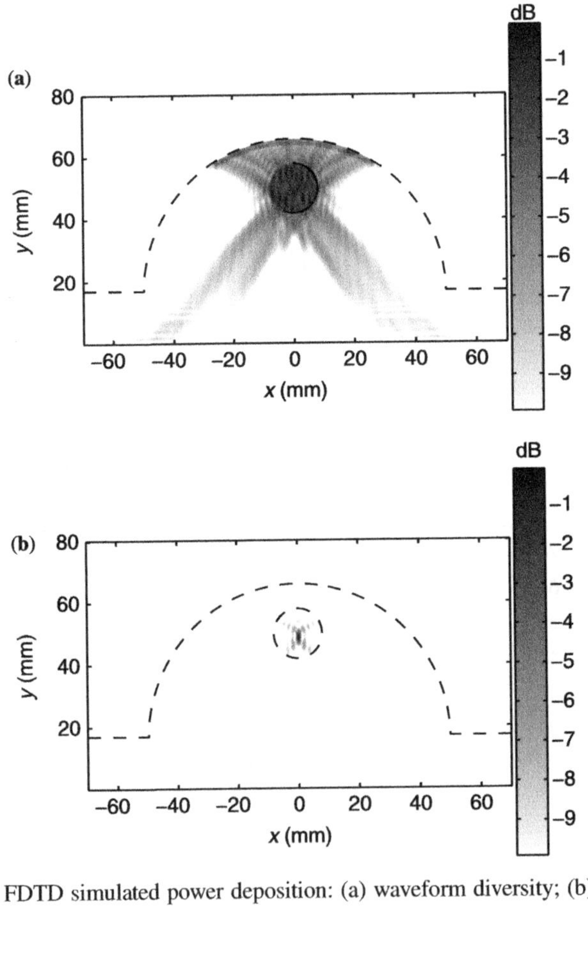

Figure 1.21 FDTD simulated power deposition: (a) waveform diversity; (b) phased array.

respectively. The convective boundary condition is used at the skin surface. We demonstrate the performance of the proposed waveform diversity-based method via several numerical examples. The conventional delay-and-sum (DAS) [i.e., the least-squares (LS) approach in Refs. 27 and 28] phased-array beamforming method is also applied to the same model for comparison purposes. The DAS-based phased-array beamformer transmits a single waveform using the following weight vector:

$$\mathbf{w} = \sqrt{\frac{c_p}{M_t}}\begin{bmatrix} e^{j2\pi f_0 \tau_1(\mathbf{r})} & e^{j2\pi f_0 \tau_2(\mathbf{r})} & \cdots & e^{j2\pi f_0 \tau_{M_t}(\mathbf{r})} \end{bmatrix}^T \tag{1.110}$$

The corresponding beampattern is

$$P(\mathbf{r}) = \left|\mathbf{a}^*(\mathbf{r})\,\mathbf{w}\right|^2 \tag{1.111}$$

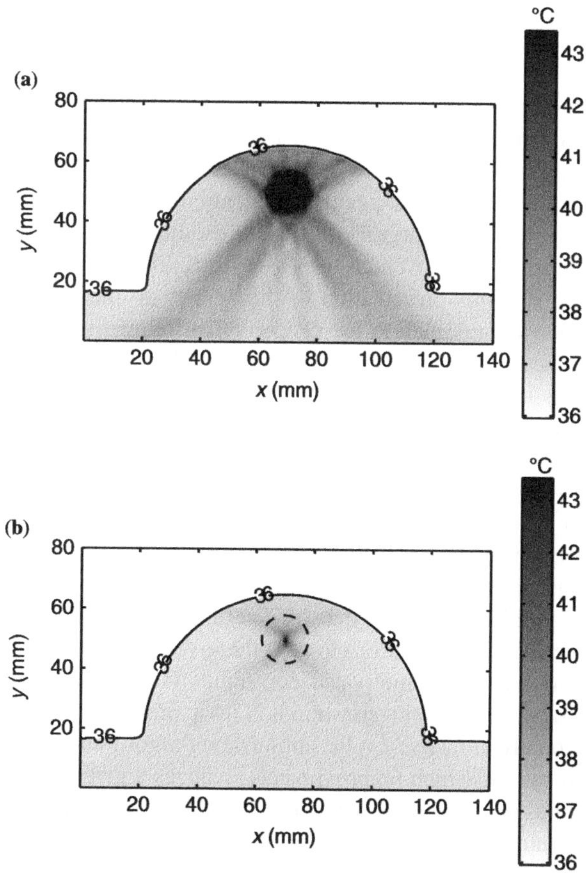

Figure 1.22 Temperature distribution: (a) waveform diversity; (b) phased array.

Figure 1.20 shows the transmit beampatterns within the breast model. Figure 1.20a is the beampattern obtained with the waveform diversity technique, which is calculated by using (1.103) with the optimal covariance matrix **R** determined by using (1.104). The figure shows that the mainbeam is matched to the tumor region well, and the sidelobe level is low. Figure 1.20b is the DAS beampattern, which is calculated using (1.111). The DAS beampattern is very narrow, and is focused on the center of the tumor. Figures 1.21a,b show the FDTD simulated acoustic power densities within the breast model for the waveform diversity technique and DAS, respectively. The acoustic power densities in Fig. 1.21 agree well with the beampatterns in Fig. 1.20; the waveform diversity allows the focal spot to be matched to the entire tumor region.

Figure 1.22 shows the temperature distribution within the breast model. Figure 1.22a is the result for waveform diversity, which shows that the entire tumor region is heated to a temperature greater than 43°C while the surrounding healthy tissues have a lower temperature (below 40°C). By comparison, DAS heats only a very small region in the center of the tumor to a temperature greater than 43°C.

1.7 CONCLUSIONS

We have presented some recent results on the emerging technology of MIMO radar with collocated antennas. We have shown that the waveform diversity offered by such a MIMO radar system enables significant superiority over its phased-array counterpart, including improved parameter identifiability, direct applicability of adaptive nonparametric techniques for parameter estimation, enhanced performance of parametric algorithms, as well as superior flexibility of transmit beampattern designs. We hope that this presentation of our results on the MIMO radar, along with the related results presented in the chapters that follow by our colleagues, will stimulate the interest deserved by this topic in both academia and government agencies as well as industry.

For the reader who wants to probe further, we mention here some ongoing research results that have not been included in this chapter. Some initial results on waveform optimization with respect to the covariance matrix of the transmitted waveforms for MIMO spacetime adaptive processing (STAP) have been presented in Refs. 19 and 29. Given a covariance matrix of the transmitted waveforms, obtained in an optimization stage such as the one described in Section 1.6, or in Refs. 19 and 29, or simply prespecified, the problem becomes that of determining a signal waveform matrix whose covariance matrix is equal or close to the given covariance matrix, *and* that also satisfies some practically motivated constraints (such as constant-modulus or low peak-to-average-power ratio constraints). In a previous paper we presented [80] a computationally attractive cyclic optimization algorithm for the synthesis of such a waveform matrix, which (approximately) realizes a given optimal covariance matrix under various practical constraints. For MIMO synthetic aperture radar (SAR) imaging, we presented [81] a related cyclic optimization algorithm for the synthesis of constant-modulus transmit signals with good auto- and cross-correlation properties. In our earlier studies cited among the references for this chapter, we also discussed the use of an instrumental variables approach to design receive filters that can be used to minimize the impact of scatterers in nearby range bins on the received signals from the range bin of interest (the so-called range compression problem).

APPENDIX IA GENERALIZED LIKELIHOOD RATIO TEST

We assume that the columns of the residual term $\mathbf{Z}$ in (1.29) are independently and identically distributed circularly symmetric complex Gaussian random vectors with zero-mean and unknown covariance matrix $\mathbf{Q}$. We derive the GLRT for each θ of interest. For notational brevity, we omit the argument θ of ρ, $\mathbf{a}$, $\mathbf{b}$, and β.

Following Kelly [82] (see also Ref. 83), we define the GLR as follows

$$\rho = 1 - \left[\frac{\max_{\mathbf{Q}} f(\mathbf{Y}|\beta = 0, \mathbf{Q})}{\max_{\beta,\mathbf{Q}} f(\mathbf{Y}|\beta, \mathbf{Q})}\right]^{1/N} \tag{1A.1}$$

where

$$f(\mathbf{Y}|\beta, \mathbf{Q}) = \pi^{-NM_r}|\mathbf{Q}|^{-N}\exp\{-\mathrm{tr}[\mathbf{Q}^{-1}(\mathbf{Y} - \mathbf{b}^c\beta\mathbf{a}^*\mathbf{X})(\mathbf{Y} - \mathbf{b}^c\beta\mathbf{a}^*\mathbf{X})^*]\} \tag{1A.2}$$

is the probability density function of the observed data matrix $\mathbf{Y}$ given the parameters β and $\mathbf{Q}$, and $\mathrm{tr}(\cdot)$ and $|\cdot|$ denote the trace and determinant of a matrix, respectively. From (1A.1), we note that the value of the GLR, ρ, lies between 0 and 1. If there is a target at a θ of interest, we usually have $\max_{\beta,\mathbf{Q}} f(\mathbf{Y}|\beta, \mathbf{Q}) \gg \max_{\mathbf{Q}} f(\mathbf{Y}|\beta = 0, \mathbf{Q})$ (i.e., $\rho \approx 1$; otherwise $\rho \approx 0$).

Solving the optimization problems in (1A.1) with respect to $\mathbf{Q}$ yields

$$\max_{\mathbf{Q}} f(\mathbf{Y}|\beta = 0, \mathbf{Q}) = (\pi e)^{-NM_r}|\hat{\mathbf{R}}_{yy}|^{-N} \tag{1A.3}$$

and

$$\max_{\beta,\mathbf{Q}} f(\mathbf{Y}|\beta, \mathbf{Q}) = (\pi e)^{-NM_r}\left\{\min_{\beta}\left|\frac{1}{N}(\mathbf{Y} - \mathbf{b}^c\beta\mathbf{a}^*\mathbf{X})(\mathbf{Y} - \mathbf{b}^c\beta\mathbf{a}^*\mathbf{X})^*\right|\right\}^{-N} \tag{1A.4}$$

Applying the technique used in Refs. 40 and 84, we obtain

$$\begin{aligned}
&\left|\frac{1}{N}(\mathbf{Y} - \mathbf{b}^c\beta\mathbf{a}^*\mathbf{X})(\mathbf{Y} - \mathbf{b}^c\beta\mathbf{a}^*\mathbf{X})^*\right| \\
&= \left|\hat{\mathbf{R}}_{yy} - \frac{\mathbf{Y}\mathbf{X}^*\mathbf{a}\beta^*\mathbf{b}^T}{N} - \frac{\mathbf{b}^c\beta\mathbf{a}^*\mathbf{X}\mathbf{Y}^*}{N} + |\beta|^2\mathbf{b}^c\mathbf{a}^*\hat{\mathbf{R}}_{xx}\mathbf{a}\mathbf{b}^T\right| \\
&= \left|\hat{\mathbf{Q}} + (\mathbf{a}^*\hat{\mathbf{R}}_{xx}\mathbf{a})\left(\mathbf{b}^c\beta - \frac{\mathbf{Y}\mathbf{X}^*\mathbf{a}}{N(\mathbf{a}^*\hat{\mathbf{R}}_{xx}\mathbf{a})}\right)\left(\mathbf{b}^c\beta - \frac{\mathbf{Y}\mathbf{X}^*\mathbf{a}}{N(\mathbf{a}^*\hat{\mathbf{R}}_{xx}\mathbf{a})}\right)^*\right|
\end{aligned}$$

$$
\begin{aligned}
&= |\hat{\mathbf{Q}}| \left| \mathbf{I} + (\mathbf{a}^*\hat{\mathbf{R}}_{xx}\mathbf{a})\hat{\mathbf{Q}}^{-1}\left(\mathbf{b}^c\beta - \frac{\mathbf{YX}^*\mathbf{a}}{N(\mathbf{a}^*\hat{\mathbf{R}}_{xx}\mathbf{a})}\right)\left(\mathbf{b}^c\beta - \frac{\mathbf{YX}^*\mathbf{a}}{N(\mathbf{a}^*\hat{\mathbf{R}}_{xx}\mathbf{a})}\right)^* \right| \\
&= |\hat{\mathbf{Q}}| \left[1 + (\mathbf{a}^*\hat{\mathbf{R}}_{xx}\mathbf{a})\left(\mathbf{b}^c\beta - \frac{\mathbf{YX}^*\mathbf{a}}{N(\mathbf{a}^*\hat{\mathbf{R}}_{xx}\mathbf{a})}\right)^* \hat{\mathbf{Q}}^{-1}\left(\mathbf{b}^c\beta - \frac{\mathbf{YX}^*\mathbf{a}}{N(\mathbf{a}^*\hat{\mathbf{R}}_{xx}\mathbf{a})}\right) \right] \\
&\geq |\hat{\mathbf{Q}}| \left[1 + \frac{\mathbf{a}^*\mathbf{XY}^*}{N^2(\mathbf{a}^*\hat{\mathbf{R}}_{xx}\mathbf{a})}\hat{\mathbf{Q}}^{-1}\left(\mathbf{I} - \frac{\mathbf{b}^c\mathbf{b}^T\hat{\mathbf{Q}}^{-1}}{\mathbf{b}^T\hat{\mathbf{Q}}^{-1}\mathbf{b}^c}\right)\mathbf{YX}^*\mathbf{a} \right]
\end{aligned}
\tag{1A.5}
$$

where we have applied the fact that $|\mathbf{I} + \mathbf{A}_1\mathbf{B}_1| = |\mathbf{I} + \mathbf{B}_1\mathbf{A}_1|$ [31], and the equality holds when $\beta = \hat{\beta}_{\text{APES}}$. Note that

$$
\begin{aligned}
&|\hat{\mathbf{Q}}| \left[1 + \frac{\mathbf{a}^*\mathbf{XY}^*}{N^2(\mathbf{a}^*\hat{\mathbf{R}}_{xx}\mathbf{a})}\hat{\mathbf{Q}}^{-1}\left(\mathbf{I} - \frac{\mathbf{b}^c\mathbf{b}^T\hat{\mathbf{Q}}^{-1}}{\mathbf{b}^T\hat{\mathbf{Q}}^{-1}\mathbf{b}^c}\right)\mathbf{YX}^*\mathbf{a} \right] \\
&= |\hat{\mathbf{Q}}| \left| \mathbf{I} + \hat{\mathbf{Q}}^{-1}\left(\mathbf{I} - \frac{\mathbf{b}^c\mathbf{b}^T\hat{\mathbf{Q}}^{-1}}{\mathbf{b}^T\hat{\mathbf{Q}}^{-1}\mathbf{b}^c}\right)\frac{\mathbf{YX}^*\mathbf{aa}^*\mathbf{XY}^*}{N^2(\mathbf{a}^*\hat{\mathbf{R}}_{xx}\mathbf{a})} \right| \\
&= \left| \hat{\mathbf{Q}} + \left(\mathbf{I} - \frac{\mathbf{b}^c\mathbf{b}^T\hat{\mathbf{Q}}^{-1}}{\mathbf{b}^T\hat{\mathbf{Q}}^{-1}\mathbf{b}^c}\right)\frac{\mathbf{YX}^*\mathbf{aa}^*\mathbf{XY}^*}{N^2(\mathbf{a}^*\hat{\mathbf{R}}_{xx}\mathbf{a})} \right| \\
&= \left| \hat{\mathbf{R}}_{yy} - \frac{\mathbf{b}^c\mathbf{b}^T\hat{\mathbf{Q}}^{-1}\mathbf{YX}^*\mathbf{aa}^*\mathbf{XY}^*}{N^2(\mathbf{b}^T\hat{\mathbf{Q}}^{-1}\mathbf{b}^c)(\mathbf{a}^*\hat{\mathbf{R}}_{xx}\mathbf{a})} \right| \\
&= |\hat{\mathbf{R}}_{yy}| \left| \mathbf{I} - \frac{\hat{\mathbf{R}}_{yy}^{-1}\mathbf{b}^c\mathbf{b}^T\hat{\mathbf{Q}}^{-1}\mathbf{YX}^*\mathbf{aa}^*\mathbf{XY}^*}{N^2(\mathbf{b}^T\hat{\mathbf{Q}}^{-1}\mathbf{b}^c)(\mathbf{a}^*\hat{\mathbf{R}}_{xx}\mathbf{a})} \right| \\
&= |\hat{\mathbf{R}}_{yy}| \left[1 - \frac{\mathbf{b}^T\hat{\mathbf{Q}}^{-1}(\hat{\mathbf{R}}_{yy} - \hat{\mathbf{Q}})\hat{\mathbf{R}}_{yy}^{-1}\mathbf{b}^c}{\mathbf{b}^T\hat{\mathbf{Q}}^{-1}\mathbf{b}^c} \right] \\
&= |\hat{\mathbf{R}}_{yy}| \frac{\mathbf{b}^T\hat{\mathbf{R}}_{yy}^{-1}\mathbf{b}^c}{\mathbf{b}^T\hat{\mathbf{Q}}^{-1}\mathbf{b}^c}
\end{aligned}
\tag{1A.6}
$$

where again we have utilized the fact that $|\mathbf{I} + \mathbf{A}_1\mathbf{B}_1| = |\mathbf{I} + \mathbf{B}_1\mathbf{A}_1|$.

From (1A.5) and (1A.6), it follows that

$$
\min_{\beta} \left| \frac{1}{N}(\mathbf{Y} - \mathbf{b}^c\beta\mathbf{a}^*\mathbf{X})(\mathbf{Y} - \mathbf{b}^c\beta\mathbf{a}^*\mathbf{X})^* \right| = |\hat{\mathbf{R}}_{yy}| \frac{\mathbf{b}^T\hat{\mathbf{R}}_{yy}^{-1}\mathbf{b}^c}{\mathbf{b}^T\hat{\mathbf{Q}}^{-1}\mathbf{b}^c} \tag{1A.7}
$$

By using (1A.3), (1A.4), and (1A.7) in (1A.1), we see that the GLR in (1.61) follows immediately.

APPENDIX 1B LEMMA AND PROOF

Lemma 1B.1 Let $\mathbf{R} \in \mathcal{C}^{M_t \times M_t}$, and let $\tilde{\mathbf{R}} \in \mathcal{R}^{2M_t \times 2M_t}$ be as defined in (1.96). Then

$$\text{rank}(\mathbf{R}) = M_t - m \Longleftrightarrow \text{rank}(\tilde{\mathbf{R}}) = 2(M_t - m) \quad \text{for} \quad m = 0, \ldots, M_t \tag{1B.1}$$

***Proof*:** Let $\mathbf{v} \in \mathcal{C}^{M_t \times 1}$, $\mathbf{v} \neq 0$, be a vector in the null space of $\mathbf{R}$, $\mathcal{N}(\mathbf{R})$:

$$\mathbf{R}\mathbf{v} = 0 \tag{1B.2}$$

This implies that

$$\tilde{\mathbf{R}}\begin{bmatrix} \text{Re}\{\mathbf{v}\} \\ \text{Im}\{\mathbf{v}\} \end{bmatrix} = 0 \tag{1B.3}$$

Moreover, since (1B.2) also implies $\text{R}(j\mathbf{v}) = 0$, we must also have

$$\tilde{\mathbf{R}}\begin{bmatrix} -\text{Im}\{\mathbf{v}\} \\ \text{Re}\{\mathbf{v}\} \end{bmatrix} = 0 \tag{1B.4}$$

The vectors appearing in (1B.3) and (1B.4) are linearly independent of each other. Indeed, if we assume that they were not, then there would exist a nonzero complex-valued scalar, say, $\zeta \neq 0$, such that

$$\begin{bmatrix} \text{Re}\{\mathbf{v}\} & -\text{Im}\{\mathbf{v}\} \\ \text{Im}\{\mathbf{v}\} & \text{Re}\{\mathbf{v}\} \end{bmatrix}\begin{bmatrix} \text{Re}\{\zeta\} \\ \text{Im}\{\zeta\} \end{bmatrix} = 0 \Longrightarrow \mathbf{v}\,\zeta = 0 \Longrightarrow \mathbf{v} = 0 \tag{1B.5}$$

which contradicts the assumption that $\mathbf{v} \neq 0$.

Thus, we have shown that from each $\mathbf{v} \in \mathcal{N}(\mathbf{R})$ we can obtain [as in (1B.3) and (1B.4)] two linearly independent vectors in $\mathcal{N}(\tilde{\mathbf{R}})$. Furthermore, we can use an argument similar to (1B.5) to show that if the vectors $\mathbf{v}_1, \mathbf{v}_2, \ldots, \in \mathcal{N}(\mathbf{R})$ are linearly independent, then so are the corresponding vectors in $\mathcal{N}(\tilde{\mathbf{R}})$. It follows from these observations that

$$\text{rank}(\mathbf{R}) = M_t - m \Longrightarrow \text{rank}(\tilde{\mathbf{R}}) \leq 2(M_t - m), \quad m \in [0, M_t] \tag{1B.6}$$

Conversely, for each $\boldsymbol{\nu} \neq 0$ satisfying $\tilde{\mathbf{R}}\boldsymbol{\nu} = 0$ [i.e., $\boldsymbol{\nu} \in \mathcal{N}(\tilde{\mathbf{R}})$], we can write $\boldsymbol{\nu}$ as $[\text{Re}\{\mathbf{v}^T\} \quad \text{Im}\{\mathbf{v}^T\}]^T$, and therefore we can build a $\mathbf{v}$ such that $\mathbf{v} \in \mathcal{N}(\mathbf{R})$. Furthermore, owing the structure of $\tilde{\mathbf{R}}$, it follows, as above, that also $[-\text{Im}\{\mathbf{v}^T\}\ \text{Re}\{\mathbf{v}^T\}]^T \in \mathcal{N}(\tilde{\mathbf{R}})$), and that $[\text{Re}\{\mathbf{v}^T\} \quad \text{Im}\{\mathbf{v}^T\}]^T$ and $[-\text{Im}\{\mathbf{v}^T\} \quad \text{Re}\{\mathbf{v}^T\}]^T$ are linearly independent of each other. Therefore, for any two such linearly independent vectors in $\mathcal{N}(\tilde{\mathbf{R}})$, there is one vector $\mathbf{v} \in \mathcal{N}(\mathbf{R})$. Again, similar to what was shown above, the linear independence of the vectors in $\mathcal{N}(\tilde{\mathbf{R}})$ implies that of the corresponding vectors in $\mathcal{N}(\mathbf{R})$. Therefore, we have shown that

$$\text{rank}(\tilde{\mathbf{R}}) = 2(M_t - m) \Longrightarrow \text{rank}(\mathbf{R}) \leq M_t - m, \quad m \in [0, M_t] \tag{1B.7}$$

The result stated in (1B.1) follows from (1B.6) and (1B.7). Indeed, if $\text{rank}(\mathbf{R}) = M_t - m$, then we must have $\text{rank}(\tilde{\mathbf{R}}) = 2(M_t - m)$ [otherwise (1B.6) and (1B.7) would imply that $\text{rank}(\tilde{\mathbf{R}}) < 2(M_t - m)$, by (1B.6), and thus that $\text{rank}(\mathbf{R}) < M_t - m$, by (1B.7), which is a contradiction]. Similarly, (1B.6) and (1B.7) can be used to conclude that $\text{rank}(\tilde{\mathbf{R}}) = 2(M_t - m) \Longrightarrow \text{rank}(\mathbf{R}) = M_t - m$. □

Lemma 1B.1 is related to a number of results reported by Goodman [85] for matrices of the form of (1.96), but it does not appear in that cited paper. In fact, while the result from Lemma 1B.1 may possibly be known, we have not been able to locate it in the literature, and this is why we provided a proof for it in this appendix.

ACKNOWLEDGMENTS

This work was supported in part by the Office of Naval Research under Grant N00014-0710293, the National Science Foundation under Grant CCF-0634786, the Defense Advanced Research Projects Agency under Grant HR0011-06-1-0031, and the Swedish Science Council (VR). Opinions, interpretations, conclusions, and recommendations are those of the authors and are not necessarily endorsed by the United States government. We are also grateful to Dr. Luzhou Xu, Ms. Yao Xie, Mr. William Roberts, Dr. Bin Guo, Dr. Xiayu Zheng, and Ms. Xumin Zhu for their contributions to the research results reported herein and for their help with putting this chapter together.

REFERENCES

1. E. Fishler, A. Haimovich, R. Blum, D. Chizhik, L. Cimini, and R. Valenzuela, MIMO radar: An idea whose time has come, *Proc. IEEE Radar Conf.* April 2004, pp. 71–78.
2. E. Fishler, A. Haimovich, R. Blum, L. Cimini, D. Chizhik, and R. Valenzuela, Performance of MIMO radar systems: Advantages of angular diversity, *Proc. 38th Asilomar Conf. Signals, Systems and Computers*, Pacific Grove, CA, Nov. 2004, Vol. 1, pp. 305–309.
3. E. Fishler, A. Haimovich, R. Blum, L. Cimini, D. Chizhik, and R. Valenzuela, Spatial diversity in radars — models and detection performance, *IEEE Trans. Signal Process.* **54**(3):823–838 (March 2006).
4. Y. Yang and R. S. Blum, MIMO radar waveform design based on mutual information and minimum mean-square error estimation, *IEEE Trans. Aerospace Electron. Syst.* **43**(1):330–343 (Jan. 2007).
5. T. Aittomaki and V. Koivunen, Low-complexity method for transmit beamforming in MIMO radars, *Proc. 2007 IEEE Int. Conf. Acoustics, Speech, and Signal Processing*, Honolulu, April 2007.
6. I. Bekkerman and J. Tabrikian, Spatially coded signal model for active arrays, *Proc. 2004 IEEE Int. Conf. Acoustics, Speech, and Signal Processing*, Montreal, Quebec, Canada, March 2004, Vol. 2, pp. ii/209–ii/212.
7. I. Bekkerman and J. Tabrikian, Target detection and localization using MIMO radars and sonars, *IEEE Trans. Signal Process.* **54**:3873–3883 (Oct. 2006).

8. I. Bekkerman and J. Tabrikian, Space-time coding for active arrays, *IEEE Trans. Signal Process.* (in press).
9. D. W. Bliss and K. W. Forsythe, Multiple-input multiple-output (MIMO) radar and imaging: Degrees of freedom and resolution, *Proc. 37th Asilomar Conf. Signals, Systems and Computers*, Pacific Grove, CA, Nov. 2003, Vol. 1, pp. 54–59.
10. K. W. Forsythe and D. W. Bliss, Waveform correlation and optimization issues for MIMO radar, *Proc. 39th Asilomar Conf. Signals, Systems and Computers*, Pacific Grove, CA, Nov. 2005, pp. 1306–1310.
11. K. W. Forsythe, D. W. Bliss, and G. S. Fawcett, Multiple-input multiple-output (MIMO) radar: Performance issues, *Proc. 38th Asilomar Conf. Signals, Systems and Computers*, Pacific Grove, CA, Nov. 2004, Vol. 1, pp. 310–315.
12. D. R. Fuhrmann and G. San Antonio, Transmit beamforming for MIMO radar systems using partial signal correlations, *Proc. 38th Asilomar Conf. Signals, Systems and Computers*, Pacific Grove, CA, Nov. 2004, Vol. 1, pp. 295–299.
13. D. R. Fuhrmann and G. San Antonio, Transmit beamforming for MIMO radar systems using signal cross-correlation, *IEEE Trans. Aerospace Electron. Syst.* (in press).
14. N. Lehmann, E. Fishler, A. M. Haimovich, R. S. Blum, D. Chizhik, L. Cimini, and R. Valenzuela, Evaluation of transmit diversity in MIMO-radar direction finding, *IEEE Trans. Signal Process.* **55**(5):2215–2225 (May 2007).
15. J. Li and P. Stoica, MIMO radar — diversity means superiority, *Proc. 14th Annual Workshop on Adaptive Sensor Array Processing* (invited), MIT Lincoln Laboratory, Lexington, MA, June 2006.
16. J. Li and P. Stoica, MIMO radar with collocated antennas: Review of some recent work, *IEEE Signal Process. Mag.* **24**(5):106–114 (Sept. 2007).
17. J. Li, P. Stoica, and Y. Xie, On probing signal design for MIMO radar, *Proc. 40th Asilomar Conf. Signals, Systems and Computers* (invited), Pacific Grove, CA, Oct. 2006.
18. J. Li, P. Stoica, L. Xu, and W. Roberts, On parameter identifiability of MIMO radar, *IEEE Signal Process. Lett.* **14**(12):968–971 (Dec. 2007).
19. J. Li, L. Xu, P. Stoica, K. W. Forsythe, and D. W. Bliss, Range compression and waveform optimization for MIMO radar: A Cramér-Rao bound based study, *IEEE Trans. Signal Process.* **56**(1):218–232 (Jan. 2008).
20. V. F. Mecca, D. Ramakrishnan, and J. L. Krolik, MIMO radar space-time adaptive processing for multipath clutter mitigation, *Proc. 4th IEEE Workshop on Sensor Array and Multi-channel Processing*, Waltham, MA, July 2006.
21. D. J. Rabideau and P. Parker, Ubiquitous MIMO multifunction digital array radar, *Proc. 37th Asilomar Conf. Signals, Systems and Computers*, Pacific Grove, CA, Nov. 2003.
22. F. C. Robey, S. Coutts, D. Weikle, J. C. McHarg, and K. Cuomo, MIMO radar theory and exprimental results, *Proc. 38th Asilomar Conf. Signals, Systems and Computers*, Pacific Grove, CA, Nov. 2004, Vol. 1, pp. 300–304.
23. P. Stoica, J. Li, and Y. Xie, On probing signal design for MIMO radar, *IEEE Trans. Signal Process.* **55**(8):4151–4161 (Aug. 2007).
24. J. Tabrikian, Barankin bounds for target localization by MIMO radars, *Proc. 4th IEEE Workshop on Sensor Array and Multi-channel Processing*, Waltham, MA, July 2006.

25. J. Tabrikian and I. Bekkerman, Transmission diversity smoothing for multi-target localization, *Proc. 2005 IEEE Int. Conf. Acoustics, Speech, and Signal Processing*, Philadelphia, PA, March 2005, Vol. 4, pp. iv/1041–iv/1044.
26. L. Xu, J. Li, and P. Stoica, Target detection and parameter estimation for MIMO radar systems, *IEEE Trans. Aerospace Electron. Syst.* (in press).
27. L. Xu, J. Li, and P. Stoica, Adaptive techniques for MIMO radar, *Proc. 4th IEEE Workshop on Sensor Array and Multi-channel Processing*, Waltham, MA, July 2006.
28. L. Xu, J. Li, and P. Stoica, Radar imaging via adaptive MIMO techniques, *Proc. 14th European Signal Processing Conf.* (invited), Florence, Italy, Sept. 2006.
29. L. Xu, J. Li, P. Stoica, K. W. Forsythe, and D. W. Bliss, Waveform optimization for MIMO radar: A Cramer-Rao bound based study, *Proc. 2007 IEEE Int. Conf. Acoustics, Speech, and Signal Processing*, Honolulu, April 2007.
30. L. Xu and J. Li, Iterative generalized likelihood ratio test for MIMO radar, *IEEE Trans. Signal Process.* **55**(6):2375–2385, (June 2007).
31. P. Stoica and R. L. Moses, *Spectral Analysis of Signals*, Prentice-Hall, Upper Saddle River, NJ, 2005.
32. H. L. Van Trees, *Optimum Array Processing: Part IV of Detection, Estimation, and Modulation Theory*, Wiley, New York, 2002.
33. A. Nehorai, D. Starer, and P. Stoica, Direction-of-arrival estimation in applications with multipath and few snapshots, *Circuits Syst. Signal Process.* **10**:327–342 (1991).
34. M. Wax and I. Ziskind, On unique localization of multiple sources by passive sensor arrays, *IEEE Trans. Acoust. Speech Signal Process.* **37**(7):996–1000 (July 1989).
35. J. R. Guerci, *Space-Time Adaptive Processing for Radar*, Artech House, Norwood, MA, 2003.
36. R. Klemm, *Principles of Space-Time Adaptive Processing*, IEE Press, London, 2002.
37. J. Ward, *Space-Time Adaptive Processing for Airborne Radar*, Technical Report 1015, MIT Lincoln Laboratory, Dec. 1994.
38. D. J. Rabideau and A. O. Steinhardt, Improved adaptive clutter cancellation through data-adaptive training, *IEEE Trans. Aerospace Electron. Syst.* **35**(3):879–891 (July 1999).
39. J. Capon, High resolution frequency-wavenumber spectrum analysis, *Proc. IEEE* **57**:1408–1418 (Aug. 1969).
40. J. Li and P. Stoica, An adaptive filtering approach to spectral estimation and SAR imaging, *IEEE Trans. Signal Process.* **44**(6):1469–1484 (June 1996).
41. P. Stoica, H. Li, and J. Li, A new derivation of the APES filter, *IEEE Signal Process. Lett.* **6**(8):205–206 (Aug. 1999).
42. E. G. Larsson, J. Li, and P. Stoica, High-resolution nonparametric spectral analysis: Theory and applications, in *High-Resolution and Robust Signal Processing*, Y. Hua, A. B. Gershman, and Q. Cheng (eds.), Marcel-Dekker, New York, 2003, pp. 151–251.
43. P. Stoica, A. Jakobsson, and J. Li, Capon, APES and matched-filterbank spectral estimation, *Signal Process.* **66**(1):45–59 (April 1998).
44. H. Li, J. Li, and P. Stoica, Performance analysis of forward-backward matched-filterbank spectral estimators, *IEEE Trans. Signal Process.* **46**:1954–1966 (July 1998).
45. A. Jakobsson and P. Stoica, Combining Capon and APES for estimation of spectral lines, *Circuits Syst. Signal Process.* **19**(2):159–169 (2000).

46. L. Xu, P. Stoica, and J. Li, A diagonal growth curve model and some signal processing applications, *IEEE Trans. Signal Process.* **54**(9):3363–3371 (Sept. 2006).

47. D. A. Harville, *Matrix Algebra from a Statistician's Perspective*, Springer, New York, 1997.

48. K. M. Buckley and L. J. Griffiths, An adaptive generalized sidelobe canceller with derivative constraints, *IEEE Trans. Anten. Propag.* **34**:311–319 (March 1986).

49. J. Li, P. Stoica, and Z. Wang, On robust Capon beamforming and diagonal loading, *IEEE Trans. Signal Process.* **51**(7):1702–1715 (July 2003).

50. P. Stoica, Z. Wang, and J. Li, Robust Capon beamforming, *IEEE Signal Process. Lett.* **10**(6):172–175 (June 2003).

51. J. Li and P. Stoica (eds.), *Robust Adaptive Beamforming*, Wiley, Hoboken, NJ, 2005.

52. T. L. Marzetta, A new interpretation for Capon's maximum likelihood method of frequency-wavenumber spectrum estimation, *IEEE Trans. Acoust. Speech Signal Process.* **31**(2):445–449 (April 1983).

53. J. Li, P. Stoica, and Z. Wang, Doubly constrained robust Capon beamformer, *IEEE Trans. Signal Process.* **52**(9):2407–2423 (Sept. 2004).

54. P. Stoica and G. Ganesan, Maximum-SNR spatial-temporal formatting for MIMO channels, *IEEE Trans. Signal Process.* **50**(12):3036–3042 (Dec. 2002).

55. P. Stoica and Y. Selen, Model-order selection: A review of information criterion rules, *IEEE Signal Process. Mag.* **21**(4):36–47 (July 2004).

56. J. Li and P. Stoica, Efficient mixed-spectrum estimation with applications to target feature extraction, *IEEE Trans. Signal Process.* **44**:281–295 (Feb. 1996).

57. J. F. Sturm, Using SeDuMi 1.02, a MATLAB toolbox for optimization over symmetric cones, *Optim. Meth. Software Online* **11–12**:625–653 (Oct. 1999) (available at http://www2.unimaas.nl/sturm/software/sedumi.html).

58. S. Boyd and L. Vandenberghe, *Convex Optimization*, Cambridge Univ. Press, UK, 2004.

59. R. Orsi, U. Helmke, and J. B. Moore, A Newton-like method for solving rank constrained linear matrix inequalities, *Proc. 43rd IEEE Conf. Decision and Control*, 2004, pp. 3138–3144.

60. E. C. Fear, P. M. Meaney, and M. A. Stuchly, Microwaves for breast cancer detection, *IEEE Potentials* **22**(1):12–18 (Feb.–March 2003).

61. B. Guo, Y. Wang, J. Li, P. Stoica, and R. Wu, Microwave imaging via adaptive beamforming methods for breast cancer detection, *J. Electromagn. Waves Appl.* **20**(1):53–63 (2006).

62. P. M. Meaney, M. W. Fanning, D. Li, S. P. Poplack, and K. D. Paulsen, A clinical prototype for active microwave imaging of the breast, *IEEE Trans. Microwave Theory Tech.* **48**(11):1841–1853 (Nov 2000).

63. F. W. Kremkau, *Diagnostic Ultrasound: Principles and Instruments*, Saunders, Philadelphia, PA, 1993.

64. T. Szabo, *Diagnostic Ultrasound Imaging: Inside Out*, Elsevier Academic Press, Burlington, MA, 2004.

65. R. A. Kruger, K. K. Kopecky, A. M. Aisen, D. R. Reinecke, G. A. Kruger, and W. L. Kiser, Thermoacoustic CT with radio waves: A medical imaging paradigm, *Radiology* **211**(1):275–278 (April 1999).

66. D. Gianfelice, A. Khiat, M. Amara, A. Belblidia, and Y. Boulanger, MR imaging-guided focused US ablation of breast cancer: Histopathologic assessment of effectiveness-initial experience, *Radiology* **227**:849–855 (April 2003).

67. M. H. Falk and R. D. Issels, Hyperthermia in oncology, *Int. J. Hypertherm.* **17**(1):1–18 (Jan. 2001).
68. C. C. Vernon, J. W. Hand, S. B. Field, D. Machin, J. B. Whaley, J. Van Der Zee, W. L. J. Van Putten, G. C. Van Rhoon, J. D. P. Van Dijk, D. G. González, F. Liu, P. Goodman, and M. Sherar, Radiotherapy with or without hyperthermia in the treatment of superficial localized breast cancer: Results from five randomized controlled trials, *Int. J. Radiation Oncol. Biol. Phys.* **35**(4):731–744 (July 1996).
69. A. J. Fenn, V. Sathiaseelan, G. A. King, and P. R. Stauffer, Improved localization of energy deposition in adaptive phased-array hyperthermia treatment of cancer, *Lincoln Lab. J.* **9**(2):187–195 (1996).
70. C. J. Diederich and K. Hynynen, Ultrasound technology for hyperthermia, *Ultrasound Med. Biol.* **25**(6):871–887 (July 1999).
71. B. Guo and J. Li, Waveform diversity based ultrasound system for hyperthermia treatment of breast cancer, *IEEE Trans. Biomed. Eng.* **55**(2):822–826 (Feb. 2008).
72. D. T. Blackstock, *Fundamentals of Physical Acoustics*, Wiley, New York, 2000.
73. A. D. Pierce, *Acoustics: An Introduction to Its Physical Principles and Applications*, McGraw-Hill, New York, 1994.
74. F. A. Duck, *Physical Properties of Tissue*, Academic Press, London, 1990.
75. F. A. Duck, A. C. Baker, and H. C. Starritt, *Ultrasound in Medicine*, Institute of Physics Publishing, Bristol, Philadelphia, PA, 1998.
76. T. K. Katsibas and C. S. Antonopoulos, A general form of perfectly matched layers for three-dimensional problems of acoustic scattering in lossless and lossy fluid media, *IEEE Trans. Ultrason. Ferroelectrics Freq. Control* **51**(8):964–972 (Aug. 2004).
77. X. Yuan, D. Borup, J. Wiskin, M. Berggren, and S. A. Johnson, Simulation of acoustic wave propagation in dispersive media with relaxation losses by using FDTD method with PML absorbing boundary condition, *IEEE Trans. Ultrason. Ferroelectrics Freq. Control* **46**(1):14–23 (Jan. 1999).
78. P. Bernardi, M. Cavagnaro, S. Pisa, and E. Piuzzi, Specific absorption rate and temperature elevation in a subject exposed in the far-field of radio-frequency sources operating in the 10-900-MHz range, *IEEE Trans. Biomed. Eng.* **50**(3):295–304 (March 2003).
79. J. Wang and O. Fujiwara, FDTD computation of temperature rise in the human head for portable telephones, *IEEE Trans. Microwave Theory Tech.* **47**(8):1528–1534 (Aug. 1999).
80. P. Stoica, J. Li, X. Zhu, and B. Guo, Waveform synthesis for diversity-based transmit beampattern design, *Proc. 14th IEEE Workshop on Statistical Signal Processing* (invited), Madison, WI, Aug. 2007.
81. J. Li, X. Zheng, and P. Stoica, MIMO SAR imaging: Signal synthesis and receiver design, *Proc. 2nd Int. Workshop on Computational Advances in Multi-Sensor Adaptive Processing* (invited), St. Thomas, US Virgin Islands, 2007.
82. E. J. Kelly, An adaptive detection algorithm, *IEEE Trans. Aerospace Electron. Syst.* **22**(1):115–127 (March 1986).
83. A. Dogandžić and A. Nehorai, Generalized multivariate analysis of variance, *IEEE Signal Process. Mag.* 39–54 (Sept. 2003).
84. Y. Jiang, P. Stoica, and J. Li, Array signal processing in the known waveform and steering vector case, *IEEE Trans. Signal Process.* **52**(1):23–35 (Jan. 2004).
85. N. R. Goodman, Statistical analysis based on a certain multi-variate complex Gaussian distribution, *Ann. Math. Stat.* **34**:152–177 (March 1963).

2

MIMO RADAR: CONCEPTS, PERFORMANCE ENHANCEMENTS, AND APPLICATIONS[1]

KEITH W. FORSYTHE AND DANIEL W. BLISS
MIT Lincoln Laboratory, Lexington, Massachusetts

Once upon a time and a very good time it was there was a moocow coming down along the road and this moocow that was coming down along the road met a nicens little boy named baby tuckoo . . . .

—James Joyce

2.1 INTRODUCTION

2.1.1 A Short History of Radar

The venerable field of radar research is far from young. Intended as an aid for naval navigation, by 1904, Christian Hulsmeyer had developed and patented a spark-gap transmitter–receiver system [1]. While very limited in its capability, Hulsmeyer's technology was an early form of radar. In 1935, Sir Robert Watson-Watt demonstrated and patented a working radar system [2]. This system had all the essential characteristics of a useful radar.

[1]This work was sponsored by the United States Air Force under United States Air Force Contract FA8721-05-C-0002. Opinions, interpretations, conclusions, and recommendations are those of the authors and are not necessarily endorsed by the United States government.

MIMO Radar Signal Processing, edited by Jian Li and Petre Stoica

During World War II, the capabilities and usefulness of radars improved significantly. Robert Watson-Watt's radar technology was quickly refined, becoming the British Chain Home air defense radar. This radar helped protect the British from air attacks during World War II. Infamously, the incoming wave of fighters during the Pearl Harbor attack was detected by an early air-surveillance radar, but this information was not used by US military commanders. This lack of appreciation for the value of radar technology changed quickly as the use of radar became pervasive by the military during the war. In fact, many of the radar concepts currently employed were developed during or just after the war, including air defense, ground-approach radar, fire control, and moving-target indication (MTI) radars.

After World War II, radar concepts continued to be developed, although not at the same rapid pace. Not until the advent of synthetic aperture radar (SAR), which became an active area of research during the 1950s [3], was there a significant change in the basic concept originally defined by Hulsmeyer. In the 1960s, the development of phased-array antennas became an important area of research, allowing radars to quickly change search direction. During the 1970s, digital signal processing became viable, and was applied to radar processing, enabling adaptive array processing and the modern radar system. Once again, driven by the advances in computational capabilities, which allow system designers to consider more complicated processing, multiple-input multiple-output (MIMO) radars have become an active area of research.

2.1.2 Definition and Characteristics of MIMO Radar

The notion of MIMO radar is simply that there are multiple radiating and receiving sites [4], as shown in Fig. 2.1. The collected information is then processed together. In some sense, MIMO radars are a generalization of multistatic radar concepts. The underlying concepts have most likely been discovered independently numerous times [5]. While not using the nomenclature MIMO radar, the RIAS and SIAR radars [6,7],

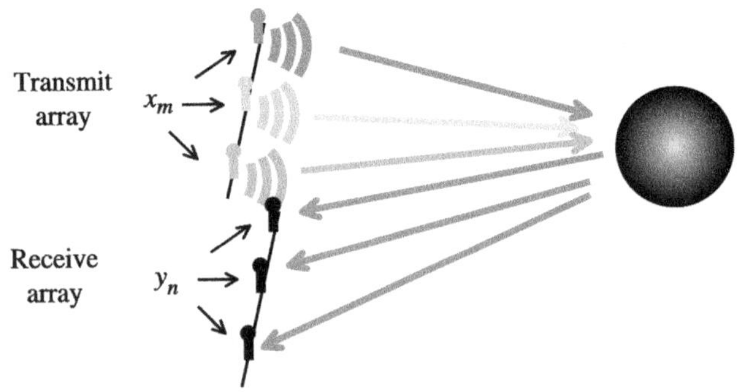

Figure 2.1 Illustration of the basic MIMO radar. The location of the mth transmitter is given by $\mathbf{x}_m$, and the location of the nth receiver is given by $\mathbf{y}_n$.

that are experimental systems developed to study air-surveillance technology, are early examples of systems that could be classified as MIMO radars.

By the most general definition, many traditional systems can be considered as special cases of MIMO radars. As an example, SAR can be considered as a form of MIMO radar. Although SAR traditionally employs a single transmit antenna and a single receive antenna, the positions of these two antennas are translated and images are formed by processing all the information jointly. The significant difference between this radar and a "typical" MIMO radar, which takes full advantage of the degrees of freedom, is that SAR does not have access to channel measurements for all transmit–receive position pairs. Equivalently, one may say that only the diagonal elements of the channel matrix are measured. Similarly, a fully polarimetric radar, that is, a radar that measures both receive polarizations for each transmit polarization, is an example of MIMO radar [4,8]. Clearly, it is a MIMO radar with a relatively small dimensionality. In addition, some spatial interpretations of MIMO radar have to be considered in a different context for polarimetric radars.

Various possible signaling techniques are used for MIMO radar. The transmit antennas radiate signals, which may or may not be correlated, and the receive antennas attempt to disentangle these signals. In much of the current literature, it is assumed that the waveforms coming from each transmit antenna are orthogonal, but this is not a requirement for MIMO radar. However, orthogonality can facilitate the processing. Two simple approaches to obtain orthogonality are to use time division or frequency division multiplexing. However, both approaches can suffer from potential performance degradation (assuming coherent operation) because of the loss of coherence of the target response. The scattering response of the target or background is commonly time-varying or frequency-selective, limiting the ability to coherently combine the information from the antenna pairs. In some applications, it is desirable to introduce correlation between the transmitted signals. For some tracking problems, optimal asymptotic angle estimation performance is given by employing strongly correlated signals [9].

There is a continuum of MIMO radar systems concepts; however, there are two basic regimes of operation considered in the current literature. In the first regime, the transmit array elements (and receive array elements) are broadly spaced, providing independent scattering responses for each antenna pairing, sometimes referred to as *statistical MIMO radar*. In the second regime, the transmit array elements (and receive array elements) are closely spaced so that the target is in the far field of the transmit–receive array, sometimes referred to as *coherent MIMO radar*. Here it is assumed that the target's scattering response is the same for each antenna pair, up to some small delay. While the answer to the question "How large must the angular separation be to get independent scattering responses?" is dependent on the details of the target, a sense of scale is provided by thinking of the target as an array of scatterers with phase responses optimized to focus energy toward one of the antennas. As shown in Fig. 2.2, if an array of appropriately phased scatterers of the physical size of the target can resolve individual locations of the antennas, then independent scattering responses would theoretically be possible. Conversely, if the overall angular antenna separation were small compared to the "beamwidth"

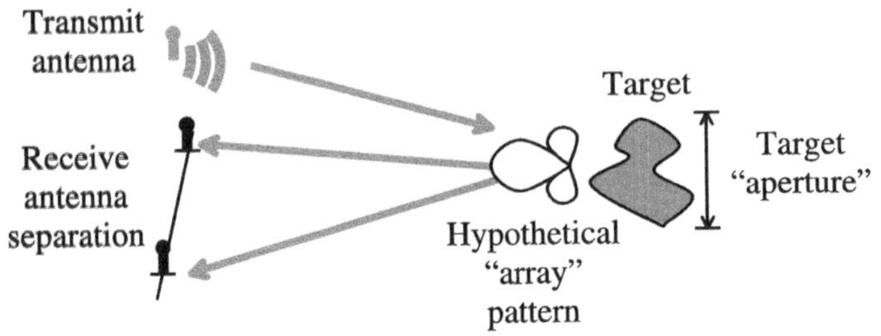

Figure 2.2 Coherent versus incoherent scattering response.

of the scatterer array, then coherent processing would be viable. This analysis is easily interpreted for air-surveillance radars for which a target is well separated from any background. While the discussion is still valid, the interpretation of a target can be less clear in the presence of a background.

2.1.3 Uses of MIMO Radar

There are a variety of potential advantages to using MIMO radar. For given system design choices, some of these advantages can be traded for others:

- Improved target detection performance
- Improved angle estimation accuracy
- Decreased minimum detectible velocity

For the first type of MIMO radar discussed in Section 2.1.2, in which the individual transmit and receive antennas are separated widely, the diversity provided by the multiplicity in transmit and receive angles can be exploited to improve the statistics of the detection performance. Similar to frequency diversity used in some traditional radars, spatial (angular) diversity can be employed to reduce the probability that a "bad" scattering response causes a target to be missed.

For the second type of MIMO radar discussed in Section 2.1.2, in which the antennas are spaced relatively closely, angle estimation performance can be improved. The estimation performance improvement can be dramatic when optimized sparse arrays are used. In some sense, the performance of the MIMO systems can be characterized by a virtual array constructed by the convolution of the locations of the transmit and receive antenna locations. Consequently, a filled virtual array can be constructed by using sparse constituent arrays. In principle, this virtual array can be much larger than the array of an equivalent traditional system; thus, the MIMO system will have much better intrinsic resolution.

Implicitly, with any array design, there is a tradeoff between spatial ambiguities caused by spatial sidelobe levels and the intrinsic spatial resolution. Specifically, in traditional antenna array processing, sparse arrays can provide a larger aperture than can filled arrays, providing improved angular resolution compared to a filled array. However, this performance improvement comes at the expense of increased sidelobe levels. Similarly, MIMO virtual arrays can be constructed to be sparse, filled, or overfilled. MIMO sidelobe levels can be reduced by decreasing the virtual array aperture size. The importance of the trade depends on the application being addressed. For upward-looking radars, such as air-surveillance radars, sidelobes are not a dominating concern. However, for airborne ground surveillance radars, ambiguities with clutter background can be a driving design constraint.

Compared to traditional systems, MIMO ground moving-target indication (GMTI) radars can be employed to improve minimum detectable velocities. Minimum detectable velocity is sensitive to both the aperture size and integration interval. Both of these characteristics can be improved using MIMO radars. There is a question of how to make a fair comparison. Performance criteria are different depending on whether the radar is doing wide-area surveillance or tracking a particular target. For the moment, we consider wide-area surveillance. In this mode, typical GMTI systems either transmit from an single element (or subarray) covering a larger area, or scan a beam from the transmit array over the area of interest. For a comparison with a traditional GMTI transmitting from a single element, the MIMO system may have n_T transmitters illuminating the same area. Assuming that the MIMO system is transmitting independent sequences simultaneously, so that the radiated power combines incoherently, the MIMO system may illuminate the ground with n_T times as much power. If the traditional GMTI system uses the entire transmit array coherently, sweeping a beam to perform its surveillance, as seen in Fig. 2.3,

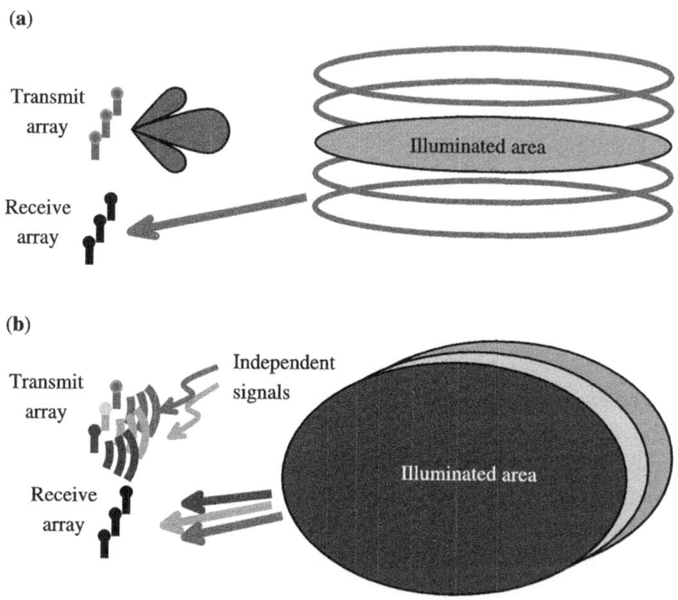

Figure 2.3 Illumination patterns of (a) traditional scanning radar and (b) MIMO radar assuming independent transmit signals.

then the traditional system illuminates the ground with n_T^2 as much power as that of a single antenna. However, the beam must be swept over the region of interest. The integration interval is about $1/n_{\text{beams}} \approx 1/n_T$. As a consequence, the integrated power is proportional to n_T^2/n_T. The MIMO system would illuminate this same total region continuously, so that the average power on the ground for the MIMO system and the swept beam is approximately the same. The combination of the longer illumination and the larger aperture of the MIMO radar provides for the possibility of improved minimum detectable velocity for GMTI systems.

2.1.4 The Current State of MIMO Radar Research

The field of MIMO radar is growing rapidly. It would be nearly impossible to cover every paper on this topic. However, here we attempt to provide a snapshot of the state of MIMO radar research.

2.1.4.1 Statistical MIMO Radar While it is not the focus of this chapter, it certainly is worth mentioning some of the research in diversity-oriented MIMO radar. Fishler [10] and Lehmann et al. [11] discuss the benefits of spatial diversity enabled by using MIMO radar. Dai et al. [12] discuss variations on this theme, in which "closer" antenna spacings are allowed. In general, these papers discuss improvements in detectability of a target because of the multiple bistatic viewing angles enabled by widely separated transmit and receive antennas.

2.1.4.2 Coherent MIMO Radar The notion of MIMO radar using closely spaced elements is discussed in Refs. 4, 14, and 15. The degrees of freedom and virtual array interpretation of MIMO radar are discussed in Refs. 4, 14, and 16, which includes a discussion of detection performance. In Ref. 4 the construction of filled or nearly filled virtual antennas arrays given sparse transmit and/or receive arrays is presented. For array design, given a certain number of antennas, there is often a tradeoff between aperture, and spatial sidelobe level. The same is true for MIMO radar. The height of these sidelobes determines the threshold point. The threshold point is the SNR at which an estimator will diverge from the Cramér–Rao bound [17], as seen in Fig. 2.4. Tools for studying this effect are introduced in Ref. 18. A study of adaptive techniques for processing MIMO radar data is considered in Ref. 19.

2.1.4.3 Waveform Optimization In general there are two views of waveform design. In the first view, the design of the details of the time series transmitted from each transmitter is considered. In Refs. 20 and 21 using simulated annealing and genetic algorithms, respectively, time series are designed for good cross-transmitter and range estimation characteristics. In Ref. 13, Doppler characteristics are added to the set of design criteria. General MIMO ambiguity function tools are considered in Ref. 22.

In the second view, which is the approach used in this chapter, the details of the time series are not considered. Rather, only the intertransmitter signal correlation is

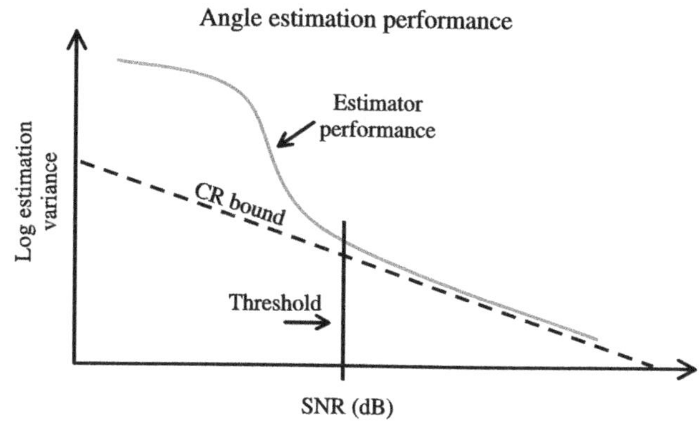

Figure 2.4 Notional performance angle estimation. Asymptotically, the estimator converges to the Cramér–Rao performance bound.

designed. Yang and Blum [23] discuss waveform optimization in terms of maximizing the mutual information, given knowledge of the wide-sense-stationary target response covariance structure. It is shown, under this assumption, that maximizing the mutual information is given by a modified water-filling solution, which provides the same solution as minimizing the minimum mean-square error. Forsythe and Bliss [17] discuss optimization of intertransmitter waveform correlation given specific channel realizations. They discuss optimization in terms of the transmit covariance structure, under a total power constraint. A related optimization, under the constraint of power radiated from each transmit element, is given in Ref. 24. San Antonio and Fuhrmann [25] discuss optimization of wideband signals for a given area illumination. As an extension to waveform construction, Donnet and Longstaff [26] consider the use of MIMO radar waveforms as communication waveforms.

2.1.4.4 MIMO Radar System Issues A number of application concepts have been considered for MIMO radar. The earliest experiments of MIMO radar were for air-surveillance systems [6,7]. These concepts were developed further in Ref. 14. Clutter mitigation has also been discussed in [27]. The performance implications for airborne ground moving-target indication (GMTI) radar is discussed in Refs. 4 and 9. These discussions include the limitation caused by clutter mitigation. The application of MIMO radar to the problem of breast cancer tumor detection is discussed in Ref. 28. For this application, self-interference mitigation is a significant concern.

2.1.5 Chapter Outline

Following this introduction, the chapter is divided into five sections. Some notation is specific to particular sections in this chapter. Notation used throughout the chapter is described in Section 2.2. In Section 2.3, we introduce the MIMO virtual aperture. In Section 2.4, we consider the performance of the radar in the presence of a single scatterer. Angle estimation performance bounds are provided, and optimal waveform correlations are introduced. In Section 2.5, we consider detection in the context of search. Using an information-theoretic approach, we investigate waveform optimization. Finally, in Section 2.6, we discuss the performance of MIMO GMTI radars. Simple characterizations of performance are introduced. We investigate waveform correlation optimization. A significant issue is the minimum detectable velocity. Detection near the zero-velocity clutter is difficult because of SNR losses.

2.2 NOTATION

We use the following notation throughout the chapter. Scalars are represented by italic lightface symbols, for example, a. Column vectors are indicated by bold lowercase variables, for example, $\mathbf{a}$, and matrices are indicated using uppercase bold characters, for example, $\mathbf{A}$. The notation $\cdot^{\dagger}$ indicates the Hermitian conjugate, while $\cdot^{*}$ or $\bar{\cdot}$ indicates element-by-element conjugation. The 2-norm is indicated using the notation $\|\cdot\|$. The Krönecker product is indicated using $\cdot \otimes \cdot$ (see Appendix 2H). The Hadamard product, which is the element-by-element matrix product, is donated $\cdot \Box \cdot$. The mth, nth entry in a matrix $\mathbf{A}$ is denoted by $A_{m,n}$ $(\mathbf{A})_{m,n}$:

$\mathbf{x}_m$	location of the mth transmit antenna (or subarray)
$\mathbf{y}_n$	location of the nth receive antenna (or subarray)
n_R	number of receiving antennas (or subarrays)
n_T	number of transmitting antennas (or subarrays)
n_S	number of samples in block
k	wavenumber ($2\pi/$wavelength)
$\mathbf{u}$	pointing vector from transmitter (or receiver) to scatterer
$\mathbf{a}_T(\mathbf{u})$	transmit steering vector
$\mathbf{a}_R(\mathbf{u})$	receive steering vector
$\mathbf{a}_D(\mathbf{u})$	pulse-domain steering vector (for MTI)
$\mathbf{a}(\mathbf{u})$	MIMO virtual array
$\mathbf{Z} \in \mathbb{C}^{n_R \times n_S}$	receive data matrix
$\mathbf{H}_\delta \in \mathbb{C}^{n_R \times n_T}$	channel matrix for delay δ
$\hat{\mathbf{H}}_\delta \in \mathbb{C}^{n_R \times n_T}$	estimate of channel matrix for delay δ
$\mathbf{S}_\delta \in \mathbb{C}^{n_T \times n_S}$	transmitted signal matrix delayed by time δ
$\mathbf{C} \in \mathbb{C}^{n_T \times n_T}$	transmit waveform correlation matrix
$\mathbf{I}_m$	$m \times m$ identity matrix
$\mathbf{I}_{m,n}$	$m \times n$ identity matrix
vec($\mathbf{X}$)	column vector formed by stacking rows of $\mathbf{X}$
$\mathbf{1}$	vector of ones

For notational simplicity, the δ subscript is sometimes suppressed. Occasionally, k is used as an index when there is little chance of confusion.

2.3 MIMO RADAR VIRTUAL APERTURE

In this section, the MIMO channel is introduced. This is followed by a discussion of the MIMO virtual array. For the analysis discussed in this section, it is assumed that the waveforms transmitted from each transmit antenna are independent.

2.3.1 MIMO Channel

Between the transmitter and receiver is the channel. In some sense, the role of radar processing is to estimate and interpret this channel. Without loss of generality, a baseband sampled signal can be considered. The $n_R \times n_S$ received data matrix $\mathbf{Z}$ is given by

$$\mathbf{Z} = \sum_{\delta} \mathbf{H}_{\delta}\mathbf{S}_{\delta} + \mathbf{N} \tag{2.1}$$

where $\mathbf{N}$ contains the sum of noise and external interference. The summation in Eq. (2.1) is over delays δ, which correspond to different range cells.

If the illuminated region contained a single simple scatterer (as is displayed in Fig. 2.1) in the far field at delay δ, then the channel matrices at all delays would be zero with the exception of $\mathbf{H}_{\delta}$, which would have the structure

$$(\mathbf{H}_{\delta})_{n,m} \propto e^{ik\mathbf{u}\cdot(\mathbf{y}_n+\mathbf{x}_m)} \tag{2.2}$$

where $k\mathbf{u}$ is the wavevector ($k \equiv 2\pi/\lambda$) and $\mathbf{x}_m$, $\mathbf{y}_n$ are 3-vectors of physical locations for the transmitter and receiver phase centers, respectively. The argument of the exponential reflects differential pathlengths between transmitter and receiver phase centers, given a far-field target in direction $\mathbf{u}$. As an example, if both the and transmitter receiver arrays have three antennas, with antenna separation d, located along a line at $\{-d,0,d\}$, in direction $\mathbf{d}$, then the channel matrix is given by

$$\mathbf{H}_{\delta} \propto \begin{pmatrix} e^{i\eta 2d} & e^{i\eta d} & e^{i\eta 0} \\ e^{i\eta d} & e^{i\eta 0d} & e^{-i\eta d} \\ e^{i\eta 0} & e^{-i\eta d} & e^{-i\eta 2d} \end{pmatrix} \tag{2.3}$$

where

$$\eta = k\mathbf{u} \cdot \frac{\mathbf{d}}{\|\mathbf{d}\|} \tag{2.4}$$

The structure of the channel matrix in Eq. (2.3) illustrates two important concepts: (1) the largest phase offsets $e^{i\eta 2d}$ and $e^{-i\eta 2d}$ are larger than those created by the real arrays (this is the motivation for the MIMO virtual array concept; in this case, the virtual array has five virtual array locations: $-2d$, $-d$, 0, d, and 2d); and (2) some entries are overrepresented. For the particular real arrays used in this example, the entries in the Hankel channel matrix are repeated. This motivates the exploration of sparse real arrays to minimize the number of repeated phase measurements.

2.3.2 MIMO Virtual Array: Resolution and Sidelobes

Considering the argument of the exponential in Eq. (2.2), it can be seen that the MIMO radar appears to have phase centers located at the virtual locations $\{\mathbf{x}_m + \mathbf{y}_n\}$. Equivalently, we say that the MIMO virtual array phase centers can be constructed by convolving the locations of the real transmitter and receiver locations.

With a single-point-scattering target, the channel matrix can be written in the form[2]

$$\mathrm{a}(\mathbf{u}) \equiv \mathrm{vec}\mathbf{H} = \mathrm{a}_R(\mathbf{u}) \otimes \mathbf{a}_T(\mathbf{u}) \tag{2.5}$$

based on the conventional response vectors

$$(\mathbf{a}_T)_m \equiv e^{ik\mathbf{u}\cdot\mathbf{x}_m} \tag{2.6}$$

and

$$(\mathbf{a}_R)_n \equiv e^{ik\mathbf{u}\cdot\mathbf{y}n} \tag{2.7}$$

The discussion of MIMO radar sidelobe performance is complicated by the differences in users' needs [9]. There is no single optimization criterion. In traditional radars, sidelobes are typically controlled by applying tapers to the antenna elements. The tapers selected depend upon the application. In order to make a comparison between the performance of traditional arrays and MIMO arrays, bounds are presented on the length of the maximum contiguous aperture of the MIMO virtual array. As a consequence, performance is at least as good as an equivalent real array achieves, assuming the same integrated SNR.

In this chapter section, aperture bounds are considered for MIMO virtual arrays for two different conditions. Under the first condition, the transmitter and receiver phase centers are constrained to occupy the same locations, denoted here as the monostatic case. Under the second condition, the constraint is relaxed so that the transmitter and receiver phase centers are not necessarily the same. This additional freedom allows construction of MIMO virtual arrays with superior performance characteristics.

For the sake of the following analysis, it is assumed that sparse physical arrays are constructed by selecting phase centers from a one-dimensional lattice with critical spacing (half-wavelength for omnidirectional antennas). The length of the largest contiguous (in terms of lattice sites) subarray of the MIMO virtual array is computed in terms of the number of phase centers in the physical array. If signal processing is restricted to this contiguous subarray, conventional tapering can be applied. This type of construction substantially increases effective aperture and guarantees sidelobe levels. In principle, the joint design of MIMO arrays and waveforms could lead to better performance.

[2]The variable vec(·) converts its matrix argument into a vector by stacking rows together. See Appendix 2H for notation.

2.3.2.1 Monostatic Contiguous Virtual Array Length In this subsection, it is assumed that the transmitter and receiver antennas occupy the same phase centers. A uniform, linear, critically sampled array with $n = n_T = n_R$ elements is used as a reference. In this array, the elements are separated by a half-wavelength (0.5λ or $\lambda/2$). In the example given in Eq. (2.3), a three-element uniform filled array was used by both the transmitter and the receiver. This real antenna array can be represented using the notation $\{1 \quad 1 \quad 1\}$, where each entry corresponds to the number of antennas at a particular location on the $\lambda/2$ linear grid. Using this array, the corresponding MIMO virtual array, which is constructed by calculating the convolution of the real array antenna positions, is given by $\{1 \quad 2 \quad 3 \quad 2 \quad 1\}$. There are two interesting implications to note here: (1) the MIMO array aperture is much larger, with length $2n - 1$; and (2) many of the virtual antenna locations are overrepresented. Consequently, by using a sparse distribution of elements in the original traditional array, a MIMO array with a larger aperture can be constructed.

As an example, by using the following sparse array

$$\{1 \quad 1 \quad 0 \quad 1 \quad 1\}$$

for the transmitter and receiver, the following virtual MIMO array can be constructed:

$$\{1 \quad 2 \quad 1 \quad 2 \quad 4 \quad 2 \quad 1 \quad 2 \quad 1\}$$

Here a four-element sparse array is used to produce a filled array of aperture 9 rather than the aperture of 7 expected from a filled real array. Similarly, the six-element sparse array

$$\{1 \quad 1 \quad 0 \quad 1 \quad 0 \quad 1 \quad 0 \quad 1 \quad 1\}$$

corresponds to the MIMO virtual array

$$\{1 \quad 2 \quad 1 \quad 2 \quad 2 \quad 2 \quad 3 \quad 2 \quad 6 \quad 2 \quad 3 \quad 2 \quad 2 \quad 2 \quad 1 \quad 2 \quad 1\}$$

which has a length of 17 rather than 11.

For the purposes of this bound, the goal is to optimize the real array such that the MIMO virtual array has a very long nonzero contiguous region. Because the virtual array is constructed from the channel matrix, a bound on the maximum virtual array length can be found. By construction, the same locations are being used by the transmitters and receivers. An implication of this is that the channel matrix is symmetric, thus limiting the number of virtual array elements available. The maximum aperture length under this constraint is

$$l \leq \frac{n(n+1)}{2} \tag{2.8}$$

In general, it may not be possible to create an array that achieves this bound.

Conversely, a lower bound on the contiguous aperture can be calculated by using a specific construction. The following construction can be employed to build real arrays that have corresponding MIMO virtual arrays with very long nonzero contiguous regions

$$\{\underbrace{111\cdots1}_{q}\;\overbrace{\underbrace{00\cdots0}_{q-1}1}^{1}\;\overbrace{\underbrace{00\cdots0}_{q-1}1}^{2}\cdots\overbrace{\underbrace{00\cdots0}_{q-1}1}^{m}\}$$

that is, q ones, followed by $q-1$ zeros, followed by a one, followed by $q-1$ zeros, followed by a one, and so on, followed by $q-1$ zeros, followed by a one. The number of repetitions, denoted m, of the sequence $0\cdots01$ is a free parameter that can be optimized. The total number of antennas n is given by the sum of the q original antennas plus the m antennas from the m copies of the $0\cdots01$ sequence: $n=q+m$. The total length of the nonzero contiguous region for the MIMO virtual array is given by

$$\begin{aligned} l &= q+mq+q-1 \\ &= 2q+(n-q)q-1 \end{aligned} \tag{2.9}$$

By maximizing the length with respect to the parameter k, the optimal length is given by

$$l(n\in\text{even})\geq\frac{n^2}{4}+n \quad \text{or} \quad l(n\in\text{odd})\geq\frac{n^2-1}{4}+n \tag{2.10}$$

The bound is valid for values of $n\geq2$.

Constructed as above, an n-element, sparsely sampled physical aperture can be used to create a MIMO virtual array that provides equivalent or better sidelobe levels and a much larger effective aperture than does an n-element critically sampled array.

2.3.2.2 Contiguous Virtual Array Length with Independent Transmitter and Receiver Positions It turns out that the largest contiguous virtual MIMO array length, assuming that the positions of the transmit and receive antennas can be controlled independently, has a much simpler story. The contiguous virtual array length is bounded from above by the number of elements in the channel matrix $n_T\times n_R$. This is an achievable bound using the following construction. For example, a 4×4 MIMO radar is considered. If the receive array is filled, given by

$$\{1\quad1\quad1\quad1\}$$

and the transmit array is sparse with antennas spaced by the length of the receive array, given by

$$\{1\quad0\quad0\quad0\quad1\quad0\quad0\quad0\quad1\quad0\quad0\quad0\quad1\}$$

then the MIMO virtual array is given by the 16-virtual-element array

$$\{1\quad1\quad1\quad1\quad1\quad1\quad1\quad1\quad1\quad1\quad1\quad1\quad1\quad1\quad1\quad1\}$$

Within this chapter, this is denoted a Nyquist virtual array. For comparison, if $n = n_R = n_T$, the virtual aperture is n times larger than a traditional system with n antennas. This is much better than the bound achieved when the common transmit–receive antenna location constraint is applied, and consequently is the preferred MIMO radar topology for many MIMO radar applications.

2.4 MIMO RADAR IN CLUTTER-FREE ENVIRONMENTS

In this section, we consider the performance of MIMO radar in clutter-free environments. To investigate this topic, we introduce a Cramér–Rao bound on angle estimation performance. The bound allows arbitrary inter-waveform correlation. Limitations on the usefulness of this bound due to near ambiguities are also discussed. Given the estimator performance bound, waveform correlation optimization is introduced, and examples are provided.

2.4.1 Limitations of Cramér–Rao Estimation Bounds

The performance of many estimators, including a MIMO radar angle estimator, often exhibit similar characteristics. As seen in Fig. 2.4, at low SNR the variance of the estimator is poor and far from the Cramér–Rao bound [29]. This is because the noise is of sufficient strength to allow the estimator to be confused by any large near ambiguities of the angle estimation statistic. As the SNR is increased to some threshold, the estimator's performance approaches the Cramér–Rao bound. While the Cramér–Rao bound is a useful tool for characterizing system performance, the threshold SNR is also important.

The basic Cramér–Rao bound for static MIMO radar angular estimation has been described previously [16,17,24]. The description is repeated here, assuming an arbitrary positive-definite transmit covariance matrix.

2.4.2 Signal Model

The accuracy of angle-of-arrival estimates for an illuminated scatterer depends on the signals transmitted by the radar. Under some constraints, these signals can be designed to enhance system performance. For our analysis, the salient characteristic of the transmitted signals is their covariance matrix. Although the general cross-correlation properties of the signals are crucial aspects of system performance, we restrict ourselves to the same-time cross-correlations, associated with a particular range cell. Cross-correlations at different times are assumed perfectly decorrelated. The same-time auto- and cross-correlations of the transmitted waveforms are contained within the transmitter's correlation matrix $\mathbf{C}$:

$$\mathbf{C} \equiv \frac{1}{n_S} \left(\mathbf{S}_\delta \mathbf{S}_\delta^\dagger \right)^T \tag{2.11}$$

The transpose is used for notational convenience later on and agrees with the definition of correlation matrix used in Section 2.6. Since $\mathbf{C} = \mathbf{C}^\dagger$, it is worth noting that $\mathbf{C}^* = \mathbf{C}^T$.

Information about the external universe is contained within the recorded data matrix $\mathbf{Z}$. However, given a single point scatterer in a range cell at delay δ and the ideal different-time cross-correlations assumed above, no information about the environment is lost if the data are correlated with respect to the transmitted signals at the matching delay. In other words, the correlator outputs form a sufficient statistic. This correlation can be used to construct estimators of the channel at a given delay. For example, consider

$$\begin{aligned}\hat{\mathbf{H}}_\delta &= \frac{1}{n_S}\mathbf{Z}\mathbf{S}_\delta^\dagger \\ &= \mathbf{H}\mathbf{C}^T + \frac{1}{n_S}\mathbf{N}\mathbf{S}_\delta^\dagger \end{aligned} \tag{2.12}$$

If $\mathbf{C} = \mathbf{I}_{n_T}$, which corresponds to orthogonal transmit waveforms, then the estimator in Eq. (2.12) is the maximum-likelihood estimator (MLE) and is efficient. Otherwise, the estimation is biased, and if the signals are strongly correlated, then it is a very poor estimator. Nonetheless, the notation $\hat{\mathbf{H}}_\delta$ is used.

For the remainder of the section, it is assumed that there is a single scatterer in a given range cell. It is also assumed that the radar will be investigating this range cell by correlating with appropriately delayed versions of the transmitted waveforms. As a consequence, the δ will be dropped for notational simplicity.

The postcorrelation noise — that is, the internal receiver noise correlated with the reference signal S — is denoted $\hat{\mathbf{N}} \equiv \mathbf{N}\mathbf{S}^\dagger/n_S$. It is assumed that the internal noise has an identity covariance matrix. Consequently, the expectation of the postcorrelation noise is given by

$$E[(\hat{\mathbf{N}})_{n,m}(\hat{\mathbf{N}}^*)_{n'm'}] = \begin{cases} \frac{1}{n_S}(\mathbf{C})_{mm'}, & \text{if } n = n' \\ 0, & \text{otherwise} \end{cases} \tag{2.13}$$

For a target at a fixed delay and bearing, indicated by the wavevector $k\,\mathbf{u}$, the MIMO array response matrix is given by

$$(\mathbf{A}(\mathbf{u}))_{n,m} = e^{ik\mathbf{u}\cdot(\mathbf{x}_m+\mathbf{y}_n)} \tag{2.14}$$

which corresponds to the the theoretical channel matrix response for a simple scatterer, up to a scale factor b. Recall that $\{\mathbf{x}_m\}$ denotes the location of transmitter phase centers while $\{\mathbf{y}_m\}$ denotes receiver phase centers. The channel, assuming a single simple scatterer, is represented by $\mathbf{H} = b\mathbf{A}(\mathbf{u})$.

It is useful to think of vector representations of $\hat{\mathbf{H}},\hat{\mathbf{N}},\mathbf{A}(\mathbf{u})$, where the indices (n, m) of $(\hat{\mathbf{H}})_{n,m}$ can be viewed lexicographically[3] as the indices of a vector. The following vector notation is employed:

$$\begin{aligned} \hat{\mathbf{h}} &\equiv \mathrm{vec}\{\hat{\mathbf{H}}\} \\ \hat{\mathbf{n}} &\equiv \mathrm{vec}\{\hat{\mathbf{N}}\} \\ \mathbf{a}(\mathbf{u}) &\equiv \mathrm{vec}\{\mathbf{A}(\mathbf{u})\} = \mathbf{a}_R(\mathbf{u}) \otimes \mathbf{a}_T(\mathbf{u}) \end{aligned} \tag{2.15}$$

Definitions are supplied in Appendix 2H. A nontraditional vector operation $\mathrm{vec}\{\cdot\}$ produces a column vector that extracts elements row-wise along the entries of the matrix argument. Then the observations can be written

$$\hat{\mathbf{h}} = b(\mathbf{I}_{n_R} \otimes \mathbf{C})\, \mathbf{a}(\mathbf{u}) + \hat{\mathbf{n}} \tag{2.16}$$

where b represents the attenuation contained in the channel matrix. Note that the observation vector has $n_R n_T$ components.

2.4.3 Fisher Information Matrix

In this section, we provide a Cramér–Rao bound for angle estimation based on Eq. (2.16). The effects of transmitter–transmitter waveform correlation are included. The signal is Gaussian with known covariance

$$E\{\hat{\mathbf{n}}\hat{\mathbf{n}}^{\dagger}\} = \frac{1}{n_S}\mathbf{I}_{n_R} \otimes \mathbf{C} \tag{2.17}$$

and unknown, whitened (using the known covariance and absorbing scalar factors into b) mean

$$b\mathbf{v} \equiv b(\mathbf{I}_{n_R} \otimes \mathbf{C})^{1/2}\mathbf{a}(\mathbf{u}) = b\mathbf{v}_R \otimes \mathbf{v}_T \tag{2.18}$$

where $\mathbf{v}_R \equiv \mathbf{a}_R$ and $\mathbf{v}_T \equiv \mathbf{C}^{1/2}\mathbf{a}_T$. The Fisher matrix is a function of derivatives of the mean only [29] and can be expressed, for some arbitrary real parameters α and β, as

$$\mathbf{F}_{\alpha,\beta} = 2\,\mathrm{Re}\left\{\left(\frac{\partial}{\partial\alpha}[b\mathbf{v}_R(\mathbf{u}) \otimes \mathbf{v}_T(\mathbf{u})]^{\dagger}\right)\left(\frac{\partial}{\partial\beta}b\mathbf{v}_R(\mathbf{u}) \otimes \mathbf{v}_T(\mathbf{u})\right)\right\} \tag{2.19}$$

The parameters in $\mathbf{u}$ are of interest, while the complex amplitude b is a nuisance parameter. This complex parameter can be viewed as two real parameters. By partioning the Fisher matrix into an upper left-hand component $\mathbf{F}_{\theta\theta}$ involving the angle parameters, a lower right-hand component $\mathbf{F}_{bb}$ involving the two amplitude parameters,

[3]Order the pairs (m, n) so that $(m, n) > (m', n')$ if and only if $m > m'$ or $m = m'$ and $n > n'$.

and off-diagonal components $\mathbf{F}_{\theta b}$ and $\mathbf{F}_{b\theta}$, we can express the angular coordinates of the inverse Fisher in terms of the Sherman–Morrison–Woodbury relations [30] as

$$\left(\mathbf{F}^{(r)}\right)^{-1} = (\mathbf{F}_{\theta\theta} - \mathbf{F}_{\theta b}\mathbf{F}_{bb}^{-1}\mathbf{F}_{b\theta})^{-1} \tag{2.20}$$

$\mathbf{F}^{(r)}$ is an effective Fisher matrix for the angular parameters alone given the uncertainty of the nusiance parameters. Sometimes $\mathbf{F}^{(r)}$ is called a "reduced Fisher matrix" [29,31].[4]

To express the reduced Fisher, first define

$$\mathbf{P}_{\mathbf{v}}^{\perp} \equiv I_{n_R n_T} - \mathbf{v}(\mathbf{v}^{\dagger}\mathbf{v})^{-1}\mathbf{v}^{\dagger} \tag{2.21}$$

Then, define the matrix $\dot{\mathbf{V}}$, which includes the derivatives of $\mathbf{v}(\mathbf{u})$ with respect to parameters in $\mathbf{u}$. These derivatives can be expressed, for example, in terms of the derivatives of two independent components of $\mathbf{u}$, denoted u_1 and u_2:

$$\begin{aligned}\dot{\mathbf{V}} &\equiv \begin{pmatrix} \dfrac{\partial \mathbf{v}(\mathbf{u})}{\partial u_1} & \dfrac{\partial \mathbf{v}(\mathbf{u})}{\partial u_2} \end{pmatrix} \\ &= \begin{pmatrix} \dfrac{\partial}{\partial u_1}[\mathbf{v}_R(\mathbf{u}) \otimes \mathbf{v}_T(\mathbf{u})] & \dfrac{\partial}{\partial u_2}[\mathbf{v}_R(\mathbf{u}) \otimes \mathbf{v}_T(\mathbf{u})] \end{pmatrix}\end{aligned} \tag{2.22}$$

Then, after a tedious calculation [31], the reduced Fisher information matrix becomes

$$\mathbf{F}^{(r)} = 2|b|^2 \operatorname{Re}\{\dot{\mathbf{V}}^{\dagger}\mathbf{P}_{\mathbf{v}}^{\perp}\dot{\mathbf{V}}\} \tag{2.23}$$

Noting that the normal chain rule applies to the Krönecker product

$$\frac{\partial}{\partial u_1}[\mathbf{v}_R(\mathbf{u}) \otimes \mathbf{v}_T(\mathbf{u})] = \frac{\partial \mathbf{v}_R(\mathbf{u})}{\partial u_1} \otimes \mathbf{v}_T(\mathbf{u}) + \mathbf{v}_R(\mathbf{u}) \otimes \frac{\partial \mathbf{v}_T(\mathbf{u})}{\partial u_1} \tag{2.24}$$

and that the projection matrix can be expanded as

$$\begin{aligned}\mathbf{P}_{\mathbf{v}}^{\perp} &= \mathbf{I}_{n_R n_T} - \mathbf{P}_{\mathbf{v}_R} \otimes \mathbf{P}_{\mathbf{v}_T} \\ &= \mathbf{I}_{n_R} \otimes \mathbf{P}_{\mathbf{v}_T}^{\perp} + \mathbf{P}_{\mathbf{v}_R}^{\perp} \otimes \mathbf{I}_{n_R} - \mathbf{P}_{\mathbf{v}_R}^{\perp} \otimes \mathbf{P}_{\mathbf{v}_T}^{\perp}\end{aligned} \tag{2.25}$$

the following relationship can be constructed:

$$\dot{\mathbf{V}}^{\dagger}\mathbf{P}_{\mathbf{v}}^{\perp}\dot{\mathbf{V}} = \|\mathbf{v}_R\|^2 \dot{\mathbf{V}}_T^{\dagger}\mathbf{P}_{\mathbf{v}_T}^{\perp}\dot{\mathbf{V}}_T + \|\mathbf{v}_T\|^2 \dot{\mathbf{V}}_R^{\dagger}\mathbf{P}_{\mathbf{v}_R}^{\perp}\dot{\mathbf{V}}_R \tag{2.26}$$

[4] In the very useful DeLong reference [31], it is worth noting that in Eq. (2.7) a factor of 2 is missing.

Consequently, the Fisher matrix in Eq. (2.23) is given by

$$\begin{aligned}\mathbf{F}^{(r)} &= 2\,|b|^2\,\mathrm{Re}\Big\{\|\mathbf{v}_R\|^2\,\dot{\mathbf{V}}_T^{\dagger}\,\mathbf{P}_{\mathbf{v}_T}^{\perp}\,\dot{\mathbf{V}}_T + \|\mathbf{v}_T\|^2\,\dot{\mathbf{V}}_R^{\dagger}\,\mathbf{P}_{\mathbf{v}_R}^{\perp}\,\dot{\mathbf{V}}_R\Big\} \\ &= 2\,|b|^2\,\mathrm{Re}\left\{n_R\dot{\mathbf{W}}_T^{\dagger}\left(\mathbf{C} - \frac{\mathbf{C}\,\mathbf{a}_T(\mathbf{u})\,\mathbf{a}_T^{\dagger}(\mathbf{u})\,\mathbf{C}}{\mathbf{a}_T^{\dagger}(\mathbf{u})\,\mathbf{C}\mathbf{a}_T(\mathbf{u})}\right)\dot{\mathbf{W}}_T\right. \\ &\quad \left. +\,\mathbf{a}_T^{\dagger}(\mathbf{u})\,\mathbf{C}\,\mathbf{a}_T(\mathbf{u})\dot{\mathbf{W}}_R^{\dagger}\left(\mathbf{I} - \frac{\mathbf{a}_R(\mathbf{u})\,\mathbf{a}_R^{\dagger}(\mathbf{u})}{n_R}\right)\dot{\mathbf{W}}_R\right\}\end{aligned} \tag{2.27}$$

where the matrix of derivatives with respect to the direction parameters is given by

$$\dot{\mathbf{W}}\alpha \equiv \begin{pmatrix}\dfrac{\partial \mathbf{a}_\alpha(\mathbf{u})}{\partial u_1} & \dfrac{\partial \mathbf{a}_\alpha(\mathbf{u})}{\partial u_2}\end{pmatrix} \tag{2.28}$$

The parameter α is replaced with either R or T for receiver or transmitter. Defining the unitary, diagonal matrices with diagonal entries

$$(\mathbf{D}_T)_{m,m} \equiv e^{-ik\mathbf{x}_m\cdot\mathbf{u}_0}, \quad \text{and} \quad (\mathbf{D}_R)_{n,n} \equiv e^{-ik\mathbf{y}_n\cdot\mathbf{u}_0} \tag{2.29}$$

and noting that $\mathbf{D}_T^{\dagger}\,\mathbf{D}_T = \mathbf{I}_{n_T}$ and $\mathbf{D}_R^{\dagger}\mathbf{D}_R = \mathbf{I}_{n_R}$, a transformed version of the transmit covariance matrix is defined:

$$\tilde{\mathbf{C}} \equiv \mathbf{D}_T\mathbf{C}\mathbf{D}_T^{-1} \tag{2.30}$$

Equation (2.27) becomes

$$\begin{aligned}\mathbf{F}^{(r)} &= 2|b|^2\,\mathrm{Re}\left\{n_R\,(\mathbf{D}_T\,\dot{\mathbf{W}}_T)^{\dagger}\left(\tilde{\mathbf{C}} - \frac{\tilde{\mathbf{C}}\mathbf{1}\mathbf{1}^{\dagger}\tilde{\mathbf{C}}}{\mathbf{1}^{\dagger}\tilde{\mathbf{C}}\mathbf{1}}\right)\mathbf{D}_T\,\dot{\mathbf{W}}_T\right. \\ &\quad \left. +\mathbf{1}^{\dagger}\tilde{\mathbf{C}}\mathbf{1}(\mathbf{D}_R\dot{\mathbf{W}}_R)^{\dagger}\left(\mathbf{I}_{n_R} - \frac{\mathbf{1}\mathbf{1}^{\dagger}}{n_R}\right)\mathbf{D}_R\,\dot{\mathbf{W}}_R\right\}\end{aligned} \tag{2.31}$$

under the assumption that $\mathbf{u}_0$ is chosen to be $\mathbf{u}$, so that $\mathbf{D}_T\mathbf{a}_T(\mathbf{u}) = 1$ and $D_R\,\mathbf{a}_R(\mathbf{u}) = 1$.

Defining transmit antenna location matrix $\mathbf{X} \in \mathbb{R}^{n_T\times 3}$ and receive antenna location matrix $\mathbf{Y} \in \mathbb{R}^{n_R\times 3}$ so that the mth row of $\mathbf{X}$ is given by $\mathbf{x}_m^T$, and the nth row of $\mathbf{Y}$ is $\mathbf{y}_n^T$, the steering vectors can be written informally as

$$\mathbf{a}_T(\mathbf{u}) = e^{ik\mathbf{X}\mathbf{u}} \quad \text{and} \quad \mathbf{a}_R(\mathbf{u}) = e^{ik\mathbf{Y}\mathbf{u}} \tag{2.32}$$

The derivative with respect to some parameter α is then given by

$$\begin{aligned}\frac{\partial}{\partial\alpha}\mathbf{a}_T(\mathbf{u}) &= ik\mathbf{X}\frac{\partial\mathbf{u}}{\partial\alpha}\,\Box\,\mathbf{a}_T(\mathbf{u}) \\ \mathbf{D}_T\frac{\partial}{\partial\alpha}\mathbf{a}_T(\mathbf{u}) &= ik\mathbf{X}\frac{\partial\mathbf{u}}{\partial\alpha}\end{aligned} \tag{2.33}$$

where □ indicates the Hadamard product. Similarly, the transformed derivative of the receive steering vector is given by

$$\mathbf{D}_R \frac{\partial}{\partial \alpha} \mathbf{a}_R(\mathbf{u}) = ik\mathbf{Y}\frac{\partial \mathbf{u}}{\partial \alpha} \tag{2.34}$$

Defining the transformed set of derivatives to be

$$\dot{\mathbf{U}} \equiv \begin{pmatrix} \frac{\partial \mathbf{u}}{\partial u_1} & \frac{\partial \mathbf{u}}{\partial u_2} \end{pmatrix} \tag{2.35}$$

the Fisher matrix is given by

$$\mathbf{F}^{(r)} = 2|b|^2k^2 \operatorname{Re}\left\{ n_R \,\dot{\mathbf{U}}^\dagger \left(\mathbf{X}^T\tilde{\mathbf{C}}\mathbf{X} - \frac{\mathbf{X}^T\tilde{\mathbf{C}}\mathbf{1}\mathbf{1}^T\tilde{\mathbf{C}}\mathbf{X}}{\mathbf{1}^T\tilde{\mathbf{C}}\mathbf{1}} \right)\dot{\mathbf{U}} \right.$$

$$\left. +(\mathbf{1}^T\tilde{\mathbf{C}}\mathbf{1})\dot{\mathbf{U}}^\dagger \left(\mathbf{Y}^T\mathbf{Y} - \frac{\mathbf{Y}^T\mathbf{1}\mathbf{1}^T\mathbf{Y}}{n_R} \right)\dot{\mathbf{U}} \right\} \tag{2.36}$$

In the special case of only a single direction parameter θ with respect to the perpendicular to a linear array of antennas, and $\mathbf{C} = p/n_T\mathbf{I}_{n_T}$, where p is the total noise-normalized transmit power, this simplifies to

$$\operatorname{var}\{\theta\} = \frac{1}{\mathbf{F}^{(r)}} = \frac{n_T}{2|b|^2pk^2\cos^2(\theta)(n_R\,\sigma_T^2 + n_T\,\sigma_R^2)} \tag{2.37}$$

Here, σ_T and σ_R are the standard deviations of antenna distance along the linear array.

2.4.4 Waveform Correlation Optimization

The Fisher matrix expressed by Eq. (2.36) can be optimized in a variety of ways. We choose to minimize the trace of the inverse of the Fisher matrix for a particular coordinate system, subject to the constraint that the signal correlation matrix $\mathbf{C}$ has a constant trace: $\operatorname{tr}\mathbf{C} = \operatorname{tr}\tilde{\mathbf{C}}$.

Let the 3×3 spatial distribution of transmit and receive antennas be characterized by

$$\mathbf{R_X} \equiv \frac{1}{n_T}\left(\mathbf{X}^T\mathbf{X} - \frac{\mathbf{X}^T\mathbf{1}\mathbf{1}^T\mathbf{X}}{n_T} \right) \tag{2.38}$$

$$\mathbf{R_Y} \equiv \frac{1}{n_R}\left(\mathbf{Y}^T\mathbf{Y} - \frac{\mathbf{Y}^T\mathbf{1}\mathbf{1}^T\mathbf{Y}}{n_R} \right) \tag{2.39}$$

In applications, the array is often one- or two-dimensional. In those cases, it is convenient to restrict dimensions to the first or to the first two coordinates; then $\mathbf{R_X}$ and $\mathbf{R_Y}$ can be viewed as 1×1 or 2×2 matrices, as is appropriate. Similar comments apply to other matrix quantities below. In addition, it is convenient to assume that the coordinate origin is chosen so that $\mathbf{X}^T\mathbf{1} = \mathbf{0}$. Note that the choice of origin does not affect the values of $\mathbf{R_X}$ and $\mathbf{R_Y}$. For the moment, assume that one can write $\tilde{\mathbf{C}}$ as

$$\tilde{\mathbf{C}} = (\mathbf{X}\ \mathbf{1})\begin{pmatrix} \mathbf{A} & \mathbf{b} \\ \mathbf{b}^\dagger & g \end{pmatrix}(\mathbf{X}\ \mathbf{1})^T \tag{2.40}$$

where the components of $\tilde{\mathbf{C}}$ are formed from a Hermitian matrix $\mathbf{A} \in \mathbb{C}^{3\times 3}$, vector $\mathbf{b} \in \mathbb{C}^{3\times 1}$, and real scalar g. Given this construction, $\tilde{\mathbf{C}}$ is not full-rank and is constrained to the subspace spanned by the column space of $\mathbf{X}$ and $\mathbf{1}$. However, it is clear that an optimal solution has, in fact, this form. Any nonzero $\mathbf{x}^\dagger\tilde{\mathbf{C}}\mathbf{x}$ with $\mathbf{x}$ outside the column span of $(\mathbf{X}\ \mathbf{1})$ increases $\mathrm{tr}(\tilde{\mathbf{C}})$ without affecting the Fisher matrix of Eq. (2.36). With the form for $\tilde{\mathbf{C}}$ given above, Eq. (2.36) can be written

$$\begin{aligned}\mathbf{F}^{(r)} &\longrightarrow 2|b|^2\, k^2\, \mathrm{Re}\left\{ n_R\dot{\mathbf{U}}^\dagger\left(\mathbf{X}^T\mathbf{X}\,\mathbf{A}\,\mathbf{X}^T\mathbf{X} - \frac{\mathbf{X}^T[\mathbf{X}\mathbf{b}\mathbf{1}^T]\mathbf{1}\mathbf{1}^T[\mathbf{1}\mathbf{b}^\dagger\mathbf{X}^T]\mathbf{X}}{\mathbf{1}^T\mathbf{1}g\mathbf{1}^T\mathbf{1}}\right)\dot{\mathbf{U}}\right. \\ &\quad \left. +\mathbf{1}^T\mathbf{1}g\ \mathbf{1}^T\mathbf{1}\dot{\mathbf{U}}^\dagger\left(\mathbf{Y}^T\mathbf{Y} - \frac{\mathbf{Y}^T\mathbf{1}\mathbf{1}^T\mathbf{Y}}{n_R}\right)\dot{\mathbf{U}}\right\} \\ &= 2|b|^2\, k^2\, n_R n_T^2\ \mathrm{Re}\left\{\dot{\mathbf{U}}^\dagger\mathbf{R_X}\left(\mathbf{A} - \frac{\mathbf{b}\mathbf{b}^\dagger}{g}\right)\mathbf{R_X}\,\dot{\mathbf{U}} + g\dot{\mathbf{U}}^\dagger\,\mathbf{R_Y}\,\dot{\mathbf{U}}\right\}\end{aligned} \tag{2.41}$$

Consider the constraint $\mathrm{tr}\ \tilde{\mathbf{C}} = p$, where p is the total noise-normalized transmit power. This becomes

$$\mathrm{tr}(\mathbf{A}\mathbf{R_X}) + g = \frac{p}{n_T} \tag{2.42}$$

Since a nonzero $\mathbf{b}$ only decreases the Fisher matrix (in the sense of the ordering of positive, semidefinite matrices), one can assume that $\mathbf{b} = \mathbf{0}$.

For convenience, we choose coordinates so that $\dot{\mathbf{U}} = \mathbf{I}_k$, where, as discussed above, k is typically 1 or 2. More explicitly, in the case of two-dimensional (2D) arrays located in the x–y plane, coordinates have the form $(u_x,\ u_y)$. It is well known that this choice of coordinates provides array patterns whose shape is independent of the steering direction. However, these coordinates do not agree with physical angular coordinates except in the important special case when the direction is $\mathbf{u} = (0,\ 0,\ 1)^T$. In our coordinates, the Fisher information matrix becomes

$$\mathbf{F}^{(r)} \longrightarrow 2|b|^2\, k^2 n_R n_T^2\ \mathrm{Re}\{\mathbf{R_X}\,\mathbf{A}\,\mathbf{R_X} + g\,\mathbf{R_Y}\} \tag{2.43}$$

The trace of the inverse of the Fisher matrix becomes

$$\mathrm{tr}\{(\mathbf{F}^{(r)})^{-1}\} = \frac{1}{2|b|^2\, k^2 n_R n_T^2}\mathrm{tr}(\mathbf{R_X}^{-1}(\mathbf{P} + \mathbf{Q})^{-1}) \tag{2.44}$$

where

$$\mathbf{P} \equiv \mathbf{R}_X^{1/2}\ \mathbf{A}\mathbf{R}_X^{1/2} \quad \text{and} \quad \mathbf{Q} \equiv g\mathbf{R}_X^{-1/2}\mathbf{R_Y}\mathbf{R}_X^{-1/2}$$

with $0 \leq \mathrm{tr}\mathbf{P} \leq p/n_T$. Now one can show that Eq. (2.44) is strictly convex in $\mathbf{P}$, provided $\mathbf{P}+\mathbf{Q} > 0$ and $\mathbf{R_X} > 0$ as Hermitian matrices. Thus, a stationary point over the convex set where $\mathrm{tr}\mathbf{P}$ is constant and $\mathbf{P}+\mathbf{Q} > 0$ is a unique minimum. Lagrange multipliers can be used to find the stationary points in the sense that

$$0 = \frac{\partial}{\partial \mathbf{P}} [\mathrm{tr}(\mathbf{R}_\mathbf{X}^{-1}(\mathbf{P}+\mathbf{Q})^{-1}) - \mu \mathrm{tr}\mathbf{P}] \tag{2.45}$$

implies $\mathbf{R}_\mathbf{X}^{-1/2} = \alpha(\mathbf{P}+\mathbf{Q})$ with $\alpha > 0$. Note that A can be assumed real without changing the Fisher matrix or the power constraint of Eq. (2.42). We have

$$\begin{aligned} \mathrm{tr}\{(\mathbf{F}^{(r)})^{-1}\} &\propto \mathrm{tr}\{\mathbf{R}_\mathbf{X}^{-1}(\mathbf{P}+\mathbf{Q})^{-1}\} \\ &= \alpha^2 \mathrm{tr}\{\mathbf{P}+\mathbf{Q}\} \\ &= \frac{\mathrm{tr}(\mathbf{R}_\mathbf{X}^{-1/2})^2}{(p/n_T)\,[\rho + (1-\rho)\mathrm{tr}(\mathbf{R}_\mathbf{X}^{-1/2}\mathbf{R_Y}\mathbf{R}_\mathbf{X}^{-1/2})]} \end{aligned} \tag{2.46}$$

for $\rho \equiv 1 - g/(p/n_T)$. Because g is bounded by 0 and p/n_T [see Eq. (2.42)], ρ is bounded by $0 \leq \rho \leq 1$. Given freedom to optimize over the value of ρ under this constraint, the optimal point exists at the boundaries: ρ is 0 or 1 depending on the value of $\mathrm{tr}\{\mathbf{R}_\mathbf{X}^{-1/2}\mathbf{R_Y}\mathbf{R}_\mathbf{X}^{-1/2}\}$. The trace expresses, in a sense, the ratio between the root-mean-square (RMS) apertures of the receiver and transmitter arrays. The solution for $\tilde{\mathbf{C}}$ is given by

$$\tilde{\mathbf{C}} = \begin{cases} \mathbf{1}\mathbf{1}^T; & \rho = 0 \\ \alpha^{-1}\mathbf{X}\mathbf{R}_\mathbf{X}^{-3/2}\mathbf{X}^T; & \rho = 1 \end{cases} \tag{2.47}$$

Thus, the trace of the reduced Cramér–Rao bound matrix is minimized when the beamforming matrix assumes the values

$$\mathbf{C} = \begin{cases} \dfrac{p}{n_T}\,\mathbf{a}_T(\mathbf{u})\,\mathbf{a}_T^\dagger(\mathbf{u}) & 1 < \mathrm{tr}\{\mathbf{R}_\mathbf{X}^{-1/2}\mathbf{R_Y}\mathbf{R}_\mathbf{X}^{-1/2}\} \\ p\dfrac{\mathbf{E_X}\mathbf{R}_\mathbf{X}^{-1/2}\mathbf{E}_\mathbf{X}^\dagger}{\mathrm{tr}\mathbf{R}_\mathbf{X}^{-1/2}} & 1 \geq \mathrm{tr}\{\mathbf{R}_\mathbf{X}^{-1/2}\mathbf{R_Y}\mathbf{R}_\mathbf{X}^{-1/2}\} \end{cases} \tag{2.48}$$

where $\mathbf{E_X} \equiv \mathbf{D}_T^\dagger\mathbf{X}(\mathbf{X}^T\mathbf{X})^{-1/2}$. These solutions have simple physical interpretations. The first form can be considered as the sum beam pointing at the target. The second form can be interpreted as a combination of difference beams. The minimized trace becomes

$$\mathrm{tr}\{(\mathbf{F}^{(r)})^{-1}\} = \begin{cases} \dfrac{1}{2k^2\,p|b|^2 n_R n_T}\,\dfrac{\mathrm{tr}(\mathbf{R}_\mathbf{X}^{-1/2})^2}{\mathrm{tr}(\mathbf{R}_\mathbf{X}^{-1/2}\mathbf{R_Y}\mathbf{R}_\mathbf{X}^{-1/2})} \\ \dfrac{1}{2k^2\,p|b|^2 n_R n_T}\,\mathrm{tr}(\mathbf{R}_\mathbf{X}^{-1/2})^2 \end{cases} \tag{2.49}$$

It should be noted that these "optimal" waveforms have not taken the threshold point, as discussed in Section 2.4.1, into consideration. Amusingly, if the difference beam is pointed directly at the target, the SNR vanishes by definition. Consequently, the high-SNR assumption of the Cramér–Rao bound is violated. This implies that the optimal (in a robust, fixed transmit power sense) waveform mixes in other components in addition to the difference beam.

2.4.5 Examples

For notional full-diversity MIMO waveforms, $\mathbf{C} = \mathbf{I}_{n_T}$, while traditional SIMO systems would use $\mathbf{C} = \mathbf{1}\mathbf{1}^T$, focusing the transmitted waveform on the scatterer. Here, the acronym SINO signifies that the signal transmitted from multiple antennas is completely coherent. It does not necessarily indicate that a single transmit element is being used. Under the further assumption that the transmitter and receiver phase centers are identical, we have seen that the Fisher information for SIMO is at least as large as that for the notional MIMO system. Other types of waveform correlation matrices can also be useful. Remove the assumption that transmitter and receiver phase centers are identical, and one sees that a difference beam at the transmitter can provide a smaller Cramér–Rao bound when the RMS transmitter aperture is larger than the RMS receiver aperture. Of course, a difference beam steered at a target results in no illumination of the target and thus poor performance. A mixture of sum and difference beams provides the best Cramér–Rao bound under a combined constraint on transmitted power and on the fraction of power in the sum beam. In this case, the sum and difference beams transmit uncorrelated waveforms. The sum beam provides some target illumination, and the difference beam

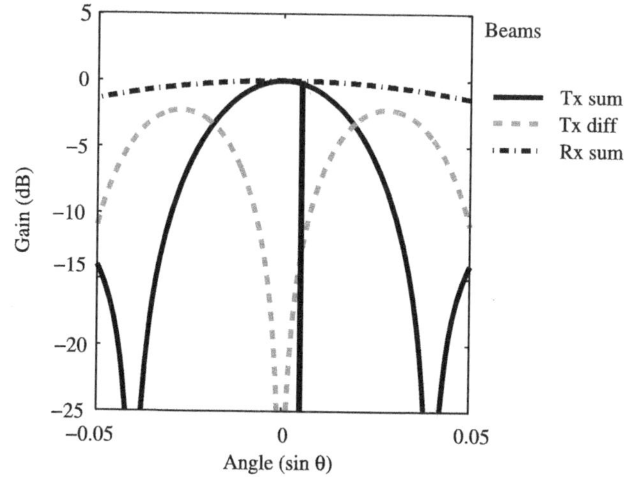

Figure 2.5 Shown are sum and difference beam patterns at the transmitter and a sum pattern at the receiver. The transmitter's aperture is larger (4×) than the receiver's aperture. The vertical line locates a target used in the figures below.

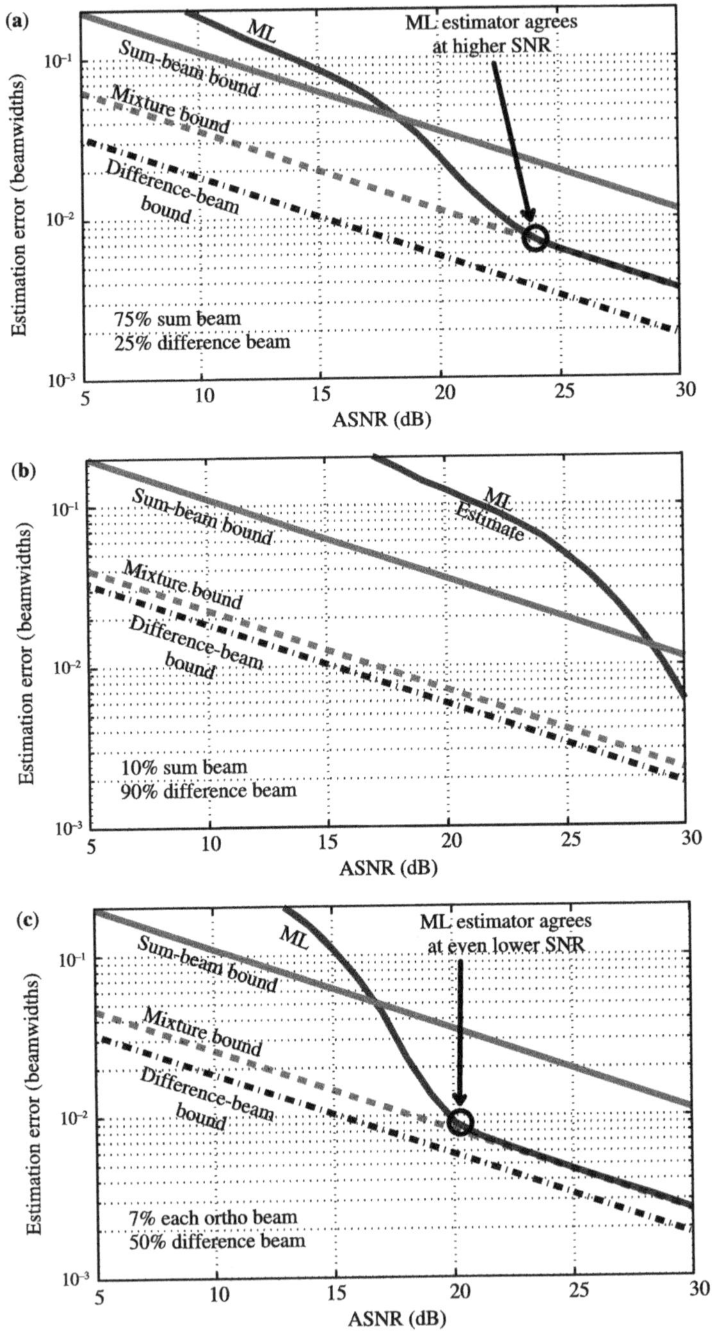

Figure 2.6 In frame (a), estimation error for mixed illumination (75% sum and 25% difference) is displayed, when the transmitter aperture is larger (6×) than the receiver aperture. In frame (b), estimation error for mixed illumination (10% sum and 90% difference) is displayed. In frame (c), estimation error for mixed illumination (50% difference and 7% in all orthogonal beams) is displayed.

leverages the larger aperture provided by the transmitters to increase accuracy. With less sum beam illumination, the Cramér–Rao bound is still approached, but at higher SNRs. However, this ignores the importance of the other possible modes in the transmit array. To avoid large sidelobes and receive full benefit of the virtual array, all modes must be exercised, which is to say that the transmit covariance should be full-rank. There is an optimization tradeoff between threshold point and the amount of energy in the difference beam. This tradeoff is not captured by the CR bound.

Figure 2.6 compares simulated estimation error with Cramér–Rao bounds for a static MIMO radar employing separate arrays for transmission and reception. Each array has eight phase centers. The receiving array has a uniform linear aperture with half-wavelength spacing; the transmitting array has a linear aperture 6 times larger with uniformly spaced phase centers. Note that this provides an oversampled MIMO virtual array. Three different mixtures of transmit beams are presented. Sum beam and difference beam are both steered at angle zero. Beam patterns for the transmitter's sum and difference beams as well as a sum beam at the receiver are shown in Fig. 2.5. Note that the target, indicated by the vertical line, is slightly off of 0° in the example.

Three Cramér–Rao bounds are shown in each panel of Fig. 2.6. One is shown for the sum beam illumination, one for difference beam illumination (in the limit), and the last (CRB) for the mixed illumination involving uncorrelated signals in the sum, difference beams and/or other beams. In principle, the difference beam bound provides the best performance when the transmitter's aperture is larger, but this accuracy cannot be achieved in practice since the target is not illuminated.

Figure 2.6a shows estimation error (ML) in the case of a mixed illumination of sum and difference beams. The power is split between the sum beam (75%) and the difference beam (25%). Note that the Cramér–Rao bound for this power splitting is close to the difference bound and significantly better than the sum bound. The estimates closely follow the mixed illumination bound at higher SNRs. In Fig. 2.6b only 10% of the transmitted power is placed in the sum beam and 90%, in the difference beam. Although the mixed-illumination Cramér–Rao bound is closer to the ideal difference bound, the estimates approach the bound only at very high SNRs, as one would expect. Finally, Fig. 2.6c shows a mixture of 50% difference beam and 7% in each orthogonal beam. This mixture clearly has the best mix of threshold and asymptotic performance out of the three examples.

2.5 OPTIMALITY OF MIMO RADAR FOR DETECTION

In the course of Section 2.4, we saw that MIMO radars can offer better parameter estimates than can SIMO radars. Before embarking on detailed examples of MIMO radar performance, we present an abstract formulation of detection and search in order to introduce some issues involving MIMO waveform design.

An abstract model of search combined with the Kullback–Leibler divergence provides a simple example of detection performance for MIMO radar. We show that the radar should form a beam on the target if the target's channel response is known. If the MIMO radar performs search, we show that the optimal beamformer depends on the strength of the target return. If the response of the target is weak, a scanned beam provides the best detection. If the response is strong, detection is best when the MIMO radar uses uncorrelated waveforms at each transmitter phase center.

2.5.1 Detection

A significant benefit of MIMO radar over SIMO radar can be illustrated using a simple model of detection for a narrowband signal. We consider a large-deviation model for detection performance. In general terms, we use the Kullback–Leibler (KL) divergence [32] to characterize the probability of detection p_d. Recall that the KL divergence is expressed by

$$I(p_2, p_1) = \int \log\left(\frac{p_2(x)}{p_1(x)}\right) p_2(x)\, dx \tag{2.50}$$

for two probability densities, $p_1(x)$ and $p_2(x)$, associated with two hypotheses, θ_1 and θ_2. Here, the signal-present hypothesis is denoted θ_1. This KL divergence provides an asymptotic exponential bound on the probability of missed detection of the form

$$1 - p_d \sim e^{-nI(p_2, p_1)} \tag{2.51}$$

where n is the number of samples. The bound is asymptotic in the number of samples n. Detection is evaluated at a fixed nontrivial (not zero or one) false-alarm probability. For MIMO radar, we can evaluate the KL divergence by using a Gaussian signal model. Specifically, let

$$\mathbf{Z} = \mathbf{HS} + \mathbf{N} \tag{2.52}$$

express the $n_R \times n_S$ matrix $\mathbf{Z}$ of observations in terms of the $n_R \times n_T$ channel matrix $\mathbf{H}$, the $n_T \times n_S$ waveform matrix $\mathbf{S}$, and the $n_R \times n_S$ noise matrix $\mathbf{N}$. Both $\mathbf{H}$ and $\mathbf{N}$ are chosen to be complex, circular, Gaussian random variates. Their covariances are given by[5]

$$\begin{aligned} \mathrm{cov}(\mathbf{H}) &= \mathbf{R} \otimes \mathbf{T}^* \\ \mathrm{cov}(\mathbf{N}) &= \mathbf{I}_{n_R} \otimes \mathbf{I}_{n_S} \end{aligned} \tag{2.53}$$

The Krönecker product stucture of cov($\mathbf{H}$) is chosen for analytical convenience; it is also capable of representing both distibuted and localized sources. For example,

[5]Let $\bar{s}$ denote conjugation of a scalar s and $\mathbf{S}^*$ denote conjugation of all entries of a matrix or vector $\mathbf{S}$. Then $E[H_{ij}\overline{H}_{Kl}] = R_{ik}\overline{T}_{jl}$, for example, where $E[\cdot]$ denotes the ensemble average of its argument. Also note $(\mathbf{T}^*)_{jk} \equiv \overline{T}_{jk}$. See Appendix 2H for Krönecker product definitions.

$\mathbf{R} \otimes \mathbf{T}^* = \mathbf{a}_R \mathbf{a}_R^\dagger \otimes \mathbf{a}_T \mathbf{a}_T^\dagger$ represents the response to a localized scatterer. The use of $\boldsymbol{T}^*$ instead of $\boldsymbol{T}$ is motivated by the transformation law of Appendix 2H.

Recall [33] that the probability distribution function of a multivariate, complex, circular Gaussian $n \times \lambda$ random matrix $\mathbf{Y}$ of zero mean and complex covariance $\boldsymbol{\Gamma} \otimes \mathbf{I}_\lambda$ is expressed by

$$p(\mathbf{Y}|\boldsymbol{\Gamma}) = \pi^{-n\lambda}|\boldsymbol{\Gamma}|^{-\lambda}e^{-\mathrm{tr}(\boldsymbol{\Gamma}^{-1}\mathbf{Y}\mathbf{Y}^\dagger)} \tag{2.54}$$

For detection, the Kullback–Leibler divergence, defined by

$$-\int \log \frac{p(\mathbf{Y}|\boldsymbol{\Gamma})}{p(\mathbf{Y}|\mathbf{I}_n)} p(\mathbf{Y}|\mathbf{I}_n)\, d_{\mathrm{vol}}\mathbf{Y} = \lambda \log|\boldsymbol{\Gamma}| + \lambda\, \mathrm{tr}(\boldsymbol{\Gamma}^{-1}) - \lambda n \tag{2.55}$$

expresses a form of differential entropy that is well defined and characterizes detection performance in the large-sample-set limit.

For the signal model in Eq. (2.52), the signal-present covariance is expressed by $\boldsymbol{\Gamma} \equiv \mathbf{R} \otimes \mathbf{T}^*$ and the dimensionality by $n = n_R n_S$. Assuming $\lambda = 1$, the Kullback–Leibler divergence becomes[6]

$$\begin{aligned}
&\log|\mathbf{R} \otimes (\mathbf{S}^\dagger\mathbf{T}\mathbf{S})^* + \mathbf{I}_{n_S n_R}| + \mathrm{tr}(\mathbf{R} \otimes (\mathbf{S}^\dagger\mathbf{T}\mathbf{S})^* + \mathbf{I}_{n_S n_R})^{-1} - n_S n_R \\
&\quad = \log|\mathbf{R} \otimes ((\mathbf{S}\mathbf{S}^\dagger)^{1/2}\mathbf{T}(\mathbf{S}\mathbf{S}^\dagger)^{1/2})^* + \mathbf{I}_{n_R n_T}| \\
&\qquad + \mathrm{tr}(\mathbf{R} \otimes ((\mathbf{S}\mathbf{S}^\dagger)^{1/2}\mathbf{T}(\mathbf{S}\mathbf{S}^\dagger)^{1/2})^* + \mathbf{I}_{n_R n_T})^{-1} - n_R n_T
\end{aligned} \tag{2.56}$$

as long as $n_T \le n_S$.

We are interested in the expected value of Eq. (2.56) over an ensemble of waveform matrices $\mathbf{S}$. This ensemble is a surrogate for the search performed by the radar. We model this search by assuming that each $\mathbf{S}$ has the singular value decomposition $\mathbf{S} = \mathbf{U}\boldsymbol{\Lambda}\mathbf{W}^\dagger$, with unitaries $\mathbf{U}$, $\mathbf{W}$, and quasidiagonal rectangular matrix $\boldsymbol{\Lambda}$ (nonzero entries only along the diagonal) with nonnegative diagonal entries. We assume that the unitaries $\mathbf{U}$ and $\mathbf{W}$ are random and uniformly distributed over all unitaries of the appropriate dimensions. The expectation over $\mathbf{U}$ will yield a function of $\mathbf{D} = \boldsymbol{\Lambda}\boldsymbol{\Lambda}^\dagger$ that can be optimized. Since Eq. (2.56) is nonconvex, we approximate the expectation of Eq. (2.56) in two regimes: that of high integrated SNRs, where the log term dominates, and that of low integrated SNR, where the KL divergence has a quadratic character. Both approximate KL divergences are convex, but in different senses. We also consider nonfading target models. The optimal beamformer is formulated for the case of a known, point scatterer when there is no requirement for search.

2.5.2 High SNR

Note that

$$(\mathbf{S}\mathbf{S}^\dagger)^{1/2}\mathbf{T}(\mathbf{S}\mathbf{S}^\dagger)^{1/2} = \mathbf{U}\mathbf{D}^{1/2}\mathbf{U}^\dagger\mathbf{T}\mathbf{U}\mathbf{D}^{1/2}\mathbf{U}^\dagger \tag{2.57}$$

[6]Observe that $\mathrm{cov}(\mathbf{A}\mathbf{Y}\mathbf{B}) = (\mathbf{A}\boldsymbol{\Gamma}\mathbf{A}^\dagger) \otimes (\mathbf{B}^\dagger\boldsymbol{\Delta}\mathbf{B})^*$ given $\mathrm{cov}(\mathbf{Y}) = \boldsymbol{\Gamma} \otimes \boldsymbol{\Delta}^*$ (see Appendix 2H).

where we define $\mathbf{D} = \mathbf{\Lambda}\mathbf{\Lambda}^{\dagger}$. Inserting this expansion into the log term of Eq. (2.56) and canceling the outer unitaries due to invariances of the determinant leads to an upper bound based on Jensen's inequality

$$E[\text{KL log term}] \leq \log\left|\mathbf{R} \otimes \left(\frac{\text{tr}(\mathbf{T})}{n_T}\mathbf{D}\right) + \mathbf{I}_{n_R n_T}\right| \tag{2.58}$$

If we denote by r_k the eigenvalues of $\mathbf{R}$ and by d_k the diagonal entries of $\mathbf{D}$, then the upper bound of Eq. (2.58) becomes (letting $\beta = \text{tr}(\mathbf{T})/n_T$)

$$f(d_1, \ldots, d_{n_T}) = \sum_{j,k} \log(1 + \beta d_j r_k) \tag{2.59}$$

Using Lagrange multipliers and imposing the constraint $\sum d_k = d$, we find, for nonzero d_k, that

$$0 = \frac{\partial f}{\partial d_k} - \mu \tag{2.60}$$

so that

$$0 = \sum_j \frac{\beta r_j}{1 + \beta r_j d_k} - \mu \tag{2.61}$$

Thus all nonzero d_k have the same value. Consider s nonzero $d_k = d/s$. Then the upper bound becomes

$$s\sum_k \log\left(1 + \frac{\beta r_k d}{s}\right) \tag{2.62}$$

with integral s, $1 \leq s \leq n_T$. A simple calculus exercise shows that the bound is maximized when $s = n_T$ and hence $\mathbf{D} = (d/n_T)\mathbf{I}n_T$. For this choice of $\mathbf{D}$, Eq. (2.58) becomes an equality that is achieved at all values of $\mathbf{U}$. In other words, uncorrelated waveforms of equal power provide a uniform upper bound on the approximate, average KL divergence in the high-SNR regime.

2.5.3 Weak-Signal Regime

We can model weak signals through covariances $\mathbf{\Gamma} = \mathbf{I}_n + \mathbf{\Sigma}$ with $0 \leq \mathbf{\Sigma} \ll \mathbf{I}_n$. We use the polynomial expansion

$$|\mathbf{I}_n + \mu\mathbf{X}| = \sum_{k=0}^{n} P_k(\mathbf{X})\mu^k \tag{2.63}$$

where each P_k is a scalar-valued polynomial in its matrix argument. Some of these terms are familiar; for example, $P_1(\mathbf{X}) = \text{tr}(\mathbf{X})$ and $P_n(\mathbf{X}) = |\mathbf{X}|$. We need the second-order term, which is expressed by

$$P_2(\mathbf{X}) = \frac{1}{2}[(\mathrm{tr}(\mathbf{X}))^2 - \mathrm{tr}(\mathbf{X}^2)] \tag{2.64}$$

Using this series expansion along with the Neumann series expansion $\boldsymbol{\Gamma}^{-1} = \mathbf{I}_n - \boldsymbol{\Sigma} + \boldsymbol{\Sigma}^2 - \ldots$, we have the weak-signal approximation of the KL divergence

$$\log|\boldsymbol{\Gamma}| + \mathrm{tr}\boldsymbol{\Gamma}^{-1} - n \approx \frac{1}{2}\mathrm{tr}\boldsymbol{\Sigma}^2 \tag{2.65}$$

Applying the weak-signal approximation to $\mathbf{R} \otimes (\mathbf{S}^{\dagger}\mathbf{T}\mathbf{S})^*$ gives us the weak-signal KL divergence

$$\frac{1}{2}\mathrm{tr}\mathbf{R}^2 \cdot \mathrm{tr}(\mathbf{T}\mathbf{S}\mathbf{S}^{\dagger}\mathbf{T}\mathbf{S}\mathbf{S}^{\dagger}) \tag{2.66}$$

As above, we let $\mathbf{S}\mathbf{S}^{\dagger} = \mathbf{U}\mathbf{D}\mathbf{U}^{\dagger}$. To evaluate the expectation over $\mathbf{U}$ of the quadratic term in $\mathbf{T}$, consider the inner product

$$\langle \mathbf{D}, \mathbf{E} \rangle \equiv \mathrm{E}[\mathrm{tr}[\mathbf{T}\mathbf{U}\mathbf{D}\mathbf{U}^{\dagger}\mathbf{T}\mathbf{U}\mathbf{E}\mathbf{U}^{\dagger}]] \tag{2.67}$$

This inner product is invariant in the sense that

$$\langle \mathbf{U}\mathbf{D}\mathbf{U}^{\dagger}, \mathbf{U}\mathbf{E}\mathbf{U}^{\dagger} \rangle = \langle \mathbf{D}, \mathbf{E} \rangle \tag{2.68}$$

It is well known[7] that such invariant inner products have the form

$$\langle \mathbf{D}, \mathbf{E} \rangle = \alpha\, \mathrm{tr}(\mathbf{D}\mathbf{E}) + \beta\, \mathrm{tr}\mathbf{D} \cdot \mathrm{tr}\mathbf{E} \tag{2.69}$$

for unknown scalars α and β. The weak-signal approximation to the KL divergence becomes

$$\mathrm{E}[\text{K.L.quad. approx.}] = \frac{1}{2}\mathrm{tr}\,\mathbf{R}^2 \cdot (\alpha\, \mathrm{tr}(\mathbf{D}^2) + \beta(\mathrm{tr}(\mathbf{D}))^2) \tag{2.70}$$

Assuming $\alpha > 0$, which is shown in Appendix 2G, we see that under a constraint on $\mathrm{tr}\mathbf{D}$, the approximation is maximized when $\mathbf{D}$ is rank 1. This corresponds to forming a beam at the transmitter (i.e., rank 1 $\mathbf{S}$) that is scanned by the unitaries $\mathbf{U}$.

[7]The mapping $i\mathbf{E} \rightarrow \mathbf{U}i\mathbf{E}\mathbf{U}^{\dagger}$ can be viewed as the adjoint action of the Lie group of unitaries $\mathbf{U}$ acting on its Lie algebra of skew-adjoint matices. Any standard reference on Lie groups and Lie algebras such as Helgason's text [34] will treat inner products that are invariant under the adjoint representation.

2.5.4 Optimal Beamforming without Search

In the special case in which there is a single target, we can write $\mathbf{R} \otimes \mathbf{T}^* = \mathbf{a}_R\mathbf{a}_R^\dagger \otimes (\mathbf{a}_T\mathbf{a}_T^\dagger)$, leading to the KL divergence [using Eq. (2.56)]

$$\log\left[1 + \|\mathbf{a}_R\|^2(\mathbf{a}_T^\dagger(\mathbf{S}\mathbf{S}^\dagger)^*\mathbf{a}_T)\right] - \frac{\|\mathbf{a}_R\|^2(\mathbf{a}_T^\dagger(\mathbf{S}\mathbf{S}^\dagger)^*\mathbf{a}_T)}{1 + \|\mathbf{a}_R\|^2(\mathbf{a}_T^\dagger(\mathbf{S}\mathbf{S}^\dagger)^*\mathbf{a}_T)} \tag{2.71}$$

which is maximized by maximizing $\mathbf{a}_T^\dagger(\mathbf{S}\mathbf{S}^\dagger)^*\mathbf{a}_T$ over $\mathbf{S}$ given a constraint on $\mathrm{tr}(\mathbf{S}\mathbf{S}^\dagger)$. A simple Lagrange multiplier argument shows that the maximum is achieved when $\mathbf{S} \propto (\mathbf{a}_T)^*\mathbf{s}^\dagger$ for some $n_S \times 1$ vector $\mathbf{s}$, which amounts to forming a beam on the target.

2.5.5 Nonfading Targets

The improvement in detection suggested by the KL divegence disappears when the target exhibits no fading. Consider, in analogy to Eq. (2.54), the probability density function

$$p(\mathbf{Y}|\mathbf{M}) = \pi^{-n\lambda}|\boldsymbol{\Gamma}|^{-\lambda}e^{-\mathrm{tr}(\mathbf{Y}-\mathbf{M})^\dagger\boldsymbol{\Gamma}^{-1}(\mathbf{Y}-\mathbf{M})} \tag{2.72}$$

based on known covariance $\boldsymbol{\Gamma}$. The mean $\mathbf{M}$ is present with the signal and absent otherwise. As a warmup to a more general result, let $n = n_T$, $\lambda = n_S$, and $\mathbf{M} = \mathbf{HS}$. The KL divergence becomes $\mathrm{tr}(\mathbf{M}^\dagger\boldsymbol{\Gamma}^{-1}\mathbf{M})$. Integrating over $\mathbf{S}$ as above leads to

$$\begin{aligned} E[\text{KL divergence}] &= \mathrm{tr}\{E[\mathbf{U}\boldsymbol{\Lambda}\boldsymbol{\Lambda}^\dagger\mathbf{U}^\dagger]\mathbf{H}^\dagger\boldsymbol{\Gamma}^{-1}\mathbf{H}\} \\ &= \frac{\mathrm{tr}(\boldsymbol{\Lambda}\boldsymbol{\Lambda}^\dagger)}{n_T}\mathrm{tr}(\mathbf{H}^\dagger\boldsymbol{\Gamma}^{-1}\mathbf{H}) = \frac{\mathrm{tr}(\mathbf{S}\mathbf{S}^\dagger)}{n_T}\mathrm{tr}(\mathbf{H}^\dagger\boldsymbol{\Gamma}^{-1}\mathbf{H}) \end{aligned}$$

since

$$E[\mathbf{U}\boldsymbol{\Lambda}\boldsymbol{\Lambda}^\dagger\mathbf{U}^\dagger] = \frac{\mathrm{tr}(\boldsymbol{\Lambda}\boldsymbol{\Lambda}^\dagger)}{n_T}\mathbf{I}_{n_T} \tag{2.73}$$

Under a constraint on $\mathrm{tr}(\mathbf{S}\mathbf{S}^\dagger)$, the type of waveform correlation does not matter. In essence, the result states that the total energy returned is the same no matter what waveforms are employed; only that energy matters for detection.

In the calculation above, we have not taken into account correlation structure in the time domain. To account for such structure, use Eq. (2.72) with $n = n_T n_S$, $\lambda = 1$, and $\mathbf{M} = \mathrm{vec}(\mathbf{HS})$.[8] Then the KL divergence becomes $\mathrm{vec}(\mathbf{HS})^\dagger\boldsymbol{\Gamma}^{-1}\mathrm{vec}(\mathbf{HS})$. Writing

[8]We use vec ($\mathbf{X}$) for a matrix $\mathbf{X}$ to denote the vector formed by ordering the entries X_{jk} lexicographically. Note that this differs from some definitions found in the literature.

$\Gamma^{-1} = \sum_k \mathbf{A}_k \otimes \mathbf{B}_k$, a calculation similar to that above shows that

$$E[\text{KL divergence}] = \text{tr}\left[\mathbf{H}^\dagger\left(\sum_k \frac{\text{tr}(\mathbf{B}_k)}{n_S}\mathbf{A}_k\right)\mathbf{H}\right]\frac{\text{tr}(\mathbf{SS}^\dagger)}{n_T} \tag{2.74}$$

Again, only the total energy in the waveforms matters.

2.5.6 Some Additional Benefits of MIMO Radar

In reality, waveform correlation has a significant impact on detection performance. Missing in the preceding analysis are two consequenes of waveform correlations: (1) employing less correlated waveforms permits the use of larger, sparsely sampled antenna apertures, which can result in better interference rejection while still avoiding the potential ambiguities associated with sparse sampling; and (2) less correlated waveforms and the different integration patterns that they allow can create additional temporal degrees of freedom that help in separating moving targets from interference or clutter. Both of these points are addressed in the context of detecting targets in clutter in Section 2.6.

2.6 MIMO RADAR WITH MOVING TARGETS IN CLUTTER: GMTI RADARS

Ground moving-target indicator (GMTI) radar offers a potentially important application for MIMO radar techniques. Sensors illuminating the ground suffer strong clutter returns that degrade detection performance. Moving targets generate responses that differ from those of stationary ground clutter returns. Adaptive filtering in Doppler and angle can mitigate the effects of ground clutter. The resulting performance is often characterized in terms of an effective SNR loss incurred as a result of the adaptive filtering. By improving the separability of targets and clutter, MIMO techniques can significantly reduce SNR loss, given equivalent physical apertures.

We present some results on the design of MIMO waveforms for optimal performance in the sense of maximizing adapted SNR without degrading area search rates. These results are based on an *ansatz* for approximating SNR loss, discussed in Section 2.6.2. We interpret the SNR loss in terms of minimal detectable velocities (MDVs) of moving targets.

We begin by introducing a signal model for the observations made by a GMTI radar in Section 2.6.1. A localization principle central to the *ansatz* for approximating SNR loss is discussed in Section 2.6.2. A detailed treatment of the *ansatz* is presented in Sections 2.6.3 and 2.6.4 for the simple case of uncorrelated transmitter waveforms. Optimal waveforms that preserve area search rates are discussed in Section 2.6.5.

2.6.1 Signal Model

Consider, at a frequency of ω radians per second, the response of a target at position $\mathbf{r}$ to a signal transmitted from a phase center at $\mathbf{x}_T$ and received at a phase center at $\mathbf{y}_R$.

The subscripts R and T index receiver and transmitter phase centers, respectively. For additional clarity, we use $\mathbf{x}$ for transmitter positions and $\mathbf{y}$ for receiver positions. The target is assumed to be moving with velocity vector $\mathbf{v}_t$, while the sensor platform moves with velocity vector $\mathbf{v}_p$. The function dist($\mathbf{r}$,$\mathbf{q}$) expresses the Euclidean distance between $\mathbf{r}$ and $\mathbf{q}$, with $\mathbf{q}$ representing a position on the sensor platform. The distance function can be linearized about ($\mathbf{r}_0$,$\mathbf{q}_0$) by the approximation

$$\text{dist}(\mathbf{r},\mathbf{q}) \approx \text{dist}(\mathbf{r}_0,\mathbf{q}_0) + [(\mathbf{r} - \mathbf{r}_0) - (\mathbf{q} - \mathbf{q}_0)] \cdot \mathbf{u}(\mathbf{r}_0,\mathbf{q}_0) \tag{2.75}$$

where $\mathbf{u}$ is a unit vector pointing from the sensor to the target. With this notation, the response vector at time t can be represented by the far-field approximation

$$\begin{aligned} &e^{-i(\omega/c)[\text{dist}(\mathbf{r}_0+\mathbf{v}_t t,\mathbf{q}_0+\mathbf{v}_p t+\mathbf{x}_T)+\text{dist}(\mathbf{r}_0+\mathbf{v}_t t,\mathbf{q}_0+\mathbf{v}_p t+\mathbf{y}_R)]} \\ &\quad \approx e^{-2i(\omega/c)\text{dist}(\mathbf{r}_0,\mathbf{q}_0)} \cdot e^{i(\omega/c)[2(\mathbf{v}_p-\mathbf{v}_t)t+\mathbf{y}_R+\mathbf{x}_T]\cdot\mathbf{u}(\mathbf{r}_0,\mathbf{q}_0)} \end{aligned} \tag{2.76}$$

where the exponent expresses phase change along the path from $\mathbf{x}_T$ to $\mathbf{y}_R$. The speed of light is denoted by c. Note that the total number of degrees of freedom in the response vector is expressed by $N = n_T n_R n_D$, where n_T represents the number of transmitter phase centers, n_R the number of receiver phase centers, and n_D the number of time samples (denoted by t). The leftmost factor in Eq. (2.76) is common to all response vectors and is ignored in the following. Of course, the target response is represented only at the center frequency. Only narrowband models are considered.

Let each sensor (i.e., each sensor capable of transmitting) transmit its own waveform denoted $s_T(\tau, t)$, where τ denotes fast time (reinitialized each pulse) and t indicates slow time (i.e., pulse number). Ignoring path delay for the waveforms $s_T(\tau, t)$, the Rth receiving sensor receives at slow time t the superimposed waveforms of all other transmitting sensors as expressed by

$$r_{Rt}(\tau) = \sum_T s_T(\tau, t) e^{(i\omega/c)[2(\mathbf{v}_p-\mathbf{v}_t)t+\mathbf{y}_R+\mathbf{x}_T]\cdot\mathbf{u}} \tag{2.77}$$

Matched filtering with the transmitted waveforms at the appropriate delay results in the observations

$$z_{TRt} = \int \bar{s}_T(\tau, t) r_{Rt}(\tau) d\tau + \int \bar{s}_T(\tau, t) n_{Rt}(\tau) d\tau \tag{2.78}$$

given the additive noise process $n_{Rt}(\tau)$. Assuming idealized temporal cross-correlation statistics [i.e., $\int \bar{s}_T(\tau, t) s_{T'}(\tau + \epsilon, t) d\tau = 0$, when $\epsilon \neq 0$][9] and white noise, the z_{TRt} form a sufficient statistic in the range cell. Denote the second

[9] If $R_{TT'}(\epsilon, t) = \int \bar{s}_T(\tau, t) s(\tau + \epsilon, t) d\tau$, we cannot have $\mathbf{R}(\epsilon, t) = r(\epsilon, t)\mathbf{R}(0, t)$ in terms of a normalized scalar autocorrelation function $r(\epsilon, t)$ unless there is a single transmitter or the equality holds only for bounded ϵ.

summand by N_{TRt}; it possesses zero mean and covariance

$$E[N_{TRt}\bar{N}_{T'R't'}] = \begin{cases} C_{TT'}(t), & \text{if } R = R' \text{ and } t = t', \\ 0, & \text{otherwise} \end{cases} \tag{2.79}$$

where

$$C_{TT'}(t) \equiv \int \bar{s}_T(\tau, t) s_{T'}(\tau, t) d\tau \tag{2.80}$$

denotes the waveform correlation matrix. It follows that Eq. (2.78) can be rewritten as

$$z_{TRt} = \sum_{T'} C_{TT'}(t) e^{i(\omega/c)[(2t\mathbf{v}_p + \mathbf{y}_R + \mathbf{x}_{T'})\cdot\mathbf{u} - 2\mathbf{t}\|\mathbf{v}_p\|\delta]} + N_{TRt} \tag{2.81}$$

The response vector of the target is a function of the unit vector $\mathbf{u}$ pointing at the target, which can be characterized locally by two components, say, u_x and u_y. It is also a function of the target velocity projected onto the same unit vector and normalized by the platform velocity; $\delta \equiv \mathbf{v}_t \cdot \mathbf{u}/\|\mathbf{v}_p\|$. Define, for a target at a fixed direction $\mathbf{u}$ and Doppler δ, the response vector $\mathbf{a}(u_x, u_y, \delta)$ with components

$$a_{TRt}(u_x, u_y, \delta) = e^{i(\omega/c)[(2t\mathbf{v}_p + \mathbf{y}_R + \mathbf{x}_T)\cdot\mathbf{u} - 2t\|\mathbf{v}_p\|\delta]} \tag{2.82}$$

The vector $\mathbf{a}(u_x, u_y, \delta)$ has the tensor product structure

$$\mathbf{a}(u_x, u_y, \delta) = \mathbf{a}_T(u_x, u_y) \otimes \mathbf{a}_R(u_x, u_y) \otimes \mathbf{a}_D(u_x, u_y, \delta) \tag{2.83}$$

with

$$(\mathbf{a}_D)_t(u_x, u_y, \delta) \equiv e^{i(\omega/c)2(\mathbf{v}_p\cdot\mathbf{u} - \|\mathbf{v}_p\|\delta)t} \tag{2.84}$$

The other factors in the tensor product are defined in Section 2.4.2. For notational convenience, this response vector, when $\delta = 0$ and when restricted to the components where $t = 0$, differs in the order of components from the vector used in Eq. (2.16). Putting notation together, the observations can be written

$$z_{TRt} = (\mathbf{D}\mathbf{a}(u_x, u_y, \delta))_{TRt} + (\mathbf{N})_{TRt} \tag{2.85}$$

with $\mathbf{D} \equiv \mathbf{C} \otimes \mathbf{I}_{n_R} \otimes \mathbf{I}_{n_D}$.

Since the noise is correlated as shown above, the data should be whitened during signal processing. Alternatively, the inner products used in signal processing and performance prediction are formed by defining $\langle \mathbf{q}, \mathbf{s} \rangle \equiv \mathbf{q}^\dagger \mathbf{D}\mathbf{s}$ with norm $\|q\|_{\mathbf{D}}^2 \equiv \langle \mathbf{q}, \mathbf{q} \rangle$. For example, the whitened array response vector becomes $(\mathbf{C}^{1/2}\mathbf{a}_T) \otimes \mathbf{a}_R \otimes \mathbf{a}_D$.

2.6.2 Localization and Adapted SNR

The performance of GMTI radar is often characterized in terms of adapted SNR. Ground clutter returns interfere with the target return and impair detection. Since the clutter exhibits correlation in angle and Doppler, as shown in Fig. 2.7, it can be suppressed when the target is separated from the clutter ridge. The figure shows the output of a nonadaptive angle–Doppler filter scanned across the angle and Doppler search region shown by the axis extents. Adaptive filtering in angle and Doppler is used to achieve the best possible SNR by adapting the filtering to the observed clutter statistics.

It is common to measure the performance of GMTI radars in terms of SNR loss, which is expressed by the ratio of adapted SNR to the SNR achieved without any clutter returns. To facilitate the calculation of adapted SNR and, more specifically, to understand the design of MIMO waveforms and arrays, we utilize an *ansatz* to approximate adapted SNR.

For simplicity, consider a one-dimensional array of elements. Conceptually, as the array aperture and the number of observations grow, the ground clutter returns are associated with the clutter ridge, a one-dimensional manifold of the angle–Doppler spectrum shown in the cartoon in Fig. 2.8. For a target near the manifold, given appropriate array geometries, a *localization principle* tells us that the clutter returns emanating from the point on the clutter ridge closest to the target are of primary importance. The appropriate metric for measuring distance is the metric determined by the statistics of the observation noise. Since whitened coordinates are used for calculations, the metric we use is Euclidean.

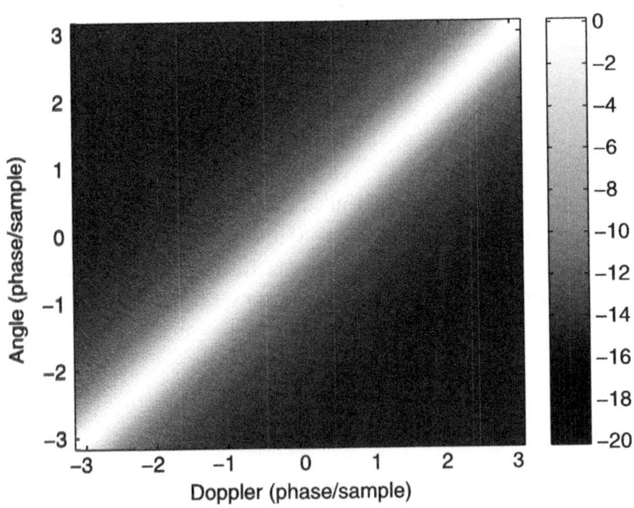

Figure 2.7 The relative power output from the beamformer of a side-looking GMTI receiver is shown as a function of angle and Doppler. Note that power decreases maximally along a gradient flowing from the clutter ridge, where the power is at its peak. In this illustration, the sample rate is fast enough to avoid Doppler ambiguities.

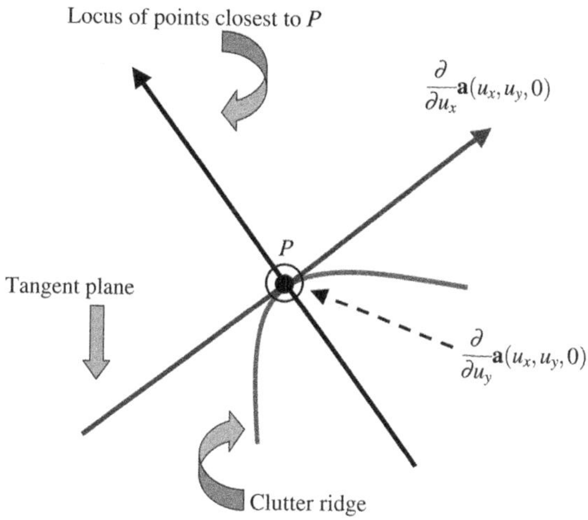

Figure 2.8 For one-dimensional arrays, in the limit of large aperture and samples, the clutter is effectively limited to a one-dimensional manifold in angle–Doppler space. For points close to the clutter ridge, adapted SNR depends only on the local region about the closest point *P*. The contribution of clutter ridge to adapted SNR is modeled by the clutter response vector at *P* as well as the tangent plane at *P*. Target response vectors have components in the tangent plane projected off, as shown. The SNR loss as a function of separation from the clutter ridge is expressed through the angular separation between the target response vectors on the normal line and the fixed clutter response vector at *P*.

The localization principle suggest an *ansatz* for approximating the adapted SNR near the clutter ridge. In outline, we project off the response vector of the closest clutter ridge point and its two angle derivatives from the target response vector. Of course, the approximation holds in the limit of strong clutter returns.

Implicit in this *ansatz* is the assumption that only the region around the closest point on the clutter ridge is important for performance. Some evidence for this localization principle is presented here. Specifically, we show that the adapted SNR for targets on the clutter ridge obeys the localization principle. While this may seem intuitively apparent for Nyquist-sampled uniform linear apertures and a conventional transmitter beamformer, it is certainly not apparent for sparsely sampled MIMO apertures and arbitrary waveforms. We present some simple conditions on the MIMO array and waveforms in order to affirm the localization principle, at least for one-dimensional arrays.

For the simple one-dimensional array used in Fig. 2.7, let $\mu = \pi u_x$, $-\pi \leq \mu \leq \pi$ represent the bearing angle and δ the normalized radial Doppler of the target. Then the adapted SNR for a target with response vector $\mathbf{a}(\mu, \delta)$ is proportional to (with fixed proportionality constant) $\mathbf{a}^\dagger(\mu, \delta)\mathcal{R}^{-1}\mathbf{a}(\mu, \delta)$, where $\mathcal{R}$ represents the covariance matrix built from response vectors $\mathbf{a}(\mu, 0)$ associated with ground clutter with angle spectrum $\mathbf{\Gamma}(\mu)$:

$$\mathcal{R} = \int_{-\pi}^{\pi} \mathbf{\Gamma}(\mu)\mathbf{a}(\mu, 0)\mathbf{a}^\dagger(\mu, 0)\frac{d\mu}{2\pi} \tag{2.86}$$

The angle spectrum weights the contributions of different clutter patches.

It is plausible to assume that adapted SNR depends most strongly on the point on the clutter ridge closest to the target, given appropriate array geometries. In fact, as the physical aperture and number of samples increase, the beampattern formed by the target response vector should localize increasingly closely to this point. A precise formulation of such an asymptotic limit would involve a parameterized target model that is beyond our scope. Instead, we will show that the adapted SNR, for targets on the clutter ridge at μ involves only the angle spectrum $\mathbf{\Gamma}(\mu)$ at μ in the asymptotic limit.

We intend to shed some light on the localization principle in the MIMO context. The arguments are only sketched in a one-dimensional example. However, a rigorous argument is presented for a basic localization principle in Appendix 2A. The key ingredient of localization for MIMO is the existence of a Nyquist virtual array for transmitters, receivers, and Doppler.

The localization principle motivates the *ansatz* mentioned above for approximating adapted SNR for targets near the clutter ridge. This *ansatz* is discussed in Section 2.6.4.

2.6.2.1 A Simplified Problem To keep the notation in check, localization is treated through a one-dimensional example that illustrates the basic principle. Arguments are only presented in outline, with rigorous results relegated to the appendixes. A complete justification of the localization principle would require an asymptotic model of the width of the clutter ridge.

Assume that the antenna array is one-dimensional along the x axis with phase centers located on a subset of a Nyquist (half-wavelength) grid. Assume in addition that a virtual sample rate is coordinated with platform motion (the distance moved by the radar between virtual samples is one-half the grid spacing) so that physical time samples are also associated with a subset of grid elements. More explicitly, if t_Δ denotes the time between virtual (i.e., potential) samples, then we require $4t_\Delta\|\mathbf{v}_p\|/\lambda = 1$. Then, from Eq. (2.85), the whitened, normalized response to a scatterer can be written in the form $(\mathbf{C}^{1/2}\mathbf{a}_T) \otimes \mathbf{a}_R \otimes \mathbf{a}_D$, where (assuming $\delta = 0$)

$$(\mathbf{a}_T)_m = e^{im\mu}, \qquad m \in M_T$$

$$(\mathbf{a}_R)_m = e^{im\mu}, \qquad m \in M_R$$

$$(\mathbf{a}_D)_m = e^{im\mu}, \qquad m \in M_D$$

and where $\mu \equiv \pi u_x$. For example, at time mt_Δ, we obtain

$$\frac{\omega}{c}(2mt_\Delta\mathbf{v}_p \cdot \mathbf{u}) = m\mu \tag{2.87}$$

The scalars M_T, M_R, and M_D are each subsets of integers that represent the phase centers used for transmitter, receiver, and pulse. The location of the scatterer is parameterized by the normalized spatial frequency μ, $-\pi \leq \mu \leq \pi$. It is important to note that spatial and temporal samples can sparsely occupy their respective apertures. Only the combined samples are required to fill the virtual array in a sense that we discuss below.

The subsets M_T, M_R, and M_D are allowed to grow in order to establish localization asymptotically. Specifically, given $\mathcal{C} = M_T \times M_R \times M_D$, we assume that

$$M_T + M_R + M_D = I_{\mathcal{C}} = \{m + j + n: (m, j, n) \in \mathcal{C}\} \tag{2.88}$$

is represented by an interval of integers

$$I_{\mathcal{C}} = \{k: k_L \leq k \leq k_U\} \tag{2.89}$$

that grows without bound in the limit. We say that M_T, M_R, M_D satisfying Eq. (2.89) form a *Nyquist virtual array*.

2.6.2.2 Covariances Localization is established by recasting the clutter covariance for the sparsely sampled aperture into a covariance for the Nyquist virtual array. The latter case is handled rigorously in Appendix 2A.

We begin by expressing a square root of the transmitter's covariance as a piece of an infinite Toeplitz matrix with stochastic, white spectrum $\hat{\mathbf{J}}(\lambda)/\sqrt{n_T}$:

$$\mathbf{J}_{pq} = \int_{-\pi}^{\pi} e^{i(p-q)\lambda} \hat{\mathbf{J}}(\lambda) \frac{d\lambda}{2\pi\sqrt{n_T}} \tag{2.90}$$

with white spectral covariance

$$E[\hat{\mathbf{J}}(\lambda_1)\hat{\mathbf{J}}^*(\lambda_2)] = 2\pi\hat{\mathbf{C}}(\lambda)\delta(\lambda_1 - \lambda_2) \tag{2.91}$$

This somewhat restrictive model of waveform correlations allows us to extend the localization principle to MIMO systems with more general waveform correlation matrices. The optimal correlation matrices discussed in Section 2.6.5 can be represented by the given form. In particular, uncorrelated MIMO waveforms of equal energy are represented by a constant $\hat{\mathbf{C}}(\lambda)$, while perfectly correlated waveforms (beamforming) are represented (formally, if not mathematically) by delta functions. The square-root representation leads to the transmitter covariance matrix

$$\begin{aligned} C_{pq} = (E[\mathbf{J}\mathbf{J}^\dagger])_{pq} &= \frac{1}{n_T}\sum_m \int_{-\pi}^{\pi} e^{i(p-m)\lambda} e^{i(m-q)\lambda} \hat{\mathbf{C}}(\lambda) \frac{d\lambda}{2\pi} \\ &= \int_{-\pi}^{\pi} e^{i(p-q)\lambda} \hat{\mathbf{C}}(\lambda) \frac{d\lambda}{2\pi} \end{aligned}$$

Since we are going to evaluate the clutter covariance matrix $\mathcal{R}$ with entries[10]

$$\mathcal{R}_{(m_1,j_1,n_1)(m_2,j_2,n_2)} = (E[(\mathbf{J}\mathbf{a}_T \otimes \mathbf{a}_R \otimes \mathbf{a}_D)(\mathbf{J}\mathbf{a}_T \otimes \mathbf{a}_R \otimes \mathbf{a}_D)^\dagger])_{(m_1,j_1,n_1)(m_2,j_2,n_2)} \tag{2.92}$$

[10]The tensor notation t_{jk} denotes a component of a vector $\mathbf{t}$ whose coordinates (jk) are lexicographically ordered (dictionary order). The matrix entry $A_{(jk)(lm)}$ is similarly indexed by row (jk) and column (lm).

we first consider

$$(E[(\mathbf{J}\mathbf{a}_T)(\mathbf{J}\mathbf{a}_T)^\dagger])_{m_1 m_2} = \int_{-\pi}^{\pi} e^{i\lambda(m_1-m_2)}\hat{\mathbf{C}}(\lambda)\frac{|\chi(\mu-\lambda)|^2}{n_T}\frac{d\lambda}{2\pi} \tag{2.93}$$

where

$$\chi(\mu) \equiv \sum_{q\in M_T} e^{i\mu q} \tag{2.94}$$

For some sparse transmitter array geometries, the function $|\chi(\mu)|^2$ is well approximated by a sum of delta functions in the limit of large n_T. For example, assume $M_T \subset n_\rho Z$. In other words, the transmitter phase centers are parameterized by multiples of a fixed integer n_ρ. If all multiples of n_ρ are involved, then in the limit of large n_T

$$\frac{|\chi(\mu-\lambda)|^2}{n_T} \longrightarrow n_\rho^{-1}\sum_k \delta(\lambda-\mu-\rho_k) \tag{2.95}$$

where $\rho_k \equiv 2\pi k/n_\rho$, $0 \le k < n_\rho$. Additions are interpreted modulo 2π so that sums lie in the interval $[-\pi, \pi)$. Thus we have the asymptotic limit

$$\begin{aligned}(E[(\mathbf{J}\mathbf{a}_T)(\mathbf{J}\mathbf{a}_T)^\dagger])_{m_1 m_2} &\longrightarrow n_\rho^{-1}\sum_k e^{i(\mu+\rho_k)(m_1-m_2)}\hat{\mathbf{C}}(\mu+\rho_k)\\ &= \left(n_\rho^{-1}\sum_k \hat{\mathbf{C}}(\mu+\rho_k)\right)e^{i\mu(m_1-m_2)}\end{aligned}$$

since $m_1 - m_2 \in n_\rho Z$. Thus the clutter covariance becomes

$$\begin{aligned}&\mathcal{R}_{(m_1,j_1,n_1)(m_2,j_2,n_2)}\\ &= (E[(\mathbf{J}\mathbf{a}_T\otimes\mathbf{a}_R\otimes\mathbf{a}_D)(\mathbf{J}\mathbf{a}_T\otimes\mathbf{a}_R\otimes\mathbf{a}_D)^\dagger])_{(m_1,j_1,n_1)(m_2,j_2,n_2)}\\ &\longrightarrow \int_{-\pi}^{\pi}\mathbf{\Gamma}(\mu)\left(n_\rho^{-1}\sum_k \hat{\mathbf{C}}(\mu+\rho_k)\right)e^{i\mu(m_1-m_2+j_1-j_2+n_1-n_2)}\frac{d\mu}{2\pi}\\ &= \mathbf{R}_{l_1 l_2}\end{aligned} \tag{2.96}$$

with $l_k = m_k + j_k + n_k$. $\mathbf{\Gamma}(\mu)$ represents the scatterer density along the clutter ridge. The MIMO radar observes an effective clutter density given by

$$\mathbf{\Gamma}(\mu)\left(n_\rho^{-1}\sum_k \hat{\mathbf{C}}(\mu+\rho_k)\right) \tag{2.97}$$

and an effective clutter covariance $\mathbf{R}$ over the indices in I_C.

2.6.2.3 Localization and the Nyquist Virtual Array Let $\mathcal{C} = \{(m, j, n)\}$ denote a fixed set of triple indices associated with transmitter, receiver, and pulse phase centers. We assume that the set of summations $I_{\mathcal{C}} \equiv \{m + j + n\colon (m, j, n) \in \mathcal{C}\}$ equals an interval of integers $\{k\colon k_L \leq k \leq k_U\}$. As the number of phase centers is scaled upward, the interval is assumed to grow without bound. Although adapted SNRs are calculated using $\mathcal{R}$ with multiple indices drawn from the set $\mathcal{C}$, these adapted SNRs are the same as those achieved with the above-defined [see Eq. (2.96)] $\mathbf{R}$ matrix with entries indexed by $I_{\mathcal{C}}$. To see this, consider vectors $\mathbf{v}$ with components v_k, $k \in I_{\mathcal{C}}$, indexed by the interval set $I_{\mathcal{C}}$. Map such a vector to a vector (really a tensor) $\mathbf{w}$ with components w_{mjn}, $(m, j, n) \in \mathcal{C}$, indexed by members of the set of triples $\mathcal{C}$ as follows: $w_{mjn} = v_{m+j+n}$. Express the linear relationship between $\mathbf{v}$ and $\mathbf{w}$ through the matrix $\mathbf{A}: \mathbf{w} = \mathbf{A}\mathbf{v}$. Note that $\mathbf{A}$ is injective (has a trivial null space) and hence $\mathbf{A}^{\dagger}$ is surjective (maps onto its range). Since $\mathcal{R} = \mathbf{A}\mathbf{R}\mathbf{A}^{\dagger}$, we have the adapted SNR

$$\max_{\rho} \frac{|\rho^{\dagger}\mathbf{w}|^2}{\rho^{\dagger}\mathcal{R}\rho} = \max_{\mathrm{r}} \frac{|\mathbf{r}^{\dagger}\mathbf{v}||^2}{\mathbf{r}^{\dagger}\mathbf{R}\mathbf{r}} \tag{2.98}$$

Thus the adapted SNR can be evaluated for the virtual array with element locations $I_{\mathcal{C}}$. This case is handled in Appendix 2A, which shows that the adapted SNR in the direction specified by μ depends only on the scatterers near μ.

2.6.3 Inner Products and Beamwidths

As a warmup to beamformer optimization, we first treat a GMTI radar employing the traditional MIMO waveform involving uncorrelated waveforms from each phase center. In the notation of Section 2.6.1, we assume $\mathbf{C} = \mathbf{I}_{n_T}$. Recall $\delta = \mathbf{v}_t \cdot \mathbf{u}/\|\mathbf{v}_p\|$, which expresses the normalized radial velocity as defined in Section 2.6.1. Note that the observation noise is white in the coordinate system associated with $\mathbf{a}$. We inntroduce a quadratic form that provides us with a shape for the angle–Doppler beam and a metric to use with our *ansatz*. First, we need to choose coordinates that simplify the notation.

In order to simplify the calculation, assume without loss of generality that the time origin is chosen so that $\sum_t t = 0$ and a spatial origin so that $\sum_{TR} \mathbf{x}_T + \mathbf{y}_R = 0$. Other choices of origin are also useful; for example, another one is used in Appendix 2F. Define

$$\mathcal{A} \equiv n_T^{-1} n_R^{-1} \frac{16}{\lambda^2} \sum_{TR} (\mathbf{x}_T + \mathbf{y}_R)(\mathbf{x}_T + \mathbf{y}_R)^T \tag{2.99}$$

where $\mathcal{A}$ expresses the mean-squared variation of the locations of phase centers. With the additional assumption that $\sum_T \mathbf{x}_T = 0$, or $\sum_R \mathbf{y}_R = 0$, or $\{\mathbf{y}_R\} = \{\mathbf{x}_T\}$, we have

$$\mathcal{A} = \frac{16}{\lambda^2} \left(n_T^{-1} \sum_T \mathbf{x}_T \mathbf{x}_T^T + n_R^{-1} \sum_R \mathbf{y}_R \mathbf{y}_R^T \right) \tag{2.100}$$

Choose coordinates so that $\mathbf{v}_p$ points along the x axis. If $\mathbf{u}(\epsilon)$ is a curve of unit vectors with real parameter ϵ, then $(\partial/\partial\epsilon)\mathbf{u}(\epsilon) = (w_x, w_y, w_z)^T$ is a tangent vector at $\mathbf{u}$. Thus $0 = u_x w_x + u_y w_y + u_z w_z$ so that $w_z = \mathbf{s}^T\mathbf{w}$ where $\mathbf{s}^T = (-(u_x/u_z), -(u_y/u_z))$ and $\mathbf{w} \equiv (w_x, w_y)^T$. Assuming that $(u_x(\epsilon), u_y(\epsilon), \delta(\epsilon))$ denotes a parameterized curve in local coordinates with tangent vector (w_x, w_y, d), define the quadratic form

$$Q((w_x, w_y, d)) \equiv \frac{4}{N\pi^2}\frac{\partial\mathbf{a}(u_x(\epsilon), u_y(\epsilon), \delta(\epsilon))^\dagger}{\partial\epsilon}\frac{\partial\mathbf{a}(u_x(\epsilon), u_y(\epsilon), \delta(\epsilon))}{\partial\epsilon} \tag{2.101}$$

where all ϵ derivatives are evaluated at $\epsilon = 0$. The constant factor $4/\pi^2$ is used to scale the inner product so that it corresponds to the definition of beamwidth used below. We can form an inner product of (w_x, w_y, d) and (w'_x, w'_y, d') from Q by using the standard polarization relation to express the inner product in terms of a matrix $\mathbf{G}$:

$$\begin{aligned}&(w'_x, w'_y, d')\mathbf{G}(w_x, w_y, d)^T\\ &\quad = \frac{1}{2}[Q((w_x, w_y, d) + (w'_x, w'_y, d')) - Q((w_x, w_y, d)) - Q((w'_x, w'_y, d'))]\end{aligned} \tag{2.102}$$

Using the facts that $(\partial\mathbf{a}/\partial\epsilon)^\dagger\mathbf{a} = 0$ and $\mathbf{a}^\dagger\mathbf{a} = \text{const.}$ given the preceding choice of coordinates, we have an implicit Taylor series approximation for the beamwidth separation $b(\epsilon)$ between the response vectors $\mathbf{a}(0)$ and $\mathbf{a}(\epsilon)$:

$$\begin{aligned}\cos\left(\frac{\pi b(\epsilon)}{2}\right)^2 &= \frac{1}{N}|\mathbf{a}(u_x(0), u_y(0), \delta(0))^\dagger\mathbf{a}(u_x(\epsilon), u_y(\epsilon), \delta(\epsilon))|^2\\ &\approx 1 - \frac{1}{N}\frac{\partial\mathbf{a}^\dagger}{\partial\epsilon}\frac{\partial\mathbf{a}}{\partial\epsilon}\epsilon^2 = 1 - \frac{\pi^2}{4}Q((w_x, w_y, d))\epsilon^2\end{aligned} \tag{2.103}$$

The normalization by $\pi/2$ in the cosine allows us to associate a one-beamwidth separation with orthogoanlity in the response vectors. This is an abtraction of the peak-to-null definition of beamwidth commonly used for antenna arrays. The inner product can be expressed in terms of $\mathcal{A}$ and

$$\rho \equiv \frac{64}{n_D\lambda^2}\sum_t t^2\|\mathbf{v}_p\|^2 \tag{2.104}$$

as follows. First note that for any real 3-vector $\mathbf{q}$, $\mathbf{q}^T(\partial/\partial\epsilon)\mathbf{u}(\epsilon) = (\mathbf{q}^{(A)} + \mathbf{q}^{(B)}\mathbf{s})^T\mathbf{w}$, where $\mathbf{q}^{(A)}$ denotes the 2-vector composed of the first two coordinates of $\mathbf{q}$ and where $\mathbf{q}^{(B)}$ is the scalar consisting of the last coordinate of $\mathbf{q}$. Partition the 3×3 matrix $\mathcal{A}$

$$\mathcal{A} = \begin{pmatrix}\mathcal{A}_{AA} & \mathcal{A}_{AB}\\ \mathcal{A}_{BA} & \mathcal{A}_{BB}\end{pmatrix} \tag{2.105}$$

into a 2×2 upper left-hand corner $\mathcal{A}_{AA}$, and so on. Define

$$\mathbf{M} \equiv \mathcal{A}_{AA} + \mathcal{A}_{AB}\mathbf{s}^T + \mathbf{s}\mathcal{A}_{BA} + \mathbf{s}\mathcal{A}_{BB}\mathbf{s}^T \tag{2.106}$$

Let $\Xi(R, T) \equiv \left(\mathbf{x}_T^{(A)} + \mathbf{x}_T^{(B)}\mathbf{s} + \mathbf{y}_R^{(A)} + \mathbf{y}_R^{(B)}\mathbf{s}\right)$. Note that

$$M = n_T^{-1} n_R^{-1} \frac{16}{\lambda^2} \sum_{T,R} \Xi(R, T)\Xi^T(R, T) \tag{2.107}$$

By evaluating the inner product $(\partial\mathbf{a}/\partial\epsilon)^\dagger(\partial\mathbf{a}/\partial\epsilon)$ in Eq. (2.103) and noting the cancellation of many terms (see Appendix 2F for a more general computation used in Section 2.6.5), it is straightforward to show that the inner product of Eq. (2.102) can be written as

$$(w_x', w_y', d')\mathbf{G}(w_x, w_y, d)^T = (w_x', w_y')\mathbf{M}(w_x, w_y)^T + \rho(d - w_x)(d' - w_x')$$

2.6.4 SNR Loss

Given a target response vector $\mathbf{a}$, a weight vector $\mathbf{w}$, and a total noise-plus-clutter covariance $\mathcal{R}$, the adapted SNR can be expressed as follows:

$$\frac{|\mathbf{w}^\dagger\mathbf{a}|^2}{\mathbf{w}^\dagger\mathcal{R}\mathbf{w}} \tag{2.108}$$

When the noise-plus-clutter covariance represents additive white noise with identity covariance, the best adapted SNR becomes $|\mathbf{a}|^2$. The addition of clutter degrades SNR. The ratio of the largest adapted SNR with noise and clutter to the SNR with noise alone is expressed by

$$\max_{\mathbf{w}} \frac{|\mathbf{w}^\dagger\mathbf{a}|^2}{|\mathbf{a}|^2\mathbf{w}^\dagger\mathcal{R}\mathbf{w}} = \frac{\mathbf{a}^\dagger\mathcal{R}^{-1}\mathbf{a}}{|\mathbf{a}|^2} \tag{2.109}$$

In GMTI applications, the clutter covariance is formed by integration over the response vectors $\mathbf{a}(u_x,u_y,0)$ for all physical (u_x,u_y), as discussed in Section 2.6.2. The localization principle of that section holds in the limit of large integration intervals and apertures. For targets near the clutter ridge and high clutter-to-noise ratios, this localization principle allows us to approximate the inverse of the clutter-plus-noise matrix by projecting onto the complement of the span of the three orthogonal (given the choice of coordinates discussed below) vectors

$$\left\{\mathbf{a}(u_x, u_y, 0), \quad \frac{\partial\mathbf{a}(u_x, u_y, 0)}{\partial u_x}, \frac{\partial\mathbf{a}(u_x, u_y, 0)}{\partial u_y}\right\}$$

Denote the projector onto the complementary span by $P_C^{\perp}$. Under this *ansatz*, SNR loss is expressed by

$$\begin{aligned}\frac{\mathbf{a}^{\dagger}\mathcal{R}^{-1}\mathbf{a}}{|\mathbf{a}|^2} &\approx \frac{\mathbf{a}^{\dagger}P_C^{\perp}\mathbf{a}}{|\mathbf{a}|^2}\\ &= N^{-1}\mathbf{a}(u_x(\epsilon),\, u_y(\epsilon),\, \delta(\epsilon))^{\dagger}P^{\perp}_{\mathbf{a}(u_x,u_y,0)}\mathbf{a}(u_x(\epsilon),\, u_y(\epsilon),\, \delta(\epsilon))\\ &= \sin^2\left(\frac{\pi b(\epsilon)}{2}\right)\end{aligned}$$

if the curve parameterized by ϵ satisfies $(\partial\mathbf{a}/\partial u_x) \perp (\partial\mathbf{a}/\partial\varepsilon) \perp (\partial\mathbf{a}/\partial u_y)$ at $(u_x,u_y,0)$ and $\epsilon = 0$. In the tangent space, clutter comes from the 2D subspace spanned by the vectors $(w_x,0,0)^T$ and $(0,w_y,0)^T$. Thus we must compute (w_x,w_y,d) orthogonal to all $(w_x{}',w_y{}',0)$. Hence

$$\begin{aligned}0 &= (w_x',\, w_y',\, 0)\mathbf{G}(w_x,\, w_y,\, d)^T\\ &= (w_x',\, w_y')\begin{pmatrix} M_{11}+\rho & M_{12}\\ M_{21} & M_{22}\end{pmatrix}(w_x,\, w_y)^T - \rho w_x' d\end{aligned}$$

The solution in (w_x, w_y) must be perpendicular to (M_{21}, M_{22}) and, hence, has the form $(w_x, w_y) \propto (-M_{22}, M_{21})$. It follows that $d = M_{22} + |M|/\rho$. The norm of this vector in the above-defined metric is easily evaluated, with the result that the unique (up to sign) orthonormal vector complementary to the clutter subspace has a d coordinate given by

$$\sqrt{\frac{M_{22}}{|\mathbf{M}|}+\rho^{-1}} \tag{2.110}$$

This coordinate corresponds (in the tangent space) to one beamwidth deviation from the clutter subspace. The SNR loss at beamwidth separation $b(\epsilon)$ is expressed by $\sin^2(\pi b(\epsilon)/2)$. Thus, the target Doppler at this SNR loss becomes approximately

$$\mathbf{v}_t \cdot \mathbf{u} \approx b(\epsilon)\|\mathbf{v}_p\|\sqrt{\frac{M_{22}}{|\mathbf{M}|}+\rho^{-1}} \tag{2.111}$$

As a funciton of SNR loss, this becomes

$$\mathbf{v}_t \cdot \mathbf{u} \approx \frac{2}{\pi}\arcsin(\sqrt{\mathrm{SNR}_{\mathrm{loss}}}) \cdot \|\mathbf{v}_p\|\sqrt{\frac{M_{22}}{|\mathbf{M}|}+\rho^{-1}} \tag{2.112}$$

This approximation is tight for losses greater than 3 dB.

For one-dimensional apertures along the x axis, $|\mathbf{M}|/M_{22}$ becomes $\mathcal{A}_{11}$, which is proportional to the mean-squared convolved aperture. Similarly, ρ is proportional to the mean-squared distance flown by the platform during the integration interval.

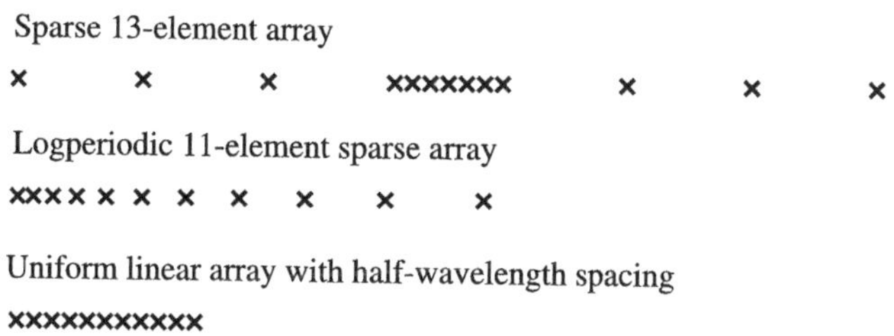

Figure 2.9 Array geometries used in the examples. Relative scales are accurate. In addition to the arrays shown, a fourth array is used. This array is based a scaled version of the logperiodic geometry. The scaling reduces its physical aperture to that of the uniform linear array.

Thus, performance is aperture-limited or Doppler-limited depending on the relative sizes of these two numbers. Although convolving increases the effective RMS aperture, the full benefits of MIMO GMTI may not be realized unless integration times are increased.

Some of the array geometries used in the examples are shown in Fig. 2.9, and the associated antenna array responses are illustrated in Fig. 2.10. Figure 2.11 shows the calculated and approximated [modeled by Eq. (2.112)] SNR loss for the uniform linear array whose pattern is shown in Fig. 2.10. The uniform array has 11 phase centers spaced apart by a half-wavelength ($\lambda/2$). Figure 2.11 also presents the same comparison for a sparse sampling of the same aperture using the logperiodic geometry in Fig. 2.9. The sensor moves a quarter-wavelength ($\lambda/4$) during a sample period. Adaptive filtering uses 32 samples. The clutter : noise ratio is 50 dB. These parameters are common to both uniformly and sparsely sampled apertures. The figure indicates that the approximation is good at SNR losses below ~3 dB, given a simple model of clutter and platform motion.

There is an additional consideration involved in the comparison of SIMO and MIMO GMTI, namely, differing SNRs on the target. This a practical issue, along

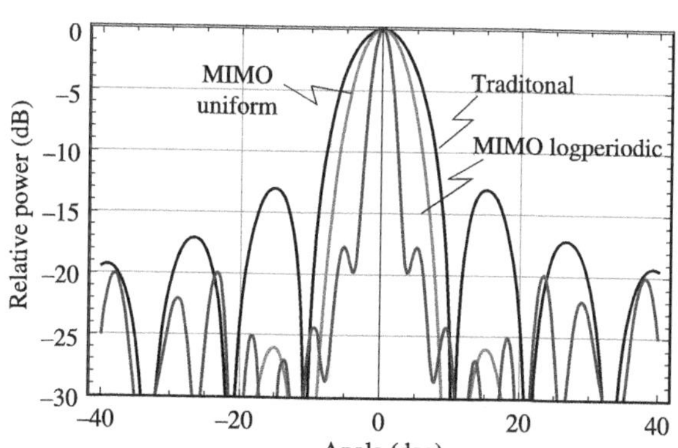

Figure 2.10 Antenna array responses for traditional uniform, MIMO uniform, and MIMO logperiodic for 11 physical antennas.

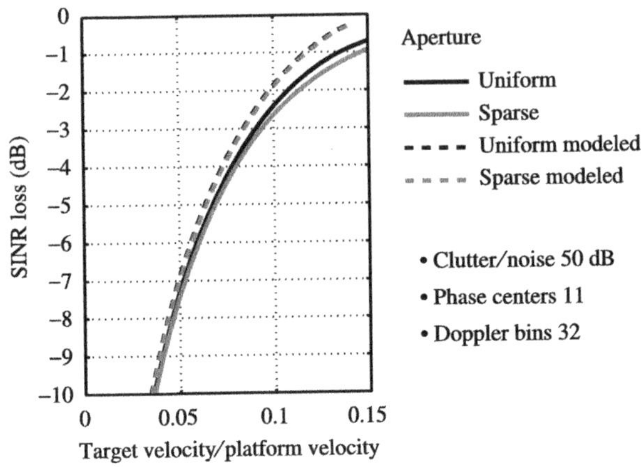

Figure 2.11 Comparison of approximated and calculated SNR loss. All arrays occupy the same aperture.

with waveform design, of some significance that is treated in Section 2.6.5. However, it is true that SIMO and MIMO radars can have the same integrated SNRs on target, given equivalent area search rates. In any case, SNR on target is not relevant to the SNR loss calculation.

Another example of SNR loss is shown in Fig. 2.12. Two 1D apertures are considered. One is a filled (half-wavelength spacing) aperture with 11 phase centers used as a conventional GMTI radar. The other aperture employs the same number of phase centers but is considerably larger and sparse. Both apertures were used in the examples shown in Fig. 2.10. Used conventionally, the sidelobe levels of the sparse aperture array are much too high even for adaptive processing. The convolved aperture used by the MIMO GMTI radar has peak sidelobes that are 13 dB down, which is good for adaptive processing and matches the peak sidelobes of the unweighted filled aperture. The figure shows an approximate 10 dB improvement

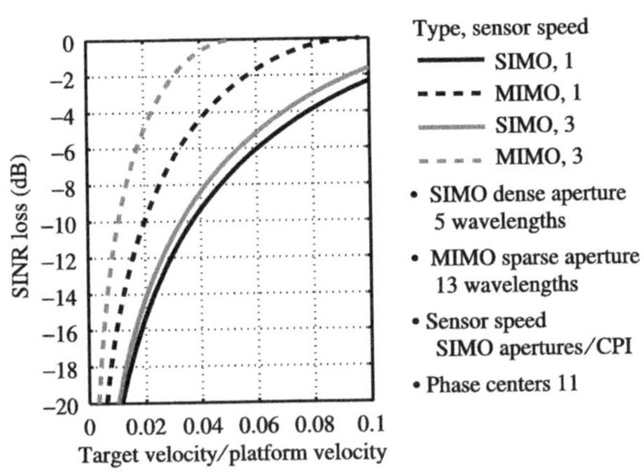

Figure 2.12 SNR loss for MIMO (sparse aperture) and SIMO (filled aperture) GMTI radars.

in SNR loss for the same level of target Doppler. As described above, the target velocity, projected on the direction vector $\mathbf{u}$, is measured as a fraction of the sensor velocity (along the array axis). Curves show the performance for a sensor that moves the length of the smaller aperture during the integration time as well as one that moves three such apertures during integration. As the equation indicates, SNR loss is improved to an asymptotic limit as the integration time is increased. Although the SIMO system is near the asymptotic limit with one aperture of integration, the MIMO system requires more integration to achieve the 10 dB improvement shown.

2.6.5 SNR Loss and Waveform Optimization

2.6.5.1 Loss Near the Clutter Ridge Thus far, GMTI MIMO radar has been assumed to employ uncorrelated waveforms. Next we consider the implications of other waveforms on adapted SNR and area search rates. We generalize the results of Section 2.6.3. The most general results apply under the additional restriction that transmitter and receiver phase centers are coincident. This does not reflect limitations on the techniques employed, but rather simplifies the statement of results.

Let $\mathbf{a}(\epsilon) \equiv \mathbf{a}(u_x(\epsilon), u_y(\epsilon), \delta(\epsilon))$ represent a parametric curve through the response vector $\mathbf{a}(0)$ corresponding to a curve $(u_x(\epsilon), u_y(\epsilon), \delta(\epsilon))$ through physical space with $\delta(0) = 0$. If the curve parameterized by ϵ satisfies $(\partial\mathbf{a}/\partial u_x)\perp(\partial\mathbf{a}/\partial\epsilon)\perp(\partial\mathbf{a}/\partial u_y)$ at $(u_x, u_y, 0)$ and $\epsilon = 0$, near $\epsilon = 0$, the SNR loss can be approximated by $\sin^2(\pi b/2)$, where

$$\cos^2\frac{\pi b(\epsilon)}{2} = \frac{|\mathbf{a}^\dagger(0)\mathbf{D}\mathbf{a}(\epsilon)|^2}{(\mathbf{a}^\dagger(0)\mathbf{D}\mathbf{a}(0))(\mathbf{a}^\dagger(\epsilon)\mathbf{D}\mathbf{a}(\epsilon))} \tag{2.113}$$

Note that orthogonality is measured with respect to the inner product $\langle\cdot,\cdot\rangle$ defined in Section 2.6.1.

The Taylor expansion of both sides of Eq. (2.113) provides a quadratic form in the tangent space at $(u_x, u_y, 0)$, expressing the shape of the resolution cell at $\epsilon = 0$ and leading to the approximation

$$\dot{b}^2 \approx \frac{4}{\pi^2}\left(\frac{\langle\dot{\mathbf{a}},\dot{\mathbf{a}}\rangle}{\|\mathbf{a}(0)\|_{\mathbf{D}}^2} - \frac{|\langle\mathbf{a},\dot{\mathbf{a}}\rangle|^2}{\|\mathbf{a}(0)\|_{\mathbf{D}}^4}\right) \tag{2.114}$$

Assume that array and Doppler coordinates are chosen as in Section 2.6.3. In addition assume that the platform velocity $\mathbf{v}_p$ is aligned with the x axis with length $v_p \equiv \|\mathbf{v}_p\|$. Letting $b \equiv b(\epsilon)|_{\epsilon=1} \approx \dot{b}\epsilon|_{\epsilon=1}$ (for small $\dot{b}$), one can write Eq. (2.114) as (see text below for variable definitions)

$$b^2 \approx \mathbf{w}^\dagger\begin{pmatrix} M_{11} & M_{12} \\ M_{21} & M_{22} \end{pmatrix}\mathbf{w} + \rho(d - w_x)^2 \tag{2.115}$$

where $(w_x, w_y, d)^T$ is the tangent vector at $\mathbf{u}$ and at $\delta = 0$. The matrix $\mathbf{M}$ can be expressed in concrete terms. First, choose coordinate origins so that $\sum_R \mathbf{y}_R = 0 = \sum_t t$. Then, defining (see Appendix 2F) the Rth column of the matrix $\mathbf{Y}$ as $\mathbf{Y}_{\cdot R} \equiv \mathbf{y}_R^{(A)} + \mathbf{y}_R^{(B)}\mathbf{s}$ and similarly the Tth column of the matrix X as $\mathbf{X}_T \equiv \mathbf{x}_T^{(A)} + \mathbf{x}_T^{(B)}\mathbf{s}$, we have (note $2\pi/\lambda = \omega/c$)

$$\mathbf{M} = \frac{16}{\lambda^2}\mathrm{Re}\left(n_R^{-1}\mathbf{Y}\mathbf{Y}^T + \frac{\mathbf{X}\tilde{\mathbf{C}}\mathbf{X}^T}{\mathbf{1}^T\tilde{\mathbf{C}}\mathbf{1}} - \frac{(\mathbf{X}\tilde{\mathbf{C}}\mathbf{1})(\mathbf{1}^T\tilde{\mathbf{C}}\mathbf{X}^T)}{|\mathbf{1}^T\tilde{C}\mathbf{1}|^2}\right) \tag{2.116}$$

where

$$\rho \equiv \frac{64}{n_D\lambda^2}v_p^2\sum_t t^2$$

$$\tilde{C}_{T'T}(\mathbf{u}) \equiv C_{T'T}(\mathbf{u})e^{i(\omega/c)(\mathbf{x}_T - \mathbf{x}_{T'})\cdot\mathbf{u}}$$

$$\mathbf{s} \equiv -\left(\frac{u_x}{u_z}, \frac{u_y}{u_z}\right)^T$$

We can interpret $\tilde{\mathbf{C}}$ as the correlation of the original waveforms after they have been phase-shifted to align coherently at the target. The variable ρ expresses the sampled mean-squared aperture (in wavelengths λ) formed by the platform motion. Note that $\mathrm{tr}\tilde{\mathbf{C}} = \mathrm{tr}\mathbf{C}$.

Another definition of SNR loss is appropriate when variations in the waveform cross-correlation matrix are considered, namely, $(\tilde{C}_{11}/\mathrm{tr}\tilde{\mathbf{C}})\sin^2(\pi b/2)$, where $\tilde{C}_{11} \equiv \mathbf{1}^T\tilde{\mathbf{C}}\mathbf{1}/n_T$. This loss is composed of gain loss due to nulling as well as loss of target energy due to defocusing of the transmitter beams. The latter mechanism was not treated explicitly in Section 2.6.4. In that section, defocusing loss is fixed by assuming $\tilde{\mathbf{C}}$ is a multiple of the identity. For example, the defocusing loss $\tilde{C}_{11}/\mathrm{tr}\tilde{\mathbf{C}}$ is unity if $\mathbf{C}$ points a beam at the target so that $\tilde{C}_{11} = n_T$, while the loss is $1/n_T$ when $\mathbf{C}$ is a multiple of the identity. It follows from the argument in Section 2.6.4 that the MDV is approximated near the clutter ridge by

$$\frac{\mathbf{v}_t\cdot\mathbf{u}}{\|\mathbf{v}_p\|} \approx \frac{2}{\pi}\sin^{-1}\left(\sqrt{\frac{\mathrm{tr}\tilde{\mathbf{C}}}{\tilde{C}_{11}}\mathrm{SNR}_{\mathrm{loss}}}\right)\cdot\sqrt{\frac{M_{22}}{|\mathbf{M}|} + \rho^{-1}} \tag{2.117}$$

Equation (2.117) can be simplified for linear arrays that are aligned along the x axis and that use the same phase centers for transmission and reception. Define the Rth component of $\mathbf{q}$ as the x component of $(4/\lambda)(\mathbf{y}_R - n_R^{-1}\sum_{R'}\mathbf{y}_{R'})$. Then, since $n_R = n = n_T$, we obtain

$$\frac{|\mathbf{M}|}{M_{22}} \longrightarrow M_{11} = \frac{\|\mathbf{q}\|^2}{n}\left(1 + \frac{\tilde{C}_{11}\tilde{C}_{22} - |\tilde{C}_{12}|^2}{\tilde{C}_{11}^2}\right) \tag{2.118}$$

where $\tilde{C}_{22} \equiv (\mathbf{q}^{\dagger}\tilde{\mathbf{C}}\mathbf{q})/\|\mathbf{q}\|^2$, $\tilde{C}_{12} = (\mathbf{1}^{\dagger}\tilde{\mathbf{C}}\mathbf{q})/(\sqrt{n}\|\mathbf{q}\|)$, and so on.

It is useful to express the reciprocal of the last factor in Eq. (2.117) in terms of an effective aperture. Given the assumptions surrounding Eq. (2.118), if ρ becomes large (e.g., with longer integration times), the reciprocal of the last factor in Eq. (2.117) approaches $\sqrt{M_{11}}$. When $\tilde{\mathbf{C}} \propto \mathbf{11}^T$, this becomes $\sqrt{M_{11}} = \sqrt{(\|\mathbf{q}\|^2/n)}$, which is proportional to the physical RMS aperture measured in wavelengths.

2.6.5.2 Minimal Detectable Velocity (MDV) We can determine the waveform cross-correlation matrix $\mathbf{C}$ that yields the smallest possible MDV, at least in the approximation shown by Eq. (2.117) and with the addtional assumptions that lead to Eq. (2.118). Note that we can assume $\tilde{C}_{12} = 0$ since that assumption only reduces MDV while leaving the $\text{tr}\mathbf{C} = \text{tr}\tilde{\mathbf{C}}$ unchanged. Let $\zeta \equiv \sqrt{\text{SNR}_{\text{loss}}\text{tr}(\tilde{\mathbf{C}})/\tilde{C}_{11}}$ and $\alpha \equiv n\text{SNR}_{\text{loss}}/\|\text{q}\|^2$. Note that $0 \leq \zeta \leq 1$. Then the minimum detectable velocity is proportional (with a positive constant) to $\sqrt{\alpha\zeta^{-2} + \rho^{-1}} \cdot \sin^{-1}(\zeta)$. Differentiating and removing common positive factors, we arrive at the lower-bound expression $(1 - \zeta^2)^{-1/2} - \zeta^{-1}\sin^{-1}(\zeta) = 1/\cos(\kappa) - \kappa/\sin(\kappa) > 0$, where $\kappa = \sin^{-1}(\zeta)$. Thus, minimizing the MDV involves minimizing $\text{tr}(\tilde{\mathbf{C}})/\tilde{C}_{11}$, with the result that $\tilde{\mathbf{C}} \propto \mathbf{11}^T$. This can be interpreted as forming a single spatial beam that is pointed at the target. However, from a system viewpoint, a cost of pointing a beam at the target is a reduced area search rate.

2.6.6 Area Search Rates

Although a single beam on transmit provides the best MDV performance on a single target, GMTI systems must search for targets first. Thus, it is fair to compare system performance on the basis of equivalent area search rates. For example, a SIMO (single transmitted waveform) system will have roughly an nth of the beam size of a MIMO system transmitting n uncorrelated waveforms. Since the area search rate is proportional to the beam size and inversely proportional to the number of pulses in a coherent processing interval (CPI), a MIMO system can afford to utilize n times as many pulses as the SIMO system for equivalent area search rates and equal integrated SNRs. Thus, these two systems could be compared using just the last factor in Eq. (2.117).

More generally, k beams can be used for search. A single beam represents the conventional SIMO approach, while an n beam search utilizes waveforms for which $\mathbf{C}$ is a multiple of the identity. One can model the search beams as a matrix $\mathbf{C}$ with $0 \leq \tilde{\mathbf{C}}/\tilde{C}_{11} \leq \mathbf{I}_m$. For example, $\mathbf{C}$ can be proportional to a rank k orthogonal projector that contains a vector (in its range) pointed at the target. As a function of k, one can replace ρ with $k^2\rho$ in Eq. (2.117). Note that $0 \leq (\tilde{C}_{11}\tilde{C}_{22} - |\tilde{C}_{12}|^2)/\tilde{C}_{11}^2 \leq \tilde{C}_{22}/\tilde{C}_{11} \leq 1$ for $\mathbf{C}$ as above. Since integrated SNRs are the same for all systems after increasing the number of pulses in proportion to k, it follows that the full MIMO system with $k = n$ offers the lowest MDV. In fact, M_{11} is increased by a factor of 2 in going from SIMO to full MIMO, and, hence, the effective aperture is increased by at least $\sqrt{2}$ (since ρ is increased by a factor of ≥ 4).

2.6.7 Some Examples

Figure 2.13 shows response patterns for three MIMO radars based on a sparsely sampled physical aperture of 13 elements with the geometry used in Section 2.3.2.

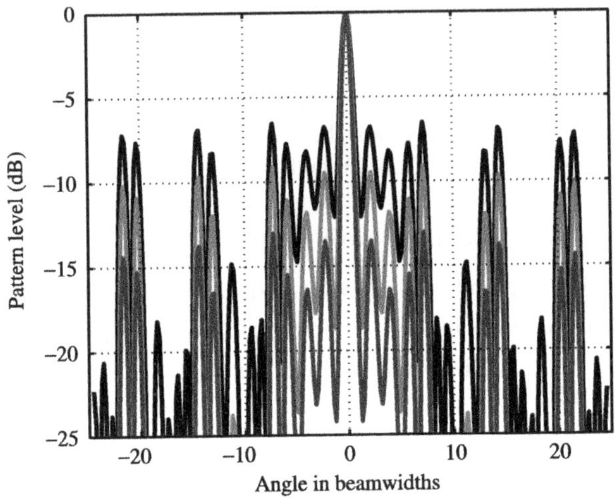

Figure 2.13 Patterns are shown for a SIMO radar (top curve) using a sparsely sampled 13-element aperture that has a uniform linear MIMO virtual subaperture, a partial MIMO radar (middle curve) using multiple covering beams and the same physical array as the SIMO radar, and a full MIMO radar (bottom curve). Peak sidelobes are reduced to levels below those of a critically sampled uniform linear array.

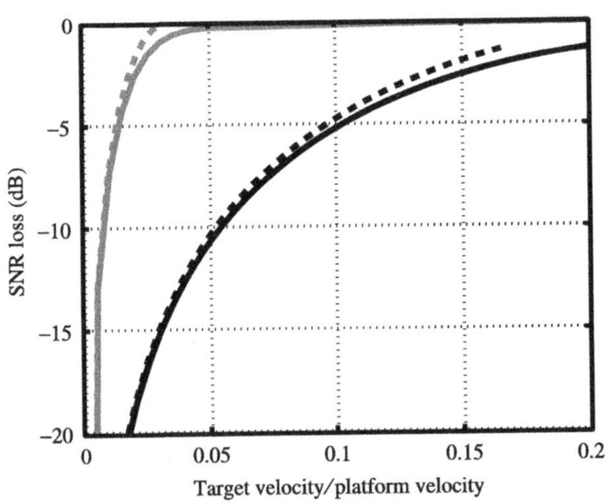

Figure 2.14 Comparison of 13-element SIMO array (right) with critically sampled aperture and 13-element sparsely sampled MIMO array (left) with larger physical aperture. MIMO array has lower effective peak sidelobes and provides equivalent area search rates with much better detectability near the clutter ridge. Performance is predicted using the *ansatz* of Section 2.6.4 (dashed curves) and simulated (solid curves).

The patterns, shown for a fixed target velocity, are based on the square of the normalized inner product of the target response vectors with a reference response vector chosen at $\mathbf{u} = (0, 1, 0)^T$. The pattern shapes are weakly dependent on target velocity. In general terms, sidelobes place a threshold on the level of radar returns required to localize weak targets.

The patterns shown in the figures are based on the same locations of physical elements but use three different waveform cross-correlation matrices (really $\tilde{\mathbf{C}}$) of the form $\mathbf{1}\mathbf{1}^T$ (SIMO), $0.2\mathbf{1}\mathbf{1}^T + 0.8\mathbf{I}_{n_T}$ (partial MIMO), and $\mathbf{I}_{n_T}$ (full MIMO). Each of these matrices is a convex combination of two cross-correlation matrices, one of which is the rank 1 matrix that steers a beam toward the target. The other matrix is the identity. Other beam mixtures are also of interest. Some obvious candidates are the beam of a single element (omnidirectional here) and the differential beam formed by q. These mixed-in beams are used to cover the sidelobes of the target response. Mixing in more of the covering beams reduces the sidelobe levels further at the cost of degraded SNR loss.

As mentioned above, equal area rates allow more slow-time integration for the full MIMO multibeam approach, resulting in the SNR loss curves shown in Fig. 2.14. This plot incorporates the additional integration gain (factor of n) that a full MIMO radar achieves over a SIMO system with the same area search rate. The figure compares a critically sampled 13-element aperture (half-wavelength element spacing) utilized by a SIMO radar to a sparsely sampled 13-element aperture utilized by a full MIMO radar. The MIMO radar has lower effective peak sidelobe levels. Moreover, since the MIMO system uses a much larger aperture, it offers much better detection performance, particularly at low target velocities.

2.7 SUMMARY

In this chapter, we discussed coherent MIMO radar concepts, performance, and applications. We provided a brief introduction to radar and discussed the current state of MIMO radar research. Coherent MIMO radar was introduced in the context of the MIMO virtual aperture. The performance of MIMO radar assuming a single scatterer in a range bin was discussed, in the context angle estimation bounds. This discussion was extended to include waveform optimization. We also discussed detection for MIMO radar in the context of search. Finally, we discussed the performance of MIMO GMTI radars, and demonstrated significant improvements in minimum defectable velocities.

APPENDIX 2A A LOCALIZATION PRINCIPLE

MIMO radar can be used for detection as well as parameter estimation. Since signal processing for MIMO radar is typically adaptive, performance is often characterized by the SNR loss suffered by adaptive beamforming. SNR loss in clutter is defined as the ratio of adapted SNR in clutter-plus-noise to SNR in noise alone, in the absence of clutter. In order to treat the performance of MIMO radars analytically, it is important to have simple, quantitative models that characterize SNR loss. One such model, the *ansatz* of Section 2.6.2, is based on approximating the effects of clutter on adapted SNR by localizing about the closest point on the clutter ridge defined in Section 2.6.2. In the next several appendices we illustrate a localization principle for adapted SNR for targets on the clutter ridge. The argument treats a receive-only problem involving one-dimensional, uniform linear arrays with half-wavelength spacing of phase centers. The results are used in Section 2.6.2. to provide an example of localization for MIMO arrays.

In the large-aperture limit, SNR loss can be localized to specific points of the spatial spectrum. To formalize this observation, consider a Toeplitz covariance based on the Fourier transform $R(n) \equiv \int_{-\pi}^{\pi} S(\omega)e^{i\omega n}d\omega/2\pi$ of the integrable, positive spectrum $S(\omega)$. Define the n-vector $\mathbf{v}^{(n)}(\omega)$ with kth component $(\mathbf{v}^{(n)}(\omega))_k \equiv e^{i\omega k}/\sqrt{n}$ and the $n \times n$ matrix $\mathbf{R}^{(n)}$ with (j, k)th entry $(\mathbf{R}^{(n)})_{jk} \equiv R(j-k)$, $1 \le j, k \le n$. In this notation, SNR loss is expressed by $\mathbf{v}^{(n)}(\omega)^{\dagger}\mathbf{R}^{(n)^{-1}}\mathbf{v}^{(n)}(\omega)$. The large n form of this expression has a simple interpretation in terms of the spectrum $S(\omega)$.

Define the matrices

$$\mathbf{A}_{jk}^{(n)} \equiv \begin{cases} R(j-k), & |j-k| \le d_n \\ 0, & |j-k| > d_n \end{cases}$$

and

$$\mathbf{B}_{jk}^{(n)} \equiv \begin{cases} R(j-k), & |j-k| > d_n \\ 0, & |j-k| \le d_n \end{cases}$$

so that $\mathbf{R}^{(n)} = \mathbf{A}^{(n)} + \mathbf{B}^{(n)}$. By definition, $\mathbf{A}^{(n)}$ is banded with bandwidth d_n. Now

$$(\mathbf{A}^{(n)}\mathbf{v}^{(n)}(\omega))_j = S^{(n)}(\omega)\frac{e^{ij\omega}}{\sqrt{n}} \tag{2A.1}$$

where

$$S^{(n)}(\omega) \equiv \sum_{|k| \le d_n} R(k)e^{-ik\omega}$$

is a truncated expansion of $S(\omega)$, as long as $d_n + 1 \le j \le n - d_n$. Thus

$$\mathbf{v}^{(n)}(\omega)^{\dagger}\mathbf{A}^{(n)}\mathbf{v}^{(n)}(\omega) = \frac{n - 2d_n}{n}S^{(n)}(\omega) + \delta_n$$

where

$$|\delta_n| \le \frac{2d_n(2d_n + 1)R(0)}{n}$$

The latter inequality is based on several observations. First, $|R(k)| \le R(0)$. Finally, there are $2d_n$ values of j that do not satisfy Eq. (2A.1), and each such j component is bounded by $R(0)(2d_n + 1)$. Since the bandwidth of $\mathbf{A}^{(n)^l}$ is at most ld_n, one has the analogous result

$$\mathbf{v}^{(n)}(\omega)^\dagger \mathbf{A}^{(n)^l} \mathbf{v}^{(n)}(\omega) = \frac{n - 2ld_n}{n} S^{(n)}(\omega)^l + \delta_n^{(l)} \tag{2A.2}$$

with

$$|\delta_n^{(l)}| \leq \frac{2ld_n(2ld_n + 1)R(0)^l}{n}$$

To proceed further, some regularity conditions are required for the spectrum $S(\omega)$. It is assumed that $S(\omega)$ is infinitely differentiable. Then one has from Appendix 2B $|R(k)| \leq C_r/k^r$ for some constant C_r independent of $k > 0$ and any integer $r > 0$. One particular consequence of this inequality is $|\mathbf{B}^{(n)}| \leq nC_r/d_n^r$, as is shown in Appendix 2C. Defining $d_n \equiv \max(\lceil n^{1/p} \rceil, 1)$, the argument in Appendix 2D shows that

$$\begin{aligned} \mathbf{v}^{(n)}(\omega)^\dagger \mathbf{R}^{(n)^l} \mathbf{v}^{(n)}(\omega) &= \mathbf{v}^{(n)}(\omega)^\dagger (\mathbf{A}^{(n)} + \mathbf{B}^{(n)})^l \mathbf{v}^{(n)}(\omega) \\ &= \mathbf{v}^{(n)}(\omega)^\dagger \mathbf{A}^{(n)^l} \mathbf{v}^{(n)}(\omega) + O(n^{1-r/p}) \end{aligned}$$

provided $r > p > 0$. Note that Eq. (2A.2) can be written

$$\mathbf{v}^{(n)}(\omega)^\dagger \mathbf{A}^{(n)^l} \mathbf{v}^{(n)}(\omega) = S^{(n)}(\omega)^l + O(n^{2/p-1})$$

since the $|S^{(n)}(\omega)|$ have a uniform upper bound. Given the differentiability of $S(\omega)$, it is standard that $S^{(n)}(\omega) \to S(\omega)$ uniformly in ω. Thus, if $p > 2$, one has $\mathbf{v}^{(n)}(\omega)^\dagger \mathbf{R}^{(n)^l}$ $\mathbf{v}^{(n)}(\omega) \to S^l(\omega)$ uniformly in ω and, hence, $\mathbf{v}^{(n)}(\omega)^\dagger P(\mathbf{R}^{(n)}) \mathbf{v}^{(n)}(\omega) \to P(S(\omega))$ uniformly in ω for any polynomial $P(x)$.

The same relation also holds for all continuous functions on a closed interval containing all the eigenvalues of $\mathbf{R}^{(n)}$ as well as all the values $S^{(n)}(\omega)$ (for sufficiently large n) and $S(\omega)$. Such an interval, fixed for all n sufficiently large, exists as discussed in Appendix 2E.

Before showing convergence, we discuss an operator norm and define $f(\mathbf{G})$ for Hermitian matrices $\mathbf{G}$. Let $|\mathbf{G}|_{\text{op}}$ denote the operator norm of $\mathbf{G}$, namely, the largest singular value of $\mathbf{G}$. Let $\mathbf{U}\Lambda\mathbf{U}^\dagger$ represent an eigendecomposition of $\mathbf{G}$ with unitary $\mathbf{U}$ and diagonal matrix Λ. Define $f(\mathbf{G}) = \mathbf{U}f(\Lambda)\mathbf{U}^\dagger$, where f is applied componentwise to the diagonal entries of Λ. Note that $f(\mathbf{G}) = P(\mathbf{G})$ when $f(x) = P(x)$ is a polynomial and $P(\mathbf{G})$ is evaluated algebraically (as a polynomial with matrix variable). If the eigenvalues of $\mathbf{G}$ are contained within an interval I, note that $|f(\mathbf{G}) - g(\mathbf{G})|_{\text{op}} \leq \sup_{x \in I}|f(x) - g(x)|$ for functions f and g. Putting these notions together, we have

$$\begin{aligned}
&|\mathbf{v}^{(n)}(\omega)^{\dagger}f(\mathbf{R}^{(n)})\mathbf{v}^{(n)}(\omega) - f(S(\omega))| \\
&\leq |\mathbf{v}^{(n)}(\omega)^{\dagger}f(\mathbf{R}^{(n)})\mathbf{v}^{(n)}(\omega) - \mathbf{v}^{(n)}(\omega)^{\dagger}P(\mathbf{R}^{(n)})\mathbf{v}^{(n)}(\omega)| \\
&+ |\mathbf{v}^{(n)}(\omega)^{\dagger}P(\mathbf{R}^{(n)})\mathbf{v}^{(n)}(\omega) - P(S(\omega))| + |P(S(\omega)) - f(S(\omega))|
\end{aligned}$$

and

$$|\mathbf{v}^{(n)}(\omega)^{\dagger}f(\mathbf{R}^{(n)})\mathbf{v}^{(n)}(\omega) - \mathbf{v}^{(n)}(\omega)^{\dagger}P(\mathbf{R}^{(n)})\mathbf{v}^{(n)}(\omega)| \leq \sup_{x \in I} |f(x) - P(x)|,$$

where I is the interval above. Then, by the Stone–Weierstrass theorem [35], the polynomials are dense in the set of continuous functions under the metric of uniform convergence on the interval I. Thus the inequality along with the previous results shows that

$$\mathbf{v}^{(n)}(\omega)^{\dagger}f(\mathbf{R}^{(n)})\mathbf{v}^{(n)}(\omega) \longrightarrow f(S(\omega)) \tag{2A.3}$$

uniformly in ω for any continuous function f on the interval I. Since, by the assumption that $S(\omega)$ is bounded away from zero, the closed interval I does not contain 0 (see Appendix 2E), and, hence, one can choose $f(x) = x^{-1}$.

To interpret this result, assume $\mathbf{R}^{(n)} = \mathbf{I}_n + (S(\omega) - 1)\mathbf{v}^{(n)}(\omega)\mathbf{v}^{(n)}(\omega)^{\dagger}$. Then the SNR loss becomes

$$\mathbf{v}^{(n)}(\omega)^{\dagger}\mathbf{R}^{(n)^{-1}}\mathbf{v}^{(n)}(\omega) = S(\omega)^{-1} \tag{2A.4}$$

Thus, comparing Eqs. (2A.3) and (2A.4), we have shown that SNR loss is approximated, in the limit of large aperture, by the SNR loss associated with a rank 1 interferer sharing the same response vector as the steering vector. The effective power of the interferer reflects the spectral density of the asymptotic interference spectrum. Multidimensional versions of Eq. (2A.3) also hold, as can be shown using essentially the same method of proof.

APPENDIX 2B BOUNDS ON $R(N)$

By the Riemann–Lesbesgue lemma $\lim_{n\to\infty} \int_{-\pi}^{\pi} f(\omega)e^{i\omega n}d\omega = 0$ for $f \in L_1(-\pi, \pi)$ (i.e., f integrable). Assume $S(\omega)$ to be infinitely differentiable. Since one can write

$$\begin{aligned}
R(n) \equiv \int_{-\pi}^{\pi} S(\omega)e^{i\omega n}\frac{d\omega}{2\pi} &= \int_{-\pi}^{\pi} S(\omega)\frac{(-i)^l}{n^l}\frac{\partial^l}{\partial\omega^l}e^{i\omega n}\frac{d\omega}{2\pi} \\
&= \frac{(i)^l}{n^l}\int_{-\pi}^{\pi}\frac{\partial^l}{\partial\omega^l}S(\omega)e^{i\omega n}\frac{d\omega}{2\pi}
\end{aligned}$$

we have $|R(n)| \leq C_l/n^l$ for any integral $l \geq 0$.

APPENDIX 2C AN OPERATOR NORM INEQUALITY

Let $|\mathbf{X}|_{\text{op}}$ denote the operator norm (largest singular value) of the matrix $\mathbf{X}$. Then

$$|\mathbf{X}|_{\text{op}} = \max_{\|\mu\|=1=\|\nu\|} \mu^{\dagger} X \nu \leq \max_{\|\mu\|=1=\|\nu\|} \sqrt{\text{tr}\mathbf{X}^{\dagger}\mathbf{X}}\sqrt{\text{tr}(\mu\nu^{\dagger})(\nu\mu^{\dagger})} = \sqrt{\text{tr}\mathbf{X}^{\dagger}\mathbf{X}}$$

Applying this inequality to $\mathbf{X} = \mathbf{B}^{(n)}$ and noting that $\text{tr}\mathbf{X}^{\dagger}\mathbf{X}$ has n^2 summands yields $|\mathbf{B}^{(n)}| \leq nC_r/d_n^r$.

APPENDIX 2D NEGLIGIBLE TERMS

Any matrix X has its eigenvalues $\{\lambda_k(X)\}$ contained in the union of Gershgorin disks: $\lambda_k(X) \in \cup_i D(X_{ii}, \sum_{j\neq i}|X_{ij}|)$ where $D(x, b)$ is the disk of radius b about x. Therefore, for the Hermitian matrices $\mathbf{R}^{(n)}$, one finds all eigenvalues in the disk (interval) $D(R(0), \sum_k |R(k)|)$.

Since the eigenvalues of all $\mathbf{R}^{(n)}$ are contained in a bounded interval, the operator norm of all $\mathbf{R}^{(n)}$ are uniformly bounded by a constant: $|\mathbf{R}^{(n)}| \leq C$. Similarly, $|\mathbf{A}^{(n)}| \leq C$. Now

$$\begin{aligned} &\mathbf{v}^{(n)}(\omega)^{\dagger}(\mathbf{A}^{(n)} + \mathbf{B}^{(n)})^{l}\mathbf{v}^{(n)}(\omega) \\ &= \mathbf{v}^{(n)}(\omega)^{\dagger}\left(\sum \mathbf{A}^{(n)^{j_1}}\mathbf{B}^{(n)^{k_1}} \cdots \mathbf{A}^{(n)^{j_s}}\mathbf{B}^{(n)^{k_s}}\right)\mathbf{v}^{(n)}(\omega) \end{aligned}$$

Using the differentiability of $S(\omega)$, one can write

$$|\mathbf{v}^{(n)}(\omega)^{\dagger}\mathbf{A}^{(n)^{j_1}}\mathbf{B}^{(n)^{k_1}} \cdots \mathbf{A}^{(n)^{j_s}}\mathbf{B}^{(n)^{k_s}}\mathbf{v}^{(n)}(\omega)| \leq C^{J}\left(\frac{nC_r}{d_n^r}\right)^{K}$$

where $J \equiv \sum_{j=1}^{s} j_s$ and $K \equiv \sum_{k=1}^{s} k_s$.

APPENDIX 2E BOUND ON EIGENVALUES

The uniform convergence of $S^{(n)}(\omega)$ to $S(\omega)$ discussed in Appendix 2A occurs over a finite, positive interval bounded away from zero as long as the function $S(\omega)$ is continuous and positive and, hence, has a positive minimum. This is an easy consequence of the series expansion for $S^{(n)}(\omega)$, which converges uniformly to $S(\omega)$. It remains to be seen that the eigenvalues of $\mathbf{R}^{(n)}$ are also contained in a positive interval bounded away from zero. We can then utilize in Appendix 2A the smallest, positive, and closed interval that encloses both the eigenvalues of $\mathbf{R}^{(n)}$ and the values of $S^{(n)}(\omega)$ and $S(\omega)$, for sufficiently large n.

It is well known (see the article by Gazzah et al. [36] and references cited therein) that both the maximum and minimum eigenvalues of $\mathbf{R}^{(n)}$ are contained in the interval $[\min_{\omega} S(\omega), \max_{\omega} S(\omega)]$. We present a sketch of a simple argument here for completeness. The details require some familiarity with Hilbert spaces and the spectral decomposition of bounded linear operators, but the argument is intuitive even without such a background.

View the infinite-dimensional matrix $R(j-k), j, k \in Z$ as a bounded linear operator on the Hilbert space $l^2(Z)$ consisting of doubly infinite sequences $\{v_j\}$ indexed by the integers with inner product $\langle \mathbf{v}, \mathbf{w}\rangle = \sum_j v_j \overline{w}_j$. Denote this Hilbert space operator as $\mathcal{R}$. Then $S(\omega)$ represents a diagonalization of $\mathcal{R}$ by unitaries via the Fourier transform. In particular, the eigenvalues [actually, spectral values, since the eigenvectors are not in $l^2(Z)$] of $\mathcal{R}$ are the values of $S(\omega)$. In particular

$$\max_{\omega} S(\omega) = \max_{\mathbf{v}} \frac{\langle \mathcal{R}\mathbf{v}, \mathbf{v}\rangle}{\langle \mathbf{v}, \mathbf{v}\rangle} \geq \max_{\mathbf{v}^{(n)}} \frac{\langle \mathcal{R}\mathbf{v}^{(n)}, \mathbf{v}^{(n)}\rangle}{\langle \mathbf{v}^{(n)}, \mathbf{v}^{(n)}\rangle} = \lambda_{\max}(\mathbf{R}^{(n)}) \tag{2E.1}$$

where $\lambda_{\max}(\cdot)$ denotes the maximum eigenvalue of its argument. The inequality in Eq. (2E.1) corresponds to the observation that $\mathbf{R}^{(n)}$ is the restriction of $\mathcal{R}$ in domain and range to a n-dimensional subspace (actually, any one of a family of such subspaces) of $l^2(Z)$. A similar inequality holds for the minimum eigenvalue, establishing our claim.

APPENDIX 2F SOME INNER PRODUCTS

The inner products in Eq. (2.114) are evaluated under the assumptions of Section 2.6.5. Specifically, coordinates are chosen so that $\sum_R \mathbf{y}_R = 0 = \sum_t t$ and the platform velocity $\mathbf{v}_p$ is aligned with the x-axis with length $v_p \equiv \|\mathbf{v}_p\|$. Then

$$\mathbf{a}(\epsilon)^{\dagger} D\mathbf{a}(\epsilon)\big|_{\epsilon=0} = n_D n_R \sum_{T',T} \tilde{C}_{T'T}(\mathbf{u}) \tag{2F.1}$$

Let $((\mathbf{x}_T^{(A)})^T (\mathbf{x}_T^{(B)})^T)^T$ partition the 3-vector $\mathbf{x}_T$ into its top two and last entries. Recalling $\mathbf{s} \equiv -((u_x/u_z), (u_y/u_z))^T$, we also have

$$\left.\frac{\partial(\mathbf{a}(0)^{\dagger} D\mathbf{a}(\epsilon))}{\partial\epsilon}\right|_{\epsilon=0} = n_D n_R \frac{i\omega}{c} \sum_{T',T} \tilde{C}_{T'T}(\mathbf{x}_T^{(A)} + \mathbf{x}_T^{(B)}\mathbf{s})^T \mathbf{w} \tag{2F.2}$$

with $\mathbf{w} \equiv (w_x, w_y)^T$. Finally, we obtain

$$\begin{aligned}
(n_D n_R)^{-1} \left.\frac{\partial \mathbf{a}(\epsilon)^{\dagger}}{\partial\epsilon} D \frac{\partial \mathbf{a}(\epsilon)}{\partial\epsilon}\right|_{\epsilon=0} &= \frac{\omega^2}{c^2}\left(\sum_{T',T} \tilde{C}_{T'T}(\mathbf{u})\right) \\
&\quad \cdot \mathbf{w}^T \left[n_R^{-1} \sum_R (\mathbf{y}_R^{(A)} + \mathbf{y}_R^{(B)}\mathbf{s})(\mathbf{y}_R^{(A)} + \mathbf{y}_R^{(B)}\mathbf{s})^T \right] \mathbf{w} \\
&\quad + 4\frac{\omega^2}{c^2}\left(\sum_{T',T} \tilde{C}_{T'T}(\mathbf{u})\right) v_p^2 \left(n_D^{-1} \sum_t t^2 \right) (d - w_x)^2 \\
&\quad + \frac{\omega^2}{c^2} \mathbf{w}^T \left[\sum_{T',T} \tilde{C}_{T'T}(\mathbf{u}) \left(\mathbf{x}_{T'}^{(A)} + \mathbf{x}_{T'}^{(B)}\mathbf{s}\right)\left(\mathbf{x}_T^{(A)} + \mathbf{x}_T^{(B)}\mathbf{s}\right)^T \right] \mathbf{w}
\end{aligned}$$

APPENDIX 2G AN INVARIANT INNER PRODUCT

Values are given for the real scalars α and β used in the expression for the invariant inner product $\langle \mathbf{D}, \mathbf{E} \rangle = \alpha\,\mathrm{tr}(\mathbf{DE}) + \beta\,\mathrm{tr}\mathbf{D} \cdot \mathrm{tr}\mathbf{E}$ of Section 2.5.3. In particular, we show $\alpha > 0$. We evaluate two inner products in order to find two equations for the two unknowns α and β. For notational convenience, let $m \equiv n_T$. First, note that

$$\mathrm{tr}(\mathbf{T}^2) = \langle \mathbf{I}_m, \mathbf{I}_m \rangle = \alpha m + \beta m^2 \tag{2G.1}$$

The second equation involves a calculation. Assume, without loss of generality, that $\mathbf{T}$ is diagonal with entries t_k. If u_k denotes the kth component of the first column of $\mathbf{U}$, then ($\mathbf{e}_k$ denotes the m-vector with all zero components except for the kth entry, which is one)

$$\langle \mathbf{e}_1\mathbf{e}_1^\dagger, \mathbf{e}_1\mathbf{e}_1^\dagger \rangle = \mathrm{E}\left[\left(\sum_k t_k |u_k|^2\right)^2\right] \tag{2G.2}$$

with expectation taken over the unitary $\mathbf{U}$. The density of the complex vector forming the first column of $\mathbf{U}$ is well known. Specifically, the magnitude squared $|u_k|^2$ of each individual component has a complex beta distribution [33] given by ($\Gamma(x)$ denotes the standard gamma function)

$$\frac{\Gamma(m)}{\Gamma(1)\Gamma(m-1)}(1-x)^{m-2} \tag{2G.3}$$

on the unit interval [0,1]. Thus

$$E\left[|u_k|^4\right] = \frac{\Gamma(m)}{\Gamma(1)\Gamma(m-1)}\int_0^1 x^2(1-x)^{m-2}\,dx = \frac{2}{m(m+1)} \tag{2G.4}$$

Furthermore, conditioned on $|u_j|^2$, $|u_k|^2$, $k \neq j$ has the conditional expectation

$$E\left[|u_k|^2 \mid |u_j|^2\right] = \frac{1-|u_j|^2}{m-1} \tag{2G.5}$$

so that

$$E\left[|u_k|^2|u_j|^2\right] = \frac{\Gamma(m)}{\Gamma(1)\Gamma(m-1)}\int_0^1 \frac{x(1-x)}{m-1}(1-x)^{m-2}\,dx = \frac{1}{m(m+1)} \tag{2G.6}$$

Putting the pieces together, we have

$$\begin{aligned}
\alpha + \beta = \langle \mathbf{e}_1\mathbf{e}_1^\dagger, \mathbf{e}_1\mathbf{e}_1^\dagger \rangle &= E\left[\left(\sum_k t_k |u_k|^2\right)^2\right] \\
&= 2\sum_{j<k} t_j t_k E\left[|u_j|^2|u_k|^2\right] + \sum_k t_k^2 E\left[|u_k|^4\right] \\
&= \frac{1}{m(m+1)}\left[(\mathrm{tr}\mathbf{T})^2 + \mathrm{tr}(\mathbf{T}^2)\right]
\end{aligned}$$

which gives us a second equation for α and β along with Eq. (2G.1). Solving this linear system gives us

$$\begin{pmatrix} \alpha \\ \beta \end{pmatrix} = \frac{1}{(m-1)m(m+1)} \begin{pmatrix} m(\mathrm{tr}\mathbf{T})^2 - \mathrm{tr}(\mathbf{T}^2) \\ -(\mathrm{tr}\mathbf{T})^2 + m\,\mathrm{tr}(\mathbf{T}^2) \end{pmatrix} \tag{2G.6}$$

Since the eigenvalues of **T** are nonnegative, it is evident that $\alpha > 0$ when $m \geq 2$.

APPENDIX 2H KRÖNECKER AND TENSOR PRODUCTS

Some of our calculations involve bookkeeping indices for various matrix and vector products. The notions and notation we use are standard, but at least one of the definitions is not. We provide a concise set of definitions and relations here.

2H.1 Lexicographical Ordering

Let

$$1 \leq i \leq m, \quad 1 \leq j \leq n, \quad 1 \leq k \leq p \tag{2H.1}$$

for integer variables. Then order the pairs (i, j) lexicographically with $(i, j) > (i', j')$ if and only if $i > i'$or $i = i'$ and $j > j'$. Thus we have the correspondence

$$(i, j) \longleftrightarrow (i-1)n + j \tag{2H.2}$$

This ordering extends naturally to integer n-tuples so that, for example

$$(i, j, k) \longleftrightarrow (i-1)np + (j-1)p + k \tag{2H.3}$$

The ordered multiindices are the basis of the definition of "vec." For any $m \times n$ matrix **A**, form the vector a whose components are given by $\mathbf{a}_{(ij)} = A_{ij}$. Note that the vector **a** has length mn. The elements of **A** are inserted row by row into the components of a. We denote this relation by

$$\mathbf{a} = \mathrm{vec}(\mathbf{A}) \tag{2H.4}$$

One should be careful to note that this definition differs in the ordering of components from that in the literature.

2H.2 Tensor and Krönecker Products

The tensor product of two vectors **a** and **b** is the vector expressed by $\mathbf{a} \otimes \mathbf{b}$ and has components

$$(\mathbf{a} \otimes \mathbf{b})_{(jk)} = a_j b_k \tag{2H.5}$$

The extension to more factors is obvious.

The Krönecker product of the $m \times n$ matrix $\mathbf{A}$ and the $m' \times n'$ matrix $\mathbf{B}$ is a $mm' \times nn'$ matrix expressed by $\mathbf{A} \otimes \mathbf{B}$, whose components are given by

$$(\mathbf{A} \otimes \mathbf{B})_{(i,j)(k,l)} = A_{ik}B_{jl} \tag{2H.6}$$

This definition agrees with what is common in the literature.

It is useful to note that

$$\text{vec}(\mathbf{ab}^T) = \mathbf{a} \otimes \mathbf{b} \tag{2H.7}$$

2H.3 Properties

A number of properties of Krönecker and tensor products can be listed. We call out two equivalent properties that take an intuitively satisfying form, thanks to our non-standard definition of "vec". First, we have

$$(\mathbf{A} \otimes \mathbf{B})(\mathbf{a} \otimes \mathbf{b}) = (\mathbf{Aa}) \otimes (\mathbf{Bb}) \tag{2H.8}$$

Then

$$\text{vec}(\mathbf{ABC}) = (\mathbf{A} \otimes \mathbf{C}^T)\text{vec}(\mathbf{B}) \tag{2H.9}$$

The equivalence between these expressions can be derived from

$$(\mathbf{A} \otimes \mathbf{B})(\mathbf{a} \otimes \mathbf{b}) = (\mathbf{A} \otimes \mathbf{B})\text{vec}(\mathbf{ab}^T) = \text{vec}(\mathbf{Aab}^T\mathbf{B}^T) = (\mathbf{Aa}) \otimes (\mathbf{Bb}) \tag{2H.10}$$

It is often useful to define the covariance of a matrix $\mathbf{Z}$ as

$$\text{cov}(\mathbf{Z}) = E[\text{vec}(\mathbf{Z})\text{vec}(\mathbf{Z})^\dagger] \tag{2H.11}$$

Assuming that $\text{cov}(\mathbf{Z}) = \mathbf{R} \otimes \mathbf{T}^*$ has a tensor product structure, we have the transformation rule

$$\text{cov}(\mathbf{AZB}) = (\mathbf{ARA}^\dagger) \otimes (\mathbf{B}^\dagger\mathbf{TB})^* \tag{2H.12}$$

ACKNOWLEDGMENTS

The authors would like to thank Glenn Fawcett and David Goldfein of MIT Lincoln Laboratory for their contributions. We would like to thank Shawn Kraut and Dorothy Ryan from MIT Lincoln Laboratory for their helpful comments. We would also like to thank Prof. Jian Li of University of Florida for her valuable suggestions.

REFERENCES

1. L. Brown, *A Radar History of World War II: Technical and Military Imperatives*, Institute of Physics Publishing, Bristol–Philadelphia, 1999.
2. A. Chodos, J. Ouellette, and E. Tretkoff, This month in physics history, *APS News*, **15**(4) (April 2006).
3. C. Sherwin, J. Ruina, and D. Rawcliffe, Some early developments in synthetic aperture radar systems, *IRE Trans. Military Electron.* **MIL-6**(26) (April 1962).
4. D. W. Bliss and K. W. Forsythe, Multiple-input multiple-output (MIMO) radar and imaging: Degrees of freedom and resolution, *Conf. Record 37th Asilomar Conf. Signals, Systems & Computers, Pacific Grove, CA*, Nov. 2003, Vol. 1, pp. 54–59.
5. R. T. Hoctor and S. A. Kassam, The unifying role of the coarray in aperture synthesis of coherent and incoherent imaging, *Proc. IEEE* **78**(4):735–752 (April 1990).
6. B. Chen, S. Zhang, Y. Wang, and J. Wang, Analysis and experimental results on sparse-array synthetic impulse and aperture radar, *Proc. 2001 CIE Int. Conf. Radar*, Oct. 2001, pp. 76–80.
7. J. Dorey, G. Garnier, and G. Auvray, RIAS, radar a impulsion et antenne synthetique, *Colloque International sur le Radar, Paris*, April 1989, pp. 556–562.
8. S. D. Howard, A. R. Calderbank, and W. Moran, A simple signal processing architecture for instantaneous radar polarimetry, *IEEE Trans. Inform. Theory* **53**(4) (April 2007).
9. K. W. Forsythe, D. W. Bliss, and G. S. Fawcett, Multiple-input multiple-output (MIMO) radar: Performance issues, *Conf. Record 38th Asilomar Conf. Signals, Systems & Computers, Pacific Grove*, CA, Nov. 2004, Vol. 1, pp. 310–315.
10. E. Fishler, A. Haimovich, and R. S. Blum, Spatial diversity in radars—models and detection performance. *IEEE Trans. Signal Process.* **54**(3) (March 2006).
11. N. H. Lehmann et al., Spatial diversity in radars—models and detection performance, *IEEE Trans. Signal Process.* **55**(5):2215–2225 (May 2007).
12. X. Dai, J. Xu, Y. Peng, and X. Xia, A new method of improving the weak target detection performance based on the MIMO radar, *Proc. 2006 CIE Int. Conf. Radar*, Oct. 2006.
13. H. A. Khan and D. J. Edwards, Doppler problems in orthogonal MIMO radars, *Proc. IEEE Conf. Radar*, April 2006.
14. F. C. Robey et al., MIMO radar theory and experimental results, *Conf. Record 38th Asilomar Conf. Signals, Systems & Computers, Pacific Grove*, CA, Nov. 2004.
15. D. Rabideau and P. Parker, Ubiquitous MIMO multifunction digital array radar, *Conf. Record 37th Asilomar Conf. Signals, Systems & Computers, Pacific Grove*, CA, Nov. 2003.
16. I. Bekkerman and J. Tabrikian, Target detection and localization using MIMO radars and sonars, *IEEE Trans. Signal Process.* **54**(10):3873–3883 (Oct. 2006).
17. K. W. Forsythe and D. W. Bliss, Waveform correlation and optimization issues for MIMO radar, *Conf. Record 39th Asilomar Conf. Signals, Systems & Computers, Pacific Grove*, CA, Oct. 2005, pp. 1306–1310.
18. J. Tabrikian, Barankin bounds for target localization by MIMO radars, *Proc. IEEE Workshop on Sensor Array and Multichannel Processing*, July 2006, pp. 287–281.
19. L. Xu, J. Li, and P. Stoica, Adaptive techniques for MIMO radar, *Proc. IEEE Workshop on Sensor Array and Multichannel Processing*, July 2006, pp. 258–262.

20. H. Deng, Polyphase code design for orthogonal netted radar systems, *IEEE Trans. Signal Process.* **52**(11):3126–3135 (Nov. 2004).
21. B. Liu, Z. He, J. Zeng, and B. Liu, Polyphase orthogonal code design for MIMO radar systems, *Proc. 2006 CIE Int. Conf. Radar*, Oct. 2006.
22. G. San Antonio, D. R. Fuhrmann, and F. C. Robey, MIMO radar ambiguity functions, *IEEE J. of Select. Topics Signal Process.* **1**(1):167–177 (June 2007).
23. Y. Yang and R. S. Blum, MIMO radar waveform design based on mutual information and minimum mean-square error estimation, *IEEE Trans. Aerospace Electron. Syst.* **43**(1):330–343 (Jan. 2007).
24. L. Xu, J. Li, P. Stoica, K. W. Forsythe, and D. W. Bliss, Waveform optimization for MIMO radar: A Cramer-Rao bound based study, *Proc. IEEE Conf. Acoustics, Speech and Signal Processing*, April 2007, Vol. 2, pp. II-917–II-920.
25. G. San Antonio and D. R. Fuhrmann, Beampattern synthesis for wideband MIMO radar systems. *Proc. 2005 Int. Workshop on Computational Advances in Multi-Sensor Adaptive Processing*, Dec. 2005, pp. 105–108.
26. B. J. Donnet and I. D. Longstaff, Combining MIMO radar with OFDM communications, *Proc. 3rd European Radar Conf.*, Sept. 2006, pp. 37–40.
27. V. F. Mecca, D. Ramakrishnan, and J. L. Krolik, MIMO radar space-time adaptive processing for multipath clutter mitigation, *Proc. IEEE Workshop on Sensor Array and Multichannel Processing*, July 2006, pp. 249–253.
28. D. W. Bliss and K. W. Forsythe, MIMO radar medical imaging: Self-interference mitigation for breast tumor detection, *Conf. Record 40th Asilomar Conf. Signals, Systems & Computers*, Oct. 2006, pp. 1558–1562.
29. S. M. Kay, *Fundamentals of Statistical Signal Processing: Estimation Theory*, Prentice-Hall, Englewood Cliffs, NJ, 1993.
30. G. H. Golub and C. F. Van Loan, *Matrix Computations*, John Hopkins University Press, Baltimore, 1996.
31. D. F. DeLong, *Multiple Signal Direction Finding with Thinned Linear Arrays*, Technical Report TST-68, MIT Lincoln Laboratory, April 13, 1983.
32. H. Chernoff, *Sequential Analysis and Optimal Design*, Society for Industrial and Applied Mathematics, 1979.
33. E. J. Kelly and K. W. Forsythe, *Adaptive Detection and Parameter Estimation for Multidimensional Signal Models*, Technical Report 848, MIT Lincoln Laboratory, April 1989.
34. Sigurdur Helgason, *Differential Geometry, Lie Groups and Symmetric Spaces*, American Mathematical Society, 2001.
35. K. Yosida, *Functional Analysis*, Springer-Verlag, New York, 1968.
36. H. Gazzah, P. A. Regalia, and J.-P. Delmas, Asymptotic eigenvalue distribution of block toeplitz matrices and application to blind SIMO channel identification, *IEEE Trans. Inform. Theory* **47**(3):1243–1251 (March 2001).

3

GENERALIZED MIMO RADAR AMBIGUITY FUNCTIONS[1]

GEOFFREY SAN ANTONIO AND DANIEL R. FUHRMANN
Department of Electrical and System Engineering, Washington University, St. Louis, Missouri
FRANK C. ROBEY
MIT Lincoln Laboratory, Lexington, Massachusetts

3.1 INTRODUCTION

Multiple-input multiple-output (MIMO) systems have gained popularity and attracted attention for their ability to enhance all areas of system performance. MIMO ideas are not new, in fact their origin can be traced to the control systems literature. Mehra [1] discussed, the idea of optimally selecting multiple system inputs to enhance parameter estimation. The early 1990s saw an emergence of MIMO ideas into the field of communication systems [2]. More recently, one will find the ideas of MIMO appearing in sensor and radar systems.

A MIMO radar system [3,4] consists of transmit and receive sensors; the transmit sensors have the ability to transmit arbitrary and independent waveforms. In many ways a MIMO radar is similar to a MIMO communication system. Although the mission of a radar system is quite different, among the many possible uses of a radar system, tracking and detecting targets, estimating target model parameters, and creating images of targets are some of the most common. Various authors have all shown how these system tasks can be enhanced by using MIMO radar.

[1]This work is sponsored in part under Air Force Contract FA8721-05-C-0002. Opinions, interpretations, recommendations, and conclusions are those of the authors and are not necessarily endorsed by the United States government.

MIMO Radar Signal Processing, edited by Jian Li and Petre Stoica

Fishler et al. [4] have shown how detection performance might be improved using MIMO techniques. In the issue of how to make use of the extra degrees of freedom (DOF) offered by MIMO has also been addressed [5–9]. Designing transmit beampatterns is significantly enhanced by using MIMO ideas. Also, adaptive techniques such as MVDR and the GLRT can be enhanced by MIMO [3,10]. Early work [11–13] emphasized some basic performance gains that might be achievable.

As mentioned above, two of the primary functions of a radar are to detect targets and estimate parameters of a model used to describe those targets. Early radars could distinguish one unambiguous parameter, *range*. Continuous-wave (CW) radars could identify only *range rate*. Pulse–Doppler radar can simultaneously identify *range* and *range rate*. Radar arrays allow for estimation of *angular* parameters. The work in this chapter shows how MIMO radar can enable the unambiguous observation of additional target parameters.

Modern radar systems are designed to be highly accurate for their intended purpose. Designers and engineers need to know the level of resolution to expect from a particular system configuration. Some of the tools used to characterize performance are statistical parameter estimation bounds and ambiguity functions. Typically, parameter estimation bounds such as the Cramér–Rao (CR) bound depend on the ambiguity function. The CR bound is a local bound; it depends on the shape of the ambiguity function in the local region surrounding the parameter estimate. The classic ambiguity function was introduced by Woodward and is used to characterize the local and global resolution properties of time delay and Doppler for narrowband waveforms. Other authors have extended Woodward's ideas to larger classes of waveforms and whole radar systems.

The purpose of this chapter is to extend the ideas of ambiguity analysis to MIMO radar systems. An ambiguity function is developed that expresses the new degrees of freedom offered by MIMO radar while reducing to Woodward's ambiguity function for a simple-single sensor narrowband system. The functions presented herein are the necessary tools for effectively evaluating whole sets or classes of waveforms to be used in a MIMO radar system. Section 3.2 presents an overview of previous work done on ambiguity functions, providing a background for establishing the MIMO ambiguity function. In Section 3.3 the signal model used to describe a MIMO radar is presented. Section 3.4 introduces a parametric model that characterizes targets using a six-parameter vector. The MIMO ambiguity function is developed in Section 3.5. Section 3.6 presents some examples of MIMO ambiguity functions and highlights some new signal design problems.

3.2 BACKGROUND

MIMO radar is in many ways a generalization of traditional radar technology. As a preliminary step to defining the MIMO radar ambiguity function, we need to establish what previous work has been done in the area of ambiguity function analysis. The results given later in this chapter reveal that the MIMO ambiguity function is a generalization of most of the previous contributions.

The earliest radar systems were designed to make simple measurements. These measurements included estimating time delay/range or velocity by using continuous wave (CW) radar. As systems became more complex and precise, the inevitable issues of accuracy and resolution arose. Early researchers introduced a function called the *ambiguity function* that captures some of the inherent resolution properties of a radar system. The ambiguity function was first introduced by Ville [14]; however, it is generally identified with Woodward because of his pioneering work [15,16]. Woodward was interested in characterizing how well one could identify the target parameters of time delay (range) and Doppler (range rate) based on the transmission of a known waveform $s(t)$. He established his ambiguity function by first noting that a good waveform is one that could be used to distinguish between radar returns with different target parameters. He defined a total mean-squared error (MSE) metric between a known waveform $s(t)$ and a frequency shifted and time-delayed version:

$$\Psi(\Delta\tau, \Delta f_v) = \int \left| s(t) - s(t - \Delta\tau)e^{-j2\pi\Delta f_v t} \right|^2 dt \tag{3.1}$$

When the square is expanded, the only remaining term depending on the parameters is a term that is the inner product between the original waveform $s(t)$ and the time-delayed/frequency-shifted version. Woodward called this the *radar ambiguity function*:

$$\chi(\Delta\tau, \Delta f_v) = \left| \int s(t)s^*(t - \Delta\tau)e^{-j2\pi\Delta f_v t} dt \right|^2 \tag{3.2}$$

A magnitude-squared operation is usually introduced to the inner product term. In order to minimize the function (3.1), the ambiguity function should be large for ($\Delta\tau = 0, \Delta f_v = 0$) and small for ($\Delta\tau \neq 0, \Delta f_v \neq 0$). An ideal ambiguity function is one that resembles a thumbtack in the ($\Delta\tau$, Δf_v) plane. A good description of the properties of this function can be found in Ref. 17. Among its more well-known properties is the fact that there is an inherent ambiguity or duality between resolution in time and resolution in frequency. For a given time–bandwidth product, targets cannot be resolved perfectly in time and frequency simultaneously. In fact, this is why (3.2) is referred to as an *ambiguity function*. In conjunction with this concept is the idea that ambiguity or energy (volume under the surface) can be moved around in the ($\Delta\tau$, Δf_v) plane but not removed. An intuitive comparison to make is to think of the ambiguity function as analogous to a probability density function. Just as the *trace* of a covariance matrix bounds the spread of the density function, the time–bandwidth product bounds the spread of the ambiguity function. It should be noted that alternative derivations of the ambiguity function exist and are equally valid. In particular, Van Trees [18] provided a derivation of the ambiguity function that starts from a statistical view of the received radar signal. In

Section 3.4 a statistical description of the received data in a MIMO radar is developed. Using this, a definition of ambiguity is proposed.

It was recognized that Woodward's ambiguity function needed modification to handle larger bandwidth signals, long duration signals, and targets with high velocity. Other authors and researchers have undertaken this task by considering waveforms with larger bandwidths and targets described by higher-order motion parameters. The hierarchy of generalizations can be confusing since many authors simply call their functions *generalized ambiguity functions*. Correct modeling of these types of scenarios has been carried out [19–21]. Those authors use parametric models that more accurately reflect the actual physical phenomena involved with moving targets and reflecting signals. As a traveling wavefield reflects off a moving target, the field either expands or compresses in time as a result of the movement of the target. When a narrowband waveform is transmitted, this compressive effect is ignored for the waveform's complex envelope and considered only for the carrier. The condition that must be met for this compressive effect to be ignored is based on the time–bandwidth product TB, target velocity v, and the propagation speed of traveling waves c:

$$\frac{2vBT}{c} \ll 1 \tag{3.3}$$

The ambiguity function derived for conditions that violate (3.3) is

$$\chi(\tau, f_v) = \left| \sqrt{\gamma} \int s(t)s^*(\gamma(t-\tau))e^{-j2\pi f_v t}\,dt \right|^2 \tag{3.4}$$

The term $\gamma = 1 + f_v/f_c$ specifically accounts the stretching/compressing in time of the reflected signal. There is a scalar term in front of the integral to account for a change in amplitude of the reflected signal as it is stretched. The amplitude scaling is necessary for the conservation of energy when the waveform is stretched in time. Several authors have suggested similar wideband ambiguity functions [22–26].

More recently, connections have been made between the wideband ambiguity function and the continuous-time wavelet transform [27–30]. While wideband waveforms have been shown to enhance resolution, the correct processing of such signals can be computationally intensive. Fast Fourier transform (FFT) can be used with narrowband signals to efficiently compute Woodward's ambiguity function. Equation (3.2) is basically a convolution integral. In contrast, the wavelet transform is the tool associated with processing wideband signals. The wavelet transform describes signals that have been shifted in time and time-dilated. Techniques for computing wavelet transforms over a continuum of time shifts and time dilations have been described [31].

This description of ambiguity functions was based on radar systems operating with a single aperture. Some researchers have also defined ambiguity functions for radar systems with multiple apertures. The first to do so was Urkowitz [32], who formulated an ambiguity function that is a function of *azimuth, elevation, range*, and *Doppler*. In his system, the waveform transmitted at each aperture is the same except for a time delay or phase shift. His focus is on waveforms that are narrowband. As a result, he shows that his ambiguity function can be factored into separable space and time components. It will be shown later that for certain situations, the MIMO ambiguity function will also be spacetime-separable. Spacetime separability has been discussed further in the literature see [33,34].

Formulations of the ambiguity function have also appeared for radars configured in a *bistatic* configuration [35]. The term *bistatic* has generally been used in the literature to refer to a radar system with a large separating baseline between transmit and receive apertures. In this configuration the resolution of target parameters such as range and Doppler becomes highly dependent on the location of the target relative to the transmit and receive apertures. We will use the term *bistatic* to refer to any system configuration with separate transmit and receive sensors.

Some authors have proposed ambiguity functions that incorporate the estimation of nuisance parameters. For example, a particular problem might include the joint estimation of delay and Doppler along with the background noise power, although this is not consistent with Woodward's original concept of ambiguity. Rendas and Moura [36] propose an ambiguity function based on the Kullback–Leibler divergence. Their approach to defining the ambiguity function is based on the general problem of parameter estimation in curved exponential families. In their paper they are able to show how their ambiguity function reduces to Woodward's function under special signal model assumptions. In a related problem, Dogandzic and Nehorai [37] derive the CR bounds for estimating range, velocity, and angle with an active array. They relate their solution to the ambiguity function proposed by Rendas and Moura [36], whose definition of ambiguity that we propose for MIMO radar is somewhat similar to ours. It should be recognized that we are not defining a notion of ambiguity that is vastly different from those of other authors. Rather, the intent of this chapter is examine the role of the different MIMO radar system components in the ambiguity function. A major difference between our definition of ambiguity and that of Rendas and Moura [36] is that our function is not normalized and does not strictly vary between 0 and 1. This normalization step is somewhat arbitrary and results in a loss of information. Traditionally, the ambiguity function has been defined as a normalized function. Most analysis focuses on the shape of the ambiguity function; however, the overall levels can provide further information, especially with regard to the CR bounds.

3.3 MIMO SIGNAL MODEL

In this section we introduce a general signal model for MIMO radar to clarify the interaction between the transmitted signals, the target, and the noise. In the following section we discuss a particular parametric model to be used in MIMO radar.

A MIMO radar system consists of N_T transmit sensors and N_R receive sensors. A series of independent signals is transmitted from each transmit sensor in a coherent fashion. The propagation of a signal from a transmit sensor to a receive sensor consists of propagation through a channel with three components: a forward-propagating channel to the target, a reflecting/scattering target, and a reverse channel to the receive sensor. Both the forward and reverse channels will be jointly parameterized by a parametric model with parameter θ.

The target will be considered point-like in nature. This means that the physical dimensions of the target are small enough that it appears to the radar as a single-point target with little or no extent. We do allow, however, for the target to possibly be composed of many smaller scattering centers. For each pair of transmit and receive sensors, the target response or scattering function is approximated as a realization of a random process. For the ith transmit sensor and jth receive sensor this scattering function will be denoted as $a_{j,i}$. The random process is such that over the time period of the transmit signal duration, the realization of the random process is constant.

Given this discussion of the target and parametric channel, the received signal at the jth receive sensor due to the ith transmit waveform can be expressed as

$$r_{j,i}(t) = s_i(t, \boldsymbol{\theta}, j)a_{j,i} + n_j(t) \tag{3.5}$$

The term $n_j(t)$ is an additive noise process independent of the target scattering function $a_{j,i}$. The term $s_i(t, \boldsymbol{\theta}, j)$ represents the ith transmitted signal modified according to the parametric channel model with parameter $\boldsymbol{\theta}$ for the jth receive sensor. The received signal will most likely be sampled; therefore, we can use matrix vector notation to represent the received signal as an $N \times 1$ vector:

$$\mathbf{r}_{j,i} = \mathbf{s}_i(\boldsymbol{\theta}, j)a_{j,i} + \mathbf{n}_j \tag{3.6}$$

Now, because there are N_T transmit signals, the received signal at the jth receive sensor is the linear combination of all such signals as in Eq. (3.5):

$$\begin{aligned} \mathbf{r}_j &= \sum_{i=1}^{N_T} \mathbf{r}_{j,i} \\ &= \sum_{i=1}^{N_T} \mathbf{s}_i(\boldsymbol{\theta}, j)a_{j,i} + \mathbf{n}_j \\ &= \mathbf{S}(\boldsymbol{\theta})\mathbf{a}_j + \mathbf{n}_j \end{aligned} \tag{3.7}$$

The term $\mathbf{S}(\boldsymbol{\theta})$ is a $N \times N_T$ matrix whose columns are the $N \times 1$ vectors $\mathbf{s}_i(\boldsymbol{\theta}, j)$:

$$\mathbf{S}(\boldsymbol{\theta}) = [\mathbf{s}_1(\boldsymbol{\theta}, j), \mathbf{s}_2(\boldsymbol{\theta}, j), \ldots, \mathbf{s}_{N_T}(\boldsymbol{\theta}, j)]$$

There are a total of N_R receive sensors; the data from each can be composed into a single vector of size $NN_R \times 1$:

$$
\begin{aligned}
\mathbf{r} &= \left[\mathbf{r}_1^T, \mathbf{r}_2^T, \ldots, \mathbf{r}_{N_R}^T\right]^T \\
&= \begin{bmatrix} \mathbf{S}(\boldsymbol{\theta}, 1) & 0 & \cdots & 0 \\ 0 & \mathbf{S}(\boldsymbol{\theta}, 2) & \cdots & 0 \\ \vdots & \vdots & \ddots & \vdots \\ 0 & 0 & 0 & \mathbf{S}(\boldsymbol{\theta}, N_R) \end{bmatrix} \begin{bmatrix} \mathbf{a}_1 \\ \mathbf{a}_2 \\ \vdots \\ \mathbf{a}_{N_R} \end{bmatrix} + \begin{bmatrix} \mathbf{n}_1 \\ \mathbf{n}_2 \\ \vdots \\ \mathbf{n}_{N_R} \end{bmatrix} \\
&= \mathbf{S}(\boldsymbol{\theta})\mathbf{a} + \mathbf{n}
\end{aligned} \tag{3.8}
$$

Equation (3.8) is the signal model used throughout the rest of this chapter to describe the received data due to a set of transmit waveforms, a point-like target with response vector $\mathbf{a}$, and a parametric channel model with parameter $\boldsymbol{\theta}$. The rest of this section describes the probability density function (pdf) for the data. A common pdf used in radar signal processing is the complex Gaussian distribution. Various authors have shown how both the noise and target response are well modeled by this distribution (e.g., see Ref. 17). The $N_T N_R \times 1$ target response vector will have the distribution

$$\mathbf{a} \sim \mathcal{CN}(0, \sigma_s^2 \boldsymbol{\Sigma}) \tag{3.9}$$

This denotes a complex Gaussian random vector with zero mean and covariance matrix $\sigma_s^2 \boldsymbol{\Sigma}$. In a later section the characteristics of the target covariance matrix play an important role in the ambiguity function. We shall encounter three types of target covariance matrices: (1) if the covariance matrix is rank 1, $\sigma_s^2 \boldsymbol{\Sigma} = \sigma_s^2 \mathbf{v}\mathbf{v}^H$, the target is called *coherent*; (2) if the covariance matrix is a multiple of the identity matrix, $\sigma_s^2 \boldsymbol{\Sigma} = \sigma_s^2 \mathbf{I}$, the target is called *noncoherent*; and (3) in cases where the target covariance matrix has rank >1 but is not a multiple of the identity matrix, the target is called *partially coherent*. The matrix $\sigma_s^2 \boldsymbol{\Sigma}$ is positive semidefinite and therefore has the eigenvalue/eigenvector decomposition

$$\sigma_s^2 \boldsymbol{\Sigma} = \sigma_s^2 \mathbf{V} \boldsymbol{\Lambda} \mathbf{V}^H$$

Also, if the rank of $\boldsymbol{\Sigma}$ is $N_s \leq N_T N_R$, then the covariance matrix can be factored as the product of two matrices:

$$\sigma_s^2 \boldsymbol{\Sigma} = \sigma_s^2 \mathbf{Q}\mathbf{Q}^H$$

The matrix $\mathbf{Q}$ has dimensions $N_T N_R \times N_s$. It is given by $\mathbf{Q} = \mathbf{V}_s \mathbf{\Lambda}_s^{1/2}$. The matrix $\mathbf{V}_s$ is the matrix of the first N_s columns of $\mathbf{V}$. The matrix $\mathbf{\Lambda}_s$ is a square $N_s \times N_s$ matrix given by the positive definite component of the matrix $\mathbf{\Lambda}$.

The noise vector will also be complex Gaussian, with the distribution

$$\mathbf{n} \sim \mathcal{CN}(0, \sigma_n^2 \mathbf{I}_{NN_R}) \tag{3.10}$$

The subscript notation on the identity matrix $\mathbf{I}_{NN_R}$ signifies an identity matrix of dimension $NN_R \times NN_R$. The received data vector has the distribution

$$\mathbf{r} \sim \mathcal{CN}\left(0, \sigma_s^2 \mathbf{S}(\boldsymbol{\theta}) \mathbf{\Sigma} \mathbf{S}^H(\boldsymbol{\theta}) + \sigma_n^2 \mathbf{I}_{NN_R}\right) \tag{3.11}$$

Later we will have occasion to use the data loglikelihood function. This is simply the natural logarithm of the data pdf $p(\mathbf{r}|\theta)$ viewed as a function of the parameter θ:

$$\begin{aligned} L(\boldsymbol{\theta}; \mathbf{r}) &= \log p(\mathbf{r}|\boldsymbol{\theta}) \\ &= -NN_R \ln(\pi) - \ln \det\left(\sigma_n^2 \mathbf{I}_{NN_R} + \sigma_s^2 \mathbf{S}(\boldsymbol{\theta}) \mathbf{\Sigma} \mathbf{S}^H(\boldsymbol{\theta})\right) \\ &\quad - \operatorname{tr}\left([\sigma_n^2 \mathbf{I}_{NN_R} + \sigma_s^2 \mathbf{S}(\boldsymbol{\theta}) \mathbf{\Sigma} \mathbf{S}^H(\boldsymbol{\theta})]^{-1} \mathbf{r}\mathbf{r}^H\right) \end{aligned} \tag{3.12}$$

Using two matrix identities, we can simplify this expression. The first is a matrix determinant identity:

$$\det(\mathbf{AB} + \mathbf{I}) = \det(\mathbf{BA} + \mathbf{I}) \tag{3.13}$$

Using this identity, the determinant in the loglikelihood function can be written as

$$\det\left(\sigma_n^2 \mathbf{I}_{NN_R} + \sigma_s^2 \mathbf{S}(\boldsymbol{\theta}) \mathbf{\Sigma} \mathbf{S}^H(\boldsymbol{\theta})\right) = (\sigma_n^2)^{NN_R} \det\left(\mathbf{I}_{N_s} + \frac{\sigma_s^2}{\sigma_n^2} \mathbf{Q}^H \mathbf{S}^H(\boldsymbol{\theta}) \mathbf{S}(\boldsymbol{\theta}) \mathbf{Q}\right)$$

The second identity is the matrix inversion lemma. If $\mathbf{A} = \mathbf{B} + \mathbf{CDC}^H$, then

$$\mathbf{A}^{-1} = \mathbf{B}^{-1} - \mathbf{B}^{-1}\mathbf{C}\left(\mathbf{D}^{-1} + \mathbf{C}^H \mathbf{B}^{-1} \mathbf{C}\right)^{-1} \mathbf{C}^H \mathbf{B}^{-1} \tag{3.14}$$

Using this identity, we can express the data covariance matrix as

$$\begin{aligned} \left[\sigma_n^2 \mathbf{I}_{NN_R} + \sigma_s^2 \mathbf{S}(\boldsymbol{\theta}) \mathbf{\Sigma} \mathbf{S}^H(\boldsymbol{\theta})\right]^{-1} &= \left(\frac{1}{\sigma_n^2}\right) \mathbf{I}_{NN_R} - \left(\frac{1}{\sigma_n^2}\right)\left(\frac{\sigma_s^2}{\sigma_n^2}\right) \mathbf{S}(\boldsymbol{\theta}) \mathbf{Q} \\ &\quad \times \left[\mathbf{I}_{N_s} + \left(\frac{\sigma_s^2}{\sigma_n^2}\right) \mathbf{Q}^H \mathbf{S}^H(\boldsymbol{\theta}) \mathbf{S}(\boldsymbol{\theta}) \mathbf{Q}\right]^{-1} \mathbf{Q}^H \mathbf{S}^H(\boldsymbol{\theta}) \end{aligned}$$

With these two simplifications the loglikelihood function may be rewritten as

$$L(\boldsymbol{\theta}; \mathbf{r}) = -NN_R \ln(\pi\sigma_n^2) - \ln\det\left(\mathbf{I}_{N_s} + \frac{\sigma_s^2}{\sigma_n^2}\mathbf{Q}^H\mathbf{S}^H(\boldsymbol{\theta})\mathbf{S}(\boldsymbol{\theta})\mathbf{Q}\right) - \left(\frac{1}{\sigma_n^2}\right)\mathrm{tr}(\mathbf{r}\mathbf{r}^H)$$
$$+ \left(\frac{\sigma_s^2}{\sigma_n^4}\right)\mathrm{tr}\left(\left[\mathbf{I}_{N_s} + \left(\frac{\sigma_s^2}{\sigma_n^2}\right)\mathbf{Q}^H\mathbf{S}^H(\boldsymbol{\theta})\mathbf{S}(\boldsymbol{\theta})\mathbf{Q}\right]^{-1}\mathbf{Q}^H\mathbf{S}^H(\boldsymbol{\theta})\mathbf{r}\mathbf{r}^H\mathbf{S}(\boldsymbol{\theta})\mathbf{Q}\right) \quad (3.15)$$

3.4 MIMO PARAMETRIC CHANNEL MODEL

In this section we introduce a parametric model that describes how the transmit signals appear at the receive sensors. The parts of this model include signal transmission, signal propagation, signal reflection, and signal reception.

3.4.1 Transmit Signal Model

Currently, proposed MIMO radars consist of coherent networks of transmit and receive sensors. These sensors could be distributed apertures or elements of a single phased array. In this development, we assume that all sensors have an isotropic radiation pattern and that no mutual coupling between sensors occurs. A MIMO radar has N_T transmit sensors and N_R receive sensors. The ith transmit sensors spatial location will have Cartesian coordinates given by the column vector $\mathbf{x}_{i,T}$ and the jth receiver will have Cartesian coordinates $\mathbf{x}_{j,R}$. These coordinates are referenced to a predefined origin shared by both the sensors and the target. The $3 \times N_T$ and $3 \times N_R$ matrices $\mathbf{X}_T$ and $\mathbf{X}_R$ denote the collection of all transmit and receive sensor locations:

$$\mathbf{X}_T = [\mathbf{x}_{1,T};\ \mathbf{x}_{2,T};\ \ldots\ ;\mathbf{x}_{N_T,T}]$$
$$\mathbf{X}_R = [\mathbf{x}_{1,R};\ \mathbf{x}_{2,R};\ \ldots\ ;\mathbf{x}_{N_R,R}]$$

Each transmit sensor has the ability to transmit an independent waveform. The actual signal transmitted from the ith sensors is

$$g_i(t) = 2\mathrm{Re}\{\hat{s}_i(t)\}$$

All sensors operate at the same carrier frequency f_c that is referenced to the same phase angle on transmit and receive for all sensors. In this sense the waveforms are phase-coherent relative to the carrier. The bandwidth B and time duration T are considered to be constant for all waveforms $\hat{s}_i(t)$. Moreover, it is assumed that the signal bandwidth satisfies the condition $B/2 < f_c$. Under this assumption, the common complex envelope notation can be applied to the form of the transmitted signal

$$g_i(t) = 2\mathrm{Re}\{s_i(t)e^{j2\pi f_c t}\} \tag{3.16}$$

where $s_i(t)$ is the complex envelope of the ith waveform.

3.4.2 Channel and Target Models

On transmission, the radar signals as given by the model in Eq. (3.16) propagate through free space, reflect off objects, and return to the radar receivers. Proper statistical modeling of the received signals requires a thorough understanding of the propagation channel characteristics and the target reflection process. For each transmit/receive sensor pair there exists a forward transmit channel to the target and a reverse receive channel from the target. These channels are modeled as lossless time delay and phase shift channels. All others losses, such as R^4 attenuation, will be assumed to be due to the target reflection process.

Several authors have considered various models for target reflection in MIMO radar systems; the two extremes are noncoherent scattering and coherent scattering. The coherence of the scattering is reflected in the covariance matrix of the target response. Proper model selection is based on a careful comparison of target complexity, signal bandwidth, signal duration, target motion, and sensor array configuration. The target model applied in this chapter is a point-like target moving with constant velocity.

Point targets will be described by a parameter vector $\boldsymbol{\theta}$ consisting of a position vector component $\mathbf{p}$ and velocity vector component $\mathbf{v}$. The position vector will be defined in the same coordinate system as the transmit and receive sensor arrays. The velocity vector will also be defined in the array coordinate system. Depending on the array configuration, each target could be identified by up to six unambiguous parameters, three for position and three for velocity. We will now show how these parameters appear in the model. In general, if the signal $h(t)$ is transmitted from sensor i, reflects off a target, and is received by sensor j, the response can be described by

$$\beta(\boldsymbol{\theta}, i, j)h\big(t - \delta(t, \boldsymbol{\theta}, \mathbf{x}_{i,T}, \mathbf{x}_{j,R})\big) \tag{3.17}$$

The function $\delta(t, \boldsymbol{\theta}, \mathbf{x}_{i,T}, \mathbf{x}_{j,R})$ is a time-varying function that depends on the target parameters and sensor locations. Using a Taylor series expansion, we find

$$\delta(t, \boldsymbol{\theta}, \mathbf{x}_{i,T}, \mathbf{x}_{j,R}) \approx \tau_{i,j}(\mathbf{p}) - \frac{f(\boldsymbol{\theta})}{f_c}(t - \tau_{i,j}(\mathbf{p})) \tag{3.18}$$

The same function can be derived using the Lorentz transformation, which specifies how wavefields and signals are related for different inertial reference frames. This particular derivation assumes that the target velocities of interest are significantly less than the speed of light. The term $\tau_{i,j}(\mathbf{p})$ is simply the two-way time delay due to a target located at $\mathbf{p}$ with transmit and receive sensors at $\mathbf{x}_{i,T}$ and $\mathbf{x}_{j,R}$:

$$\begin{aligned}\tau_{i,j}(\mathbf{p}) &= \tau_i(\mathbf{p}) + \tau_j(\mathbf{p}) \\ &= \frac{\|\mathbf{p}-\mathbf{x}_{i,T}\|}{c} + \frac{\|\mathbf{p}-\mathbf{x}_{j,R}\|}{c}\end{aligned} \tag{3.19}$$

where $\|\cdot\|$ denotes the usual Euclidean vector norm.

The term $f(\boldsymbol{\theta})$ is the frequency shift caused by a target moving with velocity vector $\mathbf{v}$ and position $\mathbf{p}$. The frequency shift is caused by an instantaneous change in pathlength between transmitter, target, and receiver. For simplification assume the radar is fixed, then $f(\boldsymbol{\theta})$ is defined as

$$f(\boldsymbol{\theta}) = \frac{1}{\lambda}\left[\frac{d}{dt}(R_T + R_R)\right] \tag{3.20}$$

where R_T and R_R are the transmit and receive pathlengths. Each derivative is the projection of the target velocity vector onto either the transmitter or receiver line of sight (LOS):

$$\begin{aligned}\frac{1}{\lambda}\frac{d}{dt}R_T &= \frac{1}{\lambda} < \frac{\mathbf{p}-\mathbf{x}_{i,T}}{\|\mathbf{p}-\mathbf{x}_{i,T}\|}, \mathbf{v} \geq f_i(\boldsymbol{\theta}) \\ \frac{1}{\lambda}\frac{d}{dt}R_R &= \frac{1}{\lambda} < \frac{\mathbf{p}-\mathbf{x}_{i,R}}{\|\mathbf{p}-\mathbf{x}_{i,R}\|}, \mathbf{v} \geq f_j(\boldsymbol{\theta})\end{aligned}$$

Now the frequency shifts can be explicitly written in terms of a transmit component and receive component:

$$f_{i,j}(\boldsymbol{\theta}) = f_i(\boldsymbol{\theta}) + f_j(\boldsymbol{\theta}) \tag{3.21}$$

Under certain assumptions concerning array geometry, simplifications can be made to both the formulas for time delay and frequency shift. In a later section, a hierarchy of assumptions is outlined that will have implications for the form of the MIMO ambiguity function. It is also convenient to define the stretch factor

$$\gamma_{i,j}(\boldsymbol{\theta}) = 1 + \frac{f_{i,j}(\boldsymbol{\theta})}{f_c}$$

3.4.3 Received Signal Parametric Model

Using the parametric model described above, the received signal at the jth receive sensor before demodulation to baseband can be written as

$$\tilde{r}_j(t, \boldsymbol{\theta}) = \sum_{i=1}^{N_T} \tilde{a}_{j,i}\sqrt{\gamma_{i,j}(\boldsymbol{\theta})}g_i\big(\gamma_{i,j}(\boldsymbol{\theta})(t - \tau_{i,j}(\mathbf{p}))\big) + \tilde{n}_j(t)$$

The quantities $\tilde{a}_{j,i}$ and $\tilde{n}_j(t)$ represent the target response and noise response before demodulation to baseband or another intermediate frequency. After complex demodulation to baseband, the received signal signal is

$$r_j(t, \boldsymbol{\theta}) = \sum_{i=1}^{N_T} a_{j,i}\sqrt{\gamma_{i,j}(\boldsymbol{\theta})}s_i\left(\gamma_{i,j}(\boldsymbol{\theta})(t - \tau_{i,j}(\mathbf{p}))\right)e^{-j2\pi\tau_{i,j}(\mathbf{p})(f_c+f_{i,j}(\boldsymbol{\theta}))}e^{j2\pi f_{i,j}(\boldsymbol{\theta})t} + n_j(t) \quad (3.22)$$

This is same data model as described by Eq. (3.7), but now with the parametric model explicitly stated. This is a continuous-time version of the received signal.

3.5 MIMO AMBIGUITY FUNCTION

In this section we develop the ambiguity function for the MIMO radar system described in the previous section. We mentioned in Section 3.2 that the ambiguity function is a tool used to indicate both the local and global resolution properties of sensing systems. It is most commonly associated with active sensing such as radar, and often is regarded as only a function of the transmit waveforms. However, the geometric sensor configuration of a MIMO radar can have as much of an impact on resolution as the waveforms. Moreover, because different sensors can observe different scattering characteristics from a single-point target, the ambiguity function is also a function of the target, or at least its statistical properties. It therefore is correct to view the ambiguity function as a function of the whole radar system: geometry, waveforms, and target. We define the generalized MIMO ambiguity function as the expected value of the data loglikelihood function for the received data with respect to the distribution $p(r|\boldsymbol{\theta}_0)$. The parameter $\boldsymbol{\theta}_0$ is considered to be the true target parameter, whereas the parameter $\boldsymbol{\theta}_1$ is the hypothesized target parameter:

$$A(\boldsymbol{\theta}_0, \boldsymbol{\theta}_1) = E_{p(\mathbf{r}|\boldsymbol{\theta}_0)}[L(\boldsymbol{\theta}_1|\mathbf{z})] \quad (3.23)$$

This definition results in an ambiguity function with properties similar to those of other well-known ambiguity functions. In Section 3.2 we mentioned that Rendas and Moura [36] propose an ambiguity function that is defined in terms of I divergence. Our definition of the MIMO radar ambiguity function is somewhat similar. Both reduce to Woodward's ambiguity function under the correct conditions, and both define functions that are possibly asymmetric in the parameters $\boldsymbol{\theta}_0$ and $\boldsymbol{\theta}_1$. Woodward's ambiguity function is a special case in which the ambiguity function is symmetric in the two parameter arguments.

As stated, ambiguity functions provide both local and global properties of parameter estimator performance. The Cramér–Rao lower bounds are the common tool used to evaluate the local performance of unbiased estimators. With the definition of ambiguity as in Eq. (3.23), the curvature of the function about a point $\boldsymbol{\theta}_0$ gives the elements of the Fisher information matrix (FIM):

$$\begin{aligned}\mathbf{J}_{ij} &= -E_{p(\mathbf{r}|\boldsymbol{\theta}_0)}\left[\frac{\partial^2 lnp(\mathbf{r}|\boldsymbol{\theta}_1)}{\partial\boldsymbol{\theta}_1^i\partial\boldsymbol{\theta}_1^j}\right]\Bigg|_{\boldsymbol{\theta}_1=\boldsymbol{\theta}_0}\\ &= -\frac{\partial^2}{\partial\boldsymbol{\theta}_1^i\partial\boldsymbol{\theta}_1^j}A(\boldsymbol{\theta}_0, \boldsymbol{\theta}_1)\Bigg|_{\boldsymbol{\theta}_1=\boldsymbol{\theta}_0}\end{aligned}$$

Using the definition of the data loglikelihood function in Eq. (3.15), we can write the ambiguity function as follows. It can be expressed in terms of two functions, an autocorrelation function, and a cross-correlation function:

$$\begin{aligned}A(\boldsymbol{\theta}_0, \boldsymbol{\theta}_1) &= E_{p(z|\boldsymbol{\theta}_0)}[L(\boldsymbol{\theta}_1|\mathbf{z})]\\ &= -NN_R\ln(\pi\sigma_n^2) - \ln\det\left(\mathbf{I}_{N_s} + \frac{\sigma_s^2}{\sigma_n^2}\phi(\boldsymbol{\theta})\right) - \left(\frac{1}{\sigma_n^2}\right)\mathrm{tr}(E_{p(\mathbf{z}|\boldsymbol{\theta}_0)}[\mathbf{r}\mathbf{r}^H])\\ &\quad + \left(\frac{1}{\sigma_n^2}\right)\left(\frac{\sigma_s^2}{\sigma_n^2}\right)\mathrm{tr}\left(\left[\mathbf{I}_{N_s} + \left(\frac{\sigma_s^2}{\sigma_n^2}\right)\phi(\boldsymbol{\theta})\right]^{-1}\mathbf{Q}^H\mathbf{S}^H(\boldsymbol{\theta})E_{p(\mathbf{z}|\boldsymbol{\theta}_0)}[\mathbf{r}\mathbf{r}^H]\mathbf{S}(\boldsymbol{\theta})\mathbf{Q}\right)\\ &= -NN_R\ln(\pi\sigma_n^2) - \ln\det\left(\mathbf{I}_{N_s} + \frac{\sigma_s^2}{\sigma_n^2}\phi(\boldsymbol{\theta}_1)\right) - NN_R - \frac{\sigma_s^2}{\sigma_n^2}\mathrm{tr}(\phi(\boldsymbol{\theta}_0))\\ &\quad + \sigma_s^2\mathrm{tr}\left(\left[\mathbf{I}_{N_s} + \left(\frac{\sigma_s^2}{\sigma_n^2}\right)\phi(\boldsymbol{\theta}_1)\right]^{-1}\phi(\boldsymbol{\theta}_1)\right)\\ &\quad + \sigma_s^2\frac{\sigma_s^2}{\sigma_n^2}\mathrm{tr}\left(\left[\mathbf{I}_{N_s} + \left(\frac{\sigma_s^2}{\sigma_n^2}\right)\phi(\boldsymbol{\theta}_1)\right]^{-1}\mathbf{Q}^H\mathbf{S}^H(\boldsymbol{\theta}_1)\mathbf{S}(\boldsymbol{\theta}_0)\mathbf{Q}\mathbf{Q}^H\mathbf{S}^H(\boldsymbol{\theta}_0)\mathbf{S}(\boldsymbol{\theta}_1)\mathbf{Q}\right)\\ &= -NN_R\ln(\pi\sigma_n^2) - \ln\det\left(\mathbf{I}_{N_s} + \frac{\sigma_s^2}{\sigma_n^2}\phi(\boldsymbol{\theta}_1)\right) - NN_R - \frac{\sigma_s^2}{\sigma_n^2}\mathrm{tr}(\phi(\boldsymbol{\theta}_0))\\ &\quad + \sigma_s^2\mathrm{tr}\left(\left[\mathbf{I}_{N_s} + \left(\frac{\sigma_s^2}{\sigma_n^2}\right)\phi(\boldsymbol{\theta}_1)\right]^{-1}\phi(\boldsymbol{\theta}_1)\right)\\ &\quad + \sigma_s^2\frac{\sigma_s^2}{\sigma_n^2}\mathrm{tr}\left(\left[\mathbf{I}_{N_s} + \left(\frac{\sigma_s^2}{\sigma_n^2}\right)\phi(\boldsymbol{\theta}_1)\right]^{-1}\psi(\boldsymbol{\theta}_1, \boldsymbol{\theta}_0)\psi^H(\boldsymbol{\theta}_1, \boldsymbol{\theta}_0)\right)\end{aligned} \tag{3.24}$$

The functions $\phi(\boldsymbol{\theta})$ and $\psi(\boldsymbol{\theta}_1, \boldsymbol{\theta}_0)$ are matrix-valued functions defined as

$$\phi(\boldsymbol{\theta}) = \mathbf{Q}^H\mathbf{S}^H(\boldsymbol{\theta})\mathbf{S}(\boldsymbol{\theta})\mathbf{Q} \tag{3.25}$$

$$\psi(\boldsymbol{\theta}_1, \boldsymbol{\theta}_0) = \mathbf{Q}^H\mathbf{S}^H(\boldsymbol{\theta}_1)\mathbf{S}(\boldsymbol{\theta}_0)\mathbf{Q} \tag{3.26}$$

Equation (3.25) is an autocorrelation-type function that is the inner product of the signal matrix $\mathbf{S}(\boldsymbol{\theta})$ and projects itself onto the weighted subspace of the target covariance matrix $\boldsymbol{\Sigma}$. Similarly, Eq. (3.26) defines a cross-correlation-type function. It is the inner product of the signal matrix $\mathbf{S}(\boldsymbol{\theta})$ under two different parameter values and is then projected onto the weighted subspace of the target covariance matrix $\boldsymbol{\Sigma}$. If the rank of the target covariance is >1, then these two functions will be matrix-valued with dimensions $N_s \times N_s$.

In our definition of the ambiguity function there is a strong dependence on both the target power level σ_s^2 and the noise power level σ_n^2 because the definition is based on the data loglikelihood function. Under some special conditions the shape of the ambiguity function is largely unaffected by these two terms. Specifically, when the SNR defined as the ratio SNR $= \sigma_s^2/\sigma_n^2$ is large, then the overall shape of the ambiguity function approaches a limiting form. It is scaled according to the actual value of σ_s^2. At a certain threshold SNR, the ambiguity function as defined begins to flatten out and the noise begins to dominate.

In the case of a single collocated transmitter and receiver, the expression in (3.24) reduces to a form that includes Woodward's ambiguity function:

$$
\begin{aligned}
A(\boldsymbol{\theta}_1, \boldsymbol{\theta}_0) &= -N \ln(\pi\sigma_n^2) - \ln\left(1 + \frac{\sigma_s^2}{\sigma_n^2}\phi(\boldsymbol{\theta}_1)\right) - N - \frac{\sigma_s^2}{\sigma_n^2}\phi(\boldsymbol{\theta}_0) \\
&\quad + \sigma_s^2 \frac{\phi(\boldsymbol{\theta}_1)}{1 + (\sigma_s^2/\sigma_n^2)\phi(\boldsymbol{\theta}_1)} + \sigma_s^2 \frac{\sigma_s^2}{\sigma_n^2} \frac{|\psi(\boldsymbol{\theta}_1, \boldsymbol{\theta}_0)|^2}{1 + (\sigma_s^2/\sigma_n^2)\phi(\boldsymbol{\theta}_1)} \\
&= -N \ln(\pi\sigma_n^2) - \ln\left(1 + \frac{\sigma_s^2}{\sigma_n^2}E\right) - N - \frac{\sigma_s^2}{\sigma_n^2}E \\
&\quad + \sigma_s^2 \frac{E}{1 + (\sigma_s^2/\sigma_n^2)E} + \sigma_s^2 \frac{\sigma_s^2}{\sigma_n^2} \frac{|\psi(\boldsymbol{\theta}_1, \boldsymbol{\theta}_0)|^2}{1 + (\sigma_s^2/\sigma_n^2)E} \\
&\propto |\psi(\boldsymbol{\theta}_1, \boldsymbol{\theta}_0)|^2 \\
&= \Bigg| \int \sqrt{\gamma(\boldsymbol{\theta}_1)}\, s(\gamma(\boldsymbol{\theta}_1)(t - \tau(\mathbf{p}_1)) \\
&\quad \cdot \sqrt{\gamma(\boldsymbol{\theta}_0)}\, s^*(\gamma(\boldsymbol{\theta}_0)(t - \tau(\mathbf{p}_0)) \\
&\quad \cdot e^{-j2\pi\tau(\mathbf{p}_1)\,(f_c + f(\boldsymbol{\theta}_1))} e^{j2\pi\tau(\mathbf{p}_0)\,(f_c + f(\boldsymbol{\theta}_0))} \\
&\quad e^{j2\pi(f(\boldsymbol{\theta}_1) - f(\boldsymbol{\theta}_0))t}\, dt \Bigg|^2 \\
&\approx \left| \int s(t - \tau(\mathbf{p}_1)) s^*(t - \tau(\mathbf{p}_0)) e^{j2\pi t(f(\boldsymbol{\theta}_1) - f(\boldsymbol{\theta}_0))}\, dt \right|^2 \qquad (3.27) \\
&= \left| \int s(t) s^*(t - \tau) e^{j2\pi t f_v}\, dt \right|^2 \\
&= \chi(\tau, f_v) \qquad (3.28)
\end{aligned}
$$

The approximation in (3.27) is the standard narrowband assumption; the frequency shift due to target motion does not affect the complex envelope for small time–bandwidth signals. We see that (3.24) does indeed reduce to (3.28) [the same as (3.2)], which is solely a function of relative time shift τ and Doppler shift f_v. Woodward's ambiguity function has been obtained; however, if no terms are dropped, (3.24) actually equals Woodward's up to an additive scalar and scalar multiplicative factor. Equations (3.27)–(3.28) have applied the fact that in this scenario the autocorrelation function (3.25) is equal to a scalar constant E (signal energy) for all θ.

In the remainder of this section we examine Eq. (3.24) further and explore aspects of signal design in determining radar resolution capability. All the expressions that follow use the coherent target model. Without loss of generality, the target covariance is given as $\Sigma = 11^T$. Additionally, we specialize the MIMO ambiguity function to a few scenarios of interest.

3.5.1 MIMO Ambiguity Function Composition

When the cross-correlation function (3.26) is expanded by substituting the definition of received signal $r_j(\theta)$, the result is a triple sum over the indices $(j, i, \hat{i})$:

$$\begin{aligned}\psi(\boldsymbol{\theta}_1, \boldsymbol{\theta}_0) &= 1^T \mathbf{S}^H(\boldsymbol{\theta}_1)\mathbf{S}(\boldsymbol{\theta}_0)1 \\ &= \sum_{j=1}^{N_R}\sum_{\hat{i}=1}^{N_T}\sum_{i=1}^{N_T}\int \sqrt{\gamma_{i,j}(\boldsymbol{\theta}_1)}\sqrt{\gamma_{\hat{i},j}(\boldsymbol{\theta}_0)} \\ &\quad s_i\big(\gamma_{i,j}(\boldsymbol{\theta}_1)(t-\tau_{i,j}(\mathbf{p}_1))\big)s_{\hat{i}}^*\Big(\gamma_{\hat{i},j}(\boldsymbol{\theta}_0)(t-\tau_{\hat{i},j}(\mathbf{p}_0))\Big) \\ &\quad \cdot e^{-j2\pi\tau_{i,j}(\mathbf{p}_1)(f_c+f_{i,j}(\boldsymbol{\theta}_1))}e^{j2\pi\tau_{\hat{i},j}(\mathbf{p}_0)(f_c+f_{\hat{i},j}(\boldsymbol{\theta}_0))} \\ &\quad \cdot e^{j2\pi(f_{i,j}(\boldsymbol{\theta}_1)-f_{\hat{i},j}(\boldsymbol{\theta}_0))t}\,dt \end{aligned} \tag{3.29}$$

One observation that can be made is that the exponential terms are components in transmit and receive array steering vectors. In its present form with no simplifications based on array geometry and target location, there is very little, if anything, that can be done to simplify this expression. One notational simplification that can be made is to collect all the inner product terms into a single matrix. Define the $N_T \times N_T$ matrix $\mathbf{R}(\boldsymbol{\theta}_1, \boldsymbol{\theta}_2, j)$, where the $(i, \hat{i})$ element is equal to

$$\begin{aligned}\mathbf{R}_{i,\hat{i}}(\boldsymbol{\theta}_1, \boldsymbol{\theta}_0, j) = \int &\sqrt{\gamma_{i,j}(\boldsymbol{\theta}_1)}s_i\big(\gamma_{i,j}(\boldsymbol{\theta}_1)(t-\tau_{i,j}(\mathbf{p}_1))\big) \\ &\cdot\sqrt{\gamma_{\hat{i},j}(\boldsymbol{\theta}_0)}\,s_{\hat{i}}^*\Big(\gamma_{\hat{i},j}(\boldsymbol{\theta}_0)(t-\tau_{\hat{i},j}(\mathbf{p}_0))\Big) \\ &\cdot e^{j2\pi(f_{i,j}(\boldsymbol{\theta}_1)-f_{\hat{i},j}(\boldsymbol{\theta}_0))t}\,dt \end{aligned} \tag{3.30}$$

Every element of the matrix has a functional dependence on $\boldsymbol{\theta}_1$, $\boldsymbol{\theta}_0$, and j. Using this notation, we can rewrite the cross-correlation function as

$$\psi(\boldsymbol{\theta}_1, \boldsymbol{\theta}_0) = \sum_{j=1}^{N_R} \sum_{\hat{i}=1}^{N_T} \sum_{i=1}^{N_T} \mathbf{R}_{i,\hat{i}}(\boldsymbol{\theta}_1, \boldsymbol{\theta}_0, j) \cdot e^{-j2\pi\tau_{i,j}(\mathbf{p}_1)(f_c+f_{i,j}(\boldsymbol{\theta}_1))} e^{j2\pi\tau_{\hat{i},j}(\mathbf{p}_0)(f_c+f_{\hat{i},j}(\boldsymbol{\theta}_0))} \tag{3.31}$$

One further simplification would be to write out a combined transmit–receive steering vector. We define the following $N_T \times 1$ vector:

$$\mathbf{a}_{TR}(\boldsymbol{\theta}, j) = \left[e^{j2\pi\tau_{1,j}(\mathbf{p})(f_c+f_{1,j}(\boldsymbol{\theta}))}; e^{j2\pi\tau_{2,j}(\mathbf{p})(f_c+f_{2,j}(\boldsymbol{\theta}))}; \ldots; e^{j2\pi\tau_{N_T,j}(\mathbf{p})(f_c+f_{N_T,j}(\boldsymbol{\theta}))}\right]^T \tag{3.32}$$

Using the notation in Eq. (3.32), we can express the cross-correlation function as a sum over a series of quadratic forms:

$$\psi(\boldsymbol{\theta}_1, \boldsymbol{\theta}_0) = \sum_{j=1}^{N_R} \mathbf{a}_{TR}^H(\boldsymbol{\theta}_1, j)\mathbf{R}(\boldsymbol{\theta}_1, \boldsymbol{\theta}_0, j)\mathbf{a}_{TR}(\boldsymbol{\theta}_0, j) \tag{3.33}$$

Some simple observations can be made about the cross-correlation function when written in this form. The transmit signals play a role only in the covariance function. By changing the signals that are sent, one can control the covariance function over the parameter space $\boldsymbol{\theta} = \boldsymbol{\theta}_1 \times \boldsymbol{\theta}_0$. In the case of transmitting arbitrary wideband signals while observing high-velocity targets using a nonsimple array geometry, all target parameters become coupled in the covariance function. As we show later, under the assumption of narrowband signals there is a space-time separability that falls out of the cross-correlation function. In the following sections a few specific geometric array configurations and signal scenarios are examined. These scenarios were chosen to highlight certain key factors at play in the cross-correlation function and thus the ambiguity function. There will be a hierarchy to the simplifications in that the progression will be from more to less complexity.

3.5.2 Cross-Correlation Function under Model Simplifications

The first simplification to be made concerns the impact of the target velocity. As was shown in a previous section, when high-velocity targets are observed with high time–bandwidth product transmit signals, the resulting compressive effect on the complex envelope in time cannot be ignored. If the target velocity is slow enough that the condition in Eq. (3.3) holds, then the signal compression can be ignored in the covariance function. In this simplification and all to follow, modified versions of various components of the cross-correlation function are denoted by superscript numbers. The covariance function elements under the abovementioned assumption can be written as follows:

$$\mathbf{R}^1_{i,\hat{i}}(\boldsymbol{\theta}_1, \boldsymbol{\theta}_0, j) = \int s_i(t - \tau_{i,j}(\mathbf{p}_1))s^*_{\hat{i}}(t - \tau_{\hat{i},j}(\mathbf{p}_0))e^{j2\pi(f_{i,j}(\boldsymbol{\theta}_1)-f_{\hat{i},j}(\boldsymbol{\theta}_0))t}dt \tag{3.34}$$

The cross-correlation function then becomes

$$\psi^1(\boldsymbol{\theta}_1, \boldsymbol{\theta}_0) = \sum_{j=1}^{N_R} \mathbf{a}^H_{TR}(\boldsymbol{\theta}_1, j)\mathbf{R}^1(\boldsymbol{\theta}_1, \boldsymbol{\theta}_0, j)\mathbf{a}_{TR}(\boldsymbol{\theta}_0, j) \tag{3.35}$$

Thus far no assumptions have been made concerning the location of the target with respect to the array. The model used is one in which components of the target velocity vector can be identified. This identifiability results from the placement of the sensors in relation to the target. The locations are such that different sensors might see different frequency shifts due to a single velocity vector. A common model simplification that can be applied is to assume that all the sensors are close enough to each other so that, for a given target velocity vector, the projection of that vector onto each sensor's LOS is nearly identical. Essentially, this assumption is valid only if the the bistatic angle β for the transmit–receive sensor pair with the largest separation falls within a predefined threshold. Theoretically, any nonzero bistatic angle would allow for the resolution of velocity vector components; however, there would appear to be a practical limit. Under this assumption, target velocity vector components cannot be resolved unambiguously; therefore, the velocity vector parameter should be reduced to a scalar parameter corresponding to the radial velocity along the radar's LOS. The frequency shift terms in the ambiguity function now become independent of individual sensors, allowing the array steering vectors to be decomposed into a transmit vector and receive vector. We can apply this simplification by first rewriting the terms in the covariance function:

$$\mathbf{R}^2_{i,\hat{i}}(\boldsymbol{\theta}_1, \boldsymbol{\theta}_0, j) = \int s_i\big(t - \tau_{i,j}(\mathbf{p}_1)\big)s^*_{\hat{i}}(t - \tau_{\hat{i},j}(\mathbf{p}_0))e^{j2\pi(f_{v_1}-f_{v_0})t}dt \tag{3.36}$$

The transmit array steering vectors are

$$\mathbf{a}_T(\boldsymbol{\theta}) = \left[e^{j2\pi\tau_1(\mathbf{p})(f_c+f_v)}; e^{j2\pi\tau_2(\mathbf{p})(f_c+f_v)}; \ldots; e^{j2\pi\tau_{N_T}(\mathbf{p})(f_c+f_v)}\right]^T \tag{3.37}$$

The receive array steering vectors are

$$\mathbf{a}_R(\boldsymbol{\theta}) = \left[e^{j2\pi\tau_1(\mathbf{p})(f_c+f_v)}; e^{j2\pi\tau_2(\mathbf{p})(f_c+f_v)}; \ldots; e^{j2\pi\tau_{N_R}(\mathbf{p})(f_c+f_v)}\right]^T \tag{3.38}$$

Now we can write the cross-correlation function as

$$\psi^2(\boldsymbol{\theta}_1, \boldsymbol{\theta}_0) = \sum_{j=1}^{N_R} \mathbf{a}_T^H(\boldsymbol{\theta}_1)\mathbf{R}^2(\boldsymbol{\theta}_1, \boldsymbol{\theta}_0, j)\mathbf{a}_T(\boldsymbol{\theta}_0) \cdot e^{-j2\pi\tau_j(\mathbf{p}_1)(f_c+f_{v_1})}e^{j2\pi\tau_j(\mathbf{p}_0)(f_c+f_{v_2})} \tag{3.39}$$

A third simplification to the model could be that the target appears in the far field of the array. Under this assumption propagating waves appear planar. Furthermore, equations derived for actual time delays due to target position can be approximated. First, the components of the position vector should be changed to (range, azimuth, elevation) from (x, y, z). Now the most sensible coordinate system is an array-centered coordinate system with one of the elements chosen as the phase center and origin. *Range* is the distance to the target as measured from the phase center. *Azimuth* and *elevation* are also referenced to the phase center. For a target in the far field, the time delays can be approximated as

$$\begin{aligned}\tau_{i,j}(\mathbf{p}) &= \tau_i(\mathbf{p}) + \tau_j(\mathbf{p}) \\ &\approx \frac{r}{c} - \frac{\mathbf{u}^T(\theta_{\mathrm{az}}, \theta_{\mathrm{el}})\mathbf{x}_{i,T}}{c} + \frac{r}{c} - \frac{\mathbf{u}^T(\theta_{\mathrm{az}}, \theta_{\mathrm{el}})\mathbf{x}_{j,R}}{c} \\ &= \frac{2r}{c} - \left[\frac{\mathbf{x}_{i,T}}{c} + \frac{\mathbf{x}_{j,R}}{c}\right]\mathbf{u}^T(\theta_{\mathrm{az}}, \theta_{\mathrm{el}})\end{aligned} \tag{3.40}$$

The function $\mathbf{u}^T(\theta_{\mathrm{az}}, \theta_{\mathrm{el}})$ is a unit vector in the direction specified by $(\theta_{\mathrm{az}}, \theta_{\mathrm{el}})$:

$$\mathbf{u}^T(\theta_{\mathrm{az}}, \theta_{\mathrm{el}}) = [\cos(\theta_{\mathrm{az}})\sin(\theta_{\mathrm{el}});\ \ \sin(\theta_{\mathrm{az}})\sin(\theta_{\mathrm{el}});\ \ \cos(\theta_{\mathrm{el}})]^T \tag{3.41}$$

At this point the cross-correlation function will not be rewritten since notationally the expression does not become much simpler. However, the cross-correlation function, denoted as $\psi^3(\boldsymbol{\theta}_1, \boldsymbol{\theta}_0)$, will refer to the case when the target is in the far field and target velocity is represented by a scalar. The target parameter vector has four components, $\theta = (r, \theta_{\mathrm{az}}, \theta_{\mathrm{el}}, v)$. Some of the examples will refer to this form of the cross-correlation function. Similarly, $\mathbf{R}^3_{i,\hat{i}}(\boldsymbol{\theta}_1, \boldsymbol{\theta}_0, j)$ will refer to elements of the covariance function in this case.

The last model simplification reduces the bandwidth significantly so that the narrowband assumption can be applied to waveforms. Under this assumption, the waveforms are considered sufficiently narrowband that actual intersensor time delays can be ignored in the delay of the complex envelope. This simplifies the covariance function. Under the narrowband assumption, we obtain

$$\begin{aligned}\mathbf{R}^4_{i,\hat{i}}(\boldsymbol{\theta}_1, \boldsymbol{\theta}_0, j) &= \int s_i(t)s_{\hat{i}}^*\left(t - \frac{2}{c}(r_1 - r_0)\right)e^{j2\pi(f_{v_1} - f_{v_0})t}dt \\ &= \int s_i(t)s_{\hat{i}}^*(t - \Delta\tau)e^{j2\pi\Delta f t}dt \qquad (3.42) \\ &\equiv \mathbf{R}(\Delta\tau, \Delta f) \qquad (3.43)\end{aligned}$$

As expected, each element of the covariance function is the simple Woodward ambiguity function depending on $\Delta\tau$ and Δf only. We call $\mathbf{R}(\Delta\tau, \Delta f)$ the *narrowband covariance function*. When this function is substituted into the cross-correlation function, one can see that there is a separability now between space (angle) and time (range):

$$\psi^4(\boldsymbol{\theta}_1, \boldsymbol{\theta}_0) = \sum_{j=1}^{N_R} e^{-j2\pi\tau_j(\mathbf{p}_1)(f_c+f_{v_1})} e^{j2\pi\tau_j(\mathbf{p}_0)(f_c+f_{v0})} \mathbf{a}_T^H(\boldsymbol{\theta}_1)\mathbf{R}^4(\boldsymbol{\theta}_1, \boldsymbol{\theta}_0, j)\mathbf{a}_T(\boldsymbol{\theta}_0)$$

$$= \sum_{j=1}^{N_R} e^{-j2\pi\tau_j(\mathbf{p}_1)(f_c+f_{v_1})} e^{j2\pi\tau_j(\mathbf{p}_2)(f_c+f_{v0})} \mathbf{a}_T^H(\boldsymbol{\theta}_1)\mathbf{R}(\Delta\tau, \Delta f)\mathbf{a}_T(\boldsymbol{\theta}_0) \qquad (3.44)$$

Careful inspection of Eq. (3.44) shows that the cross-correlation function is now independent of the absolute range terms that appear in the complex exponentials. These complex multipliers disappear when the magnitude squared is applied to the sum. Also, the steering vectors are now the same as the standard narrowband steering vectors with a frequency shift applied to the carrier frequency. Equation (3.44) can be written compactly as

$$\psi^4(\boldsymbol{\theta}_1, \boldsymbol{\theta}_0) = \hat{\mathbf{a}}_T^H(\boldsymbol{\theta}_1)\mathbf{R}(\Delta\tau, \Delta f)\hat{\mathbf{a}}_T(\boldsymbol{\theta}_0)\hat{\mathbf{a}}_R^H(\boldsymbol{\theta}_1)\hat{\mathbf{a}}_R(\boldsymbol{\theta}_0) \qquad (3.45)$$

where

$$\hat{\mathbf{a}}_T(\theta) = \left[\exp\left(j2\pi\mathbf{x}_{1,T}\frac{u^T(\theta_{az},\theta_{el})}{c}(f_c+f_v)\right);\ \exp\left(j2\pi\mathbf{x}_{2,T}\frac{u^T(\theta_{az},\theta_{el})}{c}(f_c+f_v)\right);\right.$$
$$\left.\ldots;\ \exp\left(j2\pi\mathbf{x}_{N_T,T}\frac{u^T(\theta_{az},\theta_{el})}{c}(f_c+f_v)\right)\right]^T \qquad (3.46)$$

$$\hat{\mathbf{a}}_R(\theta) = \left[\exp\left(j2\pi\mathbf{x}_{1,R}\frac{u^T(\theta_{az},\theta_{el})}{c}(f_c+f_v)\right);\ \exp\left(j2\pi\mathbf{x}_{2,R}\frac{u^T(\theta_{az},\theta_{el})}{c}(f_c+f_v)\right);\right.$$
$$\left.\ldots;\ \exp\left(j2\pi\mathbf{x}_{N_R,R}\frac{u^T(\theta_{az},\theta_{el})}{c}(f_c+f_v)\right)\right]^T \qquad (3.47)$$

3.5.3 Autocorrelation Function and Transmit Beampatterns

In previous papers [5,7,8] it was shown how the choice of transmit waveform signal correlation could affect the transmit beampattern of a MIMO radar. The major results are as follows. For an array independently transmitting narrowband wide-sense-stationary waveforms, the transmit beampattern function is

$$P_N(\boldsymbol{\theta}) = \mathbf{a}^H(\boldsymbol{\theta})\mathbf{R}\mathbf{a}(\boldsymbol{\theta}) \qquad (3.48)$$

Here $\mathbf{a}^H(\boldsymbol{\theta})$ is the narrowband transmit array steering vector parameterized by the angle θ and $\mathbf{R}$ is the zero-lag signal correlation matrix. Similarly, for an array transmitting wideband signals, the transmit beampattern is

$$P_W(\boldsymbol{\theta}) = \int_B \mathbf{a}^H(\boldsymbol{\theta}, f_c + f)\mathbf{S}(f)\mathbf{a}(\boldsymbol{\theta}, f_c + f)df \tag{3.49}$$

Again, $\mathbf{a}^H(\boldsymbol{\theta}, f_c + f)$ is the narrowband transmit steering vector parameterized by the angle $\boldsymbol{\theta}$, but now specifically calculated at the frequency $f_c + f$. The matrix $\mathbf{S}(f)$ is the cross-spectral density matrix (CSDM) of the transmitted waveforms. It is defined as the elementwise Fourier transform of the signal correlation matrix:

$$\mathbf{S}(f) = \int_0^T \mathbf{R}(\tau)e^{-j2\pi f t}dt \tag{3.50}$$

Returning to the definition of the MIMO ambiguity function, we can identify the role of the transmit beampattern. To begin, consider the second model simplification. In Eq. (3.39) the term $\mathbf{a}_T^H(\boldsymbol{\theta}_1)\mathbf{R}^2(\boldsymbol{\theta}_1, \boldsymbol{\theta}_0, j)\mathbf{a}_T(\boldsymbol{\theta}_0)$ appears. If the two target parameters are equal, $\theta_1 = \theta_0$, this term becomes the transmit beampattern. The dependence of the covariance function on j drops out:

$$\begin{aligned}
\mathbf{R}^2_{i,\hat{i}}(\boldsymbol{\theta}_1, \boldsymbol{\theta}_1, j) &= \int s_i(t - \tau_{i,j}(\mathbf{p}_1))s^*_{\hat{i}}(t - \tau_{\hat{i},j}(\mathbf{p}_1))dt \\
&= \int s_i(t)s^*_{\hat{i}}(t - [\tau_{i,j}(\mathbf{p}_1) - \tau_{\hat{i},j}(\mathbf{p}_1)])dt \\
&= \int s_i(t)s^*_{\hat{i}}(t - [\tau_i(\mathbf{p}_1) - \tau_{\hat{i}}(\mathbf{p}_1)])dt \\
&\equiv \mathbf{R}^2_{i,\hat{i}}(\tau_i(\mathbf{p}_1) - \tau_{\hat{i}}(\mathbf{p}_1)) && (3.51) \\
&= \int_B \mathbf{S}_{i,\hat{i}}(f)e^{j2\pi(\tau_i(\mathbf{p}_1) - \tau_{\hat{i}}(\mathbf{p}_1))f}df && (3.52)
\end{aligned}$$

Taking the result of Eq. (3.52) and substituting back into the quadratic form, one finds the expression for the wideband transmit beampattern:

$$\begin{aligned}
\mathbf{a}_T^H(\boldsymbol{\theta}_1)\mathbf{R}^2(\boldsymbol{\theta}_1, \boldsymbol{\theta}_1, j)\mathbf{a}_T(\boldsymbol{\theta}_1) &= \sum_{i=1}^{N_T}\sum_{\hat{i}=1}^{N_T}\int_B \mathbf{S}_{i,\hat{i}}(f)e^{j2\pi(\tau_i(\mathbf{p}_1) - \tau_{\hat{i}}(\mathbf{p}_1))f} \\
&\quad \cdot e^{j2\pi f_c(\tau_i(\mathbf{p}_1) - \tau_{\hat{i}}(\mathbf{p}_1))}df \\
&= \int_B \mathbf{a}_T^H(\mathbf{p}_1, f_c + f)\mathbf{S}(f)\mathbf{a}_T(\mathbf{p}_1, f_c + f)df. && (3.53)
\end{aligned}$$

Equation (3.53) is the same as Eq. (3.49) except for a slight notation difference. We find the presence of the parameter $\mathbf{p}_1$ instead of $\boldsymbol{\theta}$. The difference is that $\mathbf{p}_1$ is a true spatial parameter, whereas $\boldsymbol{\theta}$ is only an angle parameter. If the far-field simplifications were applied, then the angle parameter $\boldsymbol{\theta}$ would appear. Regardless, this equation is telling us how much gain one can expect for targets located at different points spatially. It is much easier to identify the transmit beampattern in the narrowband simplification of the cross-correlation function as given by Eq. (3.44). The quadratic form $\mathbf{a}_T^H(\boldsymbol{\theta}_1)\mathbf{R}(\Delta\tau, \Delta f)\mathbf{a}_T(\boldsymbol{\theta}_0)$ becomes the narrowband transmit beampattern when $\boldsymbol{\theta}_1 = \boldsymbol{\theta}_0(\Delta\tau = 0, \Delta f = 0)$.

Now, if we return to Eq. (3.39), we see that letting $\boldsymbol{\theta}_1 = \boldsymbol{\theta}_0$ results in the cross-correlation function becoming a scaled version of the transmit beampattern. This is actually just the autocorrelation function defined in Eq. (3.25):

$$
\begin{aligned}
\psi^2(\boldsymbol{\theta}_1, \boldsymbol{\theta}_1) &= \sum_{j=1}^{N_R} \mathbf{a}_T^H(\boldsymbol{\theta}_1)\mathbf{R}^2(\boldsymbol{\theta}_1, \boldsymbol{\theta}_1, j)\mathbf{a}_T(\boldsymbol{\theta}_1)e^{-j2\pi\tau_j(\mathbf{p}_1)(f_c+f_{v_1})}e^{j2\pi i\tau_j(\mathbf{p}_1)(f_c+f_{v_1})} \\
&= \sum_{j=1}^{N_R} \mathbf{a}_T^H(\boldsymbol{\theta}_1)\mathbf{R}^2(\boldsymbol{\theta}_1, \boldsymbol{\theta}_1, j)\mathbf{a}_T(\boldsymbol{\theta}_1) \\
&= N_R \int_B \mathbf{a}_T^H(\mathbf{p}_1, f_c + f)\mathbf{S}(f)\mathbf{a}_T(\mathbf{p}_1, f_c + f)df \\
&= \phi(\boldsymbol{\theta}_1)
\end{aligned} \tag{3.54}
$$

3.6 RESULTS AND EXAMPLES

In this section some visual examples of MIMO ambiguity functions are presented to illustrate new signal design possibilities that arise by using MIMO radar. It is beyond the scope of this chapter to actually undertake the design of MIMO radar waveforms; the results in this section simply provide motivation for the usefulness and necessity of the MIMO ambiguity function. They also show how the MIMO ambiguity functions can be used to evaluate sets of waveforms for use in a particular radar system configuration.

The previous sections showed how the ambiguity function is composed of two constituent functions: an autocorrelation function (equivalent to the transmit beampattern) and a cross-correlation function. In each of the following examples we begin by showing both of these functions separately. They are then be combined to show the MIMO radar ambiguity function. An interesting result that occurs is that two different choices of transmit signals may result in similar cross-correlation functions but different autocorrelation functions and thus will have greatly different ambiguity functions.

3.6.1 Orthogonal Signals

In this first example we examine the MIMO ambiguity function for a system that transmits nearly orthogonal signals from each transmit aperture. The system consists of three transmit/receive sensors located on the x axis at $[-33\lambda \;\; 0 \;\; 20\lambda]$. The operating frequency is set to $f_c = 2\,\text{GHz}$, so the spacing works out to be $[-5\,\text{m} \quad 0\,\text{m} \quad 3\,\text{m}]$. Each sensor transmits and receives with an omnidirectional pattern. A single point target is placed in the far field. The target is moving with constant velocity. Each waveform has a bandwidth $B = 500\,\text{MHz}$ and time duration $T = 20\,\mu\text{s}$; the time-bandwidth product is therefore $TB = 10{,}000$. The target is in the far field and assumed to be moving slow enough that waveform time-compression can be ignored; hence form 3 of the cross-correlation function will be used. Under all these assumptions, the ambiguity function will be a function of 3 unambiguous arguments, range, azimuth, and Doppler frequency, $\boldsymbol{\theta}_0 = \{r_0, \boldsymbol{\theta}_{\text{az},0}, f_{v_0}\}$, for a fixed set of target parameters $\boldsymbol{\theta}_1 = \{r_1, \boldsymbol{\theta}_{\text{az},1}, f_{v_1}\}$.

Figure 3.1 shows a single slice of the cross-correlation function. It is a range–angle slice taken at zero Doppler. The units used in this figure and others are dBi's. This means that the peak of the cross-correlation function is referenced to the level expected by a sensor isotropically transmitting a unit energy waveform. There are many ways to compute and visualize these functions. In this example the target has been placed in space with parameters $\theta_0 = (1000\,\text{km}, 0^\circ, 0\,\text{m/s})$, and the parameter $\boldsymbol{\theta}_1 = (r_1, \boldsymbol{\theta}_{\text{az},1}, f_{v_1})$ has been varied over the parameter space. Before proceeding further, some important comments must be made concerning the cross-correlation ambiguity function shown in Fig. 3.1. A visual inspection of this figure shows the presence of six distinct lines. Actually, there are nine lines; in three of the lines there are two overlapping lines. The reason for seeing nine lines is as follows. Three orthogonal waveforms reflect off the target and are received by the three receivers, at each receiver a filterbank is implemented in which each filter responds to only one transmitted waveform due to orthogonality, and at each receiver we see three responses: three receivers with three responses each, resulting in nine lines. These lines are ridges of ambiguity. In this figure they are range–angle

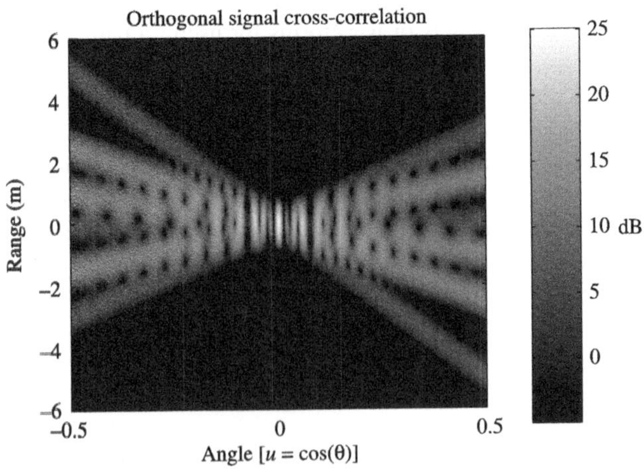

Figure 3.1 Range–angle cross-correlation — orthogonal signals — $\boldsymbol{\theta}_0 = (0\,\text{m}, 0^\circ, 0\,\text{m/s})$.

ambiguity ridges for each transmit waveform/receive sensor filter pair. The number of ridges varies depending on the type of waveform transmitted, the specific array geometry, and the target location.

Another interesting characteristic of Fig. 3.1 is the interference patterns produced when the ambiguity ridges overlap. The cause of these patterns is the unique phase ramp possessed by each ambiguity ridge. As the ridges overlap, large constructive and destructive nulls are produced. The width of each ridge is inversely proportional to the waveform bandwidth B. In a radar with closely spaced sensors the bandwidth is not large enough for the ridges to bend away from one another. Instead, the ridges would lie on top of each other, resulting in a single ridge.

In addition to the range–angle slice shown in Fig. 3.1, we can make other 2D slices or show the cross-correlation function in 3D. Figure 3.2 is a visualization of the cross-correlation function in 3D, where different 2D slices have been placed in their 3D location. One of the important properties of transmitting orthogonal signals from a MIMO radar is the resolution performance enhancement. This performance enhancement is a consequence of the ability to obtain uniform target parameter resolution simultaneously at multiple points in space and time. To illustrate this concept, the position of the point target in Fig. 3.1 is changed while the transmitted waveforms and array geometry are held fixed. Figure 3.3 shows a range–angle slice of the cross-correlation function for this scenario. As is evident from the figure, the cross-correlation function retains its previous shape, but is now shifted to the true target position. This figure illustrates the fact that with careful selection of the transmitted waveforms, good resolution can be obtained over large regions of space and time. Figure 3.4 shows the autocorrelation function for the array described above transmitting orthogonal signals and demonstrates that equal power is applied across all azimuth angles and ranges on transmit. The actual ambiguity

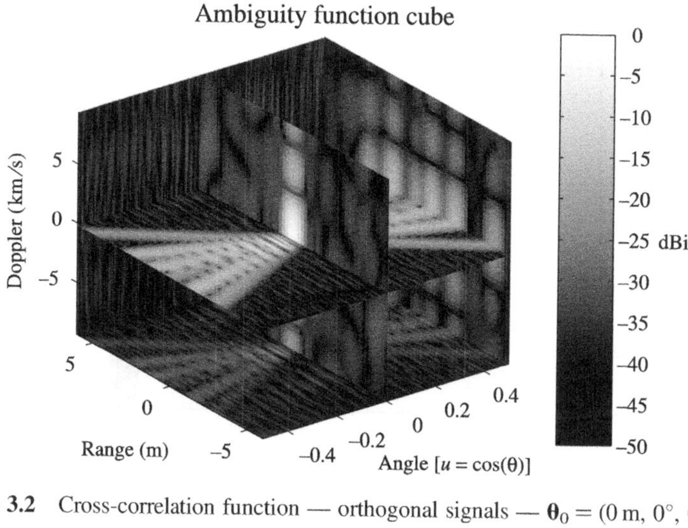

Figure 3.2 Cross-correlation function — orthogonal signals — $\boldsymbol{\theta}_0 = (0\,\text{m}, 0°, 0\,\text{m/s})$.

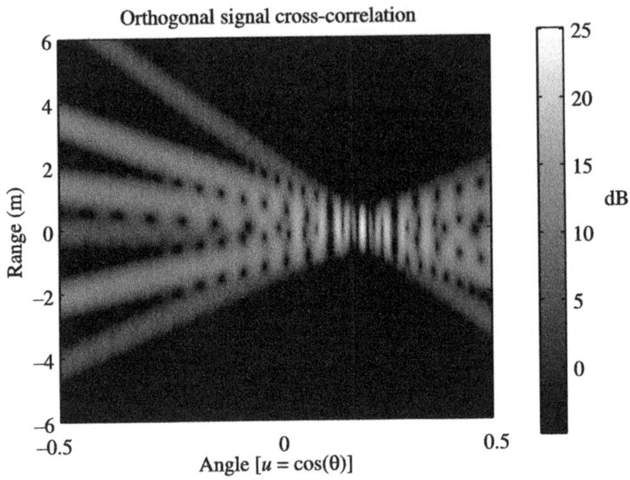

Figure 3.3 Range–angle cross-correlation — orthogonal signals — $\boldsymbol{\theta}_0 = (0\,\text{m}, 11.5°, 0\,\text{m/s})$.

function is found by computing the function given by the expression (3.24). Because the autocorrelation function is relatively invariant across the entire parameter space, the ambiguity function is simply a scaled version of the cross-correlation function. There is a slight dependence on the chosen values of σ_s^2 and σ_n^2, which were selected as $(\sigma_s^2, \sigma_n^2) = (10, 1)$. Figure 3.5 shows the ambiguity function using orthogonal signals. The -3 dB contour levels have been circled. This is done for comparison with a similar figure using coherent signals.

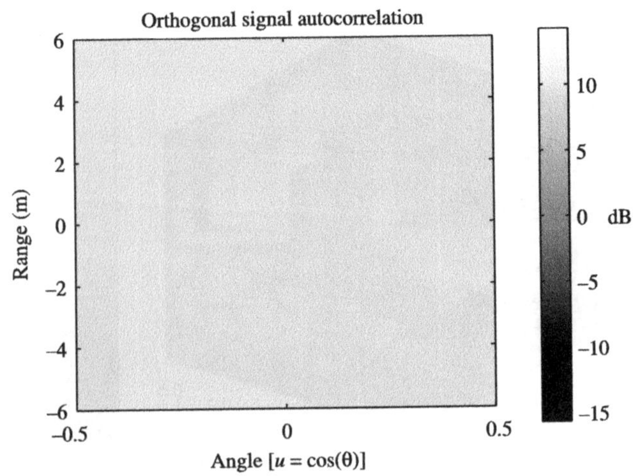

Figure 3.4 Autocorrelation — orthogonal signals.

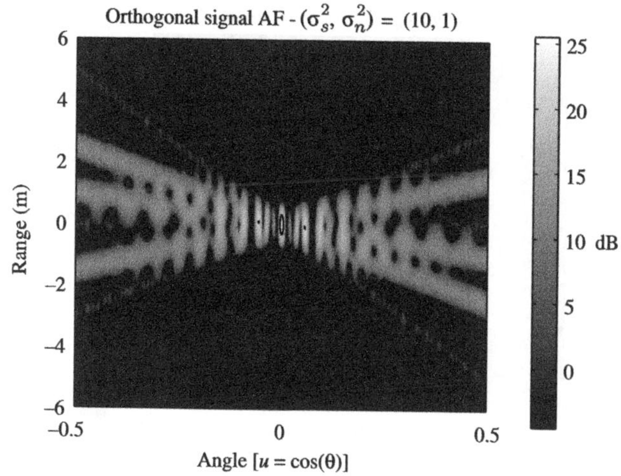

Figure 3.5 Range–angle ambiguity function — orthogonal signals — $\boldsymbol{\theta}_0 = (0\,\text{m}, 0^\circ, 0\,\text{m/s})$.

3.6.2 Coherent Signals

In the results shown above, orthogonal waveforms were transmitted from each sensor. Now the other waveform extreme is examined, transmitting coherent waveforms. These waveforms are coherent in the sense that each sensor transmits the same waveform up to a time delay or phase shift. The waveforms in this scenario focus at 0° azimuth on transmit, so no time delays or phase shifts are applied.

Figure 3.6 shows a range–angle slice of the cross-correlation function for the case were a target is located at $\boldsymbol{\theta}_0 = (1000\,\text{km}, 0^\circ, 0\,\text{m/s})$. Again, the parameter $\boldsymbol{\theta}_1$ is

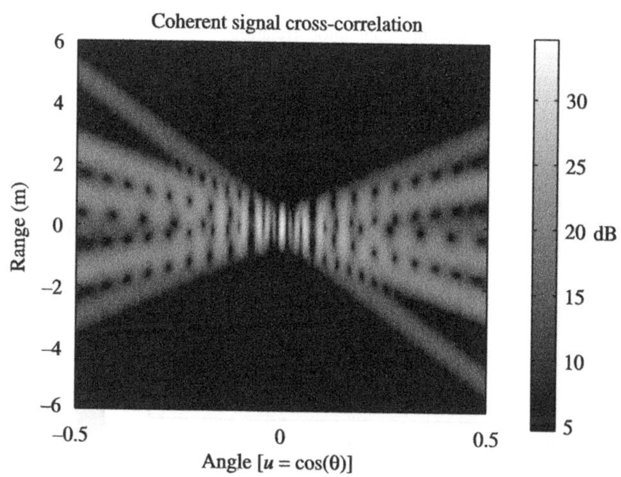

Figure 3.6 Range–angle cross-correlation function — coherent signals — $\boldsymbol{\theta}_0 = (0\,\text{m}, 0^\circ, 0\,\text{m/s})$.

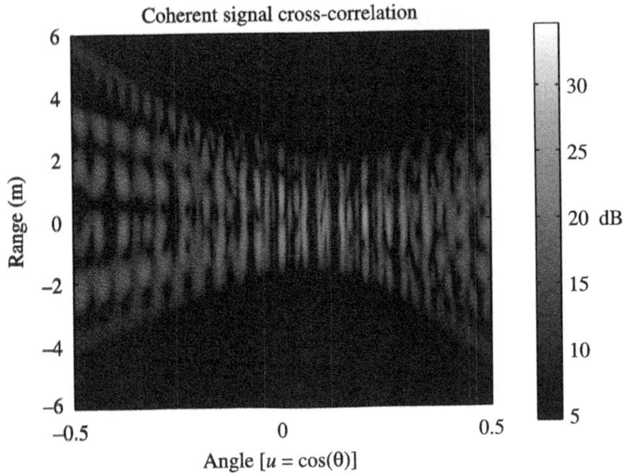

Figure 3.7 Range–angle cross-correlation — coherent signals — $\boldsymbol{\theta}_0 = (0\,\text{m}, 11.5^\circ, 0\,\text{m/s})$.

varied to produce the ambiguity function. This figure shows that when the target is in the same location as the focus point of the coherent waveforms, the same cross-correlation function is achieved as in the orthogonal waveforms case above.

Careful inspection of the figure shows that there is a 9 dB increase in the peak of the ambiguity function. This is the result of the coherent combination of the transmitted waveforms at the target. Another way to describe this result is that now there are really 27 ambiguity ridges in the MIMO ambiguity function. For every one ridge in Fig. 3.1, there are three in Fig. 3.6. As before, the target location can be varied and a new cross-correlation function can be produced. The target is

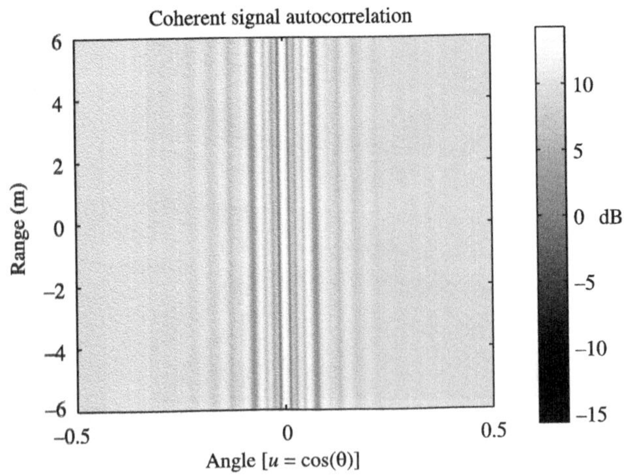

Figure 3.8 Autocorrelation function — coherent signals.

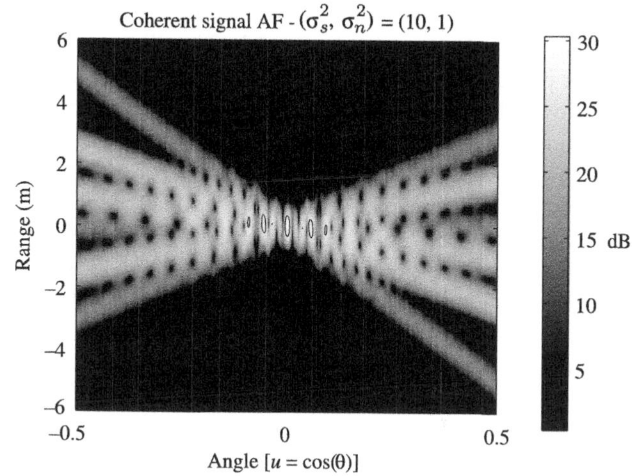

Figure 3.9 Range–angle ambiguity function — coherent signals — $\boldsymbol{\theta}_0 = (0\,\text{m}, 0^\circ, 0\,\text{m/s})$.

moved to the second position as used in the orthogonal signal case. Again, the array geometry and waveforms are left unchanged. Figure 3.7 shows a range–angle slice of the cross-correlation function. Now there are many ambiguity ridges with considerable constructive and destructive interference. This serves as an example of what coherent waveforms cannot accomplish. Specifically, when perfectly coherent waveforms are transmitted, the ability to *focus* uniformly on receive for different θ_1 values is lost. This is in contrast to the orthogonal waveform scenario where the cross-correlation function appears to simply be shifted. Similar to the orthogonal signal scenario, the autocorrelation function can be calculated. The result is the plot in Fig. 3.8. Clearly this illustrates that with the use of coherent waveforms, there is a nonuniform illumination of space with respect to angle. When the cross-correlation and autocorrelation functions are combined to form the MIMO ambiguity function, the result is as shown in Fig. 3.9. As in the case with orthogonal signals, the -3 dB contour levels have been circled. In this ambiguity function the -3 dB contours are wider and more numerous as compared to the case using orthogonal signals.

3.7 CONCLUSION

This chapter has presented a comprehensive study of ambiguity and resolution in modern MIMO radar systems and has shown how the ideas of radar ambiguity functions developed since the mid-1950s can be extended to the newly proposed class of MIMO radar systems. These systems are characterized by independent but coherent sensors possibly distributed over large baselines, transmitting waveforms with large fractional bandwidths. As a result of additional degrees of freedom, an ambiguity function is defined, which, with the proposed parametric model, is a function of 12 parameters. We have shown how this function could be simplified under various scenarios, and even reduced to Woodward's ambiguity function for simple single-sensor systems. The key result presented here is that there is a spacetime covariance function produced by the transmitted waveforms that governs the resolution performance over the parameter space. Connections were made between the MIMO ambiguity function and past work done in the area of transmit beampattern synthesis.

The results presented in this chapter show how the ideas of ambiguity functions can be applied to the new class of radars that use MIMO technology. Just as Woodward's ambiguity function is used as a tool for individual waveform design, the MIMO ambiguity function should be used for designing good MIMO waveforms. There are several areas of MIMO radar design that need further research. To date, the authors know of little, if any, work that has been done on actually producing radar waveforms capable of achieving the middle ground between orthogonality and perfect coherence. Clearly, for the results presented in this chapter to be of any practical use, this issue must be resolved. Another line of research that would complement this study would be to examine a more global parameter estimation bound such as the Barakin bound or Weiss–Weinstein bound for these MIMO systems. In many of the visual plots shown, large ambiguities were present that might cause errors when detecting targets at low SNR. These global bounds should capture the SNR threshold effects of parameter estimation problems.

REFERENCES

1. R. K. Mehra, Optimal input signals for parameter estimation in dynamic systems — survey and new results, *IEEE Trans. Automat. Control* **AC-19**(6):753–768 (Dec. 1974).
2. A. B. Gershman and N. D. Sidiropoulos (eds.), *Space-Time Processing for MIMO Communications*, Wiley, 2005.
3. F. C. Robey, Enhancing radar array performance through space-time coding, *Pre-Print* (2005).
4. E. Fishler, A. Haimovich, R. Blum, L. J. Cimini, D. Chizhik, and R. A. Valenuela, Spatial diversity in radars — models and detection performance, *IEEE Trans. Signal Process.* **3**(54):823–838 (March 2006).
5. D. R. Fuhrmann and G. San Antonio, Transmit beamforming for MIMO radar systems using partial signal correlation, *Proc. 38th Asilomar Conf. Signals, Systems, and Computers*, Monterey, CA, Nov. 2004, Vol. 1, pp. 295–299.
6. K. Forsythe and D. Bliss, Waveform correlation and optimization issues for MIMO radar, *Proc. 39th Asilomar Conf. Signals, Systems, and Computers*, Monterey, CA, 2005, pp. 1306–1310.
7. G. San Antonio and D. R. Fuhrmann, Beampattern synthesis for wideband MIMO radar systems, *Proc. 1st IEEE Int. Workshop CAMSAP*, Puerto Vallarta, Mexico, Dec. 2005, pp. 105–108.
8. D. R. Fuhrmann and G. San Antonio, Transmit beamforming for MIMO radar systems using partial signal correlation, *IEEE Trans. Aerospace Electron. Syst.* **43**(4) (Oct. 2007).

9. P. Stoica, J. Li, and Y. Xie, On probing signal design for MIMO radar, *IEEE Trans. Signal Process.* **55**(8):4151–4161 (Aug. 2007).

10. L. Xu, J. Li, and P. Stoica, Radar imaging via adaptive MIMO techniques, *Proc. EUSIPCO*, Florence, Italy, 2006.

11. D. Bliss and K. Forsythe, Multiple-input multiple-output (MIMO) radar and imaging: Degrees of freedom and resolution, *Proc. 37th Asilomar Conf. Signals, Systems, and Computers*, Monterey, CA, 2003, pp. 54–59.

12. A. S. Fletcher and F. C. Robey, Performance bounds for adaptive coherence of sparse array radar, *Proc. Adaptive Senor Array Proccesing Workshop*, MIT/Lincoln Laboratory, Lexington, MA, 2003.

13. K. Forsythe, D. Bliss, and G. Fawcett, Multiple-input multiple-output (MIMO) radar: Performance issues, *Proc. 38th Asilomar Conf. Signals, Systems, and Computers*, Monterey, CA, 2004, pp. 310–315.

14. J. Ville, Theorie et application de la notion de signal analytique, *Cables Transmission* **2**(1):61–74 (1948).

15. P. Woodward, *Probability and Information Theory, with Applications to Radar*. Pergamon Press, 1957.

16. P. Woodward, *Radar Ambiguity Analysis*, Technical Report RRE, Technical Note 731, Feb. 1967.

17. H. L. Van Trees, *Detection, Estimation, and Modulation Theory: Radar Sonar Processing and Gaussian Signals in Noise*, Part III, Wiley, 1971.

18. H. L. Van Trees, *Detection, Estimation, and Modulation Theory: Detection, Estimation, and Linear Modulation Theory*, Part I, Wiley-Interscience, 1968.

19. E. Kelly and R. Wishner, Matched-filter theory for high-velocity accelerating targets, *IEEE Trans. Military Electron.* **MIL-9**:56–69 (Jan. 1965).

20. A. Rihaczek, Radar resolution of moving targets, *IEEE Trans. Inform. Theory* **13**(1):51–56 (Jan 1967).

21. A. W. Rihaczek, Delay-Doppler ambiguity functions for wideband signals, *IEEE Trans. Aerospace Electron. Syst.* **AES-3**(4):705–711 (July 1967).

22. J. M. Speiser, Wide-band ambiguity functions, *IEEE Trans. Inform. Theory* **13**(1):122–123 (Jan. 1967).

23. R. L. Gassner and G. R. Cooper, Note on a generalized ambiguity function, *IEEE Trans. Inform. Theory* **13**:126 (Jan 1967).

24. R. J. Purdy and G. R. Cooper, A note on the volume of generalized ambiguity functions, *IEEE Trans. Inform. Theory* **14**(1):153–154 (Jan. 1968).

25. D. A. Swick, Wideband ambiguity function of pseudo-random sequences: An open problem, *IEEE Trans. Inform. Theory* 602–603 (July 1968).

26. D. C. Lush and D. A. Hudson, Ambiguity function analysis of wideband radars, *Proc. IEEE Nat. Radar Conf.*, 1991, pp. 16–20.

27. L. G. Weiss, Wavelets and wideband correlation processing, *IEEE Signal Process. Mag.* 13–32 (Jan. 1994).

28. A. Chaiyasena, L. Sibul, and A. Banyaga, Wavelet transforms, wideband ambiguity functions and group theory, *Proc. 26th ASILOMAR Conf. Signals, Systems, and Computers*, Monterey, CA, 1992, pp. 140–144.

29. Y. Fenlian and L. Maoyong, Wavelet transform with applications on radar signal synthesis, *CIE Int. Conf. Radar*, 1991, pp. 293–296.

30. L. Auslander and R. Tolimieri, Radar ambiguity functions and group theory, *SIAM Math. Anal.* **16**(3):577–601 (May 1985).

31. R. K. Young, *Wavelet Theory and Its Applications*, Kluwer Academic, 1993.

32. H. Urkowitz, C. Hauer, and J. Koval, Generalized resolution in radar systems, *Proc. IRE* **50**:2093–2105 (Oct. 1962).

33. S. Pasupathy and A. N. Venetsanopoulos, Optimum active array processing structure and space-time factorability, *IEEE Trans. Aerospace Electron. Syst.* **AES-10**(6):770–778 (Nov. 1974).

34. H. L. Van Trees, *Detection, Estimation, and Modulation Theory: Optimum Array Processing*, Part IV, Wiley Interscience, 2002.

35. T. Tsao, M. Slamani, P. Varshney, D. Weiner, and H. Schwarzlander, Ambiguity function for a bistatic radar, *IEEE Trans. Aerospace Electron. Syst.* **33**(3):1041–1050 (July 1997).

36. M. J. D. Rendas and J. M. F. Moura, Ambiguity in radar and sonar, *IEEE Trans. Signal Process.* **46**(2):294–305 (Feb. 1998).

37. A. Dogandzic and A. Nehorai, CramEr–Rao bounds for estimating range, velocity, and direction with an active array, *IEEE Trans. Signal Process.* **49**(6):1122–1137 (June 2001).

4

PERFORMANCE BOUNDS AND TECHNIQUES FOR TARGET LOCALIZATION USING MIMO RADARS

Joseph Tabrikian

Department of Electrical and Computer Engineering, Ben-Gurion University of the Negev, Beer-Sheva, Israel

4.1 INTRODUCTION

Traditional active radar systems generate spatially concentrated beams in order to achieve high gain. The target echo received by the array, can be processed using array signal processing techniques. Since the mid-1980s, array signal processing has been intensively investigated (see, e.g., Ref. 1). However, this configuration does not allow processing of the transmit signal, as it is performed on the received signal. For example, the transmit beam cannot be digitally steered.

Array processing in both transmit and receive modes is possible when the transmitted signals are spatially coded, for example, spatially orthogonal signals [2,3]. This configuration, termed *multiple-input multiple-output* (MIMO) *radars*, provides much more flexibility in many aspects, and has several advantages over traditional active radars, which do not allow array processing of the transmit signal.

MIMO configurations are commonly used in communication systems (see, e.g., Ref. 4). However, applying the idea to radar systems was raised later, probably because it contradicts the main traditional task of radar antenna arrays to provide

MIMO Radar Signal Processing, edited by Jian Li and Petre Stoica

directional beams with high gain. Nevertheless, array processing with known orthogonal signals was investigated by many researcher. Passive localization of orthogonal signals with known waveforms has been investigated [5]. It was also shown [6] that the conventional configuration of one transmitter and two receivers and an alternative configuration of two transmitters and one receiver, are equivalent in terms of Cramér–Rao bound (CRB) on bearing estimation. This configuration requires radiating two orthogonal signals from two transmitters. The potential advantage of this configuration over the conventional one is in applications where the receiving elements are to be placed on a platform of limited size. The results due to Messer et al. [6] were extended by Nielsen [7], who investigated three possible combinations of four transmitters/receivers: (1) one transmitter and three receivers, (2) two transmitters and two receivers, and (3) three transmitters and one receiver. It was found that these configurations have identical performance in terms of angle estimation accuracy, where the transmitting signals are orthogonal. Spatiotemporal coding for an antenna array was introduced by Calvary and Janer [8], who showed that a single receiver is sufficient for digital beamforming.

Orthogonal signal transmission for active radars has been proposed in [2,3]. A monostatic configuration with several transmitters and several receivers has also been proposed [3]. The properties of this configuration for target detection and localization were further analyzed [9]. The main advantages of this new configuration were found to be

- Digital beamforming of the transmitted beams in addition to the received beams, therefore avoiding beamshape loss in cases where the target is not in the center of the beam.
- Improving the angular resolution by combining the information in the transmit and the receive modes.
- Extension of the array aperture by virtual sensors, resulting in narrower beams and higher angular resolution.
- Increasing the upper limit on the number of targets that can be detected and localized by the array. This is attributed to the virtual sensors.
- Virtual spatial tapering of the extended array aperture, resulting in lower sidelobes.
- Decreasing the spatial power density of the transmit signal.

A different approach, which employs orthogonal signal transmission, is the statistical MIMO radar [10–12]. This approach considers multistatic radars in which the spacing between the radar arrays is very large compared to the signal wavelength and the transmitter and the receiver of the radar are separated such that they experience an angular spread. More recently, MIMO radars have been further investigated in various aspects [13–26].

In this chapter, the problem of target localization for MIMO radars with orthogonal and coherent signal transmission is addressed. First, some properties of orthogonal signal transmission are presented and discussed. Then, the maximum-likelihood (ML) estimator for target parameters (direction, range, and radial velocity) estimation is derived. Next, a method for decorrelating the received signals is presented. This method, *transmission diversity smoothing* (TDS), allows the use of eigenstructure-based techniques for target direction estimation. Using the CRB and the Barankin bound, the properties of the space-time coding configuration for target localization are studied. In particular, the advantages of the proposed configuration are analytically demonstrated and compared to the conventional coherent signal transmission case. This chapter is based on and extends previous works [3,9,13,14].

The chapter is organized as follows. Section 4.2 presents the problem and states the data model. Basic properties of MIMO radars for target detection and localization are presented in Section 4.3. The ML estimator and the TDS algorithm are presented in Section 4.4. The CRB and the Barankin bound for target localization are derived in Section 4.5. The derived bounds and algorithms are tested with a few examples and simulations, which are presented in Section 4.6. Finally, the main results of this chapter are discussed and concluded in Section 4.7.

4.2 PROBLEM FORMULATION

Consider a monostatic radar consisting of two colocated arrays of M_T transmitters and M_R receivers. The ith element of the transmit array transmits a bandpass signal denoted by $s_i(t)e^{j\omega_c t}$, where $s_i(t)$ is the transmitted baseband complex envelope and ω_c is the carrier angular frequency. The echo signal from a single point target at the mth receiver is given by

$$\tilde{y}_m(t) = \sum_{i=1}^{M_T} \tilde{\alpha}_{mi} s_i(t - \tilde{\tau}_{Rm}(t) - \tilde{\tau}_{Ti}(t)) e^{j\omega_c(t - \tilde{\tau}_{Rm}(t) - \tilde{\tau}_{Ti}(t))}, \quad m = 1, \ldots, M_R \tag{4.1}$$

where $\tilde{\tau}_{Ti}(t)$ is the propagation delay between the ith element in the transmit array and the target and $\tilde{\tau}_{Rm}(t)$ is the propagation delay between the target and the mth element in the receive array. The term $\tilde{\alpha}_{mi}$ is the corresponding path loss, which includes the two-way propagation loss and the reflection coefficient, which is in general a complex number. In this model we assume that the baseband signal waveform is not modified because of the Doppler effect. This assumption is justified if $T_o W v_r / c \ll 1$ where c denotes the propagation speed in the medium, v_r is the radial velocity between the target and the radar, T_o is the observation time, and W is the baseband signal bandwidth. Assuming that the target is located in the far-field at direction θ and considering constant radial velocity v_r, $\tilde{\tau}_{Rm}(t)$ and $\tilde{\tau}_{Ti}(t)$ can be written as

$$\tilde{\tau}_{Rm}(t) = \frac{r_0 + v_r t}{c} + \tau_{Rm}(\theta), \qquad m = 1, \ldots, M_R \tag{4.2}$$

$$\tilde{\tau}_{Ti}(t) = \frac{r_0 + v_r t}{c} + \tau_{Ti}(\theta), \qquad i = 1, \ldots, M_T \tag{4.3}$$

where $\tau_{Ti}(\theta)$ and $\tau_{Rm}(\theta)$ are the propagation delay of the ith transmit element and the mth receive elements, respectively, from the reference point along the direction θ. Generally, θ may be a vector including the azimuth and elevation angles of the target. In two-dimensional scenarios, in which the elevation angle cannot be resolved, θ is a scalar, denoting the azimuth angle. For simplicity of the derivations, we consider in this chapter a two-dimensional scenario, and thus θ is assumed to be a scalar. Under narrowband assumptions, that is, $W((v_r t/c) + \tau_{Ti}(\theta)) \ll 1$, the delays in $s_i(t - \tilde{\tau}_{Rm}(t) - \tilde{\tau}_{Ti}(t))$ are negligible w.r.t. (with respect to) $1/W$. After down-conversion of the received bandpass signal, the data can be expressed as

$$y_m(t) = \alpha e^{-j\omega_D t} \sum_{i=1}^{M_T} s_i(t - \tau_r) e^{-j\omega_c(\tau_{Rm}(\theta)+\tau_{Ti}(\theta))}, \quad m = 1, \ldots, M_R \tag{4.4}$$

in which $\tau_r = 2r_0/c$ and $\omega_D = 2\omega_c v_r/c$ are the two-way propagation delay between the target and the reference point and the Doppler frequency shift, respectively. The coefficient α is defined as $\alpha = \tilde{\alpha}_{mi} e^{-j\omega_c \tau_r}$, where the path losses $\tilde{\alpha}_{mi}$ are assumed to be identical for all transmit and receive elements, due to the far-field assumption. In vector notation, (4.4) can be written in the form

$$\mathbf{y}(t) = \alpha e^{-j\omega_D t} \mathbf{a}_R(\theta) \mathbf{a}_T^T(\theta) \mathbf{s}(t - \tau_r) \tag{4.5}$$

in which

$$\mathbf{a}_R(\theta) = [e^{-j\omega_c \tau_{R1}(\theta)}, \ldots, e^{-j\omega_c \tau_{RM_R}(\theta)}]^T \tag{4.6}$$

$$\mathbf{a}_T(\theta) = [e^{-j\omega_c \tau_{T1}(\theta)}, \ldots, e^{-j\omega_c \tau_{TM_T}(\theta)}]^T \tag{4.7}$$

$$\mathbf{s}(t) = [s_1(t), \ldots, s_{M_T}(t)]^T \tag{4.8}$$

Let $\mathbf{A}(\theta)$ denote the transmit–receive steering matrix

$$\mathbf{A}(\theta) = \mathbf{a}_R(\theta) \mathbf{a}_T^T(\theta) \tag{4.9}$$

Thus the signal model in (4.5) can be rewritten as

$$\mathbf{y}(t) = \alpha e^{-j\omega_D t} \mathbf{A}(\theta) \mathbf{s}(t - \tau_r) \tag{4.10}$$

In general, the transmit signals from the different elements may be correlated with correlation matrix

$$\mathbf{R_s} = \int_{T_o} \mathbf{s}(t)\mathbf{s}^H(t)dt = \begin{bmatrix} 1 & \beta_{12} & \cdots & \beta_{1M_T} \\ \beta_{21} & 1 & \cdots & \beta_{2M_T} \\ \vdots & \vdots & \ddots & \vdots \\ \beta_{M_T 1} & \beta_{M_T 2} & \cdots & 1 \end{bmatrix} \tag{4.11}$$

where β_{ij} is the complex correlation coefficient between the ith and jth signals, and the superscript $(\ )^H$ denotes the Hermitian operation. In (4.11) the transmit signal intensity is normalized, implying that the coefficient α includes both the path loss and the transmit signal intensity. In the case of phased-array radars, the transmit signals obey $\mathbf{s}(t) = \mathbf{a}_T^*(\theta_0)s(t)$, where the scalar $s(t)$ is the transmit waveform and the phases of $\mathbf{a}_T^*(\theta_0)$ steer the beam to the desired direction θ_0. In this case, it can be verified that $\mathbf{R}_s$ is given by $\mathbf{R}_s = \mathbf{a}_T^*(\theta_0)\mathbf{a}_T^T(\theta_0)$ with a unit rank. In the case of orthogonal transmit signals $\{\beta_{ij}\}_{i\neq j} = 0$, and thus the correlation matrix is an identity matrix: $\mathbf{R}_s = \mathbf{I}_{M_T}$, resulting in omnidirectional transmission.

If the transmit signal vector through the different sensors is a sequence of K modulated baseband pulses, $\mathbf{p}_0(t), \ldots, \mathbf{p}_{K-1}(t)$, transmitted in a constant pulse repetition interval (PRI) of T_{PRI}, then

$$\mathbf{s}(t) = \sum_{k=0}^{K-1} \mathbf{p}_k(t - kT_{\text{PRI}}) \tag{4.12}$$

Denoting the width of the kth modulated pulse by $T^{(k)}$, and assuming that $\omega_D T^{(k)} \ll 1, \forall k = 0, \ldots, K - 1$, then

$$e^{-j\omega_D t}\mathbf{p}_k(t - kT_{\text{PRI}} - \tau_r) \approx e^{-j\omega_D \tau_r} e^{-jk\omega_D T_{\text{PRI}}}\mathbf{p}_k(t - kT_{\text{PRI}} - \tau_r) \tag{4.13}$$

and (4.10) can be written in the form

$$\mathbf{y}(t) = \alpha e^{-j\omega_D \tau_r} \sum_{k=0}^{K-1} e^{-jk\omega_D T_{\text{PRI}}}\mathbf{A}(\theta)\mathbf{p}_k(t - kT_{\text{PRI}} - \tau_r) \tag{4.14}$$

In high-pulse-repetition-frequency (PRF) scenarios, in which $\tau_r > T_{\text{PRI}}$, the range ambiguity problem arises. An efficient solution for this problem is transmission of orthogonal waveforms over time [31,32], specifically, $\int_{T'} \mathbf{p}_k(t)\mathbf{p}_l^H(t)dt = \mathbf{R}_s\delta_{kl}$ in which $T^{(k)} = T' = \text{const}$, $k,l = 0, \ldots, K-1$, and δ_{kl} denotes the Krönecker delta.

In the presence of L_T point targets, L_I interference sources and additive measurement noise, the model in (4.10) can be extended to

$$\mathbf{y}(t) = \sum_{l=1}^{L_T} \alpha_l e^{-j\omega_{Dl}t}\mathbf{A}(\theta_l)\mathbf{s}(t - \tau_{rl}) + \sum_{l=1}^{L_I} \mathbf{a}_R(\theta_{Il})s_{Il}(t) + \mathbf{w}(t), \quad t \in [0, T_o] \tag{4.15}$$

where $\{\alpha_l, \theta_l, \tau_{rl}, \omega_{Dl}\}$ are the complex attenuation, direction, propagation delay, and Doppler frequency shift of the lth target, and θ_{Il} and $s_{Il}(t)$ are the direction and the baseband complex envelope ofthe lth interference source. The interference signals are assumed to be unknown deterministic. Let $\boldsymbol{\alpha} \triangleq [\alpha_1, \ldots, \alpha_{L_T}]^T$, and $\boldsymbol{\Phi} \triangleq [\boldsymbol{\phi}_1^T, \ldots, \boldsymbol{\phi}_{L_T}^T]^T$, where $\boldsymbol{\phi}_l \triangleq [\theta_l, \tau_{rl}, \omega_{Dl}]^T$ denotes the target location and Doppler parameters. The unknown parameters in the model stated in (4.15) are the target parameters $\boldsymbol{\xi} = [\boldsymbol{\alpha}^T, \boldsymbol{\Phi}^T]^T$, the interferences directions $\boldsymbol{\Theta}_I = [\theta_{I1}, \ldots, \theta_{IL_I}]^T$, and the interference source signals $\mathbf{s}_I(t) = [s_{I1}(t), \ldots, s_{I1}(t)]^T$. The vector $\mathbf{w}(t)$ denotes the additive noise vector, which is assumed to be temporally and spatially white circular complex Gaussian $\mathbf{R}_{\mathbf{w}}(\tau) \stackrel{\Delta}{=} E(\mathbf{w}(t)\mathbf{w}^H(t-\tau)) = \sigma_w^2 \mathbf{I}_{M_R}\delta(\tau)$, where $\delta(\cdot)$ stands for the Dirac delta function and $\mathbf{I}_{M_R}$ is an identity matrix of size M_R. If this assumption is not satisfied, the model in (4.15) can be temporally and/or spatially prewhitened. This model covers also presence of reflected paths through additional targets with different parameters. We will assume that the number of targets L_T and the number of interference sources L_I are known.

4.3 PROPERTIES

In order to illustrate the advantages of orthogonal signal transmission, in this section we consider a single range–Doppler bin. In practice, a bank of two-dimensional range–Doppler filters is used and each range–Doppler bin is processed separately. In addition, we assume an interference and clutter-free environment. In this case, the model in (4.15), after compensation for the specific range–Doppler parameters, becomes

$$\mathbf{y}(t) = \sum_{l=1}^{L_T} \alpha_l \mathbf{A}(\theta_l)\mathbf{s}(t) + \mathbf{w}(t)\,, \quad t \in [0, T_o] \tag{4.16}$$

and the unknown parameters are given by $\boldsymbol{\Theta} = [\theta_1, \ldots, \theta_{L_T}]^T$ and α.

In Appendix 4A, it is shown that the sufficient statistic for estimating $\boldsymbol{\Theta}$ and $\boldsymbol{\alpha}$ from the data $\{\mathbf{y}(t); t \in [0, T_o]\}$ is given by an $M_R \times M_T$ matrix $\mathbf{E}$

$$\mathbf{E} = \int_{T_o} \mathbf{y}(t)\mathbf{s}^H(t)dt \tag{4.17}$$

or by the vector $\boldsymbol{\eta}$ of size $M_R M_T \times 1$

$$\boldsymbol{\eta} = \text{vec}(\mathbf{E}) = \int_{T_o} \mathbf{s}^*(t) \otimes \mathbf{y}(t)dt \tag{4.18}$$

which is obtained by matching the observed data to the transmitted signals $\{\mathbf{s}(t)\}$ and where $\otimes$ denotes the Krönecker product.

It can be shown that for nonorthogonal signals with temporally and spatially white noise, the elements of the sufficient statistic vector $\boldsymbol{\eta}$ are statistically dependent. For simplicity of the algorithms we are interested in statistically independent sufficient statistic. The signal correlation matrix $\mathbf{R}_\mathbf{s}$ from (4.11) can be decomposed using singular value decomposition (SVD) as $\mathbf{R}_\mathbf{s} = \mathbf{U}\boldsymbol{\Lambda}\mathbf{U}^H$, where $\mathbf{U}$ and $\boldsymbol{\Lambda}$ are the matrices of eigenvectors and eigenvalues of $\mathbf{R}_\mathbf{s}$, respectively. Accordingly, $\mathbf{s}(t)$ is a linear transformation of a vector of orthogonal signals defined as

$$\mathbf{s}(t) \stackrel{\Delta}{=} \mathbf{U}\boldsymbol{\Lambda}^{1/2}\tilde{\mathbf{s}}(t) \tag{4.19}$$

where the signal correlation matrix of $\tilde{\mathbf{s}}(t)$ is $\mathbf{R}_{\tilde{\mathbf{s}}} = \int_{T_o} \tilde{\mathbf{s}}(t)\tilde{\mathbf{s}}^H(t)dt = \mathbf{I}_{M_T}$. Thus (4.16) can be rewritten as

$$\mathbf{y}(t) = \sum_{l=1}^{L_T} \alpha_l \mathbf{A}(\theta_l)\mathbf{U}\mathbf{\Lambda}^{1/2}\tilde{\mathbf{s}}(t) + \mathbf{w}(t), \quad t \in [0, T_o] \tag{4.20}$$

Similarly to the derivation above, the independent sufficient statistic matrix and vector can be obtained as

$$\tilde{\mathbf{E}} = \int_{T_o} \mathbf{y}(t)\tilde{\mathbf{s}}^H(t)dt \tag{4.21}$$

$$\tilde{\boldsymbol{\eta}} = \text{vec}(\tilde{\mathbf{E}}) = \int_{T_o} \tilde{\mathbf{s}}^*(t) \otimes \mathbf{y}(t)dt \tag{4.22}$$

By insertion of (4.20) into (4.21), and using the orthogonality of $\tilde{\mathbf{s}}(t)$, one obtains

$$\tilde{\mathbf{E}} = \sum_{l=1}^{L} \alpha_l \mathbf{A}(\theta_l)\mathbf{U}\mathbf{\Lambda}^{1/2} + \int_{T_o} \mathbf{w}(t)\tilde{\mathbf{s}}^H(t)dt \tag{4.23}$$

Finally, the model for the sufficient statistic $\tilde{\boldsymbol{\eta}}$ can be written in the form

$$\tilde{\boldsymbol{\eta}} = \sum_{l=1}^{L} \alpha_l \mathbf{d}(\theta_l) + \mathbf{v} \tag{4.24}$$

$$\mathbf{d}(\theta_l) \triangleq \text{vec}(\mathbf{A}(\theta_l)\mathbf{U}\mathbf{\Lambda}^{1/2}) \tag{4.25}$$

$$\mathbf{v} \triangleq \text{vec}\left(\int_{T_o} \mathbf{w}(t)\tilde{\mathbf{s}}^H(t)dt\right) \tag{4.26}$$

where $\mathbf{d}(\theta_l)$ is the equivalent transmit–receive steering vector of size $M_R M_T \times 1$ at the direction θ_l. The noise vector $\mathbf{v}$ is zero-mean, circular complex Gaussian: $\mathbf{v} \sim N^c(\mathbf{0}_{M_R M_T}, \sigma_w^2 \mathbf{I}_{M_R M_T})$, where $\mathbf{0}_M$ represents a column vector of zero elements of size M. Equation (4.24) states a model that is equivalent to (4.16), which enables one to better understand the properties of MIMO radars.

Note that the data vector in (4.25) is of size $M_R M_T$, which is expected to allow higher flexibility in the array signal processing methods. The main properties of MIMO radar with orthogonal signals are presented in the following sections.

4.3.1 Virtual Aperture Extension

In this section, it is shown that the model for the extended data vector of size $M_R M_T$ is equivalent to additional virtual sensors. The equivalent array response $\mathbf{d}(\theta)$ can be calculated for coherent and orthogonal signals. In the coherent signal case, the matrix $\mathbf{R}_s = \mathbf{u}\mathbf{u}^H$ has a single nonzero eigenvalue. Therefore, $\mathbf{U}\mathbf{\Lambda}^{1/2} = [\mathbf{u}\; 0_{M_T} \; \cdots \; 0_{M_T}]$, and the equivalent steering vector becomes

$$\mathbf{d}_{\text{coherent}}(\theta) = [\mathbf{A}(\theta)\mathbf{u};\ \mathbf{0}_{M_R(M_T-1)}] \tag{4.27}$$

where the sign ";" indicates columnwise concatenation. By substitution of (4.9) into (4.27), one obtains

$$\mathbf{d}_{\text{coherent}}(\theta) = [\mathbf{a}_R(\theta)\left(\mathbf{a}_T^T(\theta)\mathbf{u}\right);\ \mathbf{0}_{M_R(M_T-1)}] \tag{4.28}$$

Equation (4.28) shows that the equivalent transmit–receive steering vector is given by the receive steering vector $\mathbf{a}_R(\theta)$, multiplied by the gain achieved in the transmit mode, $\left(\mathbf{a}_T^T(\theta)\mathbf{u}\right) = \sum_{m=1}^{M_T} u_m a_{Tm}(\theta)$, where $\mathbf{u} = [u_1, \ldots, u_{M_T}]^T$ is the weighting vector in the transmit mode and $\mathbf{a}_T(\theta) = [a_{T1}(\theta), \ldots, a_{TM_T}(\theta)]^T$. In phased array radars, when the transmit beam is directed to θ_0, with no tapering, the correlation matrix $\mathbf{R}_s$ is given by $\mathbf{R}_s = \mathbf{a}_T^*(\theta_0)\mathbf{a}_T^T(\theta_0)$, and therefore $\mathbf{u} = (\mathbf{a}_T^*\theta_0)/\|\mathbf{a}_T(\theta_0)\|$. In this case, the gain in the transmit mode is

$$\left|\mathbf{a}_T^T(\theta)\mathbf{u}\right|^2 = \frac{\left|\mathbf{a}_T^H(\theta)\mathbf{a}_T(\theta_0)\right|^2}{\|\mathbf{a}_T(\theta_0)\|^2}$$

If the target is located in the center of the transmit beam, then the transmit gain achieves its maximum value. Otherwise, the gain is decreased as a result of beam-shape loss. Note that the effective number of elements in this case is M_R.

In the case of orthogonal signals, $\mathbf{R}_s = \mathbf{I}_{M_T}$ and $\mathbf{U}\boldsymbol{\Lambda}^{1/2} = \mathbf{I}_{M_T}$. Therefore

$$\mathbf{d}_{\text{orthogonal}}(\theta) = \text{vec}(\mathbf{A}(\theta)\mathbf{I}_{M_T}) = \mathbf{a}_T(\theta) \otimes \mathbf{a}_R(\theta) \tag{4.29}$$

The equivalent steering vector for orthogonal signals is Krönecker product of the steering vector in the receive mode $\mathbf{a}_R(\theta)$ and the steering vector in the transmit mode $\mathbf{a}_T^T(\theta)$. The equivalent steering vector for noncoherent signals includes all the $M_R M_T$ elements of $\mathbf{A}(\theta)$, which represent all the possible transmit–receive combinations. Let $\mathbf{x}_{R1}, \ldots, \mathbf{x}_{RM_R}$ and $\mathbf{x}_{T1}, \ldots, \mathbf{x}_{TM_T}$ denote the coordinates of the elements in the receive and transmit arrays, respectively. Under the far-field assumption, the expressions in (4.6) and (4.7) can be rewritten as

$$\mathbf{a}_R(\theta) = \left[e^{-j\mathbf{k}^T(\theta)\mathbf{x}_{R1}}, \ldots, e^{-j\mathbf{k}^T(\theta)\mathbf{x}_{RM_R}}\right]^T \tag{4.30}$$

$$\mathbf{a}_T(\theta) = \left[e^{-j\mathbf{k}^T(\theta)\mathbf{x}_{T1}}, \ldots, e^{-j\mathbf{k}^T(\theta)\mathbf{x}_{TM_T}}\right]^T \tag{4.31}$$

where the vector $\mathbf{k}(\theta)$ is the wavenumber vector for direction θ, which in a 2D scenario is given by $\mathbf{k}(\theta) = [\sin\theta\ \cos\theta]^T$. The mith element of $\mathbf{A}(\theta)$ is $A_{mi}(\theta) = \exp\left(-j\mathbf{k}^T(\theta)(\mathbf{x}_{Rm} + \mathbf{x}_{Ti})\right)$. Hence, the transmit–receive array response in (4.29) consists of virtual sensors located at the combinations of $\mathbf{x}_{Rm} + \mathbf{x}_{Ti}$ for

$m = 1, \ldots, M_R$, $i = 1, \ldots, M_T$. Consequently, the array aperture is virtually extended. This virtual aperture extension results in narrower beams and therefore higher angular resolution and better detection performance. Moreover, depending on the transmit and receive arrays, some of the virtual sensor locations may be identical, which can be interpreted as spatial tapering and results in lower sidelobes.

In order to illustrate these advantages, we examine an example with a $M_T = 2$ transmit element array, and $M_R = 3$ receive element array located on a two-dimensional plane. The array configurations are shown in Fig. 4.1. The receive and transmit elements are marked by circles and squares, respectively. The equivalent array structure for coherent signals is identical to the actual receive array structure, which consists of $M_R = 3$ elements. However, the equivalent array for orthogonal signals includes all the transmit–receive combinations of the elements of $\mathbf{A}(\theta)$. This is equivalent to an extended virtual array of $M_R M_T$ receive elements, located at $\mathbf{x}_{Rm} + \mathbf{x}_{Ti}$, $m = 1, \ldots, M_R$, $i = 1, \ldots, M_T$. The $M_R M_T = 6$ virtual sensors are marked by $\times$ points in Fig. 4.1. The three lower virtual elements in Fig. 4.1 correspond to the received echo signal transmitted from the lower transmit element, while the three higher virtual elements correspond to the received echo signal transmitted from the upper transmit element. By using orthogonal signals, the receiver can distinguish between the transmitted signals using matched filter with the different signals. In many cases, the locations of part of the virtual sensors are identical. For example, consider a case in which the actual locations of the receive and transmit elements are the same ($M_R = M_T$). This configuration is popular in many radar systems in which each antenna element is a transceiver. In such cases, the additional elements form a natural spatial tapering, which results in lower sidelobes [3,9]. The number of different sensor locations in this case is limited by $M_R(M_R + 1)/2$. In the case of a linear, equally spaced (LES) array of transceivers, the number of different virtual sensor locations is limited by $2M_R - 1$.

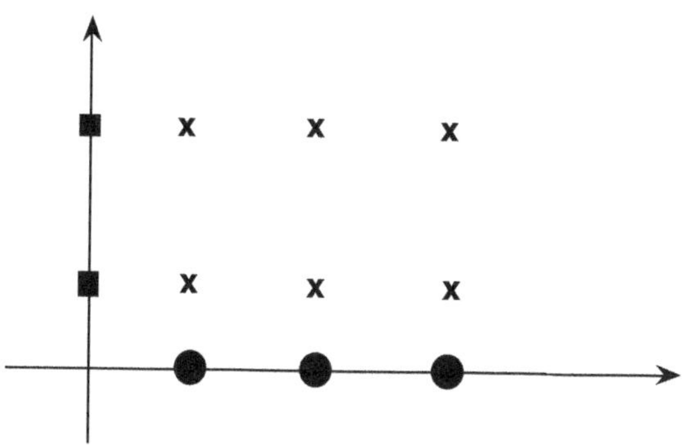

Figure 4.1 Array aperture for orthogonal signals, $M_R = 3$, $M_T = 2$. The receive and transmit elements are marked by circles and squares, respectively. The $M_R M_T = 6$ virtual sensors are denoted by $\times$.

The increased number of elements in the virtual array compared to the receive array allows more flexibility in target localization, adaptive beamforming, and null steering.

4.3.2 Spatial Coverage and Probability of Exposure

Phased-array radars transmit coherent signals in detection mode in order to form a directional beam. In the search mode, in order to scan a given region of interest (ROI), several overlapped directional beams are usually transmitted. Each directional beam is generated using coherent transmitted signals. The overlap is necessary in order to limit the maximum beamshape loss. The time on target (TOT) for each beam is equal to the total interval assigned for covering the ROI, divided by the number of beams required to cover the given ROI. In orthogonal transmit signals with omnidirectional elements, the array beam pattern is also omnidirectional. Therefore, the expected gain of phased arrays is not achieved when using orthogonal signal transmission. On the other hand, transmission of omnidirectional beams extends the spatial coverage of each beam, and therefore, it allows increase in the TOT interval for each beam. In fact, the TOT interval for orthogonal signals is equal to the interval that is assigned to scan the entire ROI. Therefore, loss of the beam gain can be compensated by increased TOT. Instead of M_T directional beams, one omnidirectional beam can be transmitted with M_T times higher TOT interval. Typically, the number of beams is close to the number of transmit elements M_T, depending on the allowed overlap between the adjacent beams, and the spatial tapering used for the beamforming. Hence, in the sequel, when comparing orthogonal and coherent signals, this TOT compensation will be considered.

The minimum TOT is determined by the required signal-to-noise ratio (SNR). In moving-target scenarios with orthogonal signal transmission, the TOT is limited in order to ensure stationarity during the observation time. In these cases, track-before-detect (TBD) techniques can be used.

A major advantage of orthogonal over coherent signal transmission is lower spatial power density, which is important if the radar is required to achieve low probability of exposure. This advantage is a direct result of lower gain in the transmit beam. Using coherent signal transmission, the transmit power is spatially concentrated and the radar signal can be detected.

In omnidirectional signal transmission, the echo signal should be processed for a larger ROI, and therefore the number of targets is statistically larger. If the targets in the given ROI are disjoint in range or Doppler, then for each range–Doppler bin, a single-target case should be considered. In practice, the probability of multiple targets in a given range–Doppler bin is low, although it is still higher than for directional signal transmission, in which only the targets within the narrow beam are illuminated.

Another advantage of orthogonal signal transmission is that it provides a uniform spatial power density with omnidirectional transmit elements. However, coherent signal transmission provides nonuniform spatial power density, which depends on the beamshape and the overlap between the adjacent beams.

4.3.3 Beampattern Improvement

As stated above, the presence of virtual sensors enables more flexibility in beamforming, and allows narrower beams to be obtained with lower sidelobes. In this section, we compare the transmit–receive patterns obtained by coherent and orthogonal signal transmission.

In digital beamforming, the pattern measures the beamformer response to a source at direction θ, when the digital beam is steered to θ_d. Thus, the transmit–receive pattern can be expressed by

$$G_{TR}(\theta, \theta_d) = C\frac{|\mathbf{d}^H(\theta)\mathbf{d}(\theta_d)|^2}{\|\mathbf{d}(\theta_d)\|^2} \tag{4.32}$$

where C is a normalization constant. In Appendix 4B, it is shown that (4.32) can be rewritten in terms of receive and transmit steering vectors as

$$G_{TR}(\theta, \theta_d) = C\frac{|\mathbf{a}_R^H(\theta)\mathbf{a}_R(\theta_d)|^2}{M_R} \cdot \frac{|\mathbf{a}_T^H(\theta)\mathbf{R}_s^T\mathbf{a}_T(\theta_d)|^2}{\mathbf{a}_T^H(\theta_d)\mathbf{R}_s^T\mathbf{a}_T(\theta_d)} \tag{4.33}$$

The first term on the right side of (4.33) is the beampattern in the receive mode and is independent of the transmit signal correlation matrix $\mathbf{R}_s$. The second term on the right side of (4.33) represents the beampattern in the transmit mode. For coherent transmit signals steered to direction θ_0, the signal correlation matrix is $\mathbf{R}_s = \mathbf{a}_T^*(\theta_0)\mathbf{a}_T^T(\theta_0)$, and the transmit–receive pattern is given by

$$G_{TR}(\theta, \theta_d) = C\frac{|\mathbf{a}_R^H(\theta)\mathbf{a}_R(\theta_d)|^2 \cdot |\mathbf{a}_T^H(\theta)\mathbf{a}_T(\theta_0)|^2}{M_R} \tag{4.34}$$

For orthonormal signals ($\mathbf{R}_s = \mathbf{I}_{M_T}$), we obtain

$$G_{TR}(\theta, \theta_d) = C\frac{|\mathbf{a}_R^H(\theta)\mathbf{a}_R(\theta_d)|^2}{M_R} \cdot \frac{|\mathbf{a}_T^H(\theta)\mathbf{a}_T(\theta_d)|^2}{M_T} \tag{4.35}$$

The term $|\mathbf{a}_T^H(\theta)\mathbf{a}_T(\theta_0)|^2$ in (4.34) introduces the attenuation in the transmit array gain due to beamshape loss, and it is constant for all values of θ_d. This attenuation does not exist for the orthogonal signals case, in which the transmit–receive pattern is the product of the transmit and the receive patterns. Moreover, the transmit–receive pattern with orthogonal signal transmission is a product of two terms, which represent the transmit and receive patterns. This result enables us to obtain narrower beams and lower sidelobes in the transmit–receive pattern. Also, in the case of ambiguity in one of the transmit or receive patterns, the other pattern may resolve the ambiguity in the overall transmit–receive pattern.

As a result of the omnidirectional signal transmission obtained by orthogonal transmit waveforms, the transmit gain is equal to one and is lower than that in the coherent signal case. However, for comparison between the coherent and orthogonal signal cases, the increase in TOT by factor M_T, obtained using coherent signal cases, should be considered.

In the following example, the advantage of orthogonal over coherent signal transmission in digital beamforming is illustrated. A scenario with uniformly linear transmit and receive arrays with $M_R = M_T = 6$ elements is considered. The two arrays are located on the same axis. Figure 4.2 presents the transmit–receive patterns for coherent (a) and orthogonal (b) signal transmission. The patterns show the array response (received power) as a function of the target angle θ for the three digitally steered beams to directions $\theta_d = 0°$, $10°$, $20°$. The transmit beam in the coherent signal case is directed to $\theta_0 = 0°$. The effect of beamshape loss is clearly seen in plot (a), when the target direction deviates from the transmit beam angle (broadside). In the case of orthogonal signals, there is no gain loss for any target direction. Another difference between the two cases is the beamwidth and sidelobe levels. In the case of orthogonal signals the beamwidth and sidelobe level are smaller than the case of coherent signals, because, according to (4.35), the transmit–receive pattern consists of the product of the receive and the transmit patterns.

Figure 4.3 compares the normalized transmit-receive patterns using coherent and orthogonal signal transmission with linear, equally spaced arrays of $M_R = 6$ receivers and $M_T = 6$ transmitters as a function of the interelement spacing in the transmit array. Both transmit and receive arrays are located on the same axis, and the interelement spacing in the receive array was fixed to half a wavelength. It can be observed

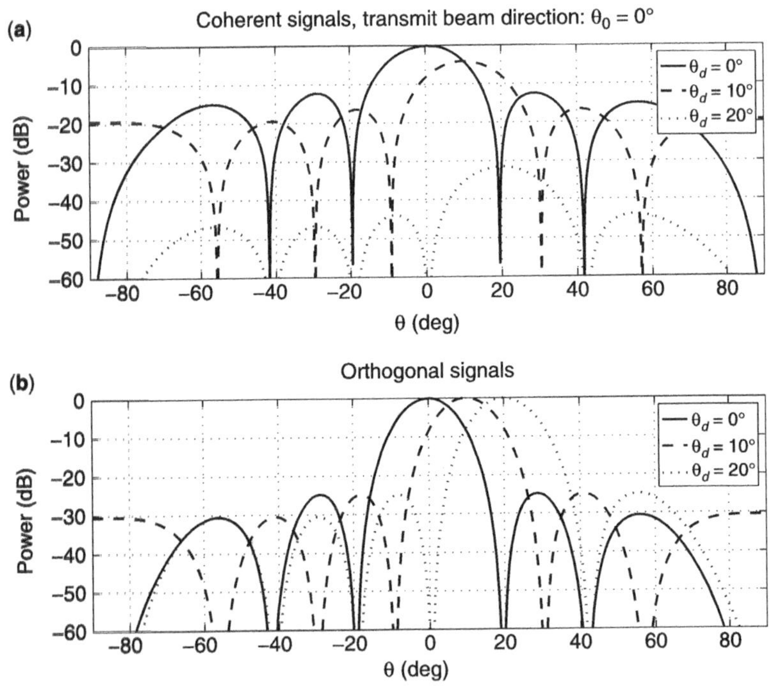

Figure 4.2 Normalized transmit–receive beampatterns using coherent and orthogonal signals for a linear array of $M_R = M_T = 6$ equally spaced transceiver elements with half-wavelength interelement spacing. The beams are digitally steered to 0°, 10°, 20°, and the transmit beam in the coherent signals case is steered to $\theta_0 = 0°$.

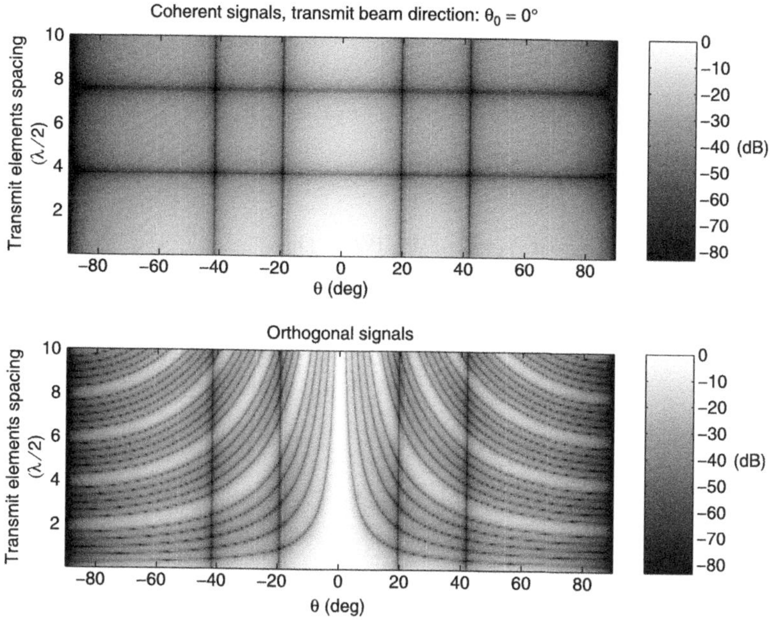

Figure 4.3 Normalized transmit–receive beampatterns using LES transmit and receive arrays with coherent and orthogonal signals as a function of transmit interelement spacing in half-wavelength units. The transmit and receive arrays are located on the same axis, and each one consists of $M_R = M_T = 6$ elements. The interelement spacing in the receive array was fixed to half a wavelength ($\lambda/2$).

that in the case of orthogonal signals, the transmit-receive main beam becomes narrower as the interelements spacing in the transmit array increases while in the case of coherent signals, the main beamwidth does not change. Furthermore, orthogonal signal transmission allows more flexibility in sidelobe level control, because of joint transmit and receive arrays in digital beamforming.

4.4 TARGET LOCALIZATION

In this section, the problem of target localization using MIMO radars is addressed. We first derive the ML estimator for the case of multiple targets in the presence of interference. This procedure is computationally expensive since it involves a multidimensional search procedure in the case of multiple targets. A preprocessing algorithm, termed *transmission diversity smoothing* (TDS), is proposed in order to decorrelate the received signals and enable the use of eigenstructure-based techniques, which are computationally attractive.

4.4.1 Maximum-Likelihood Estimation

The log-likelihood function for estimating the target parameters $\mathbf{\Phi}$ and the interference parameters $\mathbf{\Theta}_I$ for the model presented in (4.15) is derived in Appendix 4A.

It is shown that after maximization w.r.t. the nuisance parameters $\boldsymbol{\alpha}$ and $\{\mathbf{s}_I(t)\}$, the log-likelihood can be given by

$$L'_{\mathbf{y}}(\boldsymbol{\Phi}, \boldsymbol{\Theta}_I) = \text{const} - \frac{1}{\sigma_w^2}\left(E_{\mathbf{yy}} - \mathbf{e}_{\mathbf{Hy}}^H \mathbf{E}_{\mathbf{HH}}^{-1} \mathbf{e}_{\mathbf{Hy}}\right) \tag{4.36}$$

where

$$E_{\mathbf{yy}} = \int_{T_o} \left\| \mathbf{P}_{\mathbf{H}_I}^{\perp}(\boldsymbol{\Theta}_I)\mathbf{y}(t)\right\|^2 dt \tag{4.37}$$

and the lth element of the vector $\mathbf{e}_{\mathbf{Hy}}$ is given by

$$[\mathbf{e}_{\mathbf{Hy}}]_l = \mathbf{a}_R^H(\theta_l)\mathbf{P}_{\mathbf{H}_I}^{\perp}(\boldsymbol{\Theta}_I)\mathbf{E}(\tau_{rl}, \omega_{Dl})\mathbf{a}_T^*(\theta_l) \tag{4.38}$$

in which

$$\mathbf{E}(\tau_{rl}, \omega_{Dl}) = \int_{T_o} \mathbf{y}(t)\mathbf{s}^H(t-\tau_{rl})e^{j\omega_{Dl}t}\,dt \tag{4.39}$$

The lkth element of the matrix $\mathbf{E}_{\mathbf{HH}}$ is

$$[\mathbf{E}_{\mathbf{HH}}]_{lk} = \mathbf{a}_R^H(\theta_l)\mathbf{P}_{\mathbf{H}_I}^{\perp}(\boldsymbol{\Theta}_I)\mathbf{a}_R(\theta_k)\cdot \mathbf{a}_T^T(\theta_k)\mathbf{AF}_{\mathrm{ss}}(\tau_{rk}, \tau_{rl}, \omega_{Dk}, \omega_{Dl})\mathbf{a}_T^*(\theta_l) \tag{4.40}$$

in which

$$\mathbf{AF}_{\mathrm{ss}}(\tau_{rk}, \tau_{rl}, \omega_{Dk}, \omega_{Dl}) \triangleq \int_{T_o} \mathbf{s}(t-\tau_{rk})\mathbf{s}^H(t-\tau_{rl})e^{-j(\omega_{Dk}-\omega_{Dl})t}\,dt, \quad k, l = 1, \ldots, L_T \tag{4.41}$$

is the ambiguity matrix, which is a generalization of the range–Doppler ambiguity function [31] for MIMO radars [15]. For observation times T_o longer than the sum of the signal duration and the tested range τ_{rl}, the matrix $\mathbf{AF}_{\mathrm{ss}}$ becomes

$$\mathbf{AF}_{\mathrm{ss}}(\tau_{rl}, \tau_{rl}, \omega_{Dl}, \omega_{Dl}) = \mathbf{R}_{\mathbf{s}} \tag{4.42}$$

The matrix $\mathbf{P}_{\mathbf{H}_I}^{\perp}(\boldsymbol{\Theta}_I) = \mathbf{I}_{M_R} - \mathbf{H}_I(\boldsymbol{\Theta}_I)\left(\mathbf{H}_I^H(\boldsymbol{\Theta}_I)\mathbf{H}_I(\boldsymbol{\Theta}_I)\right)^{-1}\mathbf{H}_I^H(\boldsymbol{\Theta}_I)$ denotes the projection matrix into the null space spanned by the columns of $\mathbf{H}_I(\boldsymbol{\Theta}_I) = [\mathbf{a}_R(\theta_{I1}), \ldots, \mathbf{a}_R(\theta_{IL_I})]$. As expected, presence of interference in the model interference is nulling the receive term in the log-likelihood function by projection of $\mathbf{a}_R(\cdot)$ and $\mathbf{y}(t)$ into the null space of the steering vectors of the interference sources $\mathbf{H}_I(\boldsymbol{\Theta}_I)$. Note that for the case of targets in a known range–Doppler bin, the matrix $\mathbf{E}(\tau_r, \omega_D)$ is the sufficient statistic for estimating the targets and interference directions.

The ML estimator for the target and interference parameters is given by

$$(\hat{\boldsymbol{\Phi}}, \hat{\boldsymbol{\Theta}}_I)_{ML} = \arg\max_{\boldsymbol{\Phi},\boldsymbol{\Theta}_I} L'_{\mathbf{y}}(\boldsymbol{\Phi}, \boldsymbol{\Theta}_I) \tag{4.43}$$

The ML estimator in the presence of multiple targets and interference sources is impractical because of the large amount of computations required for the multidimensional search procedure.

For the case of a single target, the log-likelihood function for estimating $\boldsymbol{\phi} = [\theta, \tau_r, \omega_D]^T$ is given by

$$L'_{\mathbf{y}}(\boldsymbol{\phi}, \boldsymbol{\Theta}_I) = \text{const} - \frac{1}{\sigma_w^2}\left(E_{\mathbf{yy}} - \frac{|e_{\mathbf{Hy}}|^2}{E_{\mathbf{HH}}}\right) \tag{4.44}$$

where

$$e_{\mathbf{Hy}} = \mathbf{a}_R^H(\theta)\mathbf{P}_{\mathbf{H}_I}^{\perp}(\boldsymbol{\Theta}_I)\mathbf{E}(\tau_r, \omega_D)\mathbf{a}_T^*(\theta) \tag{4.45}$$

$$E_{\mathbf{HH}} = \mathbf{a}_R^H(\theta)\mathbf{P}_{\mathbf{H}_I}^{\perp}(\boldsymbol{\Theta}_I)\mathbf{a}_R(\theta) \cdot \mathbf{a}_T^H(\theta)\mathbf{R}_{\mathbf{s}}^T\mathbf{a}_T(\theta) \tag{4.46}$$

With no interference, the ML estimator can be simplified to

$$(\hat{\theta}, \hat{\tau}_r, \hat{\omega}_D)_{ML} = \arg\max_{\theta,\tau_r,\omega_D} \frac{|\mathbf{a}_R^H(\theta)\mathbf{E}(\tau_r, \omega_D)\mathbf{a}_T^*(\theta)|^2}{M_R\mathbf{a}_T^H(\theta)\mathbf{R}_{\mathbf{s}}^T\mathbf{a}_T(\theta)} \tag{4.47}$$

For orthogonal signal transmission, (4.47) becomes

$$(\hat{\theta}, \hat{\tau}_r, \hat{\omega}_D)_{ML} = \arg\max_{\theta,\tau_r,\omega_D} \frac{|\mathbf{a}_R^H(\theta)\mathbf{E}(\tau_r, \omega_D)\mathbf{a}_T^*(\theta)|^2}{M_R M_T} \tag{4.48}$$

and for coherent signal transmission with $\mathbf{s}(t) = s(t)\mathbf{a}_T^*(\theta_0)$, we obtain

$$(\hat{\theta}, \hat{\tau}_r, \hat{\omega}_D)_{ML} = \arg\max_{\theta,\tau_r,\omega_D} \frac{|\mathbf{a}_R^H(\theta)\mathbf{e}(\tau_r, \omega_D)|^2}{M_R} \tag{4.49}$$

where $\mathbf{e}(\tau_r, \omega_D) = \int_{T_o} \mathbf{y}(t)s^*(t-\tau_r)e^{j\omega_D t}dt$.

By comparing (4.48) and (4.49), we can see that orthogonal signal transmission enables digital steering of the transmit beam in addition to the receive beam, while in coherent signal transmission, digital beam steering is possible only in the receive beam.

This approach, which is equivalent to that for the Bartlett beamformer, can be used also for multiple-target localization by searching for the highest local maxima of the log-likelihood function. However, it is not optimal in any sense for multiple-target localization, and its performance is poor in the presence of multiple close targets. In order to reduce the effect of sidelobes, multidimensional smoothing windows can be used.

4.4.2 Transmission Diversity Smoothing

As mentioned in Section 4.4.1, the ML estimator in the presence of multiple targets involves a multidimensional search procedure, and therefore it is impractical. In this section we present an eigenstructure-based method for direction estimation of multiple targets in a given range–Doppler bin.

In many radar applications, the received echo signals from different targets is considered as coherent. The reason is that although the phases of the echo signals vary between the different pulses or snapshots because of the Doppler effect, optimal target parameters estimation requires joint estimation of the target parameters. The signal phases for a given Doppler frequency is constant, and the amplitude of the echo signal is usually constant during transmission of several pulse signals. Therefore, the echo signal from different targets is considered as coherent. This implies that eigenstructure-based methods, such as MUltiple SIgnal Classification (MUSIC) [33] or Capon's beamformer, cannot be directly implemented for multitarget localization, because these methods fail in scenarios in which the received signals are fully correlated.

In order to "decorrelate" the signals in the data autocorrelation matrix, Evans et al. [34] proposed a preprocessing technique referred to as *spatial smoothing*. Several authors (see e.g., [35–36]) investigated this method, combined with forward–backward averaging. The drawback of this approach is the reduction of the effective array aperture size, resulting in lower resolution and accuracy. An alternative "decorrelation" method is redundancy averaging [37,38]. It has been shown that this preprocessing method induces bias in the direction-of-arrival (DOA) estimates [39].

In this section, we show that the virtual sensors obtained from uncorrelated signal transmission can be grouped into subarrays with identical structures. This enables us to spatially smooth the signal autocorrelation matrix, followed by application of eigenstructure-based methods for multiple coherent target localization. Unlike the well-known spatial smoothing technique, this transmission diversity smoothing (TDS) algorithm

- is not limited to linear arrays,
- does not assume spatially white noise, and
- does not decrease the array aperture, compared to the aperture obtained in coherent signal transmission.

It should be noted that the TDS algorithm decreases the virtually extended array aperture to the case of coherent signal transmission.

Considering a given range–Doppler bin, the sufficient statistic for DOA estimation is given by (4.39). For simplicity of notation, we drop the dependence on τ_r and ω_D. By substitution of (4.9) into the model for the sufficient statistic in (4.23), one obtains

$$\tilde{\mathbf{E}} = \sum_{l=1}^{L_T} \alpha_l \mathbf{a}_R(\theta_l)\mathbf{a}_T^T(\theta_l)\mathbf{U}\boldsymbol{\Lambda}^{1/2} + \mathbf{V} \tag{4.50}$$

where

$$\mathbf{V} \triangleq [\mathbf{v}_1, \ldots, \mathbf{v}_{M_T}] = \int_{T_o} \mathbf{w}(t)\tilde{\mathbf{s}}^H(t)dt \tag{4.51}$$

It can be easily verified that the columns of $\mathbf{V}$ are zero-mean, i.i.d. with covariance matrix $\sigma_w^2 \mathbf{I}_{M_R}$. Let γ_{li} denote the ith element of the row vector $\alpha_l \mathbf{a}_T^T(\theta_l)\mathbf{U}\mathbf{\Lambda}^{1/2}$. Then, the ith column of the matrix $\tilde{\mathbf{E}}$ can be stated by

$$\tilde{\boldsymbol{\eta}}_i = \sum_{l=1}^{L_T} \mathbf{a}_R(\theta_l)\gamma_{li} + \mathbf{v}_i, \quad i = 1, \ldots, M_T. \tag{4.52}$$

Obviously, the vector $\alpha_l \mathbf{a}_T^T(\theta_l)\mathbf{U}\mathbf{\Lambda}^{1/2}$ carries some information on the target direction. However, in order to obtain a model that enables smoothing of the autocorrelation matrix, this information is ignored, and the dependence of γ_{li} on θ_l is omitted.

Equation (4.52) can be rewritten as

$$\tilde{\boldsymbol{\eta}}_i = \mathbf{A}_R(\boldsymbol{\Theta})\boldsymbol{\gamma}_i + \mathbf{v}_i, \quad i = 1, \ldots, M_T \tag{4.53}$$

where $\mathbf{A}_R(\boldsymbol{\Theta}) \triangleq [\mathbf{a}_R(\theta_1), \ldots, \mathbf{a}_R(\theta_{L_T})]$ and $\boldsymbol{\gamma}_i \triangleq [\gamma_{1i}, \ldots, \gamma_{L_T i}]^T$. As stated in Section 4.3, the equivalent model for the sufficient statistic consists of $M_R M_T$ virtual sensors. We divided these sensors into M_T subarrays of size M_R with identical structure. The measurements from the ith subarray is represented by the ith column of $\tilde{\mathbf{E}}$.

Hence, for deterministic vectors γ_i, the autocorrelation matrix of the ith subarray can be obtained as

$$\mathbf{R}_{\tilde{\eta}_i} = \mathbf{A}_R(\boldsymbol{\Theta})\boldsymbol{\gamma}_i\boldsymbol{\gamma}_i^H \mathbf{A}_R^H(\boldsymbol{\Theta}) + \underbrace{E[\mathbf{v}_i\mathbf{v}_i^H]}_{\sigma_w^2 \mathbf{I}_{M_R}}, \quad i = 1, \ldots, M_T \tag{4.54}$$

Obviously, the rank of the signal subspace for a given i is equal to one.

The matrix $\mathbf{R}_{\tilde{\eta}_i}$ represents the autocorrelation matrix of the response to the ith component of the transmitted signal $\tilde{s}_i[n]$. In this method, these matrices are smoothed, and therefore this method is termed *transmission diversity smoothing* (TDS).

By averaging the matrices $\{\mathbf{R}_{\tilde{\eta}_i}\}_{i=1}^{M_T}$, we obtain

$$\begin{aligned}\bar{\mathbf{R}}_{\tilde{\eta}} &= \frac{1}{M_T}\sum_{i=1}^{M_T} \mathbf{R}_{\tilde{\eta}_i} = \mathbf{A}_R(\boldsymbol{\Theta})\left(\frac{1}{M_T}\sum_{i=1}^{M_T} \gamma_i\gamma_i^H\right)\mathbf{A}_R^H(\boldsymbol{\Theta}) + \sigma_w^2 \mathbf{I}_{M_R} \\ &= \frac{1}{M_T}\mathbf{A}_R(\boldsymbol{\Theta})\boldsymbol{\Gamma}\boldsymbol{\Gamma}^H \mathbf{A}_R^H(\boldsymbol{\Theta}) + \sigma_w^2 \mathbf{I}_{M_R}\end{aligned} \tag{4.55}$$

where

$$\boldsymbol{\Gamma} \triangleq [\boldsymbol{\gamma}_1, \ldots, \boldsymbol{\gamma}_{M_T}] = \begin{bmatrix} \alpha_1 \mathbf{a}_T^T(\theta_1)\mathbf{U}\mathbf{\Lambda}^{1/2} \\ \vdots \\ \alpha_{L_T}\mathbf{a}_T^T(\theta_{L_T})\mathbf{U}\mathbf{\Lambda}^{1/2} \end{bmatrix} = \mathrm{diag}(\boldsymbol{\alpha})\mathbf{A}_T^T(\boldsymbol{\Theta})\mathbf{U}\mathbf{\Lambda}^{1/2} \tag{4.56}$$

in which $\mathbf{A}_T(\boldsymbol{\Theta}) \triangleq [\mathbf{a}_T(\theta_1), \ldots, \mathbf{a}_T(\theta_{L_T})]$. Therefore, by substitution of (4.56) into (4.55) and recalling that $\mathbf{R}_s = \mathbf{U}\boldsymbol{\Lambda}\mathbf{U}^H$, the smoothed autocorrelation matrix can be rewritten as

$$\bar{\mathbf{R}}_{\tilde{\eta}} = \frac{1}{M_T}\mathbf{A}_R(\boldsymbol{\Theta})\text{diag}(\boldsymbol{\alpha})\mathbf{A}_T^T(\boldsymbol{\Theta})\mathbf{R}_s\mathbf{A}_T^*(\boldsymbol{\Theta})\text{diag}(\boldsymbol{\alpha}^*)\mathbf{A}_R^H(\boldsymbol{\Theta}) + \sigma_w^2\mathbf{I}_{M_R}. \tag{4.57}$$

For coherent signal transmission, rank$(\mathbf{R}_s) = 1$, and therefore, the rank of the signal autocorrelation matrix is also equal to one. Therefore, eigenstructure-based techniques cannot be used for multitarget localization. When using orthogonal signal transmission, rank$(\mathbf{R}_s) = M_T$, and if rank$(\mathbf{A}_T(\boldsymbol{\Theta})) = L_T$ (which requires $L_T \leq M_T$), and rank(diag$(\boldsymbol{\alpha})) = L_T$ (i.e. $\alpha_l \neq 0 \; \forall l = 1, \ldots, L_T$), then rank(diag$(\boldsymbol{\alpha})\mathbf{A}_T^T(\boldsymbol{\Theta})\mathbf{R}_s$ $\mathbf{A}_T^*(\boldsymbol{\Theta})$diag$(\boldsymbol{\alpha}^*)) = L_T$. Therefore, the rank of the signal autocorrelation matrix $\mathbf{R}_{\tilde{\eta}}$ is equal to L_T if rank$(\mathbf{A}_R(\boldsymbol{\Theta})) = L_T$. To summarize, when using orthogonal signal transmission, the necessary and sufficient conditions for full rank of the signal autocorrelation matrix is rank$(\mathbf{A}_R(\boldsymbol{\Theta}))$ = rank$(\mathbf{A}_T(\boldsymbol{\Theta})) = L_T$.

This algorithm enables us to use eigenstructure-based methods such as MUSIC for multitarget localization. Note that unlike the spatial smoothing, the array aperture is not reduced. This fact results in higher performance in direction estimation. Moreover, the TDS does not constrain the transmit and/or receive array geometries to be symmetric or LES, as required in the forward–backward averaging or the spatial smoothing methods.

4.5 PERFORMANCE LOWER BOUND FOR TARGET LOCALIZATION

In this section, the CRB and the Barankin bound for target localization using MIMO radars are derived. The CRB considers local errors and is used to demonstrate the superior performance of orthogonal signal transmission resulting from (1) narrower beamwidth and (2) absence of beamshape loss as discussed above. The Barankin bound also considers the nonlocal errors and is used to demonstrate the superior performance of orthogonal signal transmission resulting from lower sidelobes as shown in Section 4.3.3. The bounds in the section are derived for the model in (4.15) without interference sources.

4.5.1 Cramér–Rao Bound

In this section, we first derive the CRB for estimating the target parameters $\tilde{\boldsymbol{\xi}} = [\tilde{\boldsymbol{\alpha}}^T, \boldsymbol{\Phi}^T]^T$, where $\tilde{\boldsymbol{\alpha}} = [\text{Re}(\alpha), \text{Im}(\alpha)]^T$, and then focus on the case of known range–Doppler parameters.

The CRB for target direction, range, and Doppler for single-input multiple-output (SIMO)/phased array radars was developed in 2001 [30]. One of the main results of this paper was that there is no coupling in the Fisher information matrix (FIM) between direction and range–Doppler parameters. In the following, we show that in the presence of incoherent transmit signals, the coupling does not necessarily vanish. The coupling can be minimized by setting some additional constraints in the processes of transmit signal design.

In the general case of L_T targets with no interference sources, the elements of the FIM for estimating $\tilde{\boldsymbol{\xi}}$ are as derived in Appendix 4C. The FIM elements for the case of a single target ($L_T = 1$) with parameters (α, θ, τ_r, ω_D) are given by

$$J_{\theta\theta} = \frac{2|\alpha|^2}{\sigma_w^2}\,\mathrm{tr}\left(\dot{\mathbf{A}}(\theta)\mathbf{R}_s\dot{\mathbf{A}}^H(\theta)\right) \tag{4.58}$$

$$J_{\tau_r\tau_r} = \frac{2|\alpha|^2}{\sigma_w^2}\,\mathrm{tr}\left(\mathbf{A}(\theta)\mathbf{AF}_{\dot{s}\dot{s}}(\tau_r, \tau_r, \omega_D, \omega_D)\mathbf{A}^H(\theta)\right) \tag{4.59}$$

$$J_{\omega_D\omega_D} = \frac{2|\alpha|^2}{\sigma_w^2}\,\mathrm{tr}\left(\mathbf{A}(\theta)\mathbf{AF}_{t^2ss}(\tau_r, \tau_r, \omega_D, \omega_D)\mathbf{A}^H(\theta)\right) \tag{4.60}$$

$$J_{\bar{\alpha}\bar{\alpha}} = \frac{2}{\sigma_w^2}\mathrm{Re}\left\{[1, j]^H\mathrm{tr}\left(\mathbf{A}(\theta)\mathbf{R}_s\mathbf{A}^H(\theta)\right)[1, j]\right\} \tag{4.61}$$

$$J_{\theta\tau_r} = -\frac{2|\alpha|^2}{\sigma_w^2}\mathrm{Re}\left\{\mathrm{tr}\left(\mathbf{A}(\theta)\mathbf{AF}_{\dot{s}s}(\tau_r, \tau_r, \omega_D, \omega_D)\dot{\mathbf{A}}^H(\theta)\right)\right\} \tag{4.62}$$

$$J_{\theta\omega_D} = -\frac{2|\alpha|^2}{\sigma_w^2}\mathrm{Re}\left\{j\,\mathrm{tr}\left(\mathbf{A}(\theta)\mathbf{AF}_{tss}(\tau_r, \tau_r, \omega_D, \omega_D)\dot{\mathbf{A}}^H(\theta)\right)\right\} \tag{4.63}$$

$$J_{\theta\bar{\alpha}} = \frac{2}{\sigma_w^2}\mathrm{Re}\left\{\alpha^*\,\mathrm{tr}\left(\mathbf{A}(\theta)\mathbf{R}_s\dot{\mathbf{A}}^H(\theta)\right)[1, j]\right\} \tag{4.64}$$

$$J_{\tau_r\bar{\alpha}} = -\frac{2}{\sigma_w^2}\mathrm{Re}\left\{\alpha^*\,\mathrm{tr}\left(\mathbf{A}(\theta)\mathbf{AF}_{s\dot{s}}(\tau_r, \tau_r, \omega_D, \omega_D)\mathbf{A}^H(\theta)\right)[1, j]\right\} \tag{4.65}$$

$$J_{\tau_r\omega_D} = \frac{2|\alpha|^2}{\sigma_w^2}\mathrm{Re}\left\{j\,\mathrm{tr}\left(\mathbf{A}(\theta)\mathbf{AF}_{ts\dot{s}}(\tau_r, \tau_r, \omega_D, \omega_D)\mathbf{A}^H(\theta)\right)\right\} \tag{4.66}$$

$$J_{\bar{\alpha}\omega_D} = -\frac{2}{\sigma_w^2}\mathrm{Re}\left\{j[1, j]^H\alpha\,\mathrm{tr}\left(\mathbf{A}(\theta)\mathbf{AF}_{tss}(\tau_r, \tau_r, \omega_D, \omega_D)\mathbf{A}^H(\theta)\right)\right\} \tag{4.67}$$

where we have used that $\mathbf{AF}_{ss}(\tau_r, \tau_r, \omega_D, \omega_D) = \mathbf{R}_s$ and

$$\mathbf{AF}_{s\dot{s}}(\tau_r, \tau_r, \omega_D, \omega_D) \stackrel{\Delta}{=} \int_{T_o} \mathbf{s}(t-\tau_r)\dot{\mathbf{s}}^H(t-\tau_r)dt = \mathbf{AF}^H_{\dot{s}s}(\tau_r, \tau_r, \omega_D, \omega_D) \tag{4.68}$$

$$\mathbf{AF}_{tss}(\tau_r, \tau_r, \omega_D, \omega_D) \stackrel{\Delta}{=} \int_{T_o} t\mathbf{s}(t-\tau_r)\mathbf{s}^H(t-\tau_r)dt \tag{4.69}$$

$$\mathbf{AF}_{ts\dot{s}}(\tau_r, \tau_r, \omega_D, \omega_D) \stackrel{\Delta}{=} \int_{T_o} t\mathbf{s}(t-\tau_r)\dot{\mathbf{s}}^H(t-\tau_r)dt \tag{4.70}$$

$$\mathbf{AF}_{\dot{\mathbf{s}}\dot{\mathbf{s}}}(\tau_r, \tau_r, \omega_D, \omega_D) \triangleq \int_{T_o} \dot{\mathbf{s}}(t-\tau_r)\dot{\mathbf{s}}^H(t-\tau_r)dt \tag{4.71}$$

$$\mathbf{AF}_{t^2\mathbf{ss}}(\tau_r, \tau_r, \omega_D, \omega_D) \triangleq \int_{T_o} t^2\mathbf{s}(t-\tau_r)\mathbf{s}^H(t-\tau_r)dt \tag{4.72}$$

It can be verified that the diagonal terms of (4.68) and (4.69) vanish; however, the off-diagonal terms are not necessarily zero, even for orthogonal signals. This implies that for the case of phased-array radars, the terms $\mathbf{AF}_{\mathbf{s}\dot{\mathbf{s}}}$ and $\mathbf{AF}_{t\mathbf{ss}}$ are equal to zero, which leads to the well-known result that there is no coupling between the pairs (α, θ) and (τ, ω). However, when using incoherent signals in MIMO radars, the mixed terms between (α, θ) and (τ, ω) are not necessarily equal to zero, unless the signals were designed to obey orthogonality in terms of (4.68) and (4.69).

Consider the problem of direction estimation of a single target assuming that there is no coupling between (α, θ) and (τ, ω), or for the case of known range–Doppler parameters. This analysis allows us to demonstrate the advantages of orthogonal signal transmission in direction estimation accuracy. In this case, the parameters of interest are given by $\tilde{\boldsymbol{\xi}} = [\,\tilde{\boldsymbol{\alpha}}^T \;\theta]^T = [\text{Re}(\alpha)\;\text{Im}(\alpha)\;\theta]^T$. The FIM [27] for estimating $\tilde{\boldsymbol{\xi}}$ can be partitioned as

$$\mathbf{J}_{\tilde{\xi}} = \begin{bmatrix} \mathbf{J}_{\tilde{\alpha}\tilde{\alpha}} & \mathbf{J}_{\theta\tilde{\alpha}}^T \\ \mathbf{J}_{\theta\tilde{\alpha}} & \mathbf{J}_{\theta\theta} \end{bmatrix} \tag{4.73}$$

and the CRB for target direction estimate can be expressed as

$$\text{CRB}(\theta) = \left[\mathbf{J}_{\theta\theta} - \mathbf{J}_{\theta\tilde{\alpha}}\mathbf{J}_{\tilde{\alpha}\tilde{\alpha}}^{-1}\mathbf{J}_{\theta\tilde{\alpha}}^T\right]^{-1} \tag{4.74}$$

where $\mathbf{J}_{\theta\theta}$, $\mathbf{J}_{\theta\tilde{\alpha}}$, and $\mathbf{J}_{\tilde{\alpha}\tilde{\alpha}}$ are as given in (4.58), (4.64), and (4.61). Thus, the CRB for target direction estimation can be expressed as

$$\begin{aligned} \text{CRB}(\theta) &= [J_{\theta\theta} - \mathbf{J}_{\theta\tilde{\alpha}}\mathbf{J}_{\tilde{\alpha}\tilde{\alpha}}^{-1}\mathbf{J}_{\theta\tilde{\alpha}}^T]^{-1} \\ &= \frac{\text{tr}(\mathbf{A}(\theta)\mathbf{R}_s\mathbf{A}^H(\theta))}{2\text{SNR}\left[\text{tr}(\dot{\mathbf{A}}(\theta)\mathbf{R}_s\dot{\mathbf{A}}^H(\theta))\text{tr}(\mathbf{A}(\theta)\mathbf{R}_s\mathbf{A}^H(\theta)) - |\text{tr}(\mathbf{A}(\theta)\mathbf{R}_s\dot{\mathbf{A}}^H(\theta))|^2\right]} \end{aligned} \tag{4.75}$$

where $\text{SNR} = |\alpha|^2/\sigma_w^2$. By substitution of (4.9) into (4.58), (4.64), and (4.61) the elements of the FIM can be expressed as

$$\begin{aligned} J_{\theta\theta} &= \frac{2|\alpha|^2}{\sigma_w^2}\text{tr}\left(\dot{\mathbf{A}}(\theta)\mathbf{R}_s\dot{\mathbf{A}}^H(\theta)\right) \\ &= \frac{2|\alpha|^2}{\sigma_w^2}\text{tr}\left(\left[\dot{\mathbf{a}}_R(\theta)\mathbf{a}_T^T(\theta) + \mathbf{a}_R(\theta)\,\dot{\mathbf{a}}_T^T(\theta)\right]\mathbf{R}_s\left[\dot{\mathbf{a}}_T^*(\theta)\mathbf{a}_R^H(\theta) + \mathbf{a}_T^*(\theta)\,\dot{\mathbf{a}}_R^H(\theta)\right]\right) \end{aligned}$$

$$
\begin{aligned}
&= \frac{2|\alpha|^2}{\sigma_w^2}\left(\mathbf{a}_T^T(\theta)\mathbf{R}_s\mathbf{a}_T^*(\theta)\,\dot{\mathbf{a}}_R^H(\theta)\,\dot{\mathbf{a}}_R(\theta) + \dot{\mathbf{a}}_T^T(\theta)\mathbf{R}_s\,\dot{\mathbf{a}}_T^*(\theta)\|\mathbf{a}_R(\theta)\|^2\right) \\
&= \frac{2|\alpha|^2}{\sigma_w^2}\left(\mathbf{a}_T^H(\theta)\mathbf{R}_s^T\mathbf{a}_T(\theta)\|\,\dot{\mathbf{a}}_R(\theta)\|^2 + \dot{\mathbf{a}}_T^H(\theta)\mathbf{R}_s^T\,\dot{\mathbf{a}}_T(\theta)M_R\right) \qquad (4.76)
\end{aligned}
$$

$$
\begin{aligned}
\mathbf{J}_{\tilde{\alpha}\tilde{\alpha}} &= \frac{2}{\sigma_w^2}\mathrm{Re}\{[1,j]^H[1,j]\mathrm{tr}\left(\mathbf{A}(\theta)\mathbf{R}_s\mathbf{A}^H(\theta)\right)\} \\
&= \frac{2}{\sigma_w^2}M_R\mathbf{a}_T^H(\theta)\mathbf{R}_s^T\mathbf{a}_T(\theta)\mathbf{I}_2 \qquad (4.77)
\end{aligned}
$$

$$
\begin{aligned}
\mathbf{J}_{\theta\tilde{\alpha}} &= \frac{2}{\sigma_w^2}\mathrm{Re}\left\{\alpha^*\mathrm{tr}\left(\mathbf{A}(\theta)\mathbf{R}_s\dot{\mathbf{A}}^H(\theta)\right)[1,j]\right\} \\
&= \frac{2M}{\sigma_w^2}\mathrm{Re}\{\alpha^*\mathbf{a}_T^H(\theta)\mathbf{R}_s^T\,\dot{\mathbf{a}}_T(\theta)[1,j]\} \qquad (4.78)
\end{aligned}
$$

where $\dot{\mathbf{a}}_R(\theta) = d\mathbf{a}_R(\theta)/d(\theta)$, $\dot{\mathbf{a}}_T(\theta) = d\mathbf{a}_T(\theta)/d(\theta)$, and it is assumed that the equality $\dot{\mathbf{a}}_R^H(\theta)\mathbf{a}_R(\theta) = 0$ is satisfied if the receive array origin is at the array centroid: $\sum_{m=1}^{M}\mathbf{x}_{Rm} = 0$. The matrix $\mathbf{I}_2$ denotes an identity matrix of size 2. Finally, the CRB given in (4.74) for direction estimation of a single target is given by

$$
\mathrm{CRB}(\theta) = \frac{1}{2\mathrm{SNR}}\left(M_R\,\dot{\mathbf{a}}_T^H(\theta)\mathbf{R}_s^T\,\dot{\mathbf{a}}_T(\theta) + \mathbf{a}_T^H(\theta)\mathbf{R}_s^T\mathbf{a}_T(\theta)\|\,\dot{\mathbf{a}}_R(\theta)\|^2 - \frac{M_R\left|\mathbf{a}_T^H(\theta)\mathbf{R}_s^T\,\dot{\mathbf{a}}_T(\theta)\right|^2}{\mathbf{a}_T^H(\theta)\mathbf{R}_s^T\mathbf{a}_T(\theta)}\right)^{-1}. \qquad (4.79)
$$

4.5.2 The Barankin Bound

In Section 4.3.3 it was shown that orthogonal signal transmission may result in lower sidelobes, depending on the transmit and receive array geometries. Lower sidelobe levels reduce the probability of nonlocal errors at low SNRs, and consequently the threshold SNR is reduced. The Barankin bound can be used in order to evaluate this advantage.

The Barankin bound [40,41] is a useful tool for predicting the threshold region in estimation problems. Identification of the threshold region enables to determine the operation conditions, such as observation time and transmission power. Zeira and Schultheiss [42,43] derived and analyzed the Barankin bound for time-delay estimation. In an earlier paper, the author used the Barankin bound for shallow-water source localization in the presence of environmental uncertainty [44]. The Barankin bound for target direction estimation is derived and analyzed for orthogonal and coherent signal transmission in the following text.

The Barankin bound on the mean-square error of any unbiased estimator of $\boldsymbol{\xi}$ from a data vector $\mathbf{y}$ is given by [41]

$$\mathrm{cov}(\hat{\boldsymbol{\xi}}) \geq \mathbf{T}(\mathbf{B} - 11^T)^{-1}\mathbf{T}^T \tag{4.80}$$

where $\hat{\boldsymbol{\xi}}$ is an unbiased estimate of $\boldsymbol{\xi}$. The matrix of test points $\mathbf{T}$ is defined as $\mathbf{T} \triangleq [\boldsymbol{\xi}_1 - \boldsymbol{\xi} \mid \boldsymbol{\xi}_2 - \boldsymbol{\xi} \mid \cdots \mid \boldsymbol{\xi}_K - \boldsymbol{\xi}]$, where $\boldsymbol{\xi}_i$, $i = 1, \ldots, K$ are test-point vectors. To maximize the bound, the test points should be chosen to maximize the right side of (4.80). The elements of the Barankin matrix $\mathbf{B}$ are given by

$$B_{ij}(\boldsymbol{\xi}) \triangleq E\left\{LR(\mathbf{y}, \boldsymbol{\xi}_i, \boldsymbol{\xi})LR(\mathbf{y}, \boldsymbol{\xi}_j, \boldsymbol{\xi})\right\}, \quad i, j = 1, \ldots, K, \tag{4.81}$$

where $LR(\mathbf{y}, \boldsymbol{\xi}_i, \boldsymbol{\xi}) \triangleq (f(\mathbf{y}|\boldsymbol{\xi}_i)/f(\mathbf{y}|\boldsymbol{\xi})$ is the likelihood ratio function.

Consider the following data model

$$\mathbf{y} = \boldsymbol{\mu}(\boldsymbol{\xi}) + \mathbf{n} \tag{4.82}$$

where $\boldsymbol{\mu}(\cdot)$ is a known deterministic function, $\mathbf{n} \sim N^c(\mathbf{0}, \mathbf{R_n})$. It has been shown in [44] that the Barankin matrix $\mathbf{B}$ for this signal model is given by

$$B_{ij}(\boldsymbol{\xi}) = \exp\left[2\mathrm{Re}\left((\boldsymbol{\mu}(\boldsymbol{\xi}) - \boldsymbol{\mu}(\boldsymbol{\xi}_i))^H \mathbf{R}_{\mathbf{n}}^{-1}(\boldsymbol{\mu}(\boldsymbol{\xi}) - \boldsymbol{\mu}(\boldsymbol{\xi}_j))\right)\right]. \tag{4.83}$$

For simplicity, we consider the case of a single target with known range and Doppler parameters. Thus, the model given above corresponds to the sufficient statistic model in (4.24). For coherent and orthogonal signal transmission, $\boldsymbol{\mu}(\cdot)$ is given by $\boldsymbol{\mu}(\boldsymbol{\xi}) = \alpha\mathbf{d}_1(\theta)$ and $\boldsymbol{\mu}(\boldsymbol{\xi}) = \alpha\mathbf{d}_0(\theta)$, respectively. The test points for $\boldsymbol{\xi}$ should be chosen to maximize the bound. Because of the linear dependence of the mean vector on α, the test points for the signal parameters are determined such that the diagonal of the Barankin matrix is minimized. It was shown [44] that the test points for α are given by $\alpha_i = \alpha(\mathbf{d}^H(\theta_i)\mathbf{d}_k(\theta))/||\mathbf{d}(\theta_i)||^2$, where θ_i is the ith test point for θ, and $k = 0,1$ is the model index.

Assuming spatially white noise, $\mathbf{R_n} = \sigma_w^2\mathbf{I}$, and the expression for α_i, the Barankin matrix elements can be written as

$$B_{ij} = \exp\left\{2\frac{|\alpha|^2}{\sigma_w^2}\mathrm{Re}\left[\mathbf{d}_k^H(\theta)(\mathbf{I} - \mathbf{P}_{ik})(\mathbf{I} - \mathbf{P}_{jk})\mathbf{d}_k(\theta)\right]\right\} \tag{4.84}$$

where $\mathbf{P}_{ik}$ is the projection matrix into the subspace spanned by $\mathbf{d}_k(\theta_i)$. The Barankin bound can now be evaluated by substituting (4.84) into (4.80).

4.6 SIMULATION RESULTS

In this section, we demonstrate via simulations the localization performances with uncorrelated signal transmission using MIMO radars. In all the examples, it is assumed that the target range and Doppler parameters are known.

Consider a LES array with a half-wavelength spacing. The array elements serve as transceivers; that is, the transmit and receive arrays are identical. In the coherent signals case, the transmit beam is steered to $\theta = 0°$. In Fig. 4.4, the CRB on target direction estimation root-mean-square-error (RMSE) in the presence of a single target ($L_T = 1$) as a function of its direction θ is plotted for coherent ($\beta = 1$) and orthonormal signals ($\beta = 0$). The array includes $M_R = M_T = 10$ elements and SNR = 20 dB. Note that for orthogonal transmitted signals the CRB is constant with respect to θ, while in the coherent transmitted signals case, the CRB increases with θ due to the transmit beamshape loss. The peak near $\theta = 12°$ in the coherent signals case corresponds to the null in the transmit beam.

Figure 4.5 presents the CRB for target direction estimation when the target is located at $\theta_1 = 0°$ in the presence of an additional target at $\theta_2 = 5°, 10°, 15°$ with an array of $M_R = M_T = 2$ elements. Obviously, the condition $L_T \leq M_R - 1$, which is required to the case of coherent signals with $|\beta| = 1$, is not satisfied. However, Fig. 4.5 shows that the CRB is finite for $|\beta| < 1$. This result is directly related to the contribution of the virtual sensors in the incoherent signals case, as mentioned in Section 4.3. In this case, the number of virtual sensors at the different locations is $M_R\ (M_R + 1)/2 = 3$, and thus localization of two targets is possible with an array of two elements. As expected, the CRB goes to infinity for $|\beta| \rightarrow 1$.

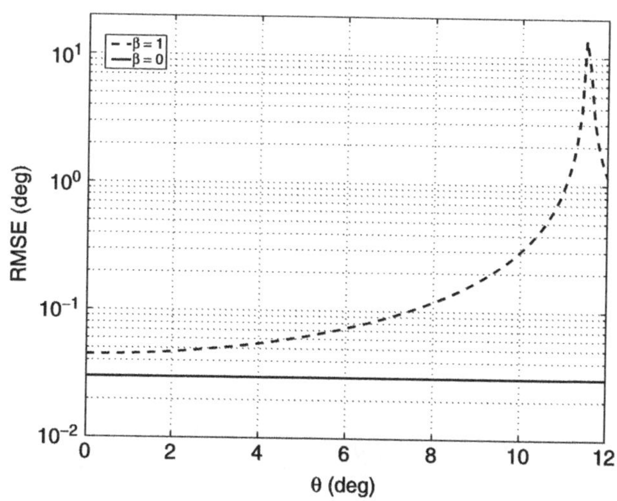

Figure 4.4 CRB on target direction estimation RMSE as a function of the target direction with $M_R = M_T = 10$, $L_T = 1$, SNR = 20 dB, and the transmit beam is steered at $\theta_0 = 0°$ for coherent signals with TOT compensation.

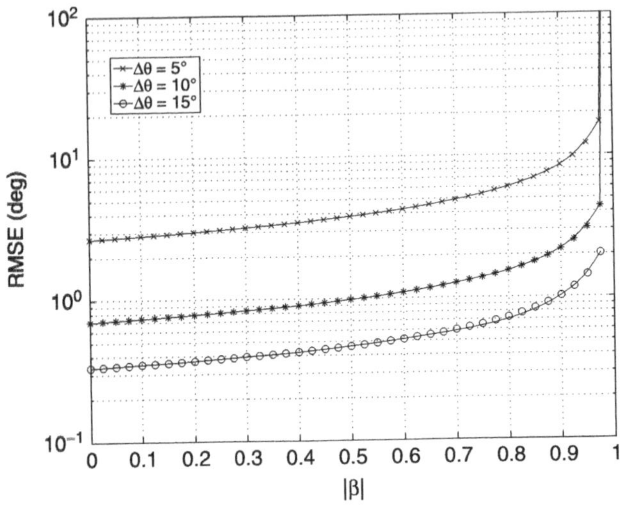

Figure 4.5 CRB on direction estimation of the first target, $L_T = 2$: $\theta_1 = 0°$, $\theta_2 = 5°$, 10°, 15°, $M_R = M_T = 2$, SNR = 0 dB. Two targets can be resolved by two elements, if the transmit signals are incoherent.

Figure 4.6 presents the CRB for a case of an array of two transceivers as a function of phase and modulus of β_{12}. It can be seen that there are two points in the image along the line $|\beta_{12}| = \text{const} \rightarrow 1$ in which the RMSE is high. This is attributed to the nonuniform spatial transmission. Such a problem does not arise in the case of orthogonal signal transmission with $\beta \rightarrow 0$.

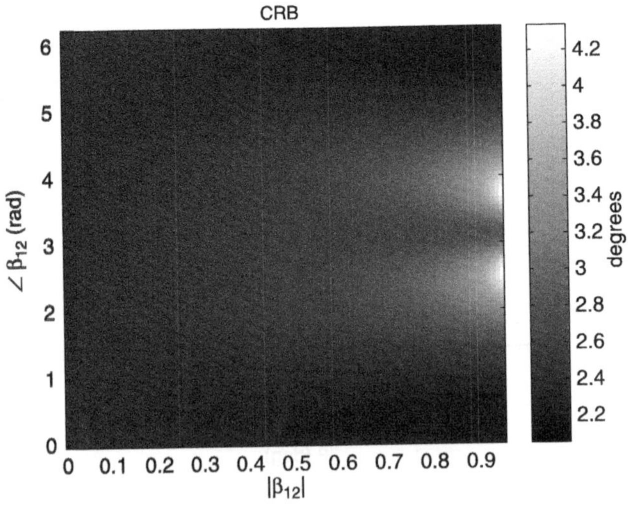

Figure 4.6 CRB on target direction estimation RMSE as a function of phase and modulus of the signal correlation β_{12}, with $M_R = M_T = 2$, $L_T = 1$, SNR = 10 dB, $\theta = 0°$.

The angular resolution with coherent and orthogonal signals is examined in Fig. 4.7. The scenario includes $M_R = M_T = 10$ elements and $L_T = 2$ targets, where the first target is located at $\theta_1 = 0°$. The CRB and performance of the ML estimator of the first target angle as a function of the direction of the second target θ_2 for both coherent and orthogonal signals are presented. The TOT compensation is factored in for the orthogonal transmitted signals. It can be noticed that the performance of the spatially coded signal model with orthogonal transmitted signals is superior to the configuration with coherent transmitted signals. At low angular separation, the RMSE of the ML estimate is lower than the CRB, because of the limited search region for estimation of θ.

The localization performance of the TDS-MUSIC algorithm with orthogonal signal transmission is evaluated and compared to the conventional model using the spatial smoothing algorithm. A LES of $M_R = M_T = 5$ sensors with half a wavelength spacing is used. The scenario includes $L_T = 2$ targets which are located at $\theta_1 = 0°$ and $\theta_2 = 15°$, respectively. In the spatial smoothing algorithm, the number of sensors in each subarray is $N_s = 4$. In Fig. 4.8, the performances of different methods for estimation of the first target localization θ_1 are presented in terms of RMSE. It can be seen that the TDS algorithm with orthogonal transmitted signals outperforms the other configurations in which the transmit signals are coherent. However, the CRB is not achieved by this configuration, because in the smoothing procedure, we dropped the information on $\boldsymbol{\Theta}$ carried by γ_{li} in (4.52). The TDS algorithm with coherent signals fails, because the rank of the signal autocorrelation matrix does not increase in the averaging process.

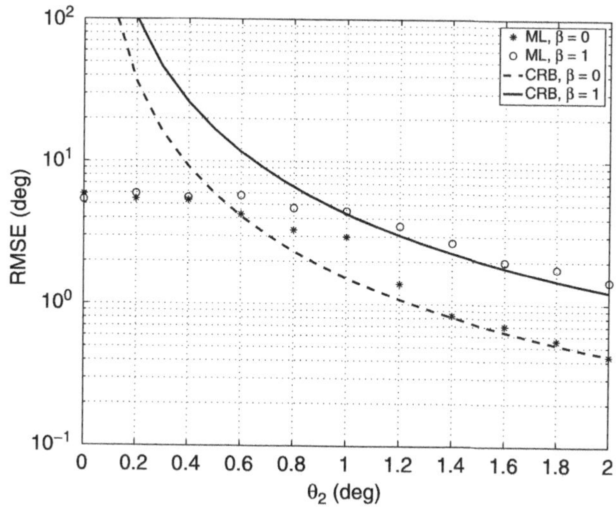

Figure 4.7 CRBs and ML performance for estimation of the first target direction with coherent and orthogonal transmitted signals; $M_R = M_T = 10$ elements, $L_T = 2$ targets: $\theta_1 = 0°$, $\theta_2 = [0:2]°$, SNR = 0 dB.

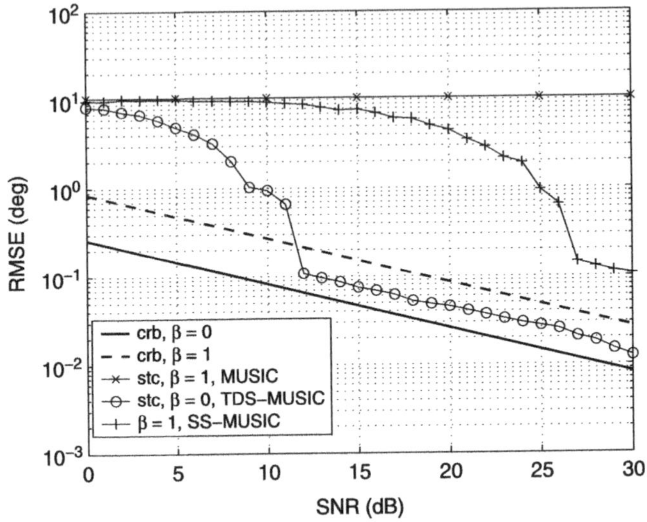

Figure 4.8 Performance of the TDS-MUSIC algorithm with orthogonal ($\beta = 0$) and coherent ($\beta = 1$) signals, compared to the spatial smoothing–MUSIC (SS-MUSIC) and CRB.

Figure 4.9 illustrates the spatial spectra of TDS-MUSIC using orthogonal and coherent signals, versus spatial smoothing combined with MUSIC, where SNR = 15 dB. The TDS-MUSIC algorithm using orthogonal transmitted signals is able to detect the two targets with a higher accuracy compared to the spatial smoothing–MUSIC, while TDS-MUSIC with coherent signals fails, as expected.

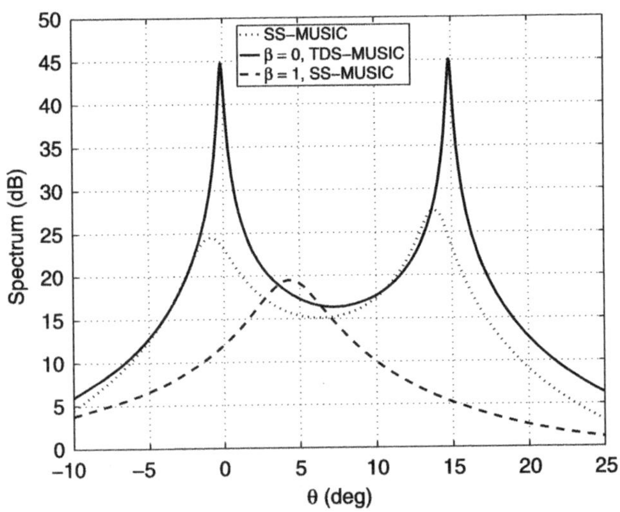

Figure 4.9 Spatial spectrum of TDS-MUSIC with $\beta = 0$ and $\beta = 1$ compared to the spatial smoothing–MUSIC; $M_R = M_T = 5$, $L_T = 2$, $\theta_1 = 0°$, $\theta_2 = 15°$, SNR = 15 dB.

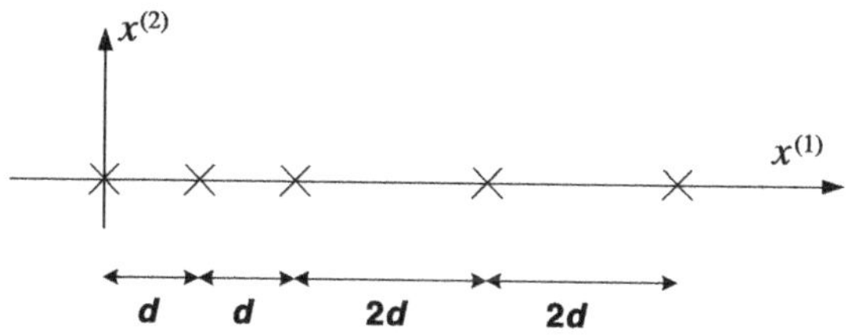

Figure 4.10 Linear, nonuniform array; $d = \lambda/2$.

The localization performance of the TDS-MUSIC with orthogonal transmitted signals with the linear, nonuniform array, shown in Fig. 4.10, is evaluated and presented in Fig. 4.11. As already discussed above, the TDS is not limited to any array geometry. The spatial smoothing algorithm cannot be implemented in this case because it requires a LES array.

Barankin bounds have been evaluated for target localization using coherent and orthogonal transmit signals. In this example, a circular array with 16 transceivers and radius of two wavelengths is used to receive the target echo signal, located at azimuth of 0°. The test points for the DOA, θ were set according to the peaks of the transmit–receive pattern (ambiguity function), given in Section 4.3.3. Figure 4.12 shows the Barankin bound for correlated and uncorrelated signals versus SNR. For comparison, the ML estimator was also implemented for the two cases. It can be observed that the Barankin bound predicts the threshold SNR location to be 5 dB below the threshold SNR exhibited by the ML. Both the Barankin bound and the ML estimator show that by orthogonal signal transmission, the threshold SNR

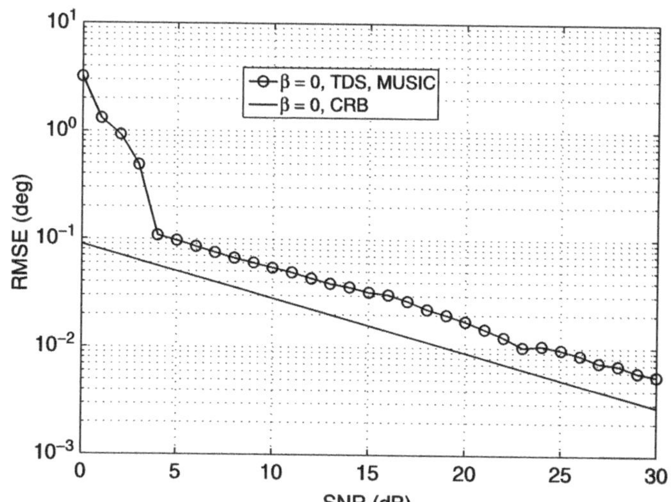

Figure 4.11 Performance of the TDS-MUSIC algorithm with linear, nonuniform array shown in Fig. 4.10, with orthogonal ($\beta = 0$) signals, compared to the CRB.

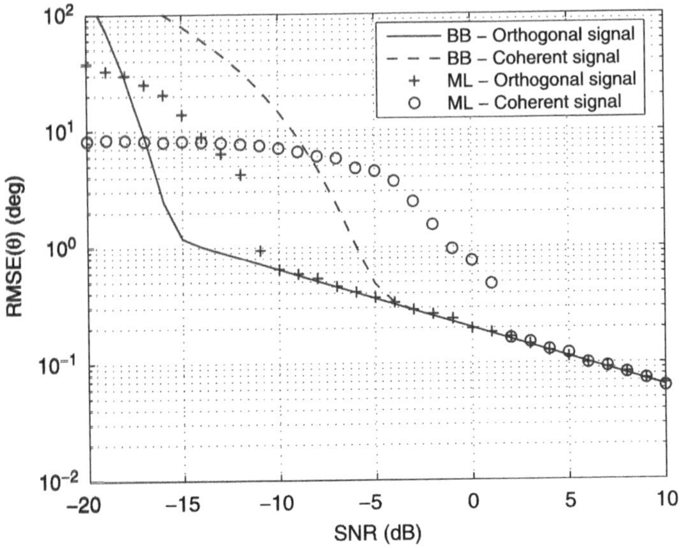

Figure 4.12 Barankin bound and ML estimation performance for target localization using coherent and orthogonal signal transmission.

is reduced by $\sim$10 dB. This improvement is achieved because of the lower sidelobes in the transmit–receive pattern. It should be noted that by uncorrelated signal transmission, the TOT is increased because of the wide spatial coverage [9]. The bound and the ML estimator performance for this case include the TOT compensation in the orthogonal signals case.

Note that it is assumed that the center of the transmitted beam is directed to the target, and thus, no "beamshape loss" is introduced in the coherent signal transmission. Presence of "beamshape loss" further increases the threshold SNR for correlated signal transmission, while it does not change the threshold SNR for the uncorrelated case.

4.7 DISCUSSION AND CONCLUSIONS

In this chapter, the problem of target localization for MIMO radars was investigated. It is shown that spatially orthogonal signal transmission enables digital beamforming and array processing of the transmit signal in addition to conventional array processing using the received signal. The ability to digitally steer the transmitted beam allows to avoid beamshape loss when the target is not located in the center of the transmitted beam. MIMO radars allow us to obtain $M_R M_T$ virtual sensors instead of M_R real sensors in the case of coherent transmit signals. These virtual sensors extend the array aperture, which results in narrower beamwidth. In array configurations, where part of the virtual sensors fall in the same locations, the virtual sensors configuration can be interpreted as virtual tapering of the array, resulting in lower sidelobes. These phenomena are directly related to higher performance in target detection, localization accuracy, and angular resolution. Another benefit from the virtual sensors is the increase of the upper limit on the number of targets that can be detected and localized, that is, $M_R M_T - 1$ targets in the cases of incoherent transmitted signals versus $M_R - 1$ in the case of coherent transmitted signals. The localization performances for orthogonal and coherent transmitted signals were evaluated using simulations and compared to the theory. The results showed superior performance of the orthogonal signals comparing to coherent signals.

Spatially orthogonal signal transmission reduces the spatial transmitted power density by M_T. This is of great importance in many military applications that require low probability of exposure of the active system. Moreover, lower spatial transmit power density reduces the interference to neighboring systems.

The ML estimator for target localization for MIMO radars was derived. The ML estimator becomes impractical in multitarget scenarios because of the large amount of computations. In nonfluctuating target scenarios, direct use of eigenstructure-based techniques is impossible, since the echo signals from the different targets in a given range–Doppler bin are fully correlated. In such cases, some preprocessing techniques such as spatial smoothing and/or forward–backward averaging can be used to decorrelate the signals. However, these techniques require specific array geometries. In addition, the spatial smoothing technique decreases the array aperture, resulting in lower performance, in terms of accuracies, resolution, and maximum number of targets, and other variables. MIMO radars with noncoherent signal transmission allow us to obtain a new preprocessing technique for decorrelating the signals. The TDS algorithm spatially smooths the received signal correlation matrix using transmission diversity, and it enables us to use eigenstructure-based techniques for multitarget localization. Unlike conventional spatial smoothing or forward–backward averaging, the TDS algorithm does not require a linear or symmetric array. In fact, no assumption of far-field targets or spatially white noise was employed. The maximum number of targets that can be localized using the TDS method is limited by $M_R - 1$, where M_R is the number of receive array elements.

The improved beampattern obtained in MIMO radars with orthogonal signals results in lower threshold SNR. The Barankin bound was derived and evaluated to study the threshold SNR. It is shown that the threshold SNR is significantly lower with orthogonal signal transmission than in the case of coherent signal transmission. This effect is a result of virtual sensors obtained in the orthogonal signal transmission. In a LES array of transceivers, the number of virtual sensors at each point vary, resulting in an equivalent triangular tapering, and therefore, lower sidelobes, and lower threshold SNR.

APPENDIX 4A LOG-LIKELIHOOD DERIVATION

In this appendix, the likelihood function for estimating the unknown parameters is derived for two models. The first model is described in (4.15), which includes multiple point targets with unknown complex attenuation, direction, propagation delay, and Doppler shift, in addition to the interference sources parameters. Next, the log-likelihood is found for a single target with known range–Doppler, with no interference sources.

4A.1 General Model

The data model described in (4.15) can be written in the form

$$\mathbf{y}(t) = \mathbf{H}(t, \boldsymbol{\Phi})\alpha + \mathbf{H}_I(\boldsymbol{\Theta}_I)\mathbf{s}_I(t) + \mathbf{w}(t)\ , \quad t \in [0, T_o] \tag{4A.1}$$

where

$$\mathbf{H}(t, \boldsymbol{\Phi}) = [\mathbf{h}(t, \phi_1), \ldots, \mathbf{h}(t, \phi_{L_T})] \tag{4A.2}$$

$$\mathbf{h}(t, \boldsymbol{\phi}_l) = \mathbf{A}(\theta_l)\mathbf{s}(t - \tau_{rl})e^{-j\omega_{Dl}t} \tag{4A.3}$$

$$\mathbf{H}_I(\boldsymbol{\Theta}_I) = [\mathbf{a}_R(\theta_{I1}), \ldots, \mathbf{a}_R(\theta_{IL_I})] \tag{4A.4}$$

$$\boldsymbol{\phi}_l = [\theta_l^T, \tau_{rl}, \omega_{Dl}]^T \tag{4A.5}$$

$$\boldsymbol{\Phi} = [\boldsymbol{\phi}_1^T, \ldots, \boldsymbol{\phi}_{L_T}^T]^T \tag{4A.6}$$

$$\boldsymbol{\Theta}_I = [\theta_{I1}^T, \ldots, \theta_{IL_I}^T]^T \tag{4A.7}$$

$$\boldsymbol{\alpha} = [\alpha_1, \ldots, \alpha_{L_T}]^T \tag{4A.8}$$

$$\mathbf{s}_I(t) = [s_{I1}(t), \ldots, s_{IL_I}(t)]^T \tag{4A.9}$$

The Karhunen–Loève expansion is used for derivation of the log-likelihood function for estimating $\boldsymbol{\Phi}$, $\boldsymbol{\alpha}$, and the interference parameters $\boldsymbol{\Theta}_I$ and $\{\mathbf{s}_I\ (t)\}$. Let Ω denote the space of the elements of $\{\mathbf{y}(t)\}$, $\{\mathbf{H}(t, \boldsymbol{\Phi})\}$, $\{\mathbf{s}_I(t)\}$, $\{\mathbf{w}(t)\}$, and let $\psi_q\ (t)$, $q = 1, 2, \ldots$ be an orthogonal basis function of Ω satisfying $< \psi_q(t), \psi_{q'}(t) >= \int_{T_o} \psi_q(t)\psi_{q'}^*(t)dt = \delta_{qq'}$ where $\delta_{qq'}$ is the Krönecker delta function. Then, the processes, $\{\mathbf{y}(t)\}$, $\{\mathbf{H}(t, \boldsymbol{\Phi})\}$, $\{\mathbf{s}_I(t)\}$, $\{\mathbf{w}(t)\}$ can be expressed by the following series

$$\mathbf{y}(t) = \sum_{q=1}^{\infty} \mathbf{y}_q\psi_q(t)$$

$$\mathbf{H}(t, \boldsymbol{\Phi}) = \sum_{q=1}^{\infty} \mathbf{H}_q(\boldsymbol{\Phi})\psi_q(t)$$

$$\mathbf{s}_I(t) = \sum_{q=1}^{\infty} \mathbf{s}_{Iq}\psi_q(t)$$

$$\mathbf{w}(t) = \sum_{q=1}^{\infty} \mathbf{w}_q\psi_q(t)$$

where $\mathbf{y}_q$, $\mathbf{H}_q(\boldsymbol{\Phi})$, $\mathbf{s}_{Iq}$, and $\mathbf{w}_q$ are the coefficients in the Karhunen–Loève expansion of $\{\mathbf{y}(t)\}$, $\{\mathbf{H}(t, \boldsymbol{\Phi})\}$, $\{\mathbf{s}_I(t)\}$, $\mathbf{w}(t)$, and they can be obtained by calculating the corresponding inner product with $\psi_q(t)$.

By Karhunen–Loève expansion of (4A.1), one obtains an equivalent discrete model for the data:

$$\mathbf{y}_q = \mathbf{H}_q(\boldsymbol{\Phi})\alpha + \mathbf{H}_I(\boldsymbol{\Theta}_I)\mathbf{s}_{Iq} + \mathbf{w}_q, \quad q = 1, 2, \ldots \tag{4A.10}$$

It can be easily shown that for a white circular complex Gaussian process, $\{\mathbf{w}(t)\}$ satisfying $E(\mathbf{w}(t)\mathbf{w}^H(t-\tau)) = \sigma_w^2 \mathbf{I}_{M_R}\delta(\tau)$, the sequence $\{\mathbf{w}_q\}$ is i.i.d. and $\mathbf{w}_q \sim N^c(\mathbf{0}_{M_R}, \sigma_w^2\mathbf{I}_{M_R})$. Thus, the log-likelihood function can be expressed as

$$L_{\mathbf{y}}(\boldsymbol{\Phi}, \boldsymbol{\alpha}, \boldsymbol{\Theta}_I, \{\mathbf{s}_{Iq}\}) = \sum_{q=1}^{\infty}\left(-M_R\log(\pi\sigma_w^2) - \frac{1}{\sigma_w^2}\left\|\mathbf{y}_q - \mathbf{H}_q(\boldsymbol{\Phi})\boldsymbol{\alpha} - \mathbf{H}_I(\boldsymbol{\Theta}_I)\mathbf{s}_{Iq}\right\|^2\right) \tag{4A.11}$$

where the first term in the right side of (4A.11) is a constant. Theoretically, this constant goes to infinity due to the summation over q. Practically, the higher-order terms in the Karhunen–Loève expansion do not contribute any significant statistical information for estimating the unknown parameters; therefore, the summation over q can be finite. After maximizing (4A.11) w.r.t. $\mathbf{s}_{Iq}$, one obtains

$$L_{\mathbf{y}}(\boldsymbol{\Phi}, \boldsymbol{\alpha}, \boldsymbol{\Theta}_I, \{\hat{\mathbf{s}}_{Iq}\}) = \text{const} - \frac{1}{\sigma_w^2}\sum_{q=1}^{\infty}\left\|\mathbf{P}_{\mathbf{H}_I}^{\perp}(\boldsymbol{\Theta}_I)\left(\mathbf{y}_q - \mathbf{H}_q(\boldsymbol{\Phi})\boldsymbol{\alpha}\right)\right\|^2 \tag{4A.12}$$

where $\mathbf{P}_{\mathbf{H}_I}^{\perp}(\boldsymbol{\Theta}_I) = \mathbf{I}_{M_R} - \mathbf{H}_I(\boldsymbol{\Theta}_I)\left(\mathbf{H}_I^H(\boldsymbol{\Theta}_I)\mathbf{H}_I(\boldsymbol{\Theta}_I)\right)^{-1}\mathbf{H}_I^H(\boldsymbol{\Theta}_I)$ is a projection matrix into the null space spanned by the columns of $\mathbf{H}_I(\boldsymbol{\Theta}_I)$. The log-likelihood function for estimating $(\boldsymbol{\Theta}, \boldsymbol{\Theta}_I)$ is obtained by maximization of (4A.12) w.r.t. $\boldsymbol{\alpha}$

$$L'_{\mathbf{y}}(\boldsymbol{\Phi}, \boldsymbol{\Theta}_I) \triangleq L_{\mathbf{y}}(\boldsymbol{\Phi}, \hat{\boldsymbol{\alpha}}, \boldsymbol{\Theta}_I, \{\hat{\mathbf{s}}_{Iq}\}) = \text{const} - \frac{1}{\sigma_w^2}\left(E_{\mathbf{yy}} - \mathbf{e}_{\mathbf{Hy}}^H\mathbf{E}_{\mathbf{HH}}^{-1}\mathbf{e}_{\mathbf{Hy}}\right) \tag{4A.13}$$

where

$$E_{\mathbf{yy}} \triangleq \sum_{q=1}^{\infty}\left\|\mathbf{P}_{\mathbf{H}_I}^{\perp}(\boldsymbol{\Theta}_I)\mathbf{y}_q\right\|^2 \tag{4A.14}$$

$$\mathbf{e}_{\mathbf{Hy}} \triangleq \sum_{q=1}^{\infty}\mathbf{H}_q^H(\boldsymbol{\Phi})\mathbf{P}_{\mathbf{H}_I}^{\perp}(\boldsymbol{\Theta}_I)\mathbf{y}_q \tag{4A.15}$$

$$\mathbf{E}_{\mathbf{HH}} \triangleq \sum_{q=1}^{\infty}\mathbf{H}_q^H(\boldsymbol{\Phi})\mathbf{P}_{\mathbf{H}_I}^{\perp}(\boldsymbol{\Theta}_I)\mathbf{H}_q(\boldsymbol{\Phi}) \tag{4A.16}$$

Using the identity

$$\int_{T_o} \mathbf{x}_1(t)\mathbf{x}_2^H(t)dt = \sum_{q=1}^{\infty} \mathbf{x}_{1q}\mathbf{x}_{2q}^H \tag{4A.17}$$

for $\mathbf{x}_i(t) = \sum_{q=1}^{\infty} \mathbf{x}_{iq}\psi_q(t),\ i = 1,2$, equations (4A.14)–(4A.16) can be rewritten as

$$E_{\mathbf{yy}} = \int_{T_o} \left\| \mathbf{P}_{\mathbf{H}_I}^{\perp}(\mathbf{\Theta}_I)\mathbf{y}(t) \right\|^2 dt \tag{4A.18}$$

$$\mathbf{e}_{\mathbf{Hy}} = \int_{T_o} \mathbf{H}^H(t, \mathbf{\Phi})\mathbf{P}_{\mathbf{H}_I}^{\perp}(\mathbf{\Theta}_I)\mathbf{y}(t)dt \tag{4A.19}$$

$$\mathbf{E}_{\mathbf{HH}} = \int_{T_o} \mathbf{H}^H(t, \mathbf{\Phi})\mathbf{P}_{\mathbf{H}_I}^{\perp}(\mathbf{\Theta}_I)\mathbf{H}(t, \mathbf{\Phi})dt \tag{4A.20}$$

Using the definitions in (4A.2) and (4A.3), the lth element of the vector $\mathbf{e}_{\mathbf{Hy}}$ is given by

$$\begin{aligned}
[\mathbf{e}_{\mathbf{Hy}}]_l &= \int_{T_o} \mathbf{h}^H(t, \phi_l)\mathbf{P}_{\mathbf{H}_I}^{\perp}(\mathbf{\Theta}_I)\mathbf{y}(t)dt \\
&= \mathrm{tr}\left[\mathbf{P}_{\mathbf{H}_I}^{\perp}(\mathbf{\Theta}_I)\left(\int_{T_o} \mathbf{y}(t)\mathbf{s}^H(t-\tau_{rl})e^{j\omega_{Dl}t}dt\right)\mathbf{A}^H(\theta_l)\right] \\
&= \mathbf{a}_R^H(\theta_l)\mathbf{P}_{\mathbf{H}_I}^{\perp}(\mathbf{\Theta}_I)\left(\int_{T_o} \mathbf{y}(t)\mathbf{s}^H(t-\tau_{rl})e^{j\omega_{Dl}t}dt\right)\mathbf{a}_T^*(\theta_l)
\end{aligned} \tag{4A.21}$$

and the lkth element of the matrix $\mathbf{E}_{\mathbf{HH}}$ is

$$\begin{aligned}
[\mathbf{E}_{\mathbf{HH}}]_{lk} &= \int_{T_o} \mathbf{h}^H(t, \phi_l)\mathbf{P}_{\mathbf{H}_I}^{\perp}(\mathbf{\Theta}_I)\mathbf{h}(t, \phi_k)dt \\
&= \mathrm{tr}\left[\mathbf{P}_{\mathbf{H}_I}^{\perp}(\mathbf{\Theta}_I)\mathbf{A}(\theta_k)\left(\int_{T_o} \mathbf{s}(t-\tau_{rk})\mathbf{s}^H(t-\tau_{rl})e^{j(\omega_{Dl}-\omega_{Dk})t}dt\right)\mathbf{A}^H(\theta_l)\right] \qquad (4A.22) \\
&= \mathbf{a}_R^H(\theta_l)\mathbf{P}_{\mathbf{H}_I}^{\perp}(\mathbf{\Theta}_I)\mathbf{a}_R(\theta_k) \\
&\quad \times \mathbf{a}_T^T(\theta_k)\left(\int_{T_o} \mathbf{s}(t-\tau_{rk})\mathbf{s}^H(t-\tau_{rl})e^{j(\omega_{Dl}-\omega_{Dk})t}dt\right)\mathbf{a}_T^*(\theta_l) \qquad (4A.23)
\end{aligned}$$

4A.2 Single Range–Doppler with No Interference

If no interference exists, then the log-likelihood function, derived above, is modified by replacing $\mathbf{P}_{\mathbf{H}_I}^{\perp}(\mathbf{\Theta}_I)$ with an identity matrix. In a single tested or known range–Doppler bin, the data can be easily compensated for propagation delay and Doppler shift. This implies that the log-likelihood function derived in (4A.13) can be used here by setting the propagation delay and Doppler shift to zero

$$L'_{\mathbf{y}}(\boldsymbol{\Theta}) = \text{const} - \frac{1}{\sigma_w^2}\left(E_{\mathbf{yy}} - \mathbf{e}_{\mathbf{Hy}}^H \mathbf{E}_{\mathbf{HH}}^{-1} \mathbf{e}_{\mathbf{Hy}}\right) \tag{4A.24}$$

where

$$E_{\mathbf{yy}} = \int_{T_o} \|\mathbf{y}(t)\|^2 dt \tag{4A.25}$$

$$\mathbf{e}_{\mathbf{Hy}} = \int_{T_o} \mathbf{H}^H(t, \boldsymbol{\Phi})\mathbf{y}(t) dt \tag{4A.26}$$

$$\mathbf{E}_{\mathbf{HH}} = \int_{T_o} \mathbf{H}^H(t, \boldsymbol{\Phi})\mathbf{H}(t, \boldsymbol{\Phi}) dt \tag{4A.27}$$

and $\mathbf{H}(t, \boldsymbol{\Phi})$ is as defined in (4A.2) with $\mathbf{h}(t, \boldsymbol{\phi}_l) = \mathbf{A}(\theta_l)\mathbf{s}(t)$. The lth element of the vector $\mathbf{e}_{\mathbf{Hy}}$ is given by

$$[\mathbf{e}_{\mathbf{Hy}}]_l = \mathbf{a}_T^H(\theta_l)\mathbf{E}^T \mathbf{a}_R(\theta_l) \tag{4A.28}$$

where

$$\mathbf{E} \triangleq \int_{T_o} \mathbf{y}(t)\mathbf{s}^H(t) dt \tag{4A.29}$$

and the lkth element of the matrix $\mathbf{E}_{\mathbf{HH}}$ is

$$[\mathbf{E}_{\mathbf{HH}}]_{lk} = \mathbf{a}_R^H(\theta_l)\mathbf{a}_R(\theta_k) \cdot \mathbf{a}_T^T(\theta_k)\left(\int_{T_o} \mathbf{s}(t)\mathbf{s}^H(t) dt\right)\mathbf{a}_T^*(\theta_l) \tag{4A.30}$$

$$= \mathbf{a}_R^H(\theta_l)\mathbf{a}_R(\theta_k) \cdot \mathbf{a}_T^H(\theta_l)\mathbf{R}_{\mathbf{s}}^T \mathbf{a}_T(\theta_k) \tag{4A.31}$$

Therefore, the sufficient statistic for estimating $\boldsymbol{\Theta}$ and α is given by $\mathbf{E}$.

APPENDIX 4B TRANSMIT–RECEIVE PATTERN DERIVATION

By inserting (4.25) into (4.32), we obtain

$$G_{TR}(\theta, \theta_d) = C\frac{\left|\text{vec}(\mathbf{A}(\theta)\mathbf{U}\boldsymbol{\Lambda}^{1/2})^H \text{vec}(\mathbf{A}(\theta_d)\mathbf{U}\boldsymbol{\Lambda}^{1/2})\right|^2}{\|\text{vec}(\mathbf{A}(\theta_d)\mathbf{U}\boldsymbol{\Lambda}^{1/2})\|^2} \tag{4B.1}$$

Using the identity [29] $\operatorname{tr}(\mathbf{XY}) = (\operatorname{vec}(\mathbf{X}^H))^H \operatorname{vec}(\mathbf{Y})$, (4B.1) can be rewritten as

$$G_{TR}(\theta, \theta_d) = C \frac{\left|\operatorname{tr}(\mathbf{\Lambda}^{1/2}\mathbf{U}^H\mathbf{A}^H(\theta)\mathbf{A}(\theta_d)\mathbf{U}\mathbf{\Lambda}^{1/2})\right|^2}{\operatorname{tr}(\mathbf{\Lambda}^{1/2}\mathbf{U}^H\mathbf{A}^H(\theta_d)\mathbf{A}(\theta_d)\mathbf{U}\mathbf{\Lambda}^{1/2})} \tag{4B.2}$$

$$= C \frac{\left|\operatorname{tr}(\mathbf{A}(\theta_d)\mathbf{R}_s\mathbf{A}^H(\theta))\right|^2}{\operatorname{tr}(\mathbf{A}(\theta_d)\mathbf{R}_s\mathbf{A}^H(\theta_d))} \tag{4B.3}$$

By substitution of (4.9) into (4B.2), the transmit–receive beampattern can be expressed as

$$G_{TR}(\theta, \theta_d) = C \frac{\left|\mathbf{a}_R^H(\theta)\mathbf{a}_R(\theta_d)\mathbf{a}_T^T(\theta_d)\mathbf{R}_s\mathbf{a}_T^*(\theta)\right|^2}{\mathbf{a}_R^H(\theta_d)\mathbf{a}_R(\theta_d)\mathbf{a}_T^T(\theta_d)\mathbf{R}_s\mathbf{a}_T^*(\theta_d)}$$

$$= C \frac{\left|\mathbf{a}_R^H(\theta)\mathbf{a}_R(\theta_d)\right|^2}{M_R} \cdot \frac{\left|\mathbf{a}_T^H(\theta)\mathbf{R}_s^T\mathbf{a}_T(\theta_d)\right|^2}{\mathbf{a}_T^H(\theta_d)\mathbf{R}_s^T\mathbf{a}_T(\theta_d)} \tag{4B.4}$$

APPENDIX 4C FISHER INFORMATION MATRIX DERIVATION

The data model in terms of Karhunen–Loève coefficients and in the absence of interference sources is given in (4A.10). Let $\tilde{\boldsymbol{\xi}}$ denote the vector of unknown parameters: $\tilde{\boldsymbol{\xi}} = [\tilde{\boldsymbol{\alpha}}^T\ \boldsymbol{\Phi}^T]^T$, where $\tilde{\boldsymbol{\alpha}} = [\tilde{\boldsymbol{\alpha}}_1^T, \ldots, \tilde{\boldsymbol{\alpha}}_{L_T}^T]^T$, and $\tilde{\boldsymbol{\alpha}}_l = [\operatorname{Re}(\alpha_l), \operatorname{Im}(\alpha_l)]^T$ for $l = 1, \ldots, L_T$. For i.i.d. noise, the FIM for estimating $\tilde{\xi}$ is given by (see, e.g., Ref. 27)

$$\mathbf{J}_{\tilde{\xi}} = \frac{2}{\sigma_w^2} \operatorname{Re}\left\{ \sum_{q=1}^{\infty} \frac{\partial(\mathbf{H}_q(\boldsymbol{\Phi})\alpha)^H}{\partial\tilde{\boldsymbol{\xi}}} \frac{\partial(\mathbf{H}_q(\boldsymbol{\Phi})\boldsymbol{\alpha})}{\partial\tilde{\boldsymbol{\xi}}} \right\} \tag{4C.1}$$

where the elements of $\partial\boldsymbol{\mu}/\partial\boldsymbol{\xi}$ are given by

$$\left[\frac{\partial\boldsymbol{\mu}}{\partial\tilde{\boldsymbol{\xi}}}\right]_{ij} \triangleq \frac{\partial\mu_i}{\partial\tilde{\xi}_j} \tag{4C.2}$$

Using the identity in (4A.17), the FIM can be expressed as

$$\mathbf{J}_{\tilde{\xi}} = \frac{2}{\sigma_w^2} \operatorname{Re}\left\{ \int_{T_o} \frac{\partial(\mathbf{H}(t, \boldsymbol{\Phi})\alpha)^H}{\partial\tilde{\boldsymbol{\xi}}} \frac{\partial(\mathbf{H}(t, \boldsymbol{\Phi})\boldsymbol{\alpha})}{\partial\tilde{\boldsymbol{\xi}}} dt \right\}. \tag{4C.3}$$

Using the definition of $\mathbf{H}(t, \mathbf{\Phi})$ from (4A.2) and (4A.3), the derivatives required for calculation of (4C.3) are given by

$$\frac{\partial \mathbf{H}(t, \mathbf{\Phi})\alpha}{\partial \theta_l} = \alpha_l e^{-j\omega_{Dl}t} \dot{\mathbf{A}}(\theta_l)\mathbf{s}(t - \tau_{rl}) \tag{4C.4}$$

$$\frac{\partial \mathbf{H}(t, \mathbf{\Phi})\alpha}{\partial \tau_{rl}} = -\alpha_l e^{-j\omega_{Dl}t} \mathbf{A}(\theta_l)\, \dot{\mathbf{s}}(t - \tau_{rl}) \tag{4C.5}$$

$$\frac{\partial \mathbf{H}(t, \mathbf{\Phi})\alpha}{\partial \omega_{Dl}} = -jt\alpha_l e^{-j\omega_{Dl}t} \mathbf{A}(\theta_l)\mathbf{s}(t - \tau_{rl}) \tag{4C.6}$$

$$\frac{\partial \mathbf{H}(t, \mathbf{\Phi})\alpha}{\partial \tilde{\boldsymbol{\alpha}}_l} = e^{-j\omega_{Dl}t} \mathbf{A}(\theta_l)\mathbf{s}(t - \tau_{rl})[1, j] \tag{4C.7}$$

where $\dot{\mathbf{A}}(\theta) = d\mathbf{A}(\theta)/d\theta$ and $\dot{\mathbf{s}}(t) = d\mathbf{s}(t)/dt$.

Accordingly, the elements of the FIM are given by

$$\begin{aligned} J_{\theta_l\theta_k} &= \frac{2}{\sigma_w^2}\mathrm{Re}\left\{\int_{T_o} \alpha_l^* \alpha_k e^{-j(\omega_{Dk}-\omega_{Dl})t}\mathbf{s}^H(t - \tau_{rl})\dot{\mathbf{A}}^H(\theta_l)\dot{\mathbf{A}}(\theta_k)\mathbf{s}(t - \tau_{rk})dt\right\} \\ &= \frac{2}{\sigma_w^2}\mathrm{Re}\left\{\alpha_l^*\alpha_k \operatorname{tr}\left(\dot{\mathbf{A}}(\theta_k)\mathbf{AF}_{\mathbf{ss}}(\tau_{rk}, \tau_{rl}, \omega_{Dk}, \omega_{Dl})\dot{\mathbf{A}}^H(\theta_l)\right)\right\} \end{aligned} \tag{4C.8}$$

$$\begin{aligned} J_{\tau_{rl}\tau_{rk}} &= \frac{2}{\sigma_w^2}\mathrm{Re}\left\{\int_{T_o} \alpha_l^* \alpha_k e^{-j(\omega_{Dk}-\omega_{Dl})t}\,\dot{\mathbf{s}}^H(t - \tau_{rl})\mathbf{A}^H(\theta_l)\mathbf{A}(\theta_k)\,\dot{\mathbf{s}}(t - \tau_{rk})dt\right\} \\ &= \frac{2}{\sigma_w^2}\mathrm{Re}\left\{\alpha_l^*\alpha_k \operatorname{tr}\left(\mathbf{A}(\theta_k)\mathbf{AF}_{\dot{\mathbf{s}}\dot{\mathbf{s}}}(\tau_{rk}, \tau_{rl}, \omega_{Dk}, \omega_{Dl})\mathbf{A}^H(\theta_l)\right)\right\} \end{aligned} \tag{4C.9}$$

$$\begin{aligned} J_{\omega_{Dl}\omega_{Dk}} &= \frac{2}{\sigma_w^2}\mathrm{Re}\left\{\int_{T_o} \alpha_l^* \alpha_k t^2 e^{-j(\omega_{Dk}-\omega_{Dl})t}\mathbf{s}^H(t - \tau_{rl})\mathbf{A}^H(\theta_l)\mathbf{A}(\theta_k)\mathbf{s}(t - \tau_{rk})dt\right\} \\ &= \frac{2}{\sigma_w^2}\mathrm{Re}\left\{\alpha_l^*\alpha_k \operatorname{tr}\left(\mathbf{A}(\theta_k)\mathbf{AF}_{t^2\mathbf{ss}}(\tau_{rk}, \tau_{rl}, \omega_{Dk}, \omega_{Dl})\mathbf{A}^H(\theta_l)\right)\right\} \end{aligned} \tag{4C.10}$$

$$\begin{aligned} J_{\tilde{\boldsymbol{\alpha}}_l\tilde{\boldsymbol{\alpha}}_k} &= \frac{2}{\sigma_w^2}\mathrm{Re}\left\{\int_{T_o} [1, j]^H e^{-j(\omega_{Dk}-\omega_{Dl})t}\mathbf{s}^H(t - \tau_{rl})\mathbf{A}^H(\theta_l)\mathbf{A}(\theta_k)\mathbf{s}(t - \tau_{rk}))dt[1, j]\right\} \\ &= \frac{2}{\sigma_w^2}\mathrm{Re}\left\{[1, j]^H \operatorname{tr}\left(\mathbf{A}(\theta_k)\mathbf{AF}_{\mathbf{ss}}(\tau_{rk}, \tau_{rl}, \omega_{Dk}, \omega_{Dl})\mathbf{A}^H(\theta_l)\right)[1, j]\right\} \end{aligned} \tag{4C.11}$$

$$\begin{aligned} J_{\theta_l\tau_{rk}} &= -\frac{2}{\sigma_w^2}\mathrm{Re}\left\{\int_{T_o} \alpha_l^* \alpha_k e^{-j(\omega_{Dk}-\omega_{Dl}t)}\mathbf{s}^H(t - \tau_{rl})\dot{\mathbf{A}}^H(\theta_l)\mathbf{A}(\theta_k)\,\dot{\mathbf{s}}(t - \tau_{rk})dt\right\} \\ &= -\frac{2}{\sigma_w^2}\mathrm{Re}\left\{\alpha_l^*\alpha_k \operatorname{tr}\left(\mathbf{A}(\theta_k)\mathbf{AF}_{\dot{\mathbf{s}}\mathbf{s}}(\tau_{rk}, \tau_{rl}, \omega_{Dk}, \omega_{Dl})\dot{\mathbf{A}}^H(\theta_l)\right)\right\} \end{aligned} \tag{4C.12}$$

$$
\begin{aligned}
J_{\theta_l \omega_{Dk}} &= -\frac{2}{\sigma_w^2}\mathrm{Re}\left\{\int_{T_o} \alpha_l^* \alpha_k j t e^{-j(\omega_{Dk}-\omega_{Dl})t} \mathbf{s}^H(t-\tau_{rl})\dot{\mathbf{A}}^H(\theta_l)\mathbf{A}(\theta_k)\mathbf{s}(t-\tau_{rk})dt\right\} \\
&= -\frac{2}{a_w^2}\mathrm{Re}\left\{ j\alpha_l^*\alpha_k \,\mathrm{tr}\left(\mathbf{A}(\theta_k)\mathbf{AF}_{t\mathbf{ss}}(\tau_{rk}, \tau_{rl}, \omega_{Dk}, \omega_{Dl})\dot{\mathbf{A}}^H(\theta_l)\right)\right\} \qquad (4C.13)
\end{aligned}
$$

$$
\begin{aligned}
J_{\theta_l \tilde{\boldsymbol{\alpha}}_k} &= \frac{2}{\sigma_w^2}\mathrm{Re}\left\{\int_{T_o} \alpha_l^* e^{-j(\omega_{Dk}-\omega_{Dl})t} \mathbf{s}^H(t-\tau_{rl})\dot{\mathbf{A}}^H(\theta_l)\mathbf{A}(\theta_k)\mathbf{s}(t-\tau_{rk})dt[1, j]\right\} \\
&= \frac{2}{\sigma_w^2}\mathrm{Re}\left\{\alpha_l^* \,\mathrm{tr}\left(\mathbf{A}(\theta_k)\mathbf{AF}_{\mathbf{ss}}(\tau_{rk}, \tau_{rl}, \omega_{Dk}, \omega_{Dl})\dot{\mathbf{A}}^H(\theta_l)\right)[1, j]\right\} \qquad (4C.14)
\end{aligned}
$$

$$
\begin{aligned}
J_{\tau_{rl} \tilde{\boldsymbol{\alpha}}_k} &= -\frac{2}{\sigma_w^2}\mathrm{Re}\left\{\int_{T_o} \alpha_l^* e^{-j(\omega_{Dk}-\omega_{Dl})t} \dot{\mathbf{s}}^H(t-\tau_{rl})\mathbf{A}^H(\theta_l)\mathbf{A}(\theta_k)\mathbf{s}(t-\tau_{rk})dt[1, j]\right\} \\
&= -\frac{2}{\sigma_w^2}\mathrm{Re}\left\{\alpha_l^* \,\mathrm{tr}\left(\mathbf{A}(\theta_k)\mathbf{AF}_{\mathbf{s}\dot{\mathbf{s}}}(\tau_{rk}, \tau_{rl}, \omega_{Dk}, \omega_{Dl})\mathbf{A}^H(\theta_l)\right)[1, j]\right\} \qquad (4C.15)
\end{aligned}
$$

$$
\begin{aligned}
J_{\tau_{rl} \omega_{Dk}} &= \frac{2}{\sigma_w^2}\mathrm{Re}\left\{\int_{T_o} \alpha_l^* \alpha_k j t e^{-j(\omega_{Dk}-\omega_{Dl})t} \dot{\mathbf{s}}^H(t-\tau_{rl})\mathbf{A}^H(\theta_l)\mathbf{A}(\theta_k)\mathbf{s}(t-\tau_{rk})dt\right\} \\
&= \frac{2}{\sigma_w^2}\mathrm{Re}\left\{ j\alpha_l^*\alpha_k \,\mathrm{tr}\left(\mathbf{A}(\theta_k)\mathbf{AF}_{t\mathbf{s}\dot{\mathbf{s}}}(\tau_{rk}, \tau_{rl}, \omega_{Dk}, \omega_{Dl})\mathbf{A}^H(\theta_l)\right)\right\} \qquad (4C.16)
\end{aligned}
$$

$$
\begin{aligned}
J_{\tilde{\boldsymbol{\alpha}}_l \omega_{Dk}} &= -\frac{2}{\sigma_w^2}\mathrm{Re}\left\{\int_{T_o} [1, j]^H \alpha_k j t e^{-j(\omega_{Dk}-\omega_{Dl})t} \mathbf{s}^H(t-\tau_{rl})\mathbf{A}^H(\theta_l)\mathbf{A}(\theta_k)\mathbf{s}(t-\tau_{rk})dt\right\} \\
&= -\frac{2}{\sigma_w^2}\mathrm{Re}\left\{ j[1, j]^H \alpha_k \mathrm{tr}\left(\mathbf{A}(\theta_k)\mathbf{AF}_{t\mathbf{ss}}(\tau_{rk}, \tau_{rl}, \omega_{Dk}, \omega_{Dl})\mathbf{A}^H(\theta_l)\right)\right\} \qquad (4C.17)
\end{aligned}
$$

where the matrix $\mathbf{AF}_{\mathbf{ss}}(\tau_{rk}, \tau_{rl}, \omega_{Dk}, \omega_{Dl})$ is as defined in (4.41), and

$$\mathbf{AF}_{\mathbf{s}\dot{\mathbf{s}}}(\tau_{rk}, \tau_{rl}, \omega_{Dk}, \omega_{Dl}) \triangleq \int_{T_o} e^{-j(\omega_{Dk}-\omega_{Dl})t} \mathbf{s}(t-\tau_{rk})\dot{\mathbf{s}}^H(t-\tau_{rl})dt \qquad (4C.18)$$

$$\mathbf{AF}_{\dot{\mathbf{s}}\mathbf{s}}(\tau_{rk}, \tau_{rl}, \omega_{Dk}, \omega_{Dl}) \triangleq \int_{T_o} e^{-j(\omega_{Dk}-\omega_{Dl})t} \dot{\mathbf{s}}(t-\tau_{rk})\mathbf{s}^H(t-\tau_{rl})dt \qquad (4C.19)$$

$$\mathbf{AF}_{t\mathbf{ss}}(\tau_{rk}, \tau_{rl}, \omega_{Dk}, \omega_{Dl}) \triangleq \int_{T_o} t e^{-j(\omega_{Dk}-\omega_{Dl})t} \mathbf{s}(t-\tau_{rk})\mathbf{s}^H(t-\tau_{rl})dt \qquad (4C.20)$$

$$\mathbf{AF}_{t\mathbf{s}\dot{\mathbf{s}}}(\tau_{rk}, \tau_{rl}, \omega_{Dk}, \omega_{Dl}) \triangleq \int_{T_o} t e^{-j(\omega_{Dk}-\omega_{Dl})t} \mathbf{s}(t-\tau_{rk})\dot{\mathbf{s}}^H(t-\tau_{rl})dt \qquad (4C.21)$$

$$\mathbf{AF}_{\dot{\mathbf{s}}\dot{\mathbf{s}}}(\tau_{rk}, \tau_{rl}, \omega_{Dk}, \omega_{Dl}) \triangleq \int_{T_o} e^{-j(\omega_{Dk}-\omega_{Dl})t} \dot{\mathbf{s}}(t-\tau_{rk})\dot{\mathbf{s}}^H(t-\tau_{rl})dt \qquad (4C.22)$$

$$\mathbf{AF}_{t^2\mathbf{ss}}(\tau_{rk}, \tau_{rl}, \omega_{Dk}, \omega_{Dl}) \triangleq \int_{T_o} t^2 e^{-j(\omega_{Dk}-\omega_{Dl})t} \mathbf{s}(t-\tau_{rk})\mathbf{s}^H(t-\tau_{rl})dt \qquad (4C.23)$$

REFERENCES

1. S. Haykin, J. Litva, and T. J. Shepherd, *Radar Array Processing*, Springer-Verlag, New York, 1993.
2. D. W. Bliss and K. W. Forsythe, Multiple-input multiple-output (MIMO) radar and imaging: Degrees of freedom and resolution, *Proc. 37th Asilomar Conf. Signals, Systems, and Computers*, Nov. 2003.
3. I. Bekkerman and J. Tabrikian, Spatially coded transmission for active arrays, *Proc. ICASSP04*, May 2004, Vol. **2**, pp. 209–212.
4. J. C. Guey and M. R. Bell, Diversity waveform sets for delay-Doppler imaging, *IEEE Trans. Inform. Theory* **44**(4):1504–1522 (Sept. 1998).
5. J. Li and R. T. Compton, Maximum likelihood angle estimation for signals with known waveforms, *IEEE Trans. Signal Process.* **41**(9):2850–2862 (Sept. 1993).
6. H. Messer, G. Sigal, and L. Bialy, On the achievable DF accuracy of two kinds of active interferometers, *IEEE Trans. Aerospace Electron. Syst.* **32**:1158–1164 (July 1996).
7. R. O. Nielsen, Performance of combinations of projectors and receivers, *IEEE J. Ocean Eng.* **27**(3) (July 2002).
8. P. Calvary and D. Janer, Spatio-temporal coding for radar array processing, *Proc. ICASSP98*, 1998, Vol. **4**, pp. 2509–2512.
9. I. Bekkerman and J. Tabrikian, Target detection and localization using MIMO radars and sonars, *IEEE Trans. Signal Process.* **54**(10):3873–3883 (Oct. 2006).
10. E. Fishler, A. Haimovich, R. Blum, D. Chizhik, L. Cimini, and R. Valenzuela, MIMO radar: An idea whose time has come, *Proc. IEEE Radar Conf. 2004*, pp. 71–78.
11. E. Fishler, A. Haimovich, R. Blum, L. Cimini, D. Chizhik, and R. Valenzuela, Spatial diversity in radars-models and detection performance, *IEEE Trans. Signal Process.* **54**(3):823–838, (March 2006).
12. N. H. Lehmann, E. Fishler, A. Haimovich, R. Blum, D. Chizhik, L. Cimini, and R. Valenzuela, Evaluation of transmit diversity in MIMO-radar direction finding, *IEEE Trans. Signal Process.* **55**(5):2215–2225 (May 2007).
13. J. Tabrikian and I. Bekkerman, Transmission diversity smoothing for multitarget localization, *Proc. ICASSP05*, March 2005, Vol. **4**, pp. 1041–1044.
14. J. Tabrikian, Barankin bounds for target localization for MIMO radars, *Proc. 4th IEEE Workshop on Sensor Array and Multi-channel Processing* (invited), July 2006, pp. 278–281.
15. G. San Antonio, D. R. Fuhrmann, and F. C. Robey, MIMO radar ambiguity functions, *IEEE J. Select. Topics Signal Process.* **1**(1):167–177 (June 2007).
16. G. San Antonio and D. R. Fuhrmann, Beampattern synthesis for wideband MIMO radar systems, *Proc. 1st IEEE Int. Workshop of Computational Advances in Multi-Sensor Adaptive Processing*, Dec. 2005.
17. G. San Antonio and D. R. Fuhrmann, Transmit beamforming for MIMO radar systems using partial signal correlation, *Proc. 38th Asilomar Conf. Signals, Systems, and Computers*, Pacific Grove, CA, Nov. 2004.
18. J. Li, P. Stoica, L. Xu, and W. Roberts, On parameter identifiability of MIMO radar, *IEEE Signal Process. Lett.* **14**(12):968–971 (Dec. 2007).
19. P. Stoica, J. Li, and Y. Xie, On probing signal design for MIMO radar, *IEEE Trans. Signal Process.* **55**(8):4151–4161 (Aug. 2007).

20. L. Xu, J. Li, and P. Stoica, Adaptive MIMO radar, *IEEE Trans. Aerospace Electron. Syst.* (in press).

21. J. Li, L. Xu, P. Stoica, K. W. Forsythe, and D. W. Bliss, Range compression and waveform optimization for MIMO radar–A Cramér-Rao bound based study, *IEEE Trans. Signal Process.* **56**(1):218–232 (Jan. 2008).

22. L. Xu, J. Li, and P. Stoica, Adaptive MIMO radar, *Proc. 4th IEEE Sensor Array and Multichannel Signal Processing Workshop*, Waltham, MA, July 2006.

23. L. Xu, J. Li, P. Stoica, K. W. Forsythe, and D. W. Bliss, Waveform optimization for MIMO radar: A Cramér–Rao bound based study, *Proc. IEEE Int. Conf. Acoustics, Speech, and Signal Processing*, Proc. ICASSP07, Honolulu, April 2007.

24. J. Li and P. Stoica, MIMO radar where diversity enables superiority, *Proc. 12th Annual Workshop on Adaptive Sensor Array Processing* (invited), Lexington, MA, June 2006.

25. L. Xu, J. Li, and P. Stoica, Radar imaging via adaptive MIMO techniques, *Proc. 14th European Signal Processing Conf.* (*EUSIPCO*) (invited), Florence, Italy, Sept. 2006.

26. J. Li, P. Stoica, and Y. Xie, On probing signal design for MIMO radar, *Proc. 40th Asilomar Conf. Signals, Systems, and Computers* (invited), Pacific Grove, CA, Oct. 2006.

27. S. M. Kay, *Fundamentals of Statistical Signal Processing: Estimation Theory,* Prentice-Hall, Englewood Cliffs, NJ, 1993.

28. A. Dogandžić and A. Nehorai, Cramér–Rao for estimating range, velocity, and direction with an active array, *IEEE Trans. Signal Process.* **49**(6):1122–1137 (June 2001).

29. A. Graham, *Kronecker Products and Matrix Calculus with Applications,* Ellis Horwood Ltd., UK, 1981.

30. A. L. Swindlehurst and P. Stoica, Maximum likelihood methods in radar array signal processing, *IEEE Proc.* **86**(2):421–441 (Feb. 1998).

31. N. Levanon and E. Mozeson, *Radar Signals*, Wiley Interscience, Hoboken, NJ, 2004.

32. A. Pinkus and J. Tabrikian, Barankin bound for range and Doppler estimation using orthogonal signal transmission, *Proc. IEEE Int. Radar Conf.*, Verona, NY, April 2006.

33. R. O. Schmidt, Multiple emitter location and signal parameter estimation, *IEEE Trans. on Anten. Propag.* **AP-34**(3):276–280 (March 1986).

34. J. E. Evans, J. R. Johnson, and D. F. Sun, High resolution angular spectrum estimation techniques for terrain scattering analysis and angle of arrival estimation, *Proc. 1st ASSP Workshop Spectral Estimation*, Hamilton, Ontario, Canada, 1981, pp. 134–139.

35. S. U. Pillai and B. H. Kwon, Forward/backward spatial smoothing techniques for coherent signal identification, *IEEE Trans. Acoust. Speech Signal Process.* **37**:8–15 (Jan. 1989).

36. J. Li, Improved angular resolution for spatial smoothing techniques, *IEEE Trans. Signal Process.* **40**:3078–3081 (Dec. 1992).

37. D. A. Linebarger and D. H. Johnson, The effect of spatial averaging on spatial correlation matrices in the presence of coherent signals, *IEEE Trans. Acoust. Speech Signal Process.* **38**:880–884 (May 1990).

38. L. C. Godara, Beamforming in the presence of correlated arrivals using structured correlation matrix, *IEEE Trans. Acoust. Speech Signal Process.* **38**:1–15 (Jan. 1990).

39. K. C. Indukumar and V. U. Reddy, A note on redundancy averaging, *IEEE Trans. Signal Process.* **40**:466–469 (Feb. 1992).

40. E. W. Barankin, Locally best unbiased estimates, *Ann. Math. Stat.* **20**:477–501 (1949).

41. R. J. McAulay and E. M. Hofstetter, Barankin bounds on parameter estimation, *IEEE Trans. Inform. Theory* **IT-17**:669–676 (Nov. 1971).

42. A. Zeira and P. M. Schultheiss, Realizable lower bounds for time delay estimation, *IEEE Trans. Signal Process.* **41**:(11)3102–3113 (Nov. 1993).

43. A. Zeira and P. M. Schultheiss, Realizable lower bounds for time delay estimation: Part 2—Threshold phenomena, *IEEE Trans. Signal Process* **42**:(5)1001–1007 (May 1994).

44. J. Tabrikian and J. L. Krolik, Barankin bounds for source localization in an uncertain ocean environment, *IEEE Trans. Signal Process* **47**:(11):2917–2927 (Nov. 1999).

5

ADAPTIVE SIGNAL DESIGN FOR MIMO RADARS

BENJAMIN FRIEDLANDER

Department of Electrical Engineering, University of California, Santa Cruz

We consider the problem of signal design for MIMO radars, where the transmit waveforms are adjusted based on target and clutter statistics. A model for the radar returns that explicitly incorporates the transmit waveforms is developed. Both estimation and detection problems are formulated for that signal model. Optimal and suboptimal algorithms are derived for designing the transmit waveforms. The performance of these algorithms is illustrated by computer simulation. These results indicate that adaptive design of the radar signal can in some cases provide improved estimation and detection performance compared to fixed designs such as the one employing a set of orthogonal waveforms.

5.1 INTRODUCTION

More recent advances in linear amplifier and waveform generation technology, and the ever-increasing processing power, have spawned interest in the development of radar systems that attempt to make full use of the spatiotemporal degrees of freedom available to the radar transmitter. These technological advances make it possible to consider the design of radar systems that allow the transmitter full flexibility in selecting the transmitted waveform (within given bandwidth and power constraints) on a pulse-by-pulse or antenna-by-antenna basis.

The flexibility of using a multiplicity of transmitted waveforms, and of adaptively adjusting these waveforms, offers significant performance advantages.

MIMO Radar Signal Processing, edited by Jian Li and Petre Stoica

Fundamentally, the additional degrees of freedom afforded by the ability to vary the transmit waveform can be used to optimize a desired performance criterion. For example, the waveform can be adapted to the target signature to enhance detectability, increase clutter or interference rejection, improve the quality of the estimated radar map, improve spatial resolution, or reduce search time.

A considerable amount of research has been done on waveform design issues such as work on optimum transmit–receive design [1–3], which assumes a deterministic target model with a range spread, using a single transmit antenna, or an antenna with multiple polarization modes [4]. In a more recent paper [5] we studied optimal optimal waveform design for a single-antenna radar. We presented a signal subspace framework that made it possible to derive the optimal radar waveform for a given scenario and evaluate the corresponding radar performance. These works involve temporal processing only.

There has been considerable interest in radar systems employing multiple antennas at both the transmitter and receiver, commonly referred to as MIMO radar [6–21]. In this case spatiotemporal processing of the radar signals is required. The waveforms transmitted by the radar may be adaptive or nonadaptive. By "adaptive" we mean that the waveforms depend on information about the particular scene being observed, such as the target and clutter statistics, or on the actual target returns. By "nonadaptive" we mean that the waveforms are designed without making use of such information. However, the waveforms may be different for different antennas and may change from pulse to pulse. We note that this terminology is by no means standard and that we use it here only in the context of the transmit waveform design. As an example, the radar may perform adaptive processing at the receiver (e.g., STAP), while using a nonadaptive method to design the transmit waveform.

Most of the work to date on signal design for MIMO radar has focused on the nonadaptive case. A widely studied approach involves the transmission of orthogonal signals on the different antennas. This makes it possible to separate the signals arriving from the different transmit antennas at the receiver, and to perform any transmit array processing functions on the receive side "after the fact." For example, one can scan the transmit beam across the illuminated area within a single dwell time, or perform adaptive beamforming to reduce interference and improve resolution [6–13]. By employing adaptive processing, it is possible to improve clutter rejection in ways that are not possible in conventional radar [14,16]. MIMO radar can also provide angular diversity, which is useful in some scenarios [22–24].

Here we consider the waveform design problem for MIMO radars in a rather general setting that allows for any waveform design procedure, either adaptive or nonadaptive. In particular, we develop a procedure to design the optimal waveform that maximizes the signal-to-interference-plus-noise ratio (SINR) at the output of the detector, as well as some suboptimal variations. The key to our approach is use of a model for the received radar signals that explicitly includes the transmitted waveforms. Most radar systems perform range processing as a first step, employing a matched filter. As is well known, the response to a single point target measured at the matched filter output does not depend on the shape of the waveform, only on its energy (however, the shape of the waveform affects the range–Doppler ambiguity function, which plays an important role when multiple targets are present). Thus, the transmitted signal does not appear explicitly in the processing subsequent to the range processing step. In Section 5.2 we develop the model for the signals received at the antenna outputs prior to any processing. This model exposes the dependence on the transmitted waveform and allows the derivation of various estimation algorithms in Section 5.3 and detection algorithms in Section 5.4.

5.2 PROBLEM FORMULATION

Consider a MIMO radar employing N_T antennas at the transmitter and N_R antennas at the receiver. The radar operates at a carrier frequency f_c and has a bandwidth B. We assume here that the two arrays are collocated, that is, represent a monostatic radar. The extension to the bistatic case is straightforward but requires a different parametrization of the scatterer locations, which entails significant changes in the equations presented here. In this section we develop a model for the signal received by the MIMO radar. We use the convention in which the time-domain variables are denoted by an overline (e.g., $\overline{\mathbf{x}}$), to distinguish them from their frequency-domain counterparts (e.g., $\mathbf{x}$).

Let $\overline{h}(\theta, \tau)$ denote the distribution of scatterer amplitudes in the area illuminated by the transmit antenna (the antenna footprint), where τ is the round-trip delay from a reference antenna to the scatterer location. The round-trip delay τ can be translated into range r by $r = c\tau/2$, where c is the speed of light. Consider the scattering from a patch of size $d\theta\, d\tau$, representing a scatterer with amplitude $\overline{h}(\theta, \tau)d\theta\, d\tau$. Let $\tau_i(\theta)$ denote the delay difference between the ith transmit antenna and the reference antenna along direction θ.

The ith transmitter transmits the signal $\overline{s}(t)e^{jw_c t}$, where $\overline{s}(t)$ is the baseband waveform. The signal reflected from the scatterer is

$$\overline{h}(\theta, \tau)d\theta\, d\tau\, \overline{s}\big(t - \tau/2 - \tau_i^t(\theta)\big)e^{jw_c(t-\tau/2-\tau_i^t(\theta))} \tag{5.1}$$

where $\tau_i^t(\theta)$ denotes the delay difference between the ith transmit antenna and the reference antenna along direction θ. This signal arrives at kth receive antenna as

$$\overline{h}(\theta, \tau)d\theta\, d\tau\, \overline{s}(t - \tau - \tau_i^t(\theta) - \tau_k^t(\theta))e^{jw_c(t-\tau-\tau_i^t(\theta)-\tau_k^r(\theta))} \tag{5.2}$$

where $\tau_k^r(\theta)$ denotes the delay difference between the kth receive antenna and the reference antenna along direction θ. The delay τ is the round-trip delay between reference antennas.

In the case where the narrowband assumption holds, we have $\bar{s}(t-\tau) \approx \bar{s}(t)$. Therefore, the signal arriving at the kth antenna is

$$\bar{h}(\theta, \tau)d\theta\, d\tau\, \bar{s}(t-\tau)e^{jw_c(t-\tau-\tau_i^t(\theta)-\tau_k^r(\theta))} \tag{5.3}$$

After downmodulating to baseband, this signal becomes

$$\bar{h}(\theta, \tau)d\theta\, d\tau\, \bar{s}(t-\tau)e^{-jw_c(\tau+\tau_i^t(\theta)+\tau_k^r(\theta))} \tag{5.4}$$

Note that so far we have considered only a single waveform $\bar{s}(t)$. Allowing for the possibility that each transmit antenna will have a different waveform, we define the vector

$$\bar{\mathbf{s}}(t) = [\bar{s}_1(t), \ldots, \bar{s}_{N_T}(t)]^T \tag{5.5}$$

consisting of the waveforms for the different antennas. Let

$$\mathbf{a}_T(\theta) = [e^{-jw_c\tau_1^t(\theta)}, \ldots, e^{-jw_c\tau_{N_T}^t(\theta)}]^T \tag{5.6}$$

denote the array manifold of the transmit array, where τ^t are the transmitter-related delay differences, and similarly

$$\mathbf{a}_R(\theta) = [e^{-jw_c\tau_1^r(\theta)}, \ldots, e^{-jw_c\tau_{N_R}^r(\theta)}]^T \tag{5.7}$$

is the array manifold of the receiver array, where τ^r are the receiver-related delay differences. The vector of received signals is then

$$\bar{\mathbf{x}}(t) = \mathbf{a}_R(\theta)\bar{h}(\theta, \tau)d\theta\, d\tau\, e^{-jw_c\tau}\mathbf{a}_T(\theta)^T\bar{\mathbf{s}}(t-\tau) \tag{5.8}$$

Without loss of generality, we absorb the phase term $e^{-jw_c\tau}$ into $\bar{h}(\theta, \tau)$. The array manifolds $\mathbf{a}_R(\theta)$ and $\mathbf{a}_T(\theta)$ were defined above for the case of identical omnidirectional antennas. In practice, these array manifolds will include the gain patterns of the antennas, effects of gain/phase mismatch, and mutual coupling, and will have a form different from the one defined above.

It is often desired to filter the received signal. Assume that a filter with impulse response $\bar{q}(t)$ is used to filter each element of $\bar{\mathbf{x}}(t)$. The output of the filter $\bar{\mathbf{y}}(t)$ is given by

$$\bar{\mathbf{y}}(t) = \mathbf{a}_R(\theta)\bar{h}(\theta, \tau)d\theta\; d\tau\mathbf{a}_T(\theta)^T\, \bar{\mathbf{s}}_q(t-\tau) \tag{5.9}$$

where $\bar{\mathbf{s}}_q(t) = q(t) * \bar{\mathbf{s}}(t)$. Note that the filtering of the received signal is equivalent to replacing the transmitted signal by its filtered version.

When the scene consists of many point scatterers, we need to integrate over all of their contributions to get

$$\overline{\mathbf{x}}(t) = \iint \mathbf{a}_R(\theta)\overline{h}(\theta, \tau)\mathbf{a}_T(\theta)^T\overline{\mathbf{s}}(t-\tau)\, d\tau\, d\theta \tag{5.10}$$

Note that we have here a convolution in the τ variable so that

$$\overline{\mathbf{x}}(t) = \int \mathbf{a}_R(\theta)\mathbf{a}_T(\theta)^T\overline{h}(\theta, t) * \overline{\mathbf{s}}(t) d\theta \tag{5.11}$$

Replacing $\overline{\mathbf{s}}(t)$ by $\overline{\mathbf{s}}_q(t)$ will give the filtered version of $\overline{\mathbf{x}}(t)$.

These continuous-time angle equations can be replaced by their sampled versions provided that the proper sampling intervals Δt and $\Delta\theta$ are used. Sampling in time and angle, we get

$$\overline{\mathbf{x}}[t_n] = \sum_m \mathbf{a}_R(\theta_m)\mathbf{a}_T(\theta_m)^T\overline{h}[\theta_m, t_n] * \overline{\mathbf{s}}[t_n] \tag{5.12}$$

where $\overline{h}[\theta_m, t_n] = \overline{h}(\theta_m, t_n)\Delta\theta\Delta t$, and where θ_m and $t_n = n\Delta t$ are the sampling points. Replacing $\overline{\mathbf{s}}[t_n]$ by $\overline{\mathbf{s}}_q\,[t_n]$ will give the filtered version of $\overline{\mathbf{x}}[t_n]$.

The sampling intervals are determined as follows. For temporal sampling we assume that $\overline{\mathbf{s}}(t)$ is bandlimited with bandwidth B. The return signal will then be similarly bandlimited. Sampling above the Nyquist rate, we get $\Delta t < 1/2B$. For spatial sampling, assume that the aperture of the composite array is D. We want to have the return from an area whose angular extent is $\Delta\theta$ to be correlated across the array aperture. The beamwidth of the radiation from the patch is $\lambda/\Delta\theta R$. We want the aperture to be within this beamwidth, therefore $D < (\lambda/\Delta\theta R)R$ or $\Delta\theta < \lambda/D$. If this condition is satisfied, we can represent the patch by a single amplitude $\overline{h}(\theta, \tau)d\theta\, d\tau$ for all the antennas.

Note the presence of the convolution $\overline{h}[\theta_m, t_n] * \overline{\mathbf{s}}[t_n]$ in this model. It is convenient to reformulate the model in the frequency domain, replacing the convolution by multiplication. Therefore we replace the sampled time sequences $\overline{\mathbf{x}}[t_n]$, $\overline{h}[\theta_m, t_n]$, and $\overline{\mathbf{s}}[t_n]$ by their discrete Fourier transforms $\mathbf{x}[f_n]$, $h[\theta_m, f_n]$, and $\mathbf{s}[f_n]$.

The received signal at frequency f_n is an $N_R \times 1$ vector given by

$$\mathbf{x}\,[f_n] = \sum_{m=1}^{N_a} \mathbf{a}_R(\theta_m)\mathbf{a}_T(\theta_m)^T h[\theta_m, f_n]\mathbf{s}[f_n] \tag{5.13}$$

where N_a is the number of azimuth cells in the antenna footprint. This can be written in a more compact matrix form as

$$\mathrm{x}\,[f_n] = \underbrace{\mathbf{A}_R\,\mathrm{diag}\{\mathbf{h}[f_n]\}\mathbf{A}_T^T}_{\mathbf{A}_s[f_n]}\,\mathrm{s}[f_n] \tag{5.14}$$

where $\mathbf{A}_R = [\mathbf{a}_R(\theta_1), \ldots, \mathbf{a}_R(\theta_{N_a})]$ is the $N_R \times N_a$ receive array manifold matrix, $\mathbf{A}_T = [\mathbf{a}_T(\theta_1), \ldots, \mathbf{a}_T(\theta_{N_a})]$ is the $N_T \times N_a$ transmit array manifold matrix, and $\mathbf{h}[f_n] = [h[\theta_1, f_n], \ldots, h[\theta_{N_a}, f_n]]^T$ is the $N_a \times 1$ vector of scatterer amplitudes at frequency f_n. Assembling these equations for all frequencies into a single matrix equation, we get

$$\mathbf{X} = [\mathbf{x}[f_1], \ldots, \mathbf{x}[f_{N_f}]] = [\mathbf{A}_s[f_1]\mathbf{s}[f_1], \ldots, \mathbf{A}_s[f_{N_f}]\mathbf{s}[f_{N_f}]] \tag{5.15}$$

where $\mathbf{X}$ is the $N_R \times N_f$ received data matrix.

An alternative way of writing the received data vector is given by

$$\mathbf{x}[f_n] = \mathbf{A}_R\mathbf{diag}\{\mathbf{h}[f_n]\}\mathbf{A}_T^T\mathbf{s}[f_n] = \underbrace{\mathbf{A}_R\,\mathbf{diag}\{\mathbf{g}[f_n]\}}_{\mathbf{A}_h[f_n]}\,\mathbf{h}[f_n] \tag{5.16}$$

where $\mathbf{g}[f_n] = \mathbf{A}_T^T\mathbf{s}[f_n]$ is the $N_a \times 1$ illumination vector at frequency f_n. Assembling these equations for all frequencies into a single matrix equation, we get

$$\mathbf{X} = [\mathbf{x}[f_1], \ldots, \mathbf{x}[f_{N_f}]] = [\mathbf{A}_h[f_1]\mathbf{h}[f_1], \ldots, \mathbf{A}_h[f_{N_f}]\mathbf{h}[f_{N_f}]] \tag{5.17}$$

Note that Eq. (5.15) expresses the received data as a linear function of the transmit waveform $\mathbf{s}[f_n]$ while Eq. (5.17) expresses the received data as a linear function of the scatterer amplitudes $\mathbf{h}[f_n]$.

It is sometimes convenient to write these equations for the $N_f N_R \times 1$ data vector $\mathbf{x} = \mathrm{vec}\{\mathbf{X}\}$, instead of for the data matrix $\mathbf{X}$. It is straightforward to show that

$$\mathbf{x} = \underbrace{\mathcal{A}_R\,\mathbf{diag}\{\mathbf{h}\}\mathcal{A}_T^T}_{\mathcal{A}_s}\,\mathbf{s} \tag{5.18}$$

where $\mathcal{A}_R$ is a $N_f N_R \times N_f N_a$ block diagonal matrix with blocks $\mathbf{A}_R$. More precisely, $\mathcal{A}_R = \mathbf{I}_{N_f} \otimes \mathbf{A}_R$, where $\mathbf{I}_{N_f}$ is an $N_f \times N_f$ identity matrix, and $\otimes$ denotes the Krönecker product. Similarly, $\mathcal{A}_T$ is a $N_f N_T \times N_f N_a$ block diagonal matrix with blocks $\mathbf{A}_T$, or $\mathcal{A}_T = \mathbf{I}_{N_f} \otimes \mathbf{A}_T$. Note that $\mathcal{A}_s$ is block diagonal with N_f blocks of size $N_R \times N_T$.

The vectors $\mathbf{h}$ and $\mathbf{s}$ are obtained by stacking the vectors $\mathbf{h}[f_n]$ and $\mathbf{s}[f_n]$, respectively, for all frequencies. More precisely, let $\mathbf{H} = [\mathbf{h}[f_1], \ldots, \mathbf{h}[f_{N_f}]]$ be the $N_a \times N_f$ matrix representing the distribution of scatters in azimuth and frequency. Then $\mathbf{h} = \mathrm{vec}\{\mathbf{H}\}$. Similarly, let $\mathbf{S} = [\mathbf{s}[f_1], \ldots, \mathbf{s}[f_{N_f}]]$ be a $N_T \times N_f$ matrix representing the transmitted waveforms. The ith row of this matrix is the DFT of the waveform transmitted by the ith antenna. Then $\mathbf{s} = \mathrm{vec}\{\mathbf{S}\}$.

The received data vector can be written alternatively as

$$\mathbf{x} = \underbrace{\mathcal{A}_R\,\mathbf{diag}\{\mathbf{g}\}}_{\mathcal{A}_h}\,\mathbf{h} \tag{5.19}$$

where $\mathbf{g} = \mathcal{A}_T^T \mathbf{s}$ is the stacked $N_a N_f \times 1$ azimuth–frequency illumination vector. Equivalently, $\mathbf{g} = \text{vec}\{\mathbf{G}\}$, where

$$\mathbf{G} = \mathbf{A}_T^T \mathbf{S} \tag{5.20}$$

is the $N_a \times N_f$ illumination function in azimuth and frequency. The matrix $\mathbf{A}_h$ is a block diagonal matrix with N_f blocks of size $N_R \times N_a$.

Using the illumination matrix $\mathbf{G}$, we have

$$\mathbf{X} = \mathbf{A}_R(\mathbf{G} \odot \mathbf{H}) \tag{5.21}$$

Equation (5.21) is the fundamental model for the received signal in terms of the scattering scene represented by $\mathbf{H}$, the radar illumination $\mathbf{G}$, and the receive array manifold matrix $\mathbf{A}_R$. This equation can be rewritten in terms of the time-domain counterparts of $\mathbf{X}$, $\mathbf{H}$, $\mathbf{G}$ as

$$\overline{\mathbf{X}} = \mathbf{X}\mathbf{F}^* = \mathbf{A}_R(\mathbf{G} \odot \mathbf{H})\mathbf{F}^* = \mathbf{A}_R((\overline{\mathbf{G}}\mathbf{F}^T) \odot (\overline{\mathbf{H}}\mathbf{F}^T))\mathbf{F}^* \tag{5.22}$$

where $\mathbf{F}$ is the $N_f \times N_f$ DFT matrix (i.e., $\mathbf{Fz}$ is the DFT of $\mathbf{z}$).

Note that if we define $\mathbf{G}_q = \mathbf{A}_T^T \mathbf{S}_q$, where $\mathbf{S}_q$ is the filtered version of $\mathbf{S}$, then we get the filtered output $\mathbf{Z} = \mathbf{A}_R(\mathbf{G}_q \odot \mathbf{H})$. The filtered version of $\mathbf{S}$ is given by $\mathbf{S}_q = \mathbf{S} \odot (1 \otimes \mathbf{q})$ where $\mathbf{q}$ is a $1 \times N_f$ vector of the filter frequency response.

5.2.1 Signal Model with Reduced Number of Range Cells

In the discussion above we assumed that the scattering map $\overline{\mathbf{H}}$ and the received data $\overline{\mathbf{X}}$ have the same number N_s of delay/range samples. In practice, the number of resolvable range cells is often smaller than the number of samples of the received data. In that case it is convenient to represent $\overline{\mathbf{H}}$ as a $N_a \times N_r$ matrix, where $N_r \leq N_s$ is the number of resolvable range cells. This requires some modification of the signal model, as is shown next.

As before, we convert $\overline{\mathbf{H}}$ into a frequency-domain $N_a \times N_f$ matrix $\mathbf{H}$, where $\mathbf{H} = \overline{\mathbf{H}}\,\overline{\mathbf{F}}^T$, with $\overline{\mathbf{F}}$ equivalent to the $N_f \times N_r$ Fourier transform matrix generating N_f frequencies from N_r time samples. The equations introduced earlier still hold, except that the square $N_f \times N_f$ DFT matrix $\mathbf{F}$ is replaced by the nonsquare DFT matrix $\overline{\mathbf{F}}$. The received data matrix $\mathbf{X}$ can be written in terms of the reduced dimension $\overline{\mathbf{H}}$ as

$$\mathbf{X} = \mathbf{A}_R(\mathbf{G} \odot (\overline{\mathbf{H}}\,\overline{\mathbf{F}}^T)) \tag{5.23}$$

Next we define the $N_a N_f \times N_a N_f$ permutation matrix $\mathbf{P}$, which converts $\text{vec}\{\mathbf{H}^T\}$ into $\text{vec}\{\mathbf{H}\}$:

$$\text{vec}\{\mathbf{H}\} = \mathbf{P}\,\text{vec}\{\mathbf{H}^T\} \tag{5.24}$$

Similarly we define the $N_aN_r \times N_aN_r$ permutation matrix $\overline{\mathbf{P}}$

$$\text{vec}\{\overline{\mathbf{H}}\} = \overline{\mathbf{P}}\,\text{vec}\{\overline{\mathbf{H}}^T\} \tag{5.25}$$

From $\mathbf{H} = \overline{\mathbf{H}}\,\overline{\mathbf{F}}^T$ it follows that $\mathbf{H}^T = \overline{\mathbf{F}}\,\overline{\mathbf{H}}^T$ and

$$\text{vec}\{\mathbf{H}^T\} = (\mathbf{I}_{N_a} \otimes \overline{\mathbf{F}})\,\text{vec}\{\overline{\mathbf{H}}^T\} \tag{5.26}$$

Multiplying by $\mathbf{P}$, we obtain

$$\text{vec}\{\mathbf{H}\} = \mathbf{P}(\mathbf{I}_{N_a} \otimes \overline{\mathbf{F}})\,\text{vec}\{\overline{\mathbf{H}}^T\} \tag{5.27}$$

Converting $\text{vec}\{\overline{\mathbf{H}}^T\}$ to $\text{vec}\{\overline{\mathbf{H}}\}$ using the permutation matrix $\overline{\mathbf{P}}$, we get

$$\text{vec}\{\mathbf{H}\} = \underbrace{\mathbf{P}(\mathbf{I}_{N_a} \otimes \overline{\mathbf{F}})\overline{\mathbf{P}}^T}_{\tilde{\mathbf{F}}}\ \text{vec}\{\overline{\mathbf{H}}\} \tag{5.28}$$

and finally

$$\mathbf{x} = \mathcal{A}_h\mathbf{h} = \underbrace{(\mathcal{A}_h\tilde{\mathbf{F}})}_{\mathcal{A}_{\bar{h}}}\,\overline{\mathbf{h}} \tag{5.29}$$

This equation relates the received data $\mathbf{x}$ to the reduced dimension $\overline{\mathbf{H}}$. Note that $\mathbf{x}$ is a $N_R\, N_f \times 1$ vector, and $\overline{\mathbf{h}}$ is a $N_a\, N_r \times 1$ vector. Thus, the scattering map $\mathbf{h}$ can be estimated by least-squares solution of the equation above, provided that $N_R\, N_f \geq N_a\, N_r$.

Note that in the model introduced earlier the data could be computed frequency by frequency, that is, that $\mathbf{X}$ can be generated column by column. Here $\mathbf{x}$ needs to be computed in its entirety, requiring significantly more computations. This added complexity is related to the fact that the reduced-dimension model involves an interpolation step. To see this more clearly note that

$$\mathbf{H}\mathbf{F}^* = \overline{\mathbf{H}}\underbrace{\overline{\mathbf{F}}^T\,\mathbf{F}^*}_{\mathbf{U}} \tag{5.30}$$

where $\mathbf{U}$ is an interpolator operating on the rows of the reduced dimension $\overline{\mathbf{H}}$, to generate N_s samples from $N_r \leq N_s$ samples.

5.2.2 Multipulse and Doppler Effects

So far we considered the radar return for a single pulse. Extending this to the multipulse case is straightforward, but requires factoring in Doppler effects.

First, consider the case where the scatters are stationary and Doppler is induced only by the motion of the platform carrying the radar system. Assume a platform moving at velocity v and that the transmit and receive arrays are aligned with the velocity vector of the platform. This assumption is not essential, and the results can be modified in an obvious manner to handle arrays with arbitrary angles relative to the platform velocity vector. A scatterer at azimuth θ relative to the line of flight will have a radial velocity of $v\cos\theta$ and a Doppler shift of $f_d = v\cos\theta f_c/c$, where f_c is the center frequency of the radar and c is the speed of light. Alternatively, $f_d = v\cos\theta/\lambda$, where λ is the radar wavelength. This Doppler frequency introduces a phase shift of $2\pi v\cos\theta p T_p/\lambda$ at pulse number p, where T_p is the pulse repetition interval. Denoting $\alpha = 2\pi v T_p/\lambda$, this phase shift can be written as $\alpha p\cos\theta$.

Let $\mathbf{A}_D$ be a $P \times N_a$ matrix representing the Doppler effect where the elements of $\mathbf{A}_D$ are given by

$$[\mathbf{A}_D]_{p,m} = e^{j\alpha p\cos\theta_m} \tag{5.31}$$

Let $\mathbf{A}_{DR}$ be the spacetime array manifold

$$\mathbf{A}_{DR} = \mathbf{A}_D \,\square\, \mathbf{A}_R \tag{5.32}$$

where $\square$ denotes the Khatri–Rao product where $\mathbf{A}\,\square\,\mathbf{B} = [\mathbf{a}_1 \otimes \mathbf{b}_1, \ldots, \mathbf{a}_N \otimes \mathbf{b}_N]$, where $\mathbf{a}_n$ and $\mathbf{b}_n$ are the columns of $\mathbf{A}$ and $\mathbf{B}$, respectively.

The spacetime manifold consists of P submatrices, each corresponding to one radar pulse. In other words

$$\mathbf{A}_{DR} = \begin{bmatrix} \mathbf{A}_{DR}[1] \\ \mathbf{A}_{DR}[2] \\ \vdots \\ \mathbf{A}_{DR}[P] \end{bmatrix} \tag{5.33}$$

It is straightforward to show that the radar return matrix for pulse p is given by

$$\mathbf{X}[p] = \mathbf{A}_{DR}[p](\mathbf{G} \odot \mathbf{H}) \tag{5.34}$$

Assembling all the radar returns into a single matrix $\mathbf{X}$

$$\mathbf{X} = \begin{bmatrix} \mathbf{X}[1] \\ \mathbf{X}[2] \\ \vdots \\ \mathbf{X}[P] \end{bmatrix} \tag{5.35}$$

we get

$$\mathbf{X} = \mathbf{A}_{DR}(\mathbf{G} \odot \mathbf{H}) \tag{5.36}$$

This equation is very similar to Eq. (5.21) for the single-pulse case, where the spatial array manifold $\mathbf{A}_R$ is replaced by the spacetime manifold $\mathbf{A}_{DR}$. In the STAP literature $\mathbf{X}$ is often called the "radar data cube."

Next we consider the case where the scatterers are moving (an airborne target, moving vehicles, etc.). In the most general case the radar return matrix is given by

$$\mathbf{X}[p] = \mathbf{A}_R(\mathbf{G} \odot \mathbf{H}[p]) \tag{5.37}$$

where $\mathbf{H}[p]$ is the scattering map and the phase of each scatterer varies from pulse to pulse in accordance with the radial velocity of the scatterer relative to the radar platform.

The case where only a single scatterer (or a rigid collection of scatterers) is in motion is of special interest because we are often interested in detecting a moving target in the presence of stationary clutter clutter. In this case the data model can be written as follows. Let $\mathbf{H}_t$ denote the zero-Doppler scattering map of a target at azimuth θ_t with velocity component v_t in the radar direction. Then the target radar return is given by

$$\mathbf{X}_t[p] = e^{j\beta_t p}\,\mathbf{A}_R(\mathbf{G} \odot \mathbf{H}_t) \tag{5.38}$$

where $\beta_t = 2\pi v_t T_p \cos\theta_t/\lambda$. Let

$$\mathbf{d}_t = [e^{j\beta_t 1}, \ldots, e^{j\beta_t P}]^T \tag{5.39}$$

be the target Doppler vector representing the target related phase shift for all pulses. Then, stacking the matrices $\mathbf{X}_t[p]$ into $\mathbf{X}_t$, we get

$$\mathbf{X}_t = \mathbf{d}_t \otimes (\mathbf{A}_R(\mathbf{G} \odot \mathbf{H}_t)) \tag{5.40}$$

Multiple targets can be handled similarly so that the total return from all targets will be a sum of terms having the form of Eq. (5.40):

$$\mathbf{X} = \sum_i \mathbf{d}_t[i] \otimes (\mathbf{A}_R(\mathbf{G} \odot \mathbf{H}_t[i])) \tag{5.41}$$

In the multipulse model discussed so far, it was assumed that the same waveforms was used on all pulses. We note that a more general formulation would allow transmission of different waveforms on different pulses, in which case the model would be

$$\mathbf{X}[p] = \mathbf{A}_{DR}[p](\mathbf{G}[p] \odot \mathbf{H}[p]) \tag{5.42}$$

where $\mathbf{G}[p] = \mathrm{A}_T^T\mathbf{s}[p]$, where $\mathbf{s}[p]$ is the waveform transmitted during pulse p. However, we do not explore the case of different waveforms for different pulses here.

5.2.3 The Complete Model

So far we considered the case where the scattering scene $\mathbf{H}$ is noise-free. In practice, these data will be contaminated by noise assumed to be zero-mean complex Gaussian and independent across measurements. Additionally, it is sometimes useful to decompose the scattering scene $\mathbf{H}$ into a component due to targets of interest $\mathbf{H}_t$ and a component due to clutter $\mathbf{H}_c$. We denote the corresponding components of the received data by $\mathbf{X}_t$ and $\mathbf{X}_c$ respectively. The complete signal model for the received radar signal $\mathbf{Y}$ is then given by

$$\mathbf{Y} = \mathbf{X}_t + \mathbf{X}_c + \mathbf{X}_n \tag{5.43}$$

where $\mathbf{X}_n$ is the noise matrix, $\mathbf{X}_t = \mathbf{A}_R(\mathbf{G} \odot \mathbf{H}_t)$ and $\mathbf{X}_c = \mathbf{A}_R(\mathbf{G} \odot \mathbf{H}_c)$. This can also be written in vectorized form as $\mathbf{y} = \mathbf{x}_t + \mathbf{x}_c + \mathbf{x}_n$, where $\mathbf{x}_n \sim \mathcal{CN}\,(0, \mathbf{R}_{x_n})$. The noise term may include receiver noise and any external interference. In the case where only receiver noise is present, the noise covariance matrix has the simpler form $\mathbf{R}_{x_n} = \sigma_n^2 \mathbf{I}$, where σ_n^2 is the noise variance.

5.2.4 The Statistical Model

The clutter return $\mathbf{x}_c$ changes randomly over time and can be represented as a random process whose statistics are determined by the characteristics of the scatterers $\mathbf{H}_c$. In the following we assume that $\mathbf{h}_c = \text{vec}\{\mathbf{H}_c\}$ is a multivariate complex Gaussian vector with zero mean and covariance $\mathbf{R}_{h_c}$, that is, $\mathbf{h}_c \sim \mathcal{CN}(0, \mathbf{R}_{h_c})$. Recalling that the clutter return is related to the scatters by $\mathbf{x}_c = \mathcal{A}_h \mathbf{h}_c$, it follows that $\mathbf{R}_{x_c} = \mathcal{A}_h \mathbf{R}_{h_c} \mathcal{A}_h^H$.

The clutter covariance matrix can always be decomposed as $\mathbf{R}_{x_c} = \mathbf{V}_{x_c} \mathbf{V}_{x_c}^H$ where $\mathbf{V}_{x_c}$ is an $N_R N_f \times r_c$ matrix, where r_c is the rank of $\mathbf{R}_{x_c}$. If all the clutter scatterers $\mathbf{h}_c$ are assumed to be uncorrelated, $\mathbf{R}_{x_c}$ will be of full rank: $r_c = N_R N_f$. If the clutter is correlated, then $r_c < N_R N_f$.

The target return $\mathbf{x}_t$ is also assumed to have a similar statistical model where $\mathbf{h}_t = \text{vec}\{\mathbf{H}_t\}$ is a multivariate complex Gaussian vector with zero mean and covariance $\mathbf{R}_{h_t}$ specifically, $\sim\mathcal{CN}(0, \mathbf{R}_{h_t})$, and $\mathbf{R}_{x_t} = \mathcal{A}_h \mathbf{R}_{h_t} \mathcal{A}_h^H$. The rank of $\mathbf{R}_{x_t}$ will be denoted r_t.

The covariance matrix of the target is usually of low rank, in which case $\mathbf{R}_{x_t} = \mathbf{V}_{x_t} \mathbf{V}_{x_t}^H$, where $\mathbf{V}_{x_t}$ is an $N_R N_f \times r_t$ matrix, where r_t is the rank of $\mathbf{R}_{x_t}$. For point targets we have $r_t = 1$, while for extended targets we may have $r_t > 1$. As seen in the following sections, this formulation allows the treatment of extended targets in a systematic way.

5.3 ESTIMATION

One of the objectives of a radar system is to estimate the scattering scene $\mathbf{H}$ in order to detect, localize, or classify targets of interest. In this section we describe two estimation techniques; the first is for the case where the transmitter uses a set of orthogonal waveforms for the different antennas, while the second can be used for arbitrary transmit waveforms.

5.3.1 Beamforming Solution

Consider a receiver where the output of each antenna is passed through a bank of filters with frequency responses $\mathbf{q}_i$, $i = 1, \ldots, N_T$, matched to the transmit waveforms. These matched filters $\mathbf{q}_i$ are the rows of $\mathbf{S}^*$. The match-filtered outputs can be written as

$$\mathbf{Z} = \begin{bmatrix} \mathbf{Z}_1 \\ \mathbf{Z}_2 \\ \vdots \\ \mathbf{Z}_{N_T} \end{bmatrix} = \begin{bmatrix} \mathbf{A}_R((\mathbf{A}_T^T\mathbf{S}_{q_1}) \odot \mathbf{H}) \\ \mathbf{A}_R((\mathbf{A}_T^T\mathbf{S}_{q_2}) \odot \mathbf{H}) \\ \vdots \\ \mathbf{A}_R((\mathbf{A}_T^T\mathbf{S}_{q_{N_T}}) \odot \mathbf{H}) \end{bmatrix} \tag{5.44}$$

where $\mathbf{Z}_i$ is an $N_R \times N_f$ matrix containing the outputs of all receive antennas filtered by $\mathbf{q}_j$, and $\mathbf{S}_{q_i} = \mathbf{S} \cdot (1_{N_T} \otimes \mathbf{q}_i)$, where 1_{N_T} is an $N_T \times 1$ vectors of $1-$ s.

Next we pass the N_TN_R outputs through a beamformer that uses a weight vector $\mathbf{V}(\theta)$ equal to the steering vector of the composite transmit and receive arrays. In other words, $\mathbf{V}(\theta) = \mathbf{a}_T(\theta) \otimes \mathbf{a}_R(\theta)$. The beamformer output $\mathbf{b}(\theta_m)$, a $1 \times N_f$ vector, can be written as

$$\mathbf{b}(\theta_m) = \mathbf{V}(\theta_m)^H\mathbf{Z} = (\mathbf{a}_R(\theta_m)^H\ \mathbf{A}_R)\sum_{n=1}^{N_T} \mathbf{a}_T[n]^*((\mathbf{A}_T^T\mathbf{S}_{q_n}) \odot \mathbf{H}) \tag{5.45}$$

where $a_T[n]$ is the nth element of $a_T(\theta)$ and θ_m is the direction at which the steering vector is pointed. Next we rewrite the illumination function as

$$\mathbf{A}_T^T\mathbf{S}_{q_n} = \mathbf{A}_T[n]^T\mathbf{S}_{q_n}[n] = \sum_{k=1,k\neq n}^{N_r} \mathbf{A}_T[k]^T\mathbf{S}_{q_n}[k] \tag{5.46}$$

where $\mathbf{A}_T$ [k] is the kth row of $\mathbf{A}_T$. Inserting this into Eq. (5.45), we get

$$\mathbf{b}(\theta_m) = (\mathbf{a}_R(\theta_m)^H\mathbf{A}_R)\sum_{n=1}^{N_T} \mathbf{a}_T[n]^*(\mathbf{A}_T[n]^T\mathbf{S}_{q_n}[n]) \odot \mathbf{H} + \tilde{\mathbf{b}}(\theta_m) \tag{5.47}$$

where

$$\tilde{\mathbf{b}}(\theta_m) = (\mathbf{a}_R(\theta_m)^H\mathbf{A}_R)\sum_{n=1}^{N_T} \mathbf{a}_T[n]^* \sum_{k=1,k\neq n}^{N_T} (\mathbf{A}_T[k]^T\mathbf{S}_{q_n}[k]) \odot \mathbf{H} \tag{5.48}$$

Note that $\mathbf{S}_{q_n}[n]$ are the transmit signals passed through their own matched filters while $\mathbf{S}_{q_n}[k]$ for $k \neq n$ are the signals passed through filters matched to the other signals. In other words, $\tilde{\mathbf{b}}(\theta_m)$ contains all the cross-terms. Assuming that the transmit waveforms are uncorrelated (or have sufficiently low correlation), the elements of $\tilde{\mathbf{b}}(\theta_m)$ will be small relative to the elements of the first term in Eq. (5.47), and can be neglected. Note also that $\mathbf{S}_{q_n}[n]$ is the range response of the nth waveform.

Assume that the waveforms are designed so they all have the same range response $\mathbf{s}_q$, namely, $\mathbf{S}_{q_n}[n] = \mathbf{s}_q$. Then Eq. (5.47) can be written as

$$\mathbf{b}(\theta_m) \approx (\mathbf{a}_R(\theta_m)^H \mathbf{A}R)(((\mathbf{a}_T(\theta_m)^H \mathbf{A}_T)^T \mathbf{s}_q) \odot \mathbf{H}) \tag{5.49}$$

We see here that the rows of $\mathbf{H}$ are weighted by the product of the transmit and receive beampatterns $\mathbf{a}_T(\theta)^H \mathbf{A}_T$, $\mathbf{a}_R(\theta)^H \mathbf{A}_R$ and the columns by the range response $\mathbf{s}_q$, as is expected. To see this more clearly, let

$$\mathbf{p}(\theta m) = (\mathbf{a}_R(\theta_m)^H \mathbf{A}_R) \odot (\mathbf{a}_T(\theta_m)^H \mathbf{A}_T) \tag{5.50}$$

be the row vector denoting the beamformer response vector when it is pointed in direction θ_m. Equation (5.49) can be rewritten as

$$\mathbf{b}(\theta_m) = (\mathbf{p}(\theta_m)\mathbf{H}) \odot \mathbf{s}_q \tag{5.51}$$

An alternative derivation of these results for the scenario with a single scatterer of unit strength at azimuth θ_m, proceeds as follows. The receive array response for transmission from antenna n at the output of the corresponding matched filter is given by $\mathbf{a}_R(\theta_m)\mathbf{a}_T(\theta_m)[n]\mathbf{s}_q$. If we assume that the signals from different transmitters are completely decoupled at the matched filter outputs, the response to this signal at the outputs of the other matched filters will be zero. Therefore the stacked response vector for all the matched filters is given by $(\mathbf{a}_T(\theta_m) \otimes \mathbf{a}_R(\theta_m))\mathbf{s}_q$. The corresponding beamformer weight vector is $\mathbf{V}(\theta_m)$, which was defined above.

Next we consider briefly the case where the transmitted signals are not orthogonal. Then the transmitted waveform matrix can be written as

$$\mathbf{S} = \mathbf{C}^T \tilde{\mathbf{S}} \tag{5.52}$$

where $\tilde{\mathbf{S}}$ is an orthogonal set of waveforms and $\mathbf{C}$ is an $N_T \times N_T$ matrix containing the coefficients of the representation of the columns of $\mathbf{S}$ in the orthogonal basis $\tilde{\mathbf{S}}$. The received data matrix for a scene with a single scatterer of unit strength at azimuth θ_m is given by

$$\mathbf{X} = \mathbf{a}_R(\theta_m)\mathbf{a}_T(\theta_m)^T \mathbf{C}^T \tilde{\mathbf{S}} = \mathbf{a}_R(\theta_m)(\mathbf{C}\mathbf{a}_T(\theta_m))^T \tilde{\mathbf{S}} \tag{5.53}$$

Passing this through a bank of filters matched to the rows of $\tilde{\mathbf{S}}$, we get

$$\mathbf{Z} = ((\mathbf{C}\mathbf{a}_T(\theta_m)) \otimes \mathbf{a}_R(\theta_m))\tilde{\mathbf{s}}_q \tag{5.54}$$

provided that all of these orthogonal signals have the *same* range response $\tilde{\mathbf{s}}_q$.

In principle, the effect of the correlation can now be removed by filtering $\mathbf{Z}$ with $\mathbf{C}^{-1} \otimes \mathbf{I}_{N_R}$

$$\begin{aligned}(\mathbf{C}^{-1} \otimes \mathbf{I}_{N_R})((\mathbf{C}\mathbf{a}_T(\theta)) \otimes \mathbf{a}_R(\theta))\tilde{\mathbf{s}}_q &= (\mathbf{C}^{-1}\mathbf{C}\mathbf{a}_T(\theta)) \otimes (\mathbf{I}_{N_R}\mathbf{a}_R(\theta))\tilde{\mathbf{s}}_q \\ &= \mathbf{a}_T(\theta) \otimes \mathbf{a}_R(\theta)\tilde{\mathbf{s}}_q \end{aligned} \tag{5.55}$$

provided that $\mathbf{C}$ is a well-conditioned nonsingular matrix.

In the more general case where the range responses corresponding to the rows of $\tilde{\mathbf{S}}$ are different, it may still be possible to decorrelate the outputs by "equalizing" the responses. This can be accomplished by dividing the rows of $\mathbf{Z}_i$ by $\tilde{s}_q/\tilde{s}_q[i]$, where $\tilde{s}_q[i]$ is the range response of the ith row of $\tilde{\mathbf{S}}$ and $\tilde{s}_q$ is a nominal range response (e.g., the average of the range responses). This will work only if the responses $\tilde{s}_q[i]$ are not too different from each other.

Note that for the purpose of the analysis above the matched filters were shown as if applied directly to the transmitted waveforms. In reality, the matched filtering is applied to the received signals, not the transmitted signals. The bank of matched filters is implemented at the receiver by

$$\tilde{\mathbf{Z}} = (1_{N_R} \otimes \mathbf{S}^*) \odot (\mathbf{X} \otimes 1_{N_T}) \tag{5.56}$$

where 1_{N_R}, 1_{N_T} are $N_R \times 1$ and $N_T \times 1$ column vectors of ones, respectively, and

$$\tilde{\mathbf{Z}} = \begin{bmatrix} \tilde{\mathbf{Z}}_1 \\ \tilde{\mathbf{Z}}_2 \\ \vdots \\ \tilde{\mathbf{Z}}_{N_R} \end{bmatrix} \tag{5.57}$$

where $\tilde{\mathbf{Z}}_i$ is the output of a matched filterbank processing the signal from the ith receive antenna (whereas $\mathbf{Z}_i$ was the ith matched filter applied to all antennas). $\tilde{\mathbf{Z}}$ is a $(N_T N_R) \times N_f$ matrix. Note that $\tilde{\mathbf{Z}}$ and $\mathbf{Z}$ are equivalent up to permutation of rows. The beamformer weight vector applied to $\tilde{\mathbf{Z}}$ is $\mathbf{a}_R(\theta) \otimes \mathbf{a}_T(\theta)$ instead of $\mathbf{a}_T(\theta) \otimes \mathbf{a}_R(\theta)$.

Computing the beam former outputs $\mathbf{b}(\theta_m)$ for all angles θ_m of interest, and arranging them in an $N_a \times N_f$ matrix $\mathbf{B}$, we get an estimate of the scattering distribution $\mathbf{H}$ in the azimuth–frequency domain. The beamformer output can be written as

$$\mathbf{B} = \mathbf{V}^H (1_{N_R} \otimes \mathbf{S}^*) \odot (\mathbf{X} \otimes 1_{N_T}) = \mathbf{V}^H \tilde{\mathbf{Z}} \tag{5.58}$$

where $\mathbf{V} = [V(\theta_1), \ldots, V(\theta_{N_a})]$.

Applying the beamformer to the time-domain received data instead of the frequency-domain data, we get an $N_a \times N_s$ matrix $\overline{\mathbf{B}}$ that provides an estimate of the scattering distribution $\overline{\mathbf{H}}$ in the azimuth–range domain.

In the multipulse case the beamformer will be applied to the receiver outputs generated for multiple pulses: $\tilde{\mathbf{Z}}[1], \ldots, \tilde{\mathbf{Z}}[P]$. These outputs will be stacked into a single data matrix $\tilde{\mathbf{Z}}$. In the case of a stationary scattering scene the signal models for the single-pulse and multipulse cases have exactly the same form, except that the spatial manifold $\mathbf{A}_R$ is replaced by the spacetime manifold $\mathbf{A}_{DR}$. It follows that the weight vectors for the multipulse case are the columns of

$$\mathbf{V} = \mathbf{A}_D \square \mathbf{A}_R \square \mathbf{A}_T \tag{5.59}$$

while in the single-pulse case they are the columns of $\mathbf{A}_R \square \mathbf{A}_T$. As before, $\mathbf{B} = \mathbf{V}^H \tilde{\mathbf{Z}}$.

In the case where there are moving targets, we need to calculate the beamformer output for all possible Doppler frequencies, because the target Dopplers are unknown. Let

$$\mathbf{a}_d(\beta) = \begin{bmatrix} e^{j2\pi\beta 0} \\ e^{j2\pi\beta 1} \\ \vdots \\ e^{j2\pi\beta(P-1)} \end{bmatrix} \tag{5.60}$$

be the Doppler manifold, where $-0.5 \leq \beta \leq 0.5$ is the normalized Doppler frequency. The spacetime steering vector now has the form

$$\mathbf{V}(\theta, \beta) = \mathbf{a}_d(\beta) \otimes \mathbf{a}_R(\theta) \otimes \mathbf{a}_T(\theta) \tag{5.61}$$

The beam former output will be calculated for a finite set of angles θ_m and Doppler frequencies β_n. Let

$$\mathbf{A}_d = [\mathbf{a}_d(\beta_1), \ldots, \mathbf{a}_d(\beta_{N_d})] \tag{5.62}$$

be the Doppler manifold matrix. Then the beam former weight vectors are the columns of

$$\mathbf{V} = \mathbf{A}_d \otimes (\mathbf{A}_R \square \mathbf{A}_T) \tag{5.63}$$

and $\mathbf{B} = \mathbf{V}^H\tilde{\mathbf{Z}}$. Comparing Eq. (5.63) to Eq. (5.59), we note that here we have all combinations of Doppler frequencies and angles, whereas before each angle had a single specific Doppler frequency associated with it. This was because the scene was stationary and motion was induced only by the motion of the radar platform.

When the scene consists of both stationary (e.g., clutter) and moving (e.g., target) components, one can choose to use the steering vector defined in either (5.59) or (5.63), depending on what is being estimated. Usually it is the moving target that is of interest. However, it is sometimes necessary to estimate the stationary background, which can be accomplished at the smaller computational cost associated with (5.59). Note that in the multipulse case the beamformer outputs can be arranged in a $N_dN_a \times N_f$ matrix $\mathbf{B}$, or equivalently in a $N_d \times N_a \times N_f$ "cube" representing an estimate of the scattering distribution in the Doppler–azimuth–frequency domain. Applying the beamform er to the time-domain received data instead of the frequency-domain data, we get a $N_d \times N_a \times N_s$ cube that provides an estimate of the scattering distribution in the Doppler–azimuth–range domain.

To illustrate the beamforming estimation, we present an example for a MIMO radar with $N_T = N_R = 10$, where the element spacing of the transmit and receive arrays was $d_T = 5\lambda$ and $d_R = 0.5\lambda$, respectively. Other system parameters were $N_s = 1024$, $N_f = 1024$, and $N_a = 256$.

Figure 5.1 depicts the beampatterns of the transmit, receive, and composite transmit-receive arrays, for rectangular windowing. Note the high angular resolution achieved here — a 3 dB beamwidth of 1° — compared to a SIMO system employing

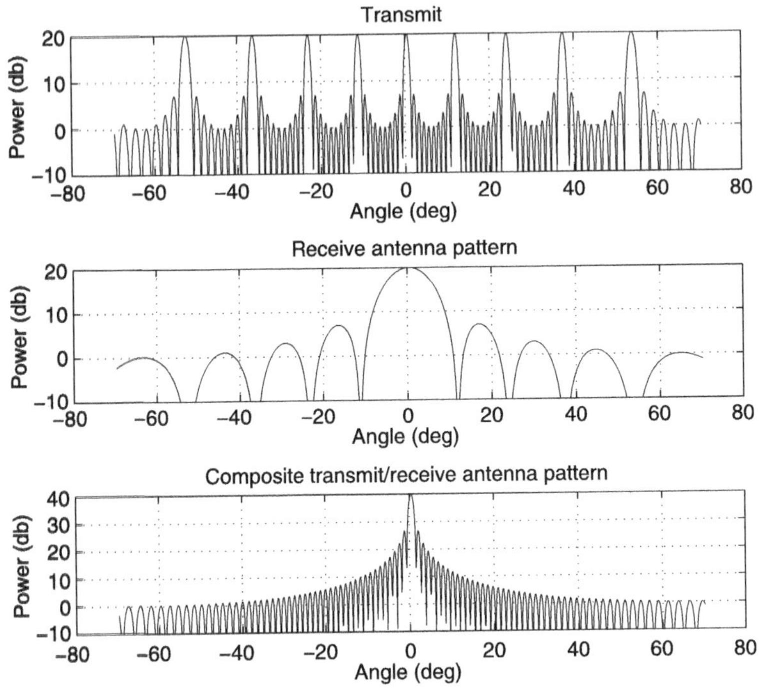

Figure 5.1 Beampatterns for $N_T = N_R = 10$ with $d_T = 5\lambda$ and $d_R = 0.5\lambda$.

a 10-element array that has a beamwidth of approximately 10°. This high resolution is a consequence of the large aperture of the transmit array where the antenna spacing was chosen to be much larger than a half-wavelength. More specifically, we let $d_T = N_R d_R$, which results in the composite transmit–receive array behaving like a $N_R N_T$-element array with $\lambda/2$ spacing. Note that the transmit beampattern has multiple beams (grating lobes). However, the receive beampattern eliminates all except one of these beams, thus resolving the possible ambiguity. The composite transmit–receive beampattern does not change if we reverse the spacing of the transmit and receive arrays, specifically, letting $d_T = 0.5\lambda$ and $d_R = 5\lambda$ in this example.

Figure 5.2 depicts the range response of the system. The transmit waveforms used here were random sequences of ± 1, independent from antenna to antenna. These are not orthogonal but have low cross-correlation. It is possible, of course, to design better sequences that are orthogonal and have good autocorrelation properties (see, e.g., Ref. 26). Figure 5.3 depicts the beamformer output matrix $\overline{\mathbf{B}}$ in azimuth–range for a simple test case with 10 point targets and SNR of 0 dB. Note the high spatial resolution achieved by the combination of the transmit and receive array beampatterns.

Repeating the experiment related in Fig. 5.3 for the multipulse case using the steering vector (5.59) yields a very similar figure except that the signal-to-noise ratio is improved by a factor of P, due to the coherent integration across pulses. In this experiment we assumed both the clutter and the targets to be stationary.

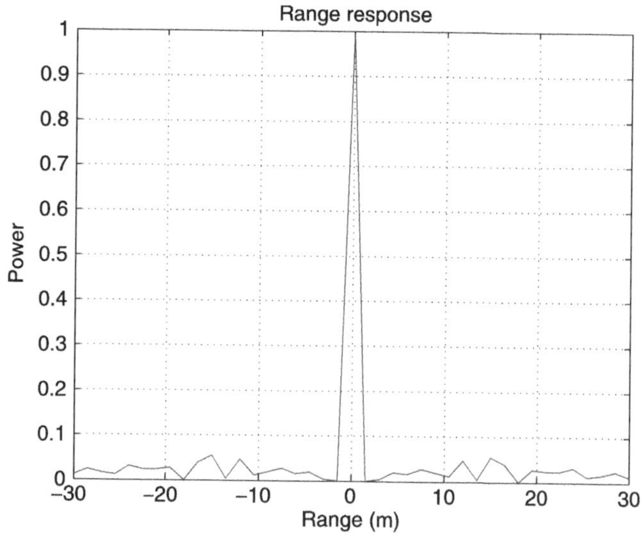

Figure 5.2 Range response using a random 1024-point transmit waveform.

Next we consider the case where the targets are moving. We used the steering vector (5.63) to estimate the Doppler–azimuth–range distribution of the scatterers. Figure 5.4 depicts an azimuth–Doppler "slice" through the range cell of a target located at azimuth of 30° and normalized Doppler frequency of 0.25. The target can be clearly seen as well as the clutter ridge.

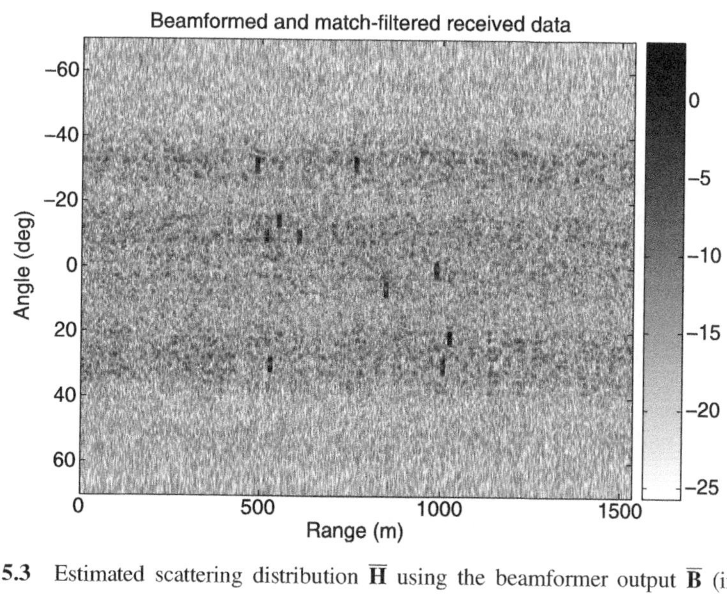

Figure 5.3 Estimated scattering distribution $\overline{\mathbf{H}}$ using the beamformer output $\overline{\mathbf{B}}$ (intensity in dB).

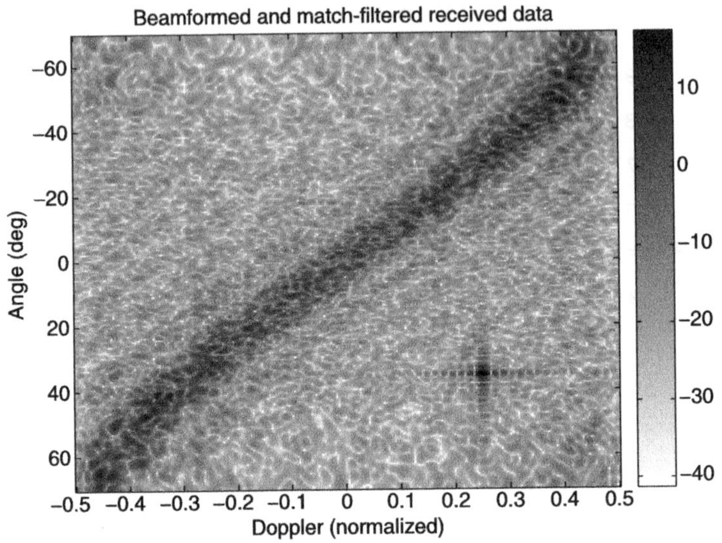

Figure 5.4 Azimuth–Doppler distribution for a range bin containing a single target and clutter, using orthogonal waveforms (intensity in dB).

5.3.2 Least-Squares Solutions

A more direct estimation of the scattering distribution follows from the fact that the received data x is a linear function of **h**, suggesting the possibility of estimating **h** using the least-squares solution of $\mathbf{x} = \mathcal{A}_{\mathrm{h}}\mathbf{h}$. Note that **x** is a $N_R N_f \times 1$ vector and **h** is an $N_a N_f \times 1$ vector. For the equation to be solvable, we must have $N_a \leq N_R$, which implies poor angular resolution. When this condition is not satisfied, we can use the reduced dimension version where $\mathbf{x} = \mathcal{A}_{\bar{\mathrm{h}}}\bar{\mathbf{h}}$, where $\bar{\mathbf{h}}$ is a $N_a N_r \times 1$ vector. If $N_a \leq N_R N_f / N_r$, we have a unique solution for $\bar{\mathbf{h}}$. Note that by decreasing N_r, we can increase the maximum value of N_a, which satisfies this inequality. In other words, we increase the number of azimuth cells at the cost of reducing the number of range cells. The least-squares solution can be used for arbitrary transmit waveforms and does not require orthogonality of the transmit waveforms.

To illustrate the least-squares solution, we present an example for a MIMO radar with $N_T = N_R = 10$, where the element spacing of the transmit and receive arrays was 5λ and 0.5λ, respectively. Other system parameters were $N_s = 128$, $N_f = 128$, $N_a = 32$, and $N_r = 32$. Note that $N_a = 32 < N_R N_f / N_r = 40$. Figure 5.5 depicts the estimated scatterer distribution $\overline{\mathbf{H}}$ in azimuth–range for a simple test case with 10 point targets and SNR of 0 dB. All 10 targets are visible in their correct locations.

5.3.3 Waveform Design for Estimation

In the absence of noise and interference, the estimation of the scatterer distribution $\overline{\mathbf{H}}$ is affected relatively little by the choice of the transmit waveform, as long as the radar

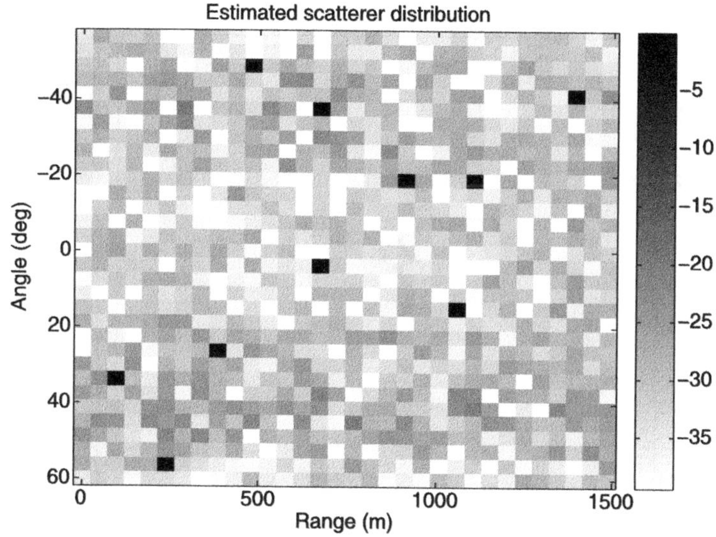

Figure 5.5 Estimated image $\overline{\mathbf{H}}$ using the least-squares solution; intensity is in decibels.

illuminates the entire area of interest, and as $\mathbf{A}_{\overline{\mathrm{h}}}$ is of sufficient rank. In the presence of noise and interference the accuracy of the estimates will be affected by the radar illumination $\mathbf{G}$, which is controlled by the transmitted waveforms $\mathbf{S}$.

Different performance criteria can be used to design the transmitted waveform, depending on the application at hand. In this section we consider two cases: (1) maximizing the total radar return and (2) matching the illumination to the scene. In the first case we design $\mathbf{S}$ so as to maximize the noise-free received signal $\mathbf{x} = \mathcal{A}_s\mathbf{s}$. In other words, we maximize $\|\mathbf{x}\|^2 = \mathbf{s}^H \mathcal{A}_s^H \mathcal{A}_s^H \mathcal{A}_s\mathbf{s}$. The solution is given by making $\mathbf{s}$ equal to the eigenvector of $\mathcal{A}_s^{\mathrm{H}}\ \mathcal{A}_s\mathbf{s}$ corresponding to its largest eigenvalue. This approach tends to focus the transmit energy on the dominant scatterer or group of scatterers, which may not be desirable, because the rest of the scene may contain features of interest that are not be adequately illuminated.

An alternative way of controlling the effective illumination is to match the illumination to the scene. In other words, we want to make the projection of $\mathbf{g}$ on $\mathbf{h}$ as large as possible. This means maximizing $\mathbf{g}^H\mathbf{P}_{\mathrm{h}}\mathbf{g} = \mathbf{s}^H \mathcal{A}_T^* \mathbf{P}_{\mathrm{h}} \mathcal{A}_T^T \mathbf{s}$. Since $\mathbf{P}_{\mathrm{h}}$ is of unit rank, the solution is to have $\mathbf{s} = (\mathcal{A}_T \mathcal{A}_T^T)^{-1}\ \mathcal{A}_T\mathbf{h}$ up to normalization of the transmit power. The total received energy is smaller than in the case described above, but it is more likely that all features of interest will be illuminated.

Note that both of these methods require knowledge of the scene $\mathbf{h}$ that they are trying to estimate and that the scene is not known a priori. This suggests an iterative procedure where we start with some initial waveform $\mathbf{s}_0$ (e.g., a set of orthogonal waveforms) and obtain an initial estimate of the scene h_0. A new waveform $\mathbf{s}_1$ is designed on the basis of the current estimate $\mathbf{h}_0$. This waveform is used to obtain a new estimate $\mathbf{h}_1$. This process may be repeated to further refine the estimate of $\mathbf{h}$. However, some care needs to be taken to make this process converge.

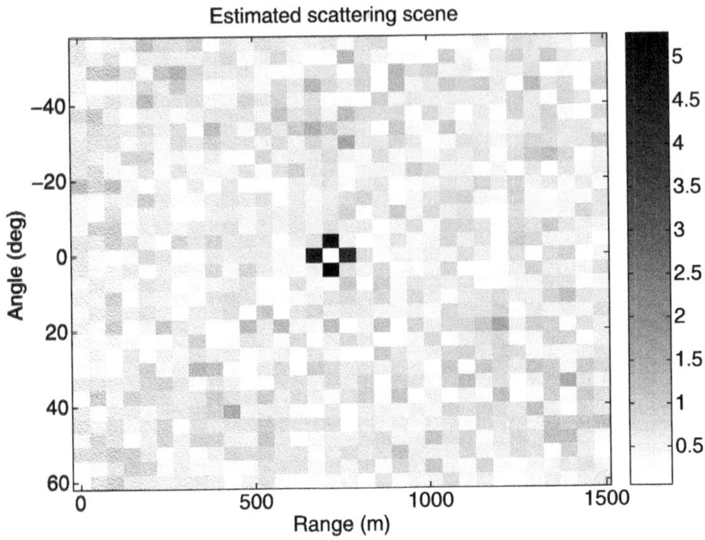

Figure 5.6 Estimated image for orthogonal waveforms.

To illustrate the matched illumination method, we present an example of a MIMO radar with $N_T = N_R = 10$, where the element spacing of the transmit and receive arrays was 5λ and 0.5λ, respectively. Other system parameters were $N_s = 128$, $N_f = 128$, $N_a = 32$, and $N_r = 32$. Figures 5.6 and 5.7 depict the estimated scatterer distribution $\overline{\mathbf{H}}$ in azimuth–range for the random waveform used in the previous images, and the matched illumination waveform discussed above. The scene consists

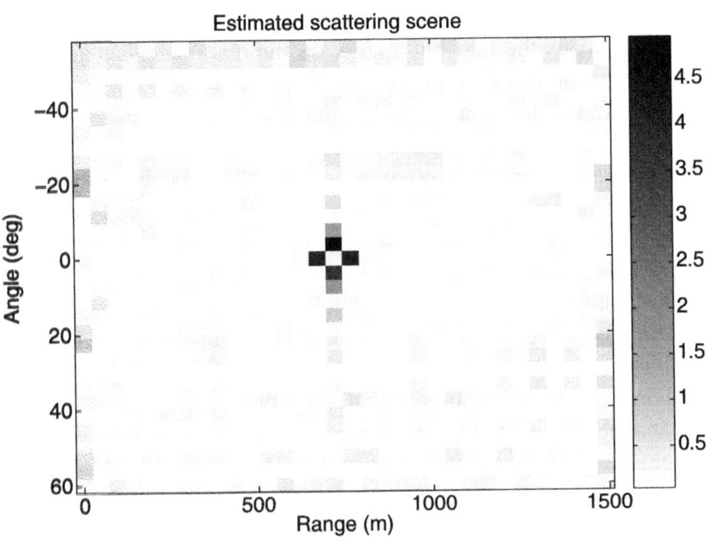

Figure 5.7 Estimated image for matched illumination waveforms.

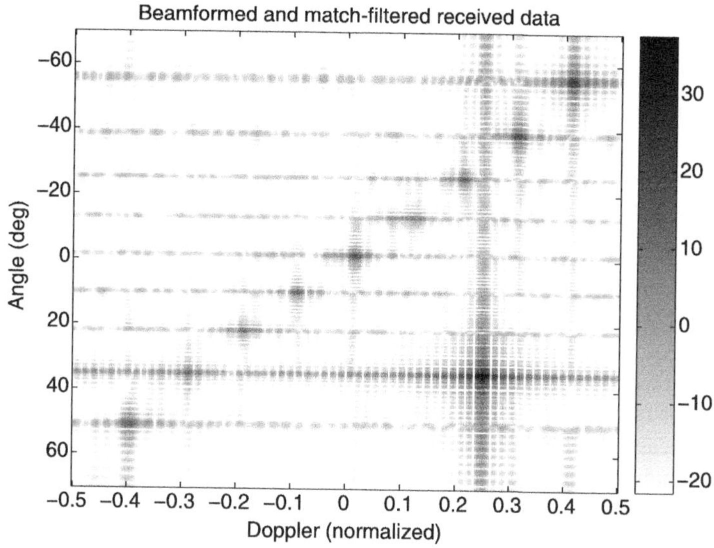

Figure 5.8 Azimuth–Doppler distribution for a range bin containing a single target and clutter, using a target-matched waveform, $d_T = 5\lambda$, $d_R = 0.5\lambda$.

of a closely spaced cluster of four point targets. The matched illumination waveform yields a less noisy image; the SNR observed in Figs. 5.6 and 5.7 is 16 and 25 dB, respectively. The effect of using different waveforms is quantified more precisely in the next section.

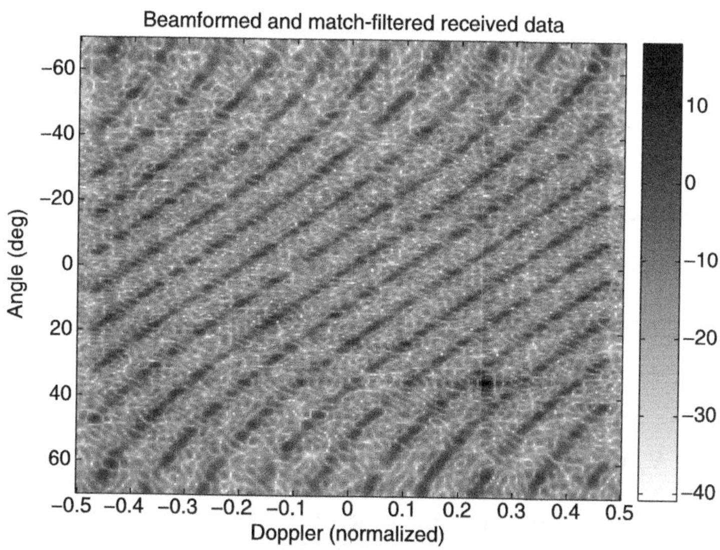

Figure 5.9 Azimuth–Doppler distribution for a range bin containing a single target and clutter, using a target-matched waveform, $d_T = 0.5\lambda$, $d_R = 5\lambda$.

Figure 5.8 depicts an example where the multiple pulses are transmitted. This is the same case depicted in Fig. 5.4, except that the target-matched waveform is used instead of the orthogonal waveform. In this case the illumination function has the shape of the transmit beampattern and the clutter is only partially illuminated, as can be clearly see from the appearance of the clutter ridge. The horizontal strips indicate the directions where the transmit beam is pointing. Figure 5.9 depicts the same example, except that the transmit and receive airays were swapped.

5.4 DETECTION

In this section we consider target detection using a MIMO radar. The radar may employ a fixed set of transmit waveforms, or it may adapt the waveforms on the basis of previously measured data. We start by developing the structure of the optimal detector under the assumption that both target and clutter statistics are known.

5.4.1 The Optimal Detector

Given the signal model presented in Section 5.2, we formulate the target detection problem as the following Gauss–Gauss binary hypothesis testing problem. We assume that the received data contain either clutter and noise, hypothesis H_0, or target clutter and noise, hypothesis H_1

$$H_0\colon \quad \mathbf{x} \sim \mathcal{CN}(0, \mathbf{R}_{x_{cn}}) \tag{5.64}$$

$$H_1\colon \quad \mathbf{x} \sim \mathcal{CN}(0, \mathbf{R}_{x_t} + \mathbf{R}_{x_{cn}} \tag{5.65}$$

where $\mathcal{CN}$ denotes the multivariate complex Gaussian distribution and $\mathbf{R}_{x_{cn}} = \mathbf{R}_{x_c} + \mathbf{R}_{x_n}$. The optimal detector for this problem is known to have the quadratic form [28]

$$d = \mathbf{x}^H \mathbf{Q} \mathbf{x} \tag{5.66}$$

where d is the detection statistic, and the kernel $\mathbf{Q}$ is given by

$$\mathbf{Q} = \mathbf{R}_{x_{cn}}^{-1} - [\mathbf{R}_{x_t} + \mathbf{R}_{x_{cn}}]^{-1} \tag{5.67}$$

This can also be written as

$$\mathbf{Q} = \mathbf{R}_{x_{cn}}^{-1} \mathbf{V}_{x_t} [\mathbf{V}_{x_t}^H \mathbf{R}_{x_{cn}}^{-1} \mathbf{V}_{x_t} + \mathbf{I}]^{-1} \mathbf{V}_{x_t}^H \mathbf{R}_{x_{cn}}^{-1} = \mathbf{W}\mathbf{W}^H \tag{5.68}$$

where $\mathbf{R}_{x_t} = \mathbf{V}_{x_t} \mathbf{V}_{x_t}^H$ and

$$\mathbf{W} = \mathbf{R}_{x_{cn}}^{-1} \underbrace{\mathbf{V}_{x_t} [\mathbf{V}_{x_t}^H \mathbf{R}_{x_{cn}}^{-1} \mathbf{V}_{x_t} + \mathbf{I}]^{-(1/2)}}_{\tilde{\mathbf{V}}_{x_t}} \tag{5.69}$$

Then the detection statistic is given by

$$d = \mathbf{x}^H \mathbf{W}\mathbf{W}^H \mathbf{x} = \|\mathbf{W}^H \mathbf{x}\|^2 \tag{5.70}$$

In the case of a unit rank target case ($r_t = 1$), $\mathbf{V}_{x_t}$ is a vector. To emphasize this fact, we denote it by the lowercase symbol $\mathbf{v}_{x_t}$. In this case

$$\mathbf{W} = \mathbf{R}_{x_{cn}}^{-1} \mathbf{v}_{x_t} \tag{5.71}$$

where we discarded a scalar multiplier term that can be absorbed into the detection threshold. This is the well-known minimum variance distortionless response (MVDR) detector [30]. The more general case of equations (5.69)–(5.70) where the rank of the target covariance is greater than unity, is the generalized MVDR (GMVDR) presented in Ref. 29.

5.4.2 The SINR

Because both the target and clutter statistics $\mathbf{R}_{x_t}$, $\mathbf{R}_{x_c}$ depend on the transmit signal **s**, the matrix **W** defining the optimal detector is a function of the transmit waveform **s**. We want to select this waveform so as to maximize the performance of the detector. This can be achieved by maxim izing the signal-to-interference-plus-noise ratio (SINR), at the output of the detector.

It follows from Eq. (5.70) that the SINR is given by

$$\text{SINR} = \frac{\text{tr}\{\mathbf{W}^H \mathbf{R}_{x_t} \mathbf{W}\}}{\text{tr}\{\mathbf{W}^H \mathbf{R}_{x_{cn}} \mathbf{W}\}} \tag{5.72}$$

or equivalently

$$\text{SINR} = \frac{\text{tr}\{\tilde{\mathbf{V}}_{x_t}^H \mathbf{R}_{x_{cn}}^{-1} \mathbf{R}_{x_t} \mathbf{R}_{x_{cn}}^{-1} \tilde{\mathbf{V}}_{x_t}\}}{\text{tr}\{\tilde{\mathbf{V}}_{x_t}^H \mathbf{R}_{x_{cn}}^{-1} \tilde{\mathbf{V}}_{x_t}\}} \tag{5.73}$$

It is sometimes convenient to rewrite this expression as

$$\text{SINR} = \frac{\text{tr}\{\mathbf{V}_{x_t}^H \mathbf{R}_{x_{cn}}^{-1} \mathbf{V}_{x_t} (\mathbf{V}_{x_t}^H \mathbf{R}_{x_{cn}}^{-1} \mathbf{V}_{x_t} + \mathbf{I})^{-1} \mathbf{V}_{x_t}^H \mathbf{R}_{x_{cn}}^{-1} \mathbf{V}_{x_t}\}}{\text{tr}\{\mathbf{V}_{x_t}^H \mathbf{R}_{x_{cn}}^{-1} \mathbf{V}_{x_t} (\mathbf{V}_{x_t}^H \mathbf{R}_{x_{cn}}^{-1} \mathbf{V}_{x_t} + \mathbf{I})^{-1}\}} \tag{5.74}$$

which makes it clear the SINR is a function of a single matrix $\mathbf{C} = \mathbf{V}_{x_t}^H \mathbf{R}_{x_{cn}}^{-1} \mathbf{V}_{x_t}$

$$\text{SINR} = \frac{\text{tr}\{\mathbf{C}(\mathbf{C} + \mathbf{I})^{-1} \mathbf{C}\}}{\text{tr}\{\mathbf{C}(\mathbf{C} + \mathbf{I})^{-1}\}} \tag{5.75}$$

This can be rewritten in terms of the eigenvalues λ_i of **C** as

$$\text{SINR} = \frac{\sum_{m=1}^{r_t} \lambda_i^2 / (\lambda_i + 1)}{\sum_{m=1}^{r_t} \lambda_i / (\lambda_i + 1)} \tag{5.76}$$

When $\mathbf{R}_{xt}$ is of unit rank, $\mathbf{C}$ is a scalar and all of these equations reduce to

$$\text{SINR} = \text{C} = \mathbf{v}_{xt}^H \mathbf{R}_{x_{cn}}^{-1} \mathbf{v}_{x_t} \tag{5.77}$$

The equations above were derived under the assumption that the detector weight vector $\mathbf{W}$ was designed using the true target vector $\mathbf{v}_{x_t}$. It is of interest to also consider the case where the detector is designed using the target vector at an assumed target location that may be different from the actual one. Therefore we consider two versions of the target signature: $\mathbf{v}_{x_t}$ — the signature corresponding to the actual target location and $\mathbf{v}_{x_t}^d$ — the signature corresponding to the target location assumed by the detector. In the case where $r_t > 1$, we have $\mathbf{V}_{x_t}$ and $\mathbf{V}_{x_t}^d$.

Recall that $\mathbf{v}_{x_t} = \mathbf{A}_{\bar{\mathbf{h}}_t} \bar{\mathbf{h}}_t$. In the case of a point target, $\bar{\mathbf{h}}_t$ is a vector of zeros with a single nonzero element corresponding to the target location, so that the assumed target signature $\mathbf{v}_{x_t}^d$ may be any one of the columns of $\mathbf{A}_{\bar{\mathbf{h}}_t}$, while the true target signature $\mathbf{v}_{x_t}$ is one particular column.

The SINR equations need to be modified as follows. For $r_t > 1$, replace $\tilde{\mathbf{V}}_{x_t}$ in equation (5.73) by

$$\tilde{\mathbf{V}}_{x_t}^d = \mathbf{V}_{x_t}^d [(\mathbf{V}_{x_t}^d)^H \mathbf{R}_{x_{cn}}^{-1} \mathbf{V}_{x_t}^d + \mathbf{I}]^{-(1/2)} \tag{5.78}$$

leaving $\mathbf{R}_{x_t}$ unchanged. In the unit rank $r_t = 1$ case

$$\text{SINR} = \frac{(\mathbf{v}_{x_t}^d)^H \mathbf{R}_{x_{cn}}^{-1} (\mathbf{v}_{x_t} \mathbf{v}_{x_t}^H) \mathbf{R}_{x_{cn}}^{-1} \mathbf{v}_{x_t}^d}{(\mathbf{v}_{x_t}^d)^H \mathbf{R}_{x_{cn}}^{-1} \mathbf{v}_{x_t}^d} \tag{5.79}$$

Both the target and clutter covariance matrices are functions of the transmitted waveform $\mathbf{s}$. It is straightforward to show that

$$\mathbf{R}_{x_t} = \underbrace{\mathcal{A}_R\, \mathbf{diag}\{\mathcal{A}_T^T \mathbf{s}\}}_{\mathbf{A}_{h_t}} \mathbf{R}_{h_t} \underbrace{\mathbf{diag}\{\mathcal{A}_T^T \mathbf{s}\}^H \mathcal{A}_R^H}_{\mathbf{A}_{h_t}^H} \tag{5.80}$$

and therefore

$$\mathbf{V}_{x_t} = \mathcal{A}_R\, \mathbf{diag}\{\mathcal{A}_T^T \mathbf{s}\} \mathbf{V}_{h_t} \tag{5.81}$$

Similarly, we have

$$\mathbf{R}_{x_c} = \underbrace{\mathcal{A}_R\, \mathbf{diag}\{\mathcal{A}_T^T \mathbf{s}\}}_{\mathbf{A}_{hc}} \mathbf{R}_{h_c} \underbrace{\mathbf{diag}\{\mathcal{A}_T^T \mathbf{s}\}^H \mathcal{A}_R^H}_{\mathbf{A}_{hc}^H} \tag{5.82}$$

Inserting these into Eq. (5.73) gives an explicit expression for the SINE in terms of $\mathbf{s}$.

In the reduced-dimension model these equations are modified as follows

$$\mathbf{V}_{x_t} = \mathcal{A}_R\, \mathbf{diag}\{\mathcal{A}_T^T \mathbf{s}\} \tilde{\mathbf{F}} \mathbf{V}_{\bar{h}_t} \tag{5.83}$$

and in equations (5.80), (5.82) we replace $\mathbf{R}_{\bar{h}_t}$ by $\tilde{\mathbf{F}}\mathbf{R}_{\bar{h}_t}, \tilde{\mathbf{F}}^H$, and $\mathbf{R}_{\bar{h}_c}$ by $\tilde{\mathbf{F}}\mathbf{R}_{\bar{h}_c}\tilde{\mathbf{F}}^H$. In the multipulse case the SINR is defined as before

$$\text{SINR} = \mathbf{v}_{x_t}^H \mathbf{R}_{x_{cn}}^{-1} \mathbf{v}_{x_t} \tag{5.84}$$

where

$$\mathbf{R}_{x_{cn}} = \mathcal{A}_h \mathbf{R}_{h_c} \mathcal{A}_h^H + \sigma^2 \mathbf{I} \tag{5.85}$$

However, $\mathcal{A}_h$ is now defined as

$$\mathcal{A}_h = \mathcal{A}_{DR}\, \mathbf{diag}\{\mathrm{g}\} \tag{5.86}$$

where $\mathcal{A}_{DR} = \mathbf{I}_{N_f} \otimes \mathbf{A}_{DR}$, and $\mathbf{v}_{x_t} = \text{vec}\{\mathbf{X}_t\}$, where

$$\mathbf{X}_t = \mathbf{d}_t \times (\mathbf{A}_R(\mathbf{G} \odot \mathbf{H}_t)) \tag{5.87}$$

Note that in the case of a point target $\mathbf{H}_t$ has a single nonzero row corresponding to the target direction.

5.4.3 Optimal Waveform Design

An optimal transmit waveform $\mathbf{s}$ can be designed by maximizing the SINR over all possible choices of $\mathbf{s}$. Because of the nonlinear dependence of SINR on $\mathbf{s}$, a closed-form solution does not seem to be available and we must resort to numerical optimization. Various optimization techniques can be used to solve for the transmit waveform $\mathbf{s}$, which maximizes the SINR. Here we consider a gradient descent optimization method [31] that requires knowledge of the derivatives of SINR with respect to the elements of $\mathbf{s}$. The update equation of the gradient descent method is given by

$$\mathbf{s} \longleftarrow \mathbf{s} + \mu \left[\frac{\partial \text{SINR}}{\partial \mathbf{s}}\right] H \tag{5.88}$$

where μ is a constant controlling the convergence rate of the algorithm. At each step of the algorithm we also rescale $\mathbf{s}$ to have a fixed norm, that is, to obey the average power constraint.

Taking the derivative of the SINR with respect to s_i, we get

$$\frac{\partial\, \text{SINR}}{\partial s_i} = 2\mathbf{s}^H \mathcal{A}_{st}^H \mathbf{R}_{x_{cn}}^{-1} \mathcal{A}_{st} \mathbf{e}_i - 2\mathbf{s}^H \mathcal{A}_{st}^H \mathbf{R}_{x_{cn}}^{-1} \frac{\partial \mathbf{R}_{x_{cn}}}{\partial s_i} \mathbf{R}_{x_{cn}}^{-1} \mathcal{A}_{st} \mathbf{s} \tag{5.89}$$

where $\mathcal{A}_{st} = \mathcal{A}_R\, \mathbf{diag}\{\mathrm{h}_t\} \mathcal{A}_T^T$, s_i is the ith entry of $\mathbf{s}$, and $\mathbf{e}_i$ is vector of zeros with a 1 at the ith position. (See Refs. 32 and 33 for examples of calculating this type of derivative.)

Recalling $\mathbf{R}_{x_c}$ from Eq. (5.82) and noting that $(\partial \mathbf{R}_{x_{cn}}/\partial s_i) = (\partial \mathbf{R}_{x_c}/\partial s_i)$, we get

$$\frac{\partial \mathbf{R}_{x_{cn}}}{\partial s_i} = \mathcal{A}_R\, \mathbf{diag}\{\mathcal{A}_T^T \mathbf{e}_i\} \mathbf{R}_{h_c} \mathbf{diag}\{\mathcal{A}_T^T \mathbf{s}\}^H \mathcal{A}_R^H \tag{5.90}$$

Note that $\mathcal{A}_T^T\mathbf{e}_i$ is the ith column of $\mathcal{A}_T^T$ or the ith row of $\mathcal{A}_T$.

Then

$$\frac{\partial\ \text{SINR}}{\partial s_i} = 2p_i - 2q_i \tag{5.91}$$

where p_i is the ith element of the row vector

$$\mathbf{p} = \mathbf{s}^H \mathcal{A}_{st}^H \mathbf{R}_{x_{cn}}^{-1} \mathcal{A}_{st} \tag{5.92}$$

and

$$q_i = \underbrace{\mathbf{s}^H \mathcal{A}_{st}^H \mathbf{R}_{x_{cn}}^{-1} \mathcal{A}_R}_{\mathbf{q}_a} \mathbf{diag}\{\mathcal{A}_T^T\mathbf{e}_i\} \underbrace{\mathbf{R}_{h_c} \mathbf{diag}\{\mathcal{A}_T^T\mathbf{s}\}^H \mathcal{A}_R^H \mathbf{R}_{x_{cn}}^{-1} \mathcal{A}_{st}\mathbf{s}}_{\mathbf{q}_b^T} \tag{5.93}$$

where $\mathbf{q}_a$ and $\mathbf{q}_b$ are row vectors defined in the equation above. Denoting $\mathbf{Q}_c = \mathcal{A}_R^H \mathbf{R}_{x_{cn}}^{-1} \mathcal{A}_{st}$, we have $\mathbf{q}_a = \mathbf{s}^H \mathbf{Q}_c^H$, $\mathbf{q}_b^T = \mathbf{R}_{h_c} \mathbf{diag}\{\mathcal{A}_T^T\mathbf{s}\}^H \mathbf{Q}_c \mathbf{s}$. Next, let $\mathcal{A}_T^T[i] = \mathcal{A}_T^T\mathbf{e}_i$ be the ithe column of $\mathcal{A}_T^T$. Then

$$q_i = \mathbf{q}_a \mathbf{diag}\{\mathcal{A}_T^T[i]\}\mathbf{q}_b^T = (\mathbf{q}_a \odot \mathbf{q}_b)\mathcal{A}_T^T[i] \tag{5.94}$$

and assembling all q_i values into a row vector $\mathbf{q}$, we get

$$\mathbf{q} = (\mathbf{q}_a \odot \mathbf{q}_b)\mathcal{A}_T^T \tag{5.95}$$

Finally

$$\frac{\partial\ \text{SINR}}{\partial s} = 2(\mathbf{p} - \mathbf{q}) \tag{5.96}$$

Equations (5.88), (5.92), (5.95), and (5.96) define the gradient descent algorithm for computing the SINR-maximizing waveform $\mathbf{s}$. To initialize the algorithm, we let $\mathbf{s}$ be one of the suboptimal waveforms described in Section 5.4.4.

Note that in the case of the reduced-dimension model $\mathbf{R}_{h_c}$ in Eq. (5.93) needs to be replaced by $\tilde{\mathbf{F}}\mathbf{R}_{\tilde{h}_c}\tilde{\mathbf{F}}^H$.

5.4.4 Suboptimal Waveform Design

In this section we present two suboptimal waveform design algorithms that are computationally less complex than the algorithms for designing the optimal waveform.

The first involves matching the radar illumination to the target in a manner similar to that considered in Section 5.3.3. Let $\mathbf{v}_{h_t}$ denote the unit rank target signature vector and design the illumination vector $\mathbf{g}$ so that it matches the target as closely as possible. In other words, let $\mathbf{g} = \mathcal{A}_T^T\mathbf{s} \approx \mathbf{v}_{h_t}$. The solution is given by $\mathbf{s} = (\mathcal{A}_T\mathcal{A}_T^T)^{-1}\mathcal{A}_T\mathbf{v}_{h_t}$ followed by normalization of $\mathbf{s}$ to meet the transmit power constraint. We will refer to the resulting waveform as the "target-matched waveform."

The second method designs the waveform using both the target signature and the clutter covariance matrix. Let $\mathbf{P}_c$ be the $N_a \times N_f$ matrix representing the power distribution of clutter in range and frequency. In vectorized form we have $\mathbf{p}_c = \mathrm{vec}\{\mathbf{P}_c\}$, which equals the diagonal of the clutter covariance matrix $\mathbf{R}_{h_c}$. The illumination will be designed so that it matches as closely as possible the target : clutter ratio. In other words, let $\mathbf{g} = \mathcal{A}_T^T \mathbf{s} \approx \mathbf{v}_{h_t} / \sqrt{\mathbf{p}_c}$, where both the division and the square-root operations are element-by-element (elementwise) operations. The solution is given by $\mathbf{s} = (A_T A_T^T)^{-1} A_T (\mathbf{v}_{h_t} / \sqrt{\mathbf{p}_c})$, followed by normalization of $\mathbf{s}$ to meet the transmit power constraint. We will refer to the resulting waveform as the "target-to-clutter-matched waveform."

5.4.5 Constrained Design

In the discussion above we allowed the waveform $\mathbf{s}$ to assume any shape, subject only to a power constraint. It is of interest to also consider the case where $\mathbf{s}$ is a linear combination of a set of L predetermined waveforms ψ_i. In other words, assume that $\mathbf{s} = \Psi \mathbf{f}$, where $\Psi = [\psi_1, \ldots, \psi_L]$ is an $N_T N_f \times L$ matrix and $\mathbf{f}$ is the $L \times 1$ vector of coefficients of the representation of $\mathbf{s}$ in this basis. It is straightforward to extend the waveform design algorithms presented here to this case, by replacing $\mathbf{s}$ with $\psi \mathbf{f}$ in all the equations.

Consider, for example, the target matched waveform. Replacing $\mathbf{s}$ with $\psi \mathbf{f}$, we get

$$g = \mathcal{A}_T^T \Psi \mathbf{f} \approx \mathbf{h}_t \tag{5.97}$$

and the least-squares solution is

$$\mathbf{f} = (\Psi^H \mathcal{A}_T^* \mathcal{A}_T^T \Psi)^{-1} \Psi^H \mathcal{A}_T^* \mathbf{h}_t \tag{5.98}$$

Different choices can be made for the spacetime basis Ψ, including

- $\Psi = \mathbf{I}_{N_T N_f}$ — this is the unconstrained case, where the spacetime basis covers the entire available space.
- $\Psi = \mathbf{I}_{N_f} \otimes \mathbf{a}_T(\theta)$ — here we force the transmitter to point a beam in direction θ. The design of the waveform will determine the temporal signal properties, while the spatial properties are fixed a priori.
- $\Psi = [\mathbf{I}_{N_f} \otimes \mathbf{a}_T(\theta_1), \ldots, \mathbf{I}_{N_f} \otimes \mathbf{a}_T(\theta_M)]$ — this is similar to the previous case, except that we predefine M different beam directions and let the design process determine the spatial distribution of the transmit energy, as well as the temporal distribution. In other words, we perform unconstrained temporal waveform design and constrained spatial design.
- $\Psi = \mathbf{q} \otimes \mathbf{I}_{N_T}$ — here we force the transmitter to use a single waveform $\mathbf{q}$ on all antennas, but allow the design procedure to determine the spatial response. In other words, we have a fixed temporal design, while the spatial design is done adaptively.

- $\Psi = [\mathbf{q}_1 \otimes \mathbf{I}_{N_T}, \ldots, \mathbf{q}_N \otimes \mathbf{I}_{N_T}]$ — this is similar to the previous case, except that we predefine N different beam temporal waveforms and let the design process determine which combination of waveforms to use. In other words, we perform unconstrained spatial design and constrained temporal design.
- $\Psi = \Psi_t \otimes \Psi_s$ — here we select any desired temporal basis function Ψ_t and a desired spatial basis function Ψ_s. In other words, we constrain both the temporal and the spatial designs.

In a separate paper [36] we studied the impact of these choices on the various waveform design procedures discussed here and the resulting performance of the estimator and detector. Our results show that proper selection of the basis is useful for obtaining stable and "well-behaved" spatial and temporal response, that is, range response and beampattern. Imposing the basis constraints does, of course, decrease performance because of the reduction in the number of degrees of freedom available in the design. However, this performance loss can be made arbitrarily small by increasing the dimension of the subspace spanned by these basis functions.

5.4.6 The Target and Clutter Models

As was shown above, the optimal detector is based on the second-order statistics of the target, clutter, and noise. Different target and clutter models arise in different applications. In this section we describe the models used in this work.

We consider both point targets — targets smaller than a single range–azimuth cell, and extended targets covering multiple range–azimuth cells. A point target has a unit rank (rank 1) covariance matrix $\mathbf{R}_{h_t}$ and $\mathbf{R}_{x_t}$. An extended target covering n_c cells has a covariance matrix with rank $1 \leq r_t \leq n_c$, depending on the degree of correlation between the radar returns from the different cells. Here we focus on the case of highly correlated returns that yield a unit rank covariance. This corresponds, for example, to a rigid target with fixed orientation relative to the radar. Motion relative to the radar will introduce the same random phase fluctuations to the returns from all of the scatterers constituting the target so that the target scatters can be represented by $\mathbf{v}_{h_t} e^{j\phi}$ where $\mathbf{v}_{h_t}$ is a deterministic vector and ϕ is random phase. In this case $\mathbf{R}_{h_t} = \mathbf{v}_{h_t}\mathbf{v}_{h_t}^H$ where $\mathbf{v}_{h_t}$ equals $\mathbf{h}_t$ up to an arbitrary unit magnitude complex scalar. A target that is not a rigid collection of point scatterers, or is one that is rotating relative to the radar, will induce different random variations along the vector $\mathbf{h}_t$, and the corresponding covariance matrix $\mathbf{R}_{h_t}$ will have an effective rank $r_t > 1$.

We assume that the clutter is represented by a random scattering map $\mathbf{H}_c$ where clutter returns from different range–azimuth cells are uncorrelated. The clutter covariance matrix is therefore diagonal and the elements of the diagonal represent the clutter power distribution $\mathbf{p}_c$. The rank of the clutter covariance matrix $\mathbf{R}_{h_c}$ is relatively large, and the matrix may in general be of full rank.

In order to implement the detector, knowledge of the target and clutter covariance matrices $\mathbf{R}_{h_t}$ and $\mathbf{R}_{h_c}$ is required. The clutter covariance can be estimated by collecting multiple radar returns, estimating $\mathbf{H}_c$ using the methods described in Section 5.3, and forming the sample covariance matrix.

The target statistics are computed for an assumed target position and spectral characteristics. In the case of a point target the signature $\mathbf{v}_{h_t}$ is a function of the target azimuth. In this case we design the transmit waveform using the constrained approach discussed earlier with $\Psi = \mathbf{I}_{N_f} \otimes \mathbf{a}_T(\theta)$. In other words, we force the transmitter to point a beam in an assumed direction θ. The assumed target direction is scanned across the area of interest over subsequent transmission. In the case of an extended target, the waveform can be designed to match its spectral signature, provided that this signature is known a priori.

5.4.7 Numerical Examples

To illustrate the performance of the various detectors, we calculated the SINR for a MIMO radar with $N_T = N_R = 10$, where the element spacing of the transmit and receive arrays was $d_T = 5\lambda$ and $d_R = 0.5\lambda$, respectively. Other system parameters were $N_s = N_f = N_a = 128$. The clutter power $\mathbf{p}_c$ is assumed to be uniform over all range/azimuth cells in the scene. The SINR values are summarized in Tables 5.1 and 5.2 for an orthogonal set of waveforms and a target-to-clutter-matched waveform. The theoretical SINR results, also shown in these tables, are derived in Appendix 5A, where it is shown that for uniform illumination (single

TABLE 5.1 Experimental and Theoretical[a] SINR (in dB) for a MIMO Radar with Orthogonal Waveforms

	SNR = 0 dB	SNR = 20 dB
CNR = −20 dB	10.8 [10.0]	30.8 [30.0]
CNR = 0 dB	10.7 [10.0]	30.7 [30.0]
CNR = 20 dB	7.0 [7.5]	27.0 [27.5]
CNR = 40 dB	−10.1 [−9.0]	9.8 [11.0]

[a]Theoretical SINR values are given in square brackets.

TABLE 5.2 Experimental and Theoretical[a] SINR (in dB) for a MEMO Radar with Target-to-Clutter-Matched Waveforms

	SNR = 0 dB	SNR = 20 dB
CNR = 20 dB	20.0 [20.0]	40.0 [40.0]
CNR = 0 dB	20.0 [20.0]	40.0 [40.0]
CNR = 20 dB	16.8 [17.5]	36.8 [37.5]
CNR = 40 dB	0.3 [1.0]	20.3 [21.0]

[a]Theoretical SINR values are given in square brackets.

transmit antenna, or MIMO with orthogonal waveforms)

$$\mathrm{SINR} = \frac{N_R \mathrm{SNR}}{\mathrm{CNR}/N_f + 1} \tag{5.99}$$

and for nonuniform illumination produced by the adaptive waveforms discussed earlier (target-matched, target-to-clutter matched, optimal waveforms),

$$\mathrm{SINR} = \frac{N_R N_T \mathrm{SNR}}{\mathrm{CNR}/N_f + 1} \tag{5.100}$$

The CNR (SNR) are defined as the total clutter (target) power received during the collection time, divided by the noise power.

A number of observations can be made from these tables. There is a good match between the theoretical and experimental results. The factor of N_R ($N_R N_T$) is due to the receive (transmit and receive) array gain. As expected, the transmitter array gain is lost when using orthogonal waveforms that provide uniform illumination of the scene. The adaptively designed waveforms provide a nonuniform illumination and recapture the lost gain. Note, however, that the orthogonal waveforms make it possible to do the transmit beam forming at the receiver, after the fact. This makes it possible to scan the entire area in the antenna footprint in a single pulse period. The adaptive waveforms illuminate only a portion of the radar footprint and will generally require N_R pulses to scan the entire area.

The SINR discussed above was the SINR when the detector uses the target location, or $\mathbf{v}_{x_t}^d = \mathbf{v}_{x_t}$. It is of interest to plot the SINR at the detector output as the assumed target location is scanned over the entire range/azimuth scene, for a given target. Figures 5.10 and 5.11 depict this SINR for a MIMO radar with

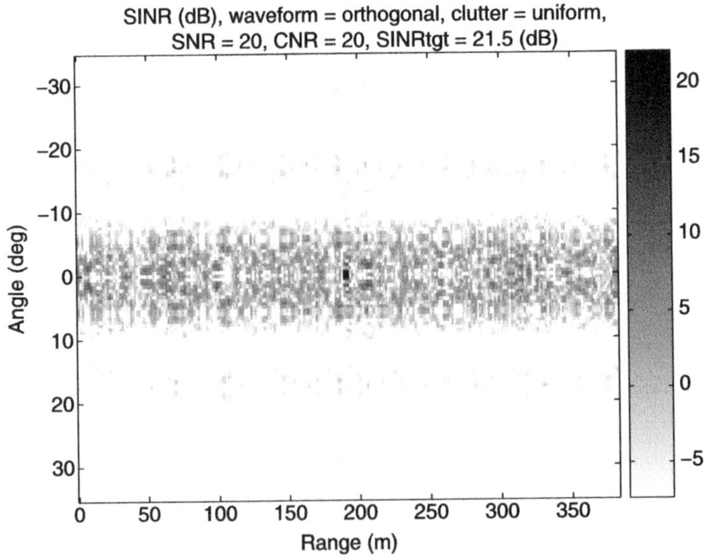

Figure 5.10 SINR for orthogonal waveforms.

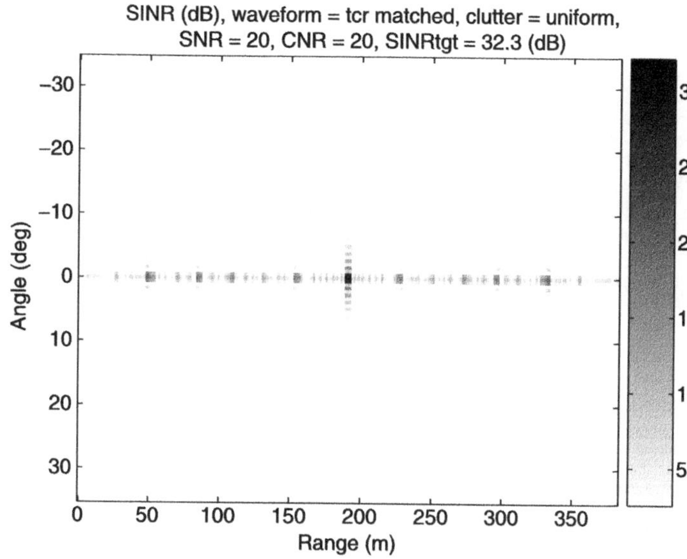

Figure 5.11 SINR for target-to-clutter-matched waveforms.

$N_T = N_R = 10$, where the element spacing of the transmit and receive arrays was $d_T = 5\lambda$ and $d_R = 0.5\lambda$, respectively. Other system parameters were $N_s = N_f = 256$, $N_a = 128$. In Fig. 5.10 the transmitter employed a set of orthogonal waveforms, while in Fig. 5.11 a target-to-clutter-matched waveform was used. The highly concentrated shape of the SINR surface makes it possible to localize the target with high resolution. The improved SINR when using an adaptively designed waveform can be observed by comparing the two figures.

To gain some insight into the shape of the waveform designed by the adaptive design procedure, we present two examples for the same MIMO radar considered above. Figure 5.12 depicts the illumination function **G** for four waveforms: orthogonal, target-matched, target-to-clutter-matched, and optimal, for the case of a point target.

Note that the adaptive waveforms and the corresponding illumination function are spectrally flat, corresponding to a narrow pulse in the time domain. This results in a narrow range response. In the spatial domain the waveforms form a narrow transmit beam focused on the target. The processing gain results from the high spatial resolution and the small range cell that contain the target, but contains only a fraction of the clutter energy. In the case of uniform clutter the SINR at the detector output is given by Eq. (5.100) for the adaptive waveforms, and by Eq. (5.99) for the orthogonal waveform.

Figure 5.13 depicts the illumination function for the case of an extended target, where the target occupies a single azimuth cell, but is spread along all range cells. In this case the adaptive waveforms are spectrally narrow, resulting with a wide range response so as to contain the extended target. In the spatial domain the waveforms form a narrow transmit beam focused on the target. In the case of uniform

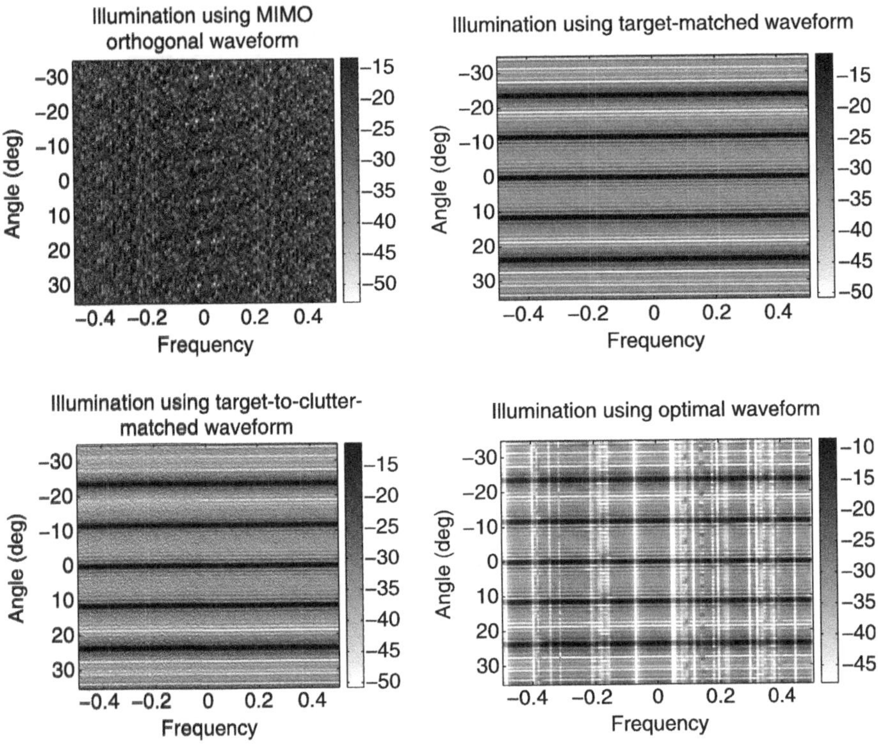

Figure 5.12 Illumination function **G** for four waveforms: orthogonal, target-matched, target-to-clutter-matched, and optimal, for the case of a point target.

clutter the SINR is approximately the same as in the point target case described above. The processing gain results from the high spatial resolution and the range response, which is narrow in the frequency domain and thus picks up only a small fraction of the clutter energy.

So far we have considered examples where the clutter was uniform. When the clutter is nonuniform, there are opportunities for additional temporal processing gain beyond the gains captures by Eqs. (5.99) and (5.100). As an example, consider the case of an extended target with nonuniform clutter. By designing the shape of the waveform in frequency (or time) to illuminate parts of the target where the clutter is relatively small, we get potentially significant improvement relative to uniform illumination (see e.g., Refs. 34 and 35).

Next we present an example of the multipulse case. Figure 5.14 depicts the SINR for different Doppler frequencies in a single range cell containing a target. Results are shown for the orthogonal waveform and the target-matched waveform. As expected, the SINR for the target-matched waveform is larger by (approximately) a factor of N_T than the SINR for the orthogonal waveform because of the transmit array gain.

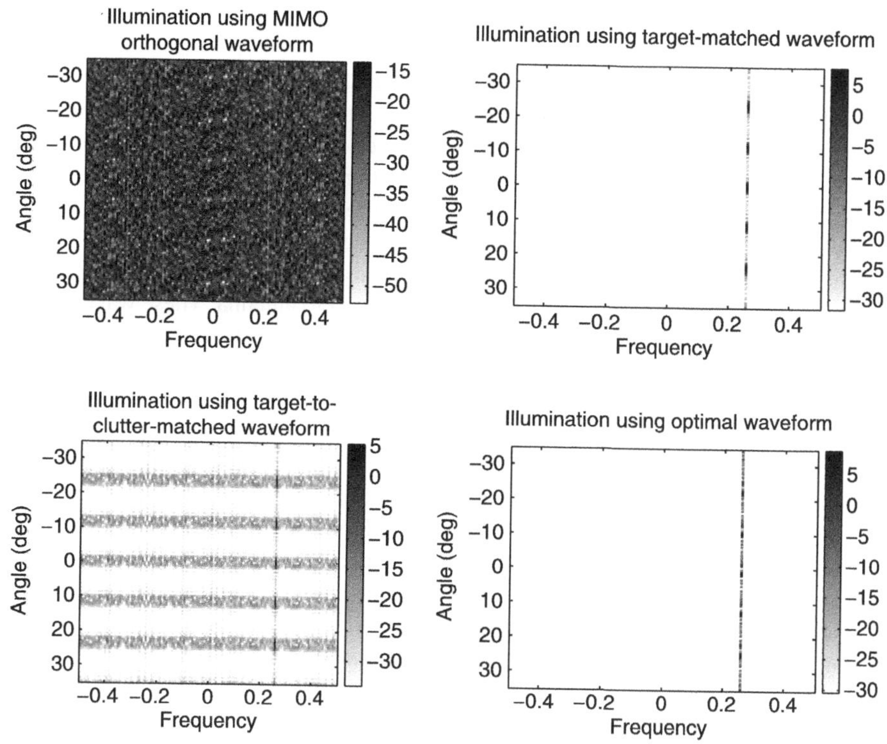

Figure 5.13 Illumination function **G** for four waveforms: orthogonal, target-matched, target-to-clutter-matched, and optimal, for the case of an extended target.

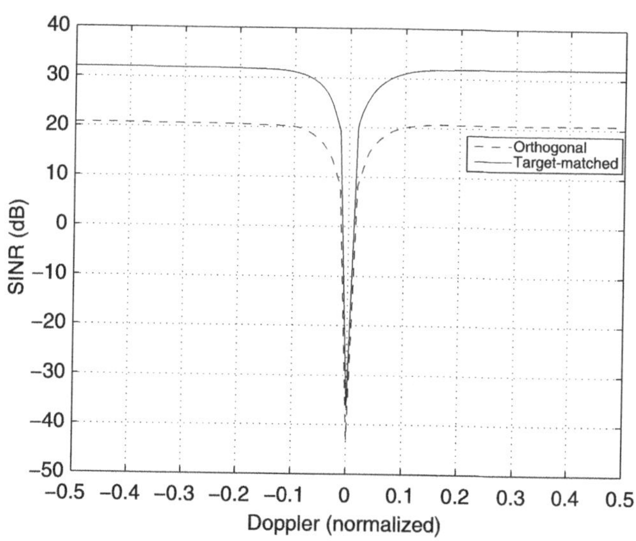

Figure 5.14 SINR versus Doppler for the orthogonal waveform and the target-matched waveform.

5.5 MIMO RADAR AND PHASED ARRAYS

The use of multiple antennas on both transmit and receive is, of course, not new. Phased arrays that form beams on both transmit and receive have been around for a long time. (See Refs. 37 and 38 and references cited therein for a historical overview of phased-array radar technology.) Phased arrays transmit a single waveform that is fed to the different antennas with different phases (delays). In other words, the waveforms at the different antennas are perfectly correlated.

The more recent use of the term "MIMO radar" refers to the case where the transmit array uses waveforms that may differ from antenna to antenna. These waveforms are usually assumed to be orthogonal or uncorrelated [39]. In fact, some references *define* MIMO radar as a radar employing orthogonal waveforms at the transmitter. In this chapter we considered a more general class of transmit waveforms. However, for the moment we focus on the case of orthogonal waveforms.

The current literature on MIMO radar discusses its various performance characteristics such as having a high angular resolution and the ability to search an extended area with that high resolution in a single dwell [7]. It is useful to note that many of the capabilities of MIMO radar can be replicated by a phased-array radar, which uses the same physical antenna array and the same transmit power. The spatial resolution of the radar system relates primarily to the array geometry and the directivity of the antenna elements that it employs. The combined transmit–receive array operates as a virtual array with $N_R N_T$ elements, as discussed in Section 5.3.1, resulting in higher resolution than a radar with a single transmit antenna. In other words, both the MIMO radar and the phasedarray have the same combined transmit–receive array manifold $a_T(\theta) \otimes a_R(\theta)$.

To see this more clearly, we consider the signal received from a point target with unit radar cross section located at azimuth θ_t and delay (range) τ_t. In the case of a phased-array radar, the received signal matrix $\mathbf{X}$ is given by

$$\mathbf{X} = \mathbf{a}_R(\theta_t)\mathbf{a}_T^T\mathbf{S} \tag{5.101}$$

where the transmitted signal consists of a single waveform $\mathbf{s}$ fed through a transmit beamformer with weight vector $\mathbf{W}_T$:

$$\mathbf{S} = \mathbf{W}_T^*\mathbf{s} \tag{5.102}$$

The transmit power is normalized to unity so we assume that $\|\mathbf{s}\|^2 = 1$ and $\|\mathbf{W}_T\|^2 = 1$. The rows of $\mathbf{X}$ are passed through a matched filter to produce the match-filtered signal matrix $\mathbf{Z}$

$$\mathbf{Z} = \mathbf{a}_R(\theta_t)\mathbf{a}_T(\theta_t)^T\mathbf{W}_T^*\mathbf{s}_q \tag{5.103}$$

where $\mathbf{s}_q$ is the match-filtered waveform $\mathbf{s}$. Assuming that the matched filter is normalized to unit power gain, we have $\|\mathbf{s}_q\|^2 = 1$.

The filtered signal is then beamformed using a beamformer whose weight vector is $\mathbf{W}_R$:

$$\mathbf{y} = \mathbf{W}_R^H \mathbf{Z} = (\mathbf{W}_R^H \mathbf{a}_R(\theta_t))(\mathbf{W}_T^H \mathbf{a}_T(\theta_t))\mathbf{s}_q \tag{5.104}$$

We assume that the beamformer is normalized to have unit noise power gain, specifically, $\|\mathbf{W}_R\|^2 = 1$.

We note that the beamformer weight vectors $\mathbf{W}_R, \mathbf{W}_T$ can be designed in different ways. These could be fixed weights designed to achieve a desired beampattern, or they may be dependent on the data, leading to an adaptive beamformer on receive, transmit, or both.

Next consider a MIMO radar that makes use of the same transmit and receive arrays and therefore has the same array manifolds $\mathbf{a}_R(\theta), \mathbf{a}_T(\theta)$. The receive signal matrix $\mathbf{X}$ is given by

$$\mathbf{X} = \mathbf{a}_R(\theta_t)\mathbf{a}_T(\theta_t)^T \mathbf{S} \tag{5.105}$$

where the rows of $\mathbf{S}$ are orthogonal waveforms. The total transmit power is unity as before so that $\|\mathbf{S}\|^2 = 1$. This means that each row of $\mathbf{S}$ has norm $1/N_T$. Passing this through a filter matched to the ith transmit waveform, we get

$$\mathbf{Z}_i = \mathbf{a}_R(\theta_t)\mathbf{a}_T(\theta_t)[i]\mathbf{s}_q \tag{5.106}$$

where $\mathbf{a}_T(\theta_t)[i]$ is the ith element of the vector $\mathbf{a}_T(\theta_t)$. Assuming that the matched filter is normalized to unit power gain, we have $\|\mathbf{s}_q\|^2 = 1/N_T$. We assume that the contributions of the other waveforms can be neglected because of orthogonality. Assembling the outputs of the different matched filters into an output matrix $\mathbf{Z}$, we get

$$\mathbf{Z} = (\mathbf{a}_T(\theta_t) \otimes \mathbf{a}_R(\theta_t))\mathbf{s}_q \tag{5.107}$$

The filtered signal is then beamformed using a beamformer whose weight vector is $\mathbf{W}_T \otimes \mathbf{W}_R$:

$$\mathbf{y} = (\mathbf{W}_T^H \otimes \mathbf{W}_R^H)\mathbf{Z} = (\mathbf{W}_R^H \mathbf{a}_R(\theta_t))(\mathbf{W}_T^H \mathbf{a}_T(\theta_t))\mathbf{s}_q \tag{5.108}$$

Comparing equations (5.104) and (5.108), we conclude that both systems produce exactly the same outputs if the same beamformer weights $\mathbf{W}_T, \mathbf{W}_R$ are used, except that the signal power in the phasedarray is N_T times larger than in the MIMO radar. As we have noted before, the MIMO radar has an SNR that is N_T times smaller than that of the phased-array system, all other things being equal.

Next we consider the multipulse case, which leads to spacetime adaptive processing (STAP). STAP radar has been extensively studied for airborne applications [27].

MIMO radar can be used in STAP as was discussed by various authors (see, e.g., Refs. 14 and 15). A comparison of STAP using transmit beamforming to STAP that uses orthogonal waveforms follows closely the earlier discussion. In the case of a phasedarray the received data matrix $\mathbf{X}$, obtained by stacking the data matrices for a sequence of pulses, is given by

$$\mathbf{X} = (\mathbf{a}_d(\beta_t) \otimes \mathbf{a}_R(\theta_t))\mathbf{a}_T^T \mathbf{W}_T^* \mathbf{s} \tag{5.109}$$

where $\mathbf{a}_d(\beta_t)$ is the target Doppler manifold [see Eq. (5.60)]. After matched filtering, we have

$$\mathbf{Z} = (\mathbf{a}_d(\beta_t) \otimes \mathbf{a}_R(\theta_t))\mathbf{a}_T(\theta_t)^T \mathbf{W}_T^* \mathbf{s}_q \tag{5.110}$$

Applying a Doppler–azimuth weight vector $\mathbf{W}_d \otimes \mathbf{W}_R$, we get

$$\mathbf{y} = (\mathbf{W}_d^H \otimes \mathbf{W}_R^H)\mathbf{Z} = (\mathbf{W}_d^H \mathbf{a}_d(\beta_t))(\mathbf{W}_R^H \mathbf{a}_R(\theta_t))(\mathbf{W}_T^H \mathbf{a}_T(\theta_t))\mathbf{s}_q \tag{5.111}$$

In the case of a MIMO radar, the received data matrix $\mathbf{X}$ is given by

$$\mathbf{X} = (\mathbf{a}_d(\beta_t) \otimes \mathbf{a}_R(\theta_t))\mathbf{a}_T(\theta_t)^T \mathbf{S} \tag{5.112}$$

and after passing it through a bank of matched filters we get

$$\mathbf{Z} = (\mathbf{a}_d(\beta_t) \otimes \mathbf{a}_R(\theta_t) \otimes \mathbf{a}_R(\theta_t))\mathbf{s}_q \tag{5.113}$$

Applying a weight vector $\mathbf{W}_d \otimes W_T \otimes \mathbf{W}_R$, we get

$$\mathbf{y} = (\mathbf{W}_d^H \otimes \mathbf{W}_T^H \otimes \mathbf{W}_R^H)\mathbf{Z} = (\mathbf{W}_d^H \mathbf{a}_d(\beta_t))(\mathbf{W}_R^H \mathbf{a}_R(\theta_t))(\mathbf{W}_T^H \mathbf{a}_T(\theta_t))\mathbf{s}_q \tag{5.114}$$

Comparing Eqs. (5.111) and (5.114), we again conclude that the two systems generate the same outputs except for an SNR difference.

Even though the two systems produce the same outputs for the same set of beamformer weight vectors (with the exception of the SNR issue), there is an important difference. In the phased-array system the weight vector $\mathbf{W}_T$ is applied on the transmit side, whereas in the MIMO radar system it is applied on the receive side. This seemingly small difference has some important implications, as discussed next.

5.5.1 Scan Transmit Beam after Receive

A typical choice for the transmit beamformer weight vector is $\mathbf{W}_T = \mathbf{a}_T(\theta_t)/\|\mathbf{a}_T(\theta_t)\|$ (possible tapering may be included). This makes the transmit array point a beam in the assumed target direction. In the phased-array implementation the transmitter requires multiple dwells to cover the area of interest because in each dwell it illuminates only a segment of the area that falls within the transmit beam. The transmit array can form N_T independent beams and will require (approximately) N_T dwell times to scan the entire area.

The situation is different when using a MIMO radar with orthogonal waveform. During the transmission no beam is formed and the entire area is illuminated during each dwell. The transmit beam will be formed during the processing of the received signal. This beam can be scanned over the entire area of interest. Thus, unlike the phased-array radar, the MIMO radar is able to cover the entire area during a single dwell. This comes at the cost of an N_T-fold reduction of the received SNR. Note that this SNR reduction may not have a significant effect on system performance if, for example, the scenario is clutter-dominated (*after* clutter reduction). However, if it is desired to compare the two systems under the same SNR conditions, the SNR reduction must be considered. In order to have the same SNR as the phased-array radar, the MIMO radar will have to integrate the radar returns from N_T dwells. Thus, in an equal SNR comparison, both systems will require a comparable number of dwells to cover a given area.

5.5.2 Adaptation of Transmit Beampattern

In some applications it is desirable to adjust the transmit beam on the basis of observed data. Consider, for example, a situation where it is desired to steer the nulls of the transmit beampattern so as to avoid illuminating directions that might give rise to multipath clutter. Information about the desired null steering directions is not available at the transmitter, and is obtained only by analyzing the received data. This is made possible in the MIMO radar where the transmit beampattern is determined "after the fact." Accomplishing this objective using a phased-array radar would be more difficult. It would require a two-step approach: a probing phase where the transmitter illuminates the area of interest to determine the directions in which the nulls should be steered, and a second phase where the transmit beampattern is configured with the appropriate nulls.

In other words, MIMO radar accomodates processing schemes that require data-dependent adaptation of the transmit beampattern. Such schemes are more difficult to implement using a phased-array radar.

5.5.3 Combined Transmit–Receive Beamforming

In a phased-array radar, beamforming is applied separately for the transmit array and the receive array, using the weight vectors $\mathbf{W}_T$ and $\mathbf{W}_R$, respectively. In the case of MIMO radar we applied the weight vector $\mathbf{W}_T \otimes \mathbf{W}_R$ to the virtual transmit/receive array, which again corresponds to beamforming separately the transmit and receive array. Note, however, that in this case it is possible to apply an arbitrary beamforming weight vector $\mathbf{W}$ to the entire virtual $N_TN_R \times 1$ array where $\mathbf{W} \neq \mathbf{W}_T \otimes \mathbf{W}_R$. In other words, in the phased-array radar we must design separately the transmit and receive beamformers using the $N_T + N_R$ degrees of freedom afforded by the number of coefficients in $\mathbf{W}_T$ and $\mathbf{W}_R$. However, in MIMO radar we can design the beamformer $\mathbf{W}$ using the corresponding N_TN_R degrees of freedom. In a phased-array radar the composite transmit–receive beampattern is constrained to be the product of the transmit and receive patterns. In a MIMO radar the composite pattern can be an arbitrary pattern of an $N_T N_R$-element array. Note that to ensure the full dimension of this virtual array, the transmit and receive arrays must be properly designed, as was discussed earlier (e.g., one is a critically sampled array and the other is a thinned array).

The discussion above focused on the case of MIMO radar with orthogonal waveforms. Earlier in this chapter we considered the use of a very general class of waveforms designed to achieve a desired illumination of the radar scene. Assume that we specify the desired illumination $\mathbf{g}_n$ at N different angles

$$\mathbf{a}_T^T(\theta_n)\mathbf{S} = \mathbf{g}_n \tag{5.115}$$

where $\mathbf{g}_n$ is a $1 \times N_f$ row vector. Assembling the N equations into a single matrix equation, we get

$$\begin{bmatrix} \mathbf{a}_T^T(\theta_1) \\ \mathbf{a}_T^T(\theta_2) \\ \vdots \\ \mathbf{a}_T^T(\theta_N) \end{bmatrix} \mathbf{S} = \begin{bmatrix} \mathbf{g}_1 \\ \mathbf{g}_2 \\ \vdots \\ \mathbf{g}_N \end{bmatrix} \tag{5.116}$$

which can be solved for the transmitted signal matrix $\mathbf{S}$. Assuming that the angles θ_n are approximately a beamwidth apart, the vectors $a_T(\theta_n)$ are approximately orthogonal so that

$$\mathbf{S} \approx \sum_{n=1}^{N} \mathbf{a}_T^*(\theta_n)\mathbf{g}_n / \|\mathbf{a}(\theta_n)\|^2 \tag{5.117}$$

Consider next a phasedarray that scans the area of interest during N dwells, pointing the transmit/receive beams at directions θ_n. The desired illumination function at the nth dwell is

$$\mathbf{a}_T^T(\theta)\mathbf{W}_T\mathbf{s} = \mathbf{g}_n \tag{5.118}$$

Letting $\mathbf{W}_T = \mathbf{a}_T^*(\theta_n)$, we get

$$\mathbf{s} = \frac{\mathbf{g}_n}{\|\mathbf{a}(\theta_n)\|^2} \tag{5.119}$$

or

$$\mathbf{S} = \mathbf{W}_T\mathbf{s} = \frac{\mathbf{a}_T^*(\theta_n)\mathbf{g}_n}{\|\mathbf{a}(\theta_n)\|^2} \tag{5.120}$$

Comparing the illumination produced by the two systems, we note that they have the same spatiotemporal characteristics, except that the MIMO radar illuminates the entire area at each dwell, while the phased-array scans the illumination across the area during multiple dwells. For this reason, the intensity of the illumination of the phased-array radar exceeds that of the MIMO radar. Note that here the phasedarray uses a different waveform at each dwell time. We conclude that both the phased-array radar and the MIMO radar can create a general spacetime illumination of the area of interest.

Another type of MIMO radar that has been proposed is the "statistical MIMO radar" described in Refs. 22–25. In this system the transmit antennas are assumed to be sufficiently far apart that transmit beamforming of the type used in a phasedarray is impossible because of the lack of phase coherence between antennas; this type of radar system is quite different from the one considered in this chapter.

APPENDIX 5A THEORETICAL SINR CALCULATION

Evaluating the SINR generally requires a numerical computation. However, for the case of uniform clutter and uniform illumination, the evaluation is straightforward. Recall that $\text{SINR} = \mathbf{v}_{x_t}^H \mathbf{R}_{x_{cn}}^{-1} \mathbf{v}_{x_t}$, where $\mathbf{R}_{x_c} = \mathbf{A}_h \mathbf{R}_h \mathbf{A}_h^H$. Assuming uniform clutter we have $\mathbf{R}_h = (\text{CNR}/N_a N_f)\, \mathbf{I}_{N_a N_f}$ so that trace $\{\mathbf{R}_h\} = \text{CNR}$, where CNR is the clutter : noise ratio. In this case

$$\mathbf{R}_{x_c} = \frac{\text{CNR}}{N_a N_f} \mathcal{A}_R \, \mathbf{diag}\{\mathbf{g} \odot \mathbf{g}^*\} \mathcal{A}_R^H \tag{5A.1}$$

Assuming uniform illumination $\mathbf{diag}\{\mathbf{g} \odot \mathbf{g}^*\} = c_g \mathbf{I} N_a N_f$, where c_g is a constant to be determined. Note that $\mathbf{A}_R \mathbf{A}_R^H \approx N_a \mathbf{I}_{N_R}$, which leads to

$$\mathbf{R}_{x_c} = \frac{\text{CNR} c_g}{N_f} \mathbf{I}_{N_R N_f} \tag{5A.2}$$

Also note that

$$\|\mathbf{g}\|^2 = \mathbf{s}^H \mathbf{A}_T^* \mathbf{A}_T^T \mathbf{s} = N_a \|\mathbf{s}\|^2 = N_a \tag{5A.3}$$

Therefore we have

$$\mathbf{tr}\{\mathbf{diag}\{\mathbf{g} \odot \mathbf{g}^*\}\} = \|\mathbf{g}\|^2 = N_a = c_g N_a N_f \tag{5A.4}$$

and we conclude that $c_g = 1/N_f$. It follows that

$$\mathbf{R}_{x_{cn}} = \frac{\text{CNR}}{N_f^2} \mathbf{I}_{N_R N_f} + \frac{1}{N_f} \mathbf{I}_{N_R N_f} \tag{5A.5}$$

We can now evaluate the SINR as

$$\mathrm{SINR} = \mathbf{v}_{x_t}^H \mathbf{R}_{x_{cn}}^{-1} \mathbf{v}_{x_t} = \frac{\|\mathbf{v}_{x_t}\|^2}{\mathrm{CNR}/N_f^2 + 1/Nf} \tag{5A.6}$$

The norm of the target signature is given by

$$\begin{aligned}\|\mathbf{v}_{x_t}\|^2 &= \mathbf{h}_t^H \mathbf{A}_h^H \mathbf{A}_h \mathbf{h}_t = \mathbf{h}_t^H \mathbf{diag}\{\mathbf{g}^H\} \mathcal{A}_R^H \mathcal{A}_R \, \mathbf{diag}\{\mathbf{g}^H\} \mathbf{h}_t \\ &= N_R \mathbf{h}_t^H \mathbf{diag}\{\mathbf{g} \odot \mathbf{g}^*\} \mathbf{h}_t\end{aligned} \tag{5A.7}$$

Note that $\mathbf{A}_h^H \mathbf{A}_h \approx N_R \mathbf{I}$ and therefore

$$\|\mathbf{v}_{x_t}\|^2 = \frac{N_R}{N_f} \|\mathbf{h}_t\|^2 = \frac{N_R}{N_f} \mathrm{SNR} \tag{5A.8}$$

where SNR is the signal-to-noise ratio. Finally we conclude that

$$\mathrm{SINR} = \frac{N_R \mathrm{SNR}}{\mathrm{CNR}/N_f + 1} \tag{5A.9}$$

This result has the following natural interpretation. If N_f is interpreted as the number of range cells, then CNR/N_f is the amount of clutter in the given range cell. The receive array gain is N_R. Because we have uniform illumination, the total received clutter power in the range cell is independent of the shape of the receive beampattern. In other words, the CNR is unaffected by spatial processing at the receiver. However, the target signal is increased by the array gain, providing an increase of the SNR by the factor N_R.

The uniform illumination corresponds to the case where orthogonal signals are used in the different antennas, and also to the case where only a single antenna is being used. If any of the adaptive waveforms described earlier is used, the resulting waveforms create a narrow beam in the direction of the target providing an additional array gain of N_T. Therefore

$$\mathrm{SINR} = \frac{N_R N_T \mathrm{SNR}}{\mathrm{CNR}/N_f + 1} \tag{5A.10}$$

REFERENCES

1. J. R. Guerci and S. U. Pillai, Theory and application of optimum transmit-receive radar, *Proc. IEEE 2000 Radar Conf.*, Alexandria, VA, 2000, pp. 705–710.
2. S. U. Pillai, D. C. Youla, S. Oh, and J. R. Guerci, Optimum transmit-receiver design in the presence of signal-dependent interference and channel noise, *Proc. 33rd Asilomar Conf. Signals, Systems, and Computers*, Pacific Grove, CA, Oct. 24–27, 1999, pp. 870–875.
3. S. U. Pillai, D. C. Youla, H. S. Oh, and J. R. Guerci, Matched transmit-receiver design in the presence of signal dependent noise, *IEEE Trans. Inform. Theory* **46**(2) (March 2000).

4. S. U. Pillai, H. S. Oh, and J. R. Guerci, Multichannel matched transmit-receiver design in presence of signal dependent interference and noise, *Proc. 1st IEEE Sensor Array and Multichannel Signal Processing Workshop*, Cambridge, MA, March 16–17, 2000.
5. B. Friedlander, A subspace framework for adaptive radar waveform design, *Proc. Asilomar Conf. Signal, Systems and Computers*, Pacific Grove, CA, Oct. 31–Nov. 2, 2005.
6. D. W. Bliss and K. W. Forsythe, Multiple-input multiple-output (MIMO) radar: Degrees of freedom and resolution, *Proc. 37th Asilomar Conf. Signals, Systems and Computers*, Pacific Grove, CA, Nov. 2003, Vol. 1, pp. 54–59.
7. K. Forsythe, D. Bliss, and G. Fawcett, Multiple-input multiple-output (MIMO) radar: Performance issues, *Proc. 38th Asilomar Conf. Signals, Systems and Computers*, 2004, pp. 310–315.
8. K. Forsythe and D. Bliss, Waveform correlation and optimization issues for MIMO radar, *Proc. 39th Asilomar Conf. Signals, Systems and Computers*, Oct. 28–Nov. 1, 2005, pp. 1306–1310.
9. D. Rabideau and P. Parker, Ubiquitous MIMO multifunction digital array radar, *Proc. 37th Conf. Signals, Systems and Computers*, Pacific Grove, CA, Nov. 2003.
10. F. C. Robey, S. Coutts, D. D. Weikle, J. C. McHarg, and K. Cuomo, MIMO radar theory and exprimental results, *Proc. 38th Asilomar Conf. Signals, Systems and Computers*, Nov. 7–10, 2004, Vol. 1, pp. 300–304.
11. L. B. White and P. S. Ray, Receiver design for MIMO tracking radars, *Proc. 2004 Waveform Diversity and Design Conf.*, Edinburgh, UK, Nov. 2004.
12. D. R. Fuhrmann and G. S. Antonio, Transmit beam-forming for MIMO radar systems using partial signal correlation, *Proc. 38th Asilomar Conf. Signals, Systems and Computers*, Nov. 2004, Vol. 1, pp. 295–299.
13. D. Fuhrmann and G. San Antonio, Transmit beamforming for MIMO radar systems using signal cross-correlation, *IEEE Trans. Aerospace Electron. Syst.* **44**(1):171–186 (Jan. 2008).
14. V. Mecca, D. Ramakrishnan, and J. Krolik, MIMO radar space-time adaptive processing for multipath clutter mitigation, *Proc. IEEE Workshop on Sensor Array and Multichannel Signal Processing*, July 12–14, 2006, pp. 249–253.
15. V. F. Mecca, D. Ramakrishnan, and J. L. Krolik, MIMO radar space-time adaptive processing for multipath clutter mitigation, *Proc. IEEE Sensor Array and Multichannel Signal Processing Workshop*, 2006, pp. 249–253.
16. L. Xu, J. Li, and P. Stoica, Adaptive techniques for MIMO radar, *Proc. IEEE Workshop on Sensor Array and Multichannel Signal Processing*, July 2006, pp. 258–262.
17. B. J. Donnet and I. D. Longstaff, MIMO radar, techniques and opportunities, *Proc. 3rd European Radar Conf.*, Sept. 2006, pp. 112–115.
18. P. Stoica, J. Li, and Y. Xie, On probing signal design for MIMO radar, *IEEE Trans. Signal Process.* **55**(8):5395–5406 (Nov. 2007).
19. I. Bekkerman and J. Tabrikian, Target detection and localization using MIMO radars and sonars, *IEEE Trans. Signal Process.* **54**(10):3873–3883 (Oct. 2006).
20. G. San Antonio, D. R. Fuhrmann, and F. C. Robey, MIMO radar ambiguity functions, *IEEE J. Select. Topics Signal Process.* **1**(1):167–177 (June 2007).
21. N. H. Lehmann, A. M. Haimovich, R. S. Blum, and L. Cimini, High resolution capabilities of MIMO radar, *Proc. 40th Asilomar Conf. Signals, Systems and Computers*, Oct.–Nov. 2006, pp. 25–30.

22. E. Fishler, A. Haimovich, R. Blum, D. Chizhik, L. Cimini, and R. Valenzuela, MIMO radar: An idea whose time has come, *Proc. IEEE Radar Conf.*, April 2004.
23. E. Fishler, A. Haimovich, R. Blum, L. Cimini, D. Chizhik, and R. Valenzuela, Statistical MIMO radar, *Proc. 12th Conf. Adaptive Sensors Array Processing*, March 2004.
24. E. Fishler, A. Haimovich, R. Blum, L. Cimini, D. Chizhik, and R. Valenzuela, Performance of MIMO radar systems: Advantages of angular diversity, *Proc. 38th Asilomar Conf. Signals, Systems and Computers*, 2004, pp. 305–309.
25. E. Fishler, A. Haimovich, R. Blum, L. Cimini, D. Chizhik, and R. Valenzuela, Performance of MIMO radar systems: Advantages of angular diversity, *Proc. 38th Asilomar Conf. Signals, Systems and Computers*, 2004, Vol. 1, pp. 305–309.
26. H. A. Khan, Y. Zhang, C. Ji, C. J. Stevens, D. J. Edwards, and D. O'Brien, Optimizing polyphase sequences for orthogonal netted radar, *IEEE Signal Process. Lett.* **13**(10):589–592 (Oct. 2006).
27. R. Klemm, *Principles of Space-Time Adaptive Processing*, Institution of Electrical Engineers, London, 2002.
28. L. L. Scharf, *Statistical Signal Processing*, Addison-Wesley, Reading, MA, 1991.
29. Y. Jin and B. Friedlander, Detection of distributed sources using sensor arrays, *IEEE Trans. Signal Process.* **52**(6):1537–1548 (June 2004).
30. S. Haykin (ed.), *Array Signal Processing*, Prentice-Hall, Englewood Cliffs, NJ, 1985.
31. L. Ljung and T. Söderström, *Theory and Practice of Recursive Identification*, MIT Press, Cambridge, MA, 1983.
32. S. Haykin, *Adaptive Filter Theory*, 4th ed., Prentice-Hall, Englewood Cliffs, NJ, 2004.
33. T. K. Moon and W. C. Stirling, *Mathematical Methods and Algorithms for Signal Processing*, Prentice-Hall, Englewood Cliffs, NJ, 1999.
34. B. Friedlander, Adaptive waveform design for a multi-antenna radar system, *Proc. Asilomar Conf. Signal, Systems and Computers*, Pacific Grove, CA, Oct. 29–Nov. 1, 2006.
35. B. Friedlander, Waveform design for MIMO radars, *IEEE Trans. Aerospace Electron. Syst.* **43**(3):1227–1238 (July 2007).
36. B. Friedlander, Waveform design for MIMO radar with space-time constraints, *Proceedings of the Asilomar Conference on Signal, Systems and Computers*, Nov. 4–7, 2007, Pacific Grove, CA, pp. 2168–2172.
37. A. J. Fenn, D. H. Temme, W. P. Delaney, and W. E. Courtney, The development of phased-array radar technology, *Lincoln Lab. J.* **12**(2):321–340 (2000).
38. C. A. Fowler, Old radar types never die; they just phased array, or . . . 55 years of trying to avoid mechanical scan, *IEEE AES Syst. Mag.* 24A–24L (Sept. 1998).
39. B. Liu, Z. He, J. Zeng, and B. Liu, Polyphase orthogonal code design for MIMO radar systems, *Proc. Int. Conf. Radar*, CIE '06, Oct. 2006, pp. 1–4.

6

MIMO RADAR SPACETIME ADAPTIVE PROCESSING AND SIGNAL DESIGN

CHUN-YANG CHEN AND P. P. VAIDYANATHAN

Department of Electrical Engineering, California Institute of Technology, Pasadena

In the traditional transmitting beamforming radar system, the transmitting antennas send coherent waveforms that form a highly focused beam. In the multiple-input multiple-output (MIMO) radar system, the transmitter sends noncoherent (possibly orthogonal) broad (possibly omnidirectional) waveforms. These waveforms can be extracted at the receiver by a matched filterbank. The extracted signals can be used to obtain more diversity or to improve the spatial resolution for clutter. This chapter focuses on spacetime adaptive processing (STAP) for MIMO radar systems, which improves the spatial resolution for clutter.

With a slight modification, STAP methods developed originally for the single-input multiple-output (SIMO) radar system (conventional radar) can also be used in MIMO radar. However, in the MIMO radar, the rank of the jammer–clutter subspace becomes very large, especially the jammer subspace. It affects both the complexity and the convergence of the STAP algorithm. In this chapter, the clutter space and its rank in the MIMO radar are explored. By using the geometry of the problem rather than data, the clutter subspace can be represented using prolate spheroidal wavefunctions (PSWFs). A new STAP algorithm is also proposed. It computes the clutter space using the PSWF and utilizes the block diagonal property of the jammer covariance matrix. Because of full utilization of the geometry and the structure of the covariance matrix, the method has very good SINR performance and low computational complexity.

MIMO Radar Signal Processing, edited by Jian Li and Petre Stoica

Besides the STAP algorithms, this chapter also discusses some design issues for the transmitted waveforms. Orthogonal waveforms can increase spatial resolution. However, the transmitted waveforms affect not only the spatial resolution but also the Doppler resolution and range resolution. The resolution of the system can be characterized by the MIMO ambiguity function. Several properties of the MIMO ambiguity function are proposed in this chapter. These properties provide some insights into the MIMO ambiguity function. The MIMO ambiguity function for the periodic pulse signal and some of its special cases are also derived in this chapter.

6.1 INTRODUCTION

The concept of MIMO radars has drawn considerable attention [1–18]. MIMO radars emit orthogonal waveforms [1–13] or noncoherent [14–18] waveforms instead of transmitting coherent waveforms, which form a focused beam in traditional transmitter-based beamforming. In the MIMO radar receiver, a matched filterbank is used to extract the orthogonal waveform components. There are two different kinds of approaches for using the noncoherent waveforms:

1. Increased spatial diversity can be obtained [4,5]. In this scenario, the transmitting antenna elements are far enough from each other relative to the distance from the target that the target radar cross sections (RCSs) are independent random variables for different transmitting paths. When the orthogonal components are transmitted from different antennas, each orthogonal waveform will carry independent information about the target. This spatial diversity can be utilized to perform better detection [4,5].
2. A better spatial resolution for clutter can be obtained. In this scenario, the distances between transmitting antennas are small enough compared to the distance between the target and the radar station that the target RCS is identical for all transmitting paths. The phase differences caused by different transmitting antennas along with the phase differences caused by different receiving antennas can form a new *virtual array* steering vector. With judiciously designed antenna positions, one can create a very long array steering vector with a small number of antennas. Thus the spatial resolution for clutter can be dramatically increased at a small cost [1,2]. In this chapter, we focus on this second advantage.

The adaptive techniques for processing the data from airborne antenna arrays are called *spacetime adaptive processing* (STAP) techniques. The basic theory of STAP for the traditional single-input multiple-output (SIMO) radar has been well developed [42,43]. Many algorithms have been proposed (e.g., see Refs. 32–43 and references cited therein), for improving the complexity and convergence of the STAP in the SIMO radar. With a slight modification, these methods can also be applied to the MIMO radar case. The MIMO extension of STAP can be found in

Ref. 2. The MIMO radar STAP for multipath clutter mitigation can be found in Ref. 11. However, in the MIMO radar, spacetime adaptive processing (STAP) becomes even more challenging because of the extra dimension created by the orthogonal waveforms. On one hand, the extra dimension increases the rank of the jammer and clutter subspace, especially the jammer subspace. This makes the STAP more complex. On the other hand, the extra degrees of freedom created by the MIMO radar allows us to filter out more clutter subspace with little effect on SINR.

In this chapter, we explore the clutter subspace and its rank in MIMO radar. Using the geometry of the MIMO radar and the prolate spheroidal wavefunction (PSWF), a method for computing the clutter subspace is developed. Then we develop a STAP algorithm that computes the clutter subspace using the geometry of the problem rather than data and utilizes the block diagonal structure of the jammer covariance matrix. Because of its full utilization of the geometry and the structure of the covariance matrix, our method has very good SINR performance and significantly lower computational complexity compared to fully adaptive methods (Section 6.5.2). We also discuss the design of signals used in transmission.

In practice, the clutter subspace might change because of effects such as the internal clutter motion (ICM), velocity misalignment, array manifold mismatch, and channel mismatch [42]. In this chapter, we consider an "ideal model," which does not account for these effects. When this model is not valid, the performance of the algorithm will degrade. One way to overcome this problem might be to estimate the clutter subspace by using a combination of both the assumed geometry and the received data. Another way might be to develop a more robust algorithm against the clutter subspace mismatch. These ideas will be explored in the future. The term "adaptive" in the STAP literature is meant to imply that quantities such as weight vectors depend on signal statistics. Real-time adaptation is not the topic of concern in this chapter.

The MIMO radar emits orthogonal waveforms to increase the spatial resolution. However, the transmitted waveforms also affect the Doppler resolution and the range resolution. In SIMO radar the resolution of the radar system can be characterized by ambiguity function [49]. The idea of ambiguity functions has been extended to the MIMO case [37]. The MIMO ambiguity function characterizes the resolution of the MIMO radar system. In this chapter, we focus on the study of STAP systems that use periodic pulse signals. Therefore, we derive the MIMO ambiguity function for the periodic pulse signals and some of its special cases. These results can be used to design the transmitted waveforms. Some properties of the MIMO ambiguity functions are also derived in this chapter. These properties provide some insights into the MIMO ambiguity function.

The remainder of the chapter is organized as follows. In Section 6.2, the concept of MIMO radar is briefly reviewed. In Section 6.3, we formulate the STAP approach for MIMO radar. In Section 6.4, we explore the clutter subspace and its rank in the MIMO radar. Using prolate spheroidal wavefunctions (PSWFs), we construct a data-independent basis for clutter signals. In Section 6.5, we propose a new STAP method for MIMO radar. This method utilizes the technique proposed in

Section 6.4 to find the clutter subspace and estimates the jammer-plus-noise covariance matrix separately. Finally, the beamformer is calculated by using matrix inversion lemma. As we will see later, this method has very satisfactory SINR performance. In Section 6.6, we compare the SINR performance of different STAP methods based on numerical simulations. In Section 6.7, we study ambiguity functions in the context of MIMO radar. We review the so-called cross-ambiguity functions. The idea of these functions was introduced in Ref. 37. We also derive a number of new properties of these functions. Finally, Section 6.8 concludes the chapter. Some parts of the results in this chapter were reported by the authors in other publications [12,13].

6.1.1 Notations

Matrices are denoted by capital letters in boldface (e.g., **A**). Vectors are denoted by lowercase letters in boldface (e.g., **x**). The superscript $\dagger$ denotes transpose conjugation. The notation diag($\mathbf{A}, \mathbf{A}, \ldots, \mathbf{A}$) denotes a block diagonal matrix whose diagonal blocks are **A**. The notation $\lceil a \rceil$ is defined as the smallest integer larger than a. The notation $\lfloor a \rfloor$ is defined as the largest integer smaller than a. The function sinc(x) is defined as $\text{sinc}(x) = \sin(\pi x)/\pi x$.

6.2 THE VIRTUAL ARRAY CONCEPT

One of the main advantages of MIMO radar is that the degrees of freedom can be greatly increased by the concept of virtual array. In this section, we briefly review this concept. More detailed reviews can be found in the literature [1,2,6,7]. Consider an arbitrary transmitting array with M antenna elements and an arbitrary receiving array with N antenna elements. The mth transmitting antenna is located at $\mathbf{x}_{T,m} \in \mathbb{R}^3$ and the nth receiving antenna is located at $\mathbf{x}_{R,n} \in \mathbb{R}^3$. Figure 6.1a shows an example with $M = 3$ and $N = 4$. The mth transmitting antenna emits the waveform $\phi_m(t)$. The emitted waveforms are orthogonal:

$$\int \phi_m(\tau)\phi_k^*(\tau)d\tau = \delta_{mk}$$

In each receiving antenna, these orthogonal waveforms are extracted by M matched filters. Therefore, the total number of extracted signal equals NM. Consider a far-field point target. The target response in the mth matched-filter output of the nth receiving antenna can be expressed as

$$y_{n,m}^{(t)} = \rho_t \exp\left(j\frac{2\pi}{\lambda}\mathbf{u}_t^T(\mathbf{x}_{T,m} + \mathbf{x}_{T,n})\right) \tag{6.1}$$

where $\mathbf{u}_t \in \mathbb{R}^3$ is a unit vector pointing toward the target from the radar station and ρ_t is the amplitude of the signal reflected by the target. One can see that the phase differences are created by both the transmitting and the receiving antenna locations.

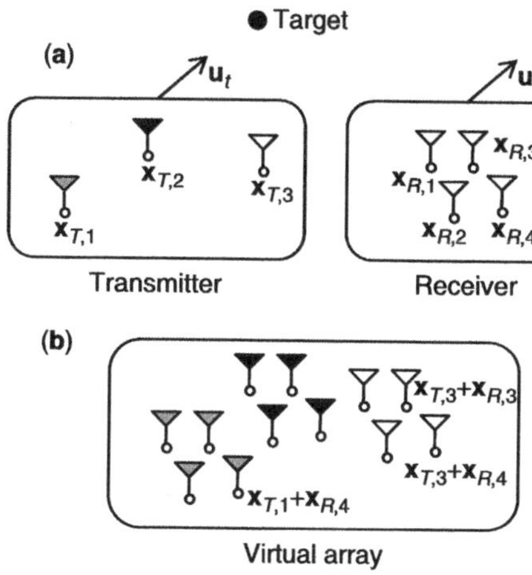

Figure 6.1 (a) Illustration of a MIMO radar system with $M = 3$ and $N = 4$; (b) corresponding virtual array.

The target response in (6.1) is the same as the target response received by a receiving array with NM antenna elements located at

$$\{\mathbf{x}_{T,m} + \mathbf{x}_{R,n} | \ n = 0, 1, \ldots, N-1, \ m = 0, 1, \ldots, M-1\}.$$

We call this NM-element array a *virtual array*. Figure 6.2b shows the corresponding virtual array of the MIMO radar system illustrated in (a). Thus, we can create an NM-element virtual array by using only $N + M$ physical antenna elements. Note

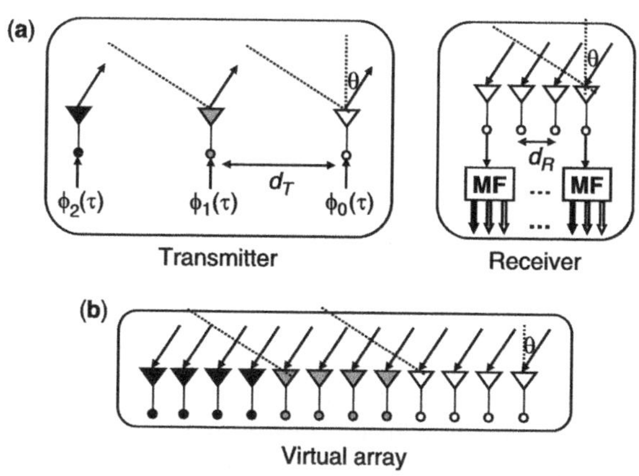

Figure 6.2 (a) Illustration of a ULA MIMO radar system with $M = 3$ and $N = 4$; (b) corresponding virtual array.

that in this example, the transmitters are sufficiently far apart that there is no overlap of the virtual elements in Fig. 1.1b.

The relation between the transmitting array, receiving array, and the virtual array can be further characterized by a convolution [7]. Define

$$g_T(\mathbf{x}) = \sum_{m=0}^{M-1} \delta(\mathbf{x} - \mathbf{x}_{T,m}) \tag{6.2}$$

and

$$g_R(\mathbf{x}) = \sum_{n=0}^{N-1} \delta(\mathbf{x} - \mathbf{x}_{R,n}) \tag{6.3}$$

These functions characterize the antenna locations in the transmitter and receiver. Because the virtual array has NM virtual elements located at $\{x_{T,m} + x_{R,n}\}$, the corresponding function that characterizes the antenna location of the virtual array can be expressed as

$$g_V(\mathbf{x}) = \sum_{m=0}^{M-1} \sum_{n=0}^{N-1} \delta(\mathbf{x} - (\mathbf{x}_{T,m} + \mathbf{x}_{R,n})) \tag{6.4}$$

Comparing (6.2)–(6.4), one can see that

$$g_V(\mathbf{x}) = (g_T * g_R)(\mathbf{x}), \tag{6.5}$$

where $*$ denotes convolution. One can observe this relation from Fig. 1.1. The array in Fig. 1.1b can be obtained by performing convolution of the arrays in Fig. 1.1a. This relation was observed in an earlier study [7].

An idea somewhat related to the virtual array concept is called *sum coarray* [38,39]. The main difference is that the sum coarray concept is applicable only to the SIMO system. In the SIMO system, the overall beampattern is the product (composite) of the transmit and receive beampatterns. The overall beampattern is therefore related to a weight vector $\mathbf{w}_{tr}$ that equals the convolution of the transmit beamformer $\mathbf{w}_t$ and the receive beamformer $\mathbf{w}_r$:

$$\mathbf{w}_{tr} = \mathbf{w}_t * \mathbf{w}_r \tag{6.6}$$

This new weight vector $\mathbf{w}_{tr}$ can be viewed as a beamformer of a longer array called *coarray*. In terms of the array geometry, this coarray is exactly the virtual array. However, these two approaches are completely different because of the difference between SIMO and MIMO systems. In the MIMO virtual array, the weight vector has a total of NM degrees of freedom. However, in coarray, the weight vector has only $N+M$ degrees of freedom because of (6.6). Also, the virtual array beamforming is performed in the receiver only, but the coarray beamforming is performed in both sides of the transmitter and receiver.

Example 6.1. Uniform Linear Virtual Array In this chapter, we focus only on the STAP method for the uniform linear arrays (ULA). In this case, the antenna locations $x_{T,m}$ and $x_{R,n}$ reduce to scalars and

$$\begin{aligned} x_{R,n} &= nd_R, \qquad n = 0, 1, \ldots, N-1 \\ x_{T,m} &= md_T, \qquad m = 0, 1, \ldots, M-1 \end{aligned}$$

where d_R is the spacing between the receiving antennas and d_T is the spacing between the transmitting antennas. Figure 6.2 shows an example with $M=3$ and $N=4$. Similar to the arbitrary antenna case, the target response in the mth matched filter of the nth receiving antenna can be expressed as

$$\rho_t \exp\left(j\frac{2\pi}{\lambda}(nd_R \sin\theta + md_T \sin\theta)\right) \tag{6.7}$$

where θ is the viewing direction of the target. The phase differences are created by both transmitting and receiving antenna locations. Define

$$f_s \triangleq \frac{d_R}{\lambda}\sin\theta, \quad \text{and} \quad \gamma \triangleq \frac{d_T}{d_R}$$

Equation (6.7) can be further simplified as

$$\rho_t \exp(j2\pi f_s(n+\gamma m))$$

If we choose

$$\gamma = N \tag{6.8}$$

the set $\{n+\gamma m\}$ becomes $\{0, 1, \ldots, NM-1\}$. Thus the NM signals in (6.7) can be viewed as the signals received by a virtual array with NM elements [2] as shown in Fig. 1.2b. It is as if we have a uniform linear receiving array with NM elements. Thus NM degrees of freedom can be obtained with only $N+M$ physical array elements. Similarly, we can obtain this result by using the convolution described in (6.5). From this perspective, one can see that the choice of $\gamma = N$ results a uniform virtual array. One can view the antenna array as a way to sample the electromagnetic wave in the spatial domain. The MIMO radar idea allows "sampling" in both transmitter and receiver and creates a total of NM "samples." Utilizing these extra samples in spatial domain, a better spatial resolution can be obtained.

Example 6.2. Overlapped Linear Virtual Array Instead of choosing $\gamma = N$ in (6.8), one can choose $\gamma = 1$. In this case, the target response in the mth antenna of

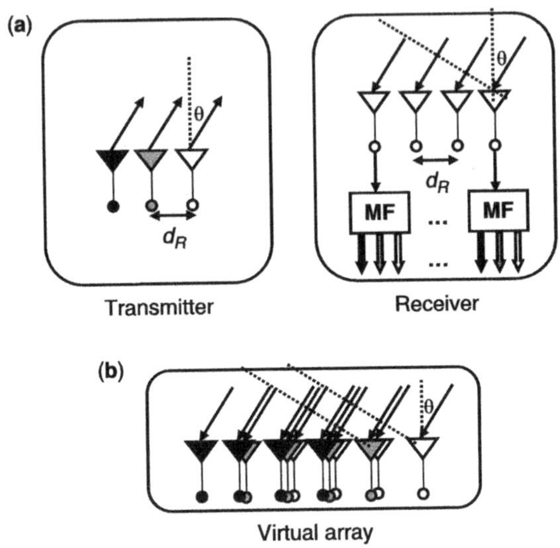

Figure 6.3 (a) Illustration of a MIMO radar system with $M = 3$, $N = 4$ and $d_T = d_R$; (b) corresponding virtual array.

the nth receiver can be expressed as

$$\rho_t \exp(j2\pi f_s(n + m))$$

Figure 6.3 shows an example of the transmitter, the receiver, and their corresponding virtual array. In this case, the virtual array is more complicated; it has several virtual elements that are at the same locations. In a certain sense, we can regard this as a nonuniform virtual array. The advantage of choosing $\gamma = 1$ is that the radar station can form a focused beam by emitting correlated waveforms $\{\phi_m(t)\}$ [14]. The transmit beamforming cannot be done in the case $\gamma = N$, because the sampling rate in the spatial domain is too low to prevent aliasing. However, the advantage of choosing $\gamma = N$ is that the virtual array is longer as shown in Fig. 1.2b which results in a better spatial resolution. The performance of the system with $\gamma = 1$ and $\gamma = N$ will be compared later in an example in Section 6.3.4. Intermediate values $1 < \gamma < N$ are possible, but are not considered here.

6.3 SPACETIME ADAPTIVE PROCESSING IN MIMO RADAR

In this section, we formulate the STAP problem in MIMO radar. The MIMO extension for STAP first appeared in an earlier paper [2]. We focus here on the idea of using the extra degrees of freedom to increase the spatial resolution for clutter.

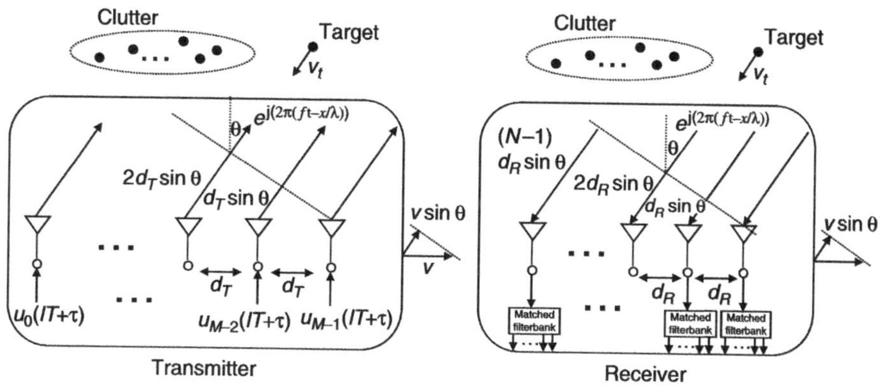

Figure 6.4 Illustration of a MIMO radar system with M transmitting antennas and N receiving antennas. The radar station is moving at speed v.

6.3.1 Signal Model

Figure 6.4 shows the geometry of the MIMO radar STAP with uniform linear arrays (ULA), where

1. d_T is the spacing of the transmitting antennas.
2. d_R is the spacing of the receiver antennas.
3. M is the number of transmitting antennas.
4. N is the number of the receiving antennas.
5. T is the radar pulse period.
6. v_t is the target speed toward the radar station.
7. v is the speed of the radar station.

We assume the two antenna arrays to be linear and parallel. The transmitter and the receiver are close enough that they share the same angle variable θ. The radar station movement is assumed to be parallel to the linear antenna array. This assumption has been made in most of the airborne ground moving-target indicator (GMTI) systems. Each array is composed of omnidirectional elements. The transmitted signals of the mth antenna can be expressed as

$$u_m(lT+\tau) = \sqrt{E}\phi_m(\tau)e^{j2\pi f(lT+\tau)} \tag{6.9}$$

for $m = 1, 2, \ldots, M-1$, where

1. $\phi_m(t)$ is the baseband pulse waveform.
2. l indicates the index of radar pulse (slow time).
3. τ represents the time within the pulse (fast time).

4. T is the radar pulse period.
5. f is the carrier frequency.
6. E is the transmitted energy for the pulse.

In pulse radar systems, the support of the pulse $\phi_m(t)$ is usually very small. The transmitted signal $u_m(t)$ consists of a chain of repeated short pulses $\phi_m(t)$. The demodulated received signal of the nth antenna can be expressed as

$$\begin{aligned} y_n(lT+\tau) &= \sum_{m=0}^{M-1} \rho_t \phi_m^{\nu_t}\left(lT+\tau-\frac{2r}{c}-nd_R\sin\theta_t - md_T\sin\theta_t\right) \\ &\quad + \sum_{i=0}^{N_c-1}\sum_{m=0}^{M-1} \rho_i \phi_m^{\nu_i}\left(lT+\tau+\frac{2r}{c}-nd_R\sin\theta_i - md_T\sin\theta_i\right) \\ &\quad + y_n^{(J)}(lT+\tau)+y_n^{(w)}(lT+\tau) \end{aligned} \tag{6.10}$$

where

$$\begin{aligned} \phi_m^{\nu}(t) &\triangleq \phi_m(t)e^{j2\pi\nu t} \\ \nu_t &= \frac{2(\nu_t+\nu\sin\theta_t)}{c}f \\ \nu_i &= \frac{2(\nu\sin\theta_i)}{c}f \end{aligned}$$

and

1. r is the distance of the range bin of interest.
2. c is the speed of light.
3. ρ_t is the amplitude of the signal reflected by the target.
4. ρ_i is the amplitude of the signal reflected by the ith clutter.
5. θ_t is the viewing direction of the target.
6. θ_i is the viewing direction of the ith clutter.
7. ν_t is the Doppler frequency of the target.
8. ν_i is the Doppler frequency of the ith clutter.
9. N_c is the number of clutter signals.
10. $y_n^{(J)}$ is the jammer signal in the nth antenna output.
11. $y_n^{(w)}$ is the white noise in the nth antenna output.

The first term in (6.10) represents the signal reflected by the target. The second term is the signal reflected by the clutter. The last two terms represent the jammer signal and white noise. We assume that there is no internal clutter motion (ICM) or antenna array misalignment [42]. The phase differences in the reflected signals are caused by the Doppler shift, the differences of the receiving antenna locations, and the differences of the transmitting antenna locations. Assume that the baseband pulses $\{\phi_m(t)\}$ are narrowband and the duration of $\{\phi_m(t)\}$, namely, T_ϕ, is small enough such that $T_\phi \nu_t \approx 0$ and $T_\phi\ \nu_i \approx 0$. Then (6.10) can be simplified as

$$y_n\left(lT + \tau + \frac{2r}{c}\right) \approx \sum_{m=0}^{M-1} \rho_t \phi_m(\tau) e^{j(2\pi/\lambda)(\sin\theta_t(2vTl + d_R n + d_T m) + 2v_t Tl)} + \sum_{i=0}^{N_c-1}\sum_{m=0}^{M-1} \rho_i \phi_m(\tau) e^{j(2\pi/\lambda)(\sin\theta_i(2vTl + d_R n + d_T m))} + y_n^{(J)}\left(lT + \tau + \frac{2r}{c}\right) + y_n^{(w)}\left(lT + \tau + \frac{2r}{c}\right) \tag{6.11}$$

In the MIMO radar, the transmitting waveforms $\phi_m(\tau)$ satisfy orthogonality:

$$\int \phi_m(\tau)\phi_k^*(\tau)d\tau = \delta_{mk} \tag{6.12}$$

The sufficient statistics for the range of interest r can be extracted by a bank of matched filters as shown in Fig. 6.4. The extracted signals can be expressed as

$$\begin{aligned} y_{n,m,l} &\triangleq \int y_n\left(lT + \tau + \frac{2r}{c}\right)\phi_m^*(\tau)d\tau \\ &= \rho_t e^{j(2\pi/\lambda)(\sin\theta_t(2vTl + d_R n + d_T m) + 2v_t Tl)} \\ &\quad + \sum_{i=0}^{N_c-1} \rho_i e^{j(2\pi/\lambda)(\sin\theta_i(2vTl + d_R n + d_T m))} + y_{n,m,l}^{(J)} + y_{n,m,l}^{(w)} \end{aligned} \tag{6.13}$$

for $n = 0, 1, \ldots, N-1$, $m = 0, 1, \ldots, M-1$, and $l = 0, 1, \ldots, L-1$, where $y_{n,m,l}^{(J)}$ is the corresponding jammer signal, $y_{n,m,l}^{(w)}$ is the corresponding white noise, and L is the number of the pulses in a coherent processing interval (CPI). To simplify this equation, we define the following normalized spatial and Doppler frequencies:

$$\begin{aligned} f_s &\triangleq \frac{d_R}{\lambda}\sin\theta_t, \quad f_{s,i} \triangleq \frac{d_R}{\lambda}\sin\theta_i \\ f_D &\triangleq \frac{2(v\sin\theta_t + v_t)}{\lambda}T \end{aligned} \tag{6.14}$$

One can observe that the normalized Doppler frequency of the target is a function of both target viewing direction and speed. Throughout this chapter we shall assume that $d_R = \lambda/2$, to avoid spatial aliasing is avoided. Using the definition above, we can rewrite the extracted signal in (6.13) as

$$y_{n,m,l} = \rho_t e^{j2\pi f_s(n+\gamma m)} e^{j2\pi f_D l} + \sum_{i=0}^{Nc-1} \rho_i e^{j2\pi f_{s,i}(n+\gamma m+\beta l)} + y_{n,m,l}^{(J)} + y_{n,m,l}^{(w)} \tag{6.15}$$

for $n = 0, 1, \ldots, N-1$, $m = 0, 1, \ldots, M-1$, and $l = 0, 1, \ldots, L-1$, where

$$\gamma \triangleq \frac{d_T}{d_R} \quad \text{and} \quad \beta \triangleq \frac{2vT}{d_R} \tag{6.16}$$

6.3.2 Fully Adaptive MIMO-STAP

The goal of spacetime adaptive processing (STAP) is to find a linear combination of the extracted signals so that the SINR can be maximized. Thus the target signal can be extracted from the interferences, clutter, and noise to perform the detection. Stacking the MIMO STAP signals in (6.15), we obtain the NML vector

$$\mathbf{y} = (y_{0,0,0} \quad y_{1,0,0} \quad \cdots \quad y_{N-1,M-1,L-1})^T \tag{6.17}$$

Then the linear combination can be expressed as $\mathbf{w}^\dagger\mathbf{y}$, where $\mathbf{w}$ is the weight vector for the linear combination. The SINR maximization can be obtained by minimizing the total variance under the constraint that the target response is unity. It can be expressed as the following optimization problem

$$\begin{aligned} &\min_{\mathbf{w}} \mathbf{w}^\dagger \mathbf{R} \mathbf{w} \\ &\text{subject to } \mathbf{w}^\dagger \mathbf{s}(f_s, f_D) = 1, \end{aligned} \tag{6.18}$$

where $\mathbf{R} \triangleq E[\mathbf{y}\mathbf{y}^\dagger]$ and $\mathbf{s}(f_s, f_D)$ is the size-NML MIMO spacetime steering vector, which consists of the elements

$$e^{j2\pi f_s(n+\gamma m)} e^{j2\pi f_D l} \tag{6.19}$$

for $n = 0, 1, \ldots, N-1$, $m = 0, 1, \ldots, M-1$, and $l = 0, 1, 1, \ldots, L-1$. This $\mathbf{w}$ is called a *minimum variance distortionless response* (MVDR) beamformer [25]. The covariance matrix $\mathbf{R}$ can be estimated by using the neighboring range bin cells. In practice, in order to prevent self-nulling, a target-free covariance matrix can be estimated by using guard cells [42]. The well-known solution to the problem stated above is [25]

$$\mathbf{w} = \frac{\mathbf{R}^{-1}\mathbf{s}(f_s, f_D)}{\mathbf{s}(f_s, f_D)^\dagger \mathbf{R}^{-1} \mathbf{s}(f_s, f_D)} \tag{6.20}$$

However, the covariance matrix **R** is $NML \times NML$; it is much larger than in the SIMO case because of the extra dimension. The complexity of the inversion of such a large matrix is high. The estimation of such a large covariance matrix also converges slowly. To overcome these problems, partially adaptive techniques can be applied. The methods described in Section 6.6 are examples of such partially adaptive techniques. In SIMO radar literature such partially adaptive methods are commonly used as well [42,43].

6.3.3 Comparison with SIMO System

In the traditional transmit beamforming, or single-input multiple-output (SIMO) radar, the transmitting antennas send coherent waveforms that form a highly focused beam. The waveforms can be expressed as

$$\phi_m(\tau) = \phi(\tau) w_{Tm}$$

for $m = 1, 2, \ldots, M-1$, where $\{w_{Tm}\}$ are the transmit beamforming weights. The sufficient statistics can be extracted by a single matched filter for every receiving antenna. The extracted signal can be expressed as

$$\begin{aligned}
y_{n,l} &\triangleq \int y_n\left(lT + \tau + \frac{2r}{c}\right)\phi^*(\tau)d\tau \\
&= \rho_t e^{j2\pi f_s n} e^{j2\pi f_D l} \sum_{m=0}^{M-1} w_{Tm} e^{j2\pi f_s \gamma m} \\
&\quad + \sum_{i=0}^{N_c-1} \rho_i e^{j2\pi f_{s,i}(n+\beta l)} \sum_{m=0}^{M-1} w_{Tm} e^{j2\pi f_{s,i}\gamma m} + y_{n,l}^{(J)} + y_{n,l}^{(w)}
\end{aligned} \tag{6.21}$$

for $n = 0, 1, \ldots, N-1$, and $l = 0, 1, \ldots, L-1$, where $y_{n,l}^{(J)}$ is the corresponding jammer signal and $y_{n,l}^{(w)}$ is the corresponding white noise. Comparing the MIMO signals in (6.15) and the SIMO signals in (6.21), one can see that a linear combination with respect to m has been performed on the SIMO signal in the target term and the clutter term. The MIMO radar, however, leaves all degrees of freedom to the receiver. Note that in the receiver, one can perform the same linear combination with respect to m on the MIMO signal in (6.15) to create the SIMO signal in (6.21). The only difference is that the transmitting power for the SIMO signal is less because of the focused beam used in the transmitter. For the SIMO radar, the number of degrees of freedom is M in the transmitter and NL in the receiver. The total number of degrees of freedom is $M + NL$. However, for the MIMO radar, the number of degrees of freedom is NML, which is much larger than $M + NL$. These extra degrees of freedom can be used to obtain a better spatial resolution for clutter.

The MIMO radar transmits omnidirectional orthogonal waveforms from each antenna element; therefore, it illuminates all angles. The benefit of SIMO radar is that it transmits a focused beam, which conserves transmitting power. Therefore, for a articular angle of interest, the SIMO radar enjoys a processing gain of M compared to the MIMO radar. However, for some applications such as scanning or imaging, it is necessary to illuminate all angles. In this case, the benefit of focused beam no longer exists because both systems need to consume the same energy for illuminating all angles. The SIMO system will need to steer the focused transmit beam to illuminate all angles.

A second point is that, for computation of the MIMO beamformer in (6.20), the matrix inversion $\mathbf{R}^{-1}$ needs to be computed only once and can be applied for all angles. The transmitting array in a MIMO radar does not have a focused beam; thus all the ground points within a range bin are uniformly illuminated. The clutter covariance seen by the receiving antenna array is, therefore, the same for all angles. In the SIMO case, the matrix inversions need to be computed for different angles because the clutter signal changes as the beam is steered through all angles.

6.3.4 The Virtual Array in STAP

Observing the MIMO spacetime steering vector defined in (6.19), one can view the first term $e^{j2\pi f_s(n+\gamma m)}$ as a sampled version of the sinusoidal function $e^{j2\pi f_s x}$. Recall that γ is defined in (6.15) as the ratio of the antenna spacing of the transmitter and receiver. To obtain a good spatial frequency resolution, these signals should be critically sampled and be of sufficiently long duration. One can choose $\gamma = N$ because it maximizes the time duration while maintaining critical sampling [2] as shown in Fig. 6.2. Sorting the sample points $n + \gamma m$ for $n = 0, 1, \ldots, N-1$, and $m = 0, 1, \ldots, M-1$, we obtain the permuted sample points $k = 0, 1, \ldots, NM-1$. Thus the target response in (6.19) can be rewritten as

$$e^{j2\pi f_s k} e^{j2\pi f_D l}$$

for $k = 0, 1, \ldots, NM-1$, and $l = 0, 1, \ldots, L-1$. It is as if we have a virtual receiving array with NM antennas. However, the resolution is actually obtained by only M antennas in the transmitter and N antennas in the receiver. Figure 6.5 compares the SINR performance of the MIMO system and the SIMO system in the viewing direction of zero degree. The optimal spacetime beamformer described in (6.20) is used. The parameter L equals 16, and β equals 1.5 in this example. In all plots it is assumed that the energy transmitted by any single antenna element to illuminate all angles is fixed.

The SINR plot as a function of Doppler frequency is a good way of measuring radar performance for moving targets. In Fig. 6.5 the SINR drops near zero-Doppler frequency because it is not easy to distinguish the slowly moving target from the still ground clutter. A sharper SINR curve means that the corresponding system is more capable of detecting a slower-moving target. The MIMO system with $\gamma = 1$ has a slightly better performance than the SIMO system with the same antenna structure. For the virtual array structure where $\gamma = N$, the MIMO system

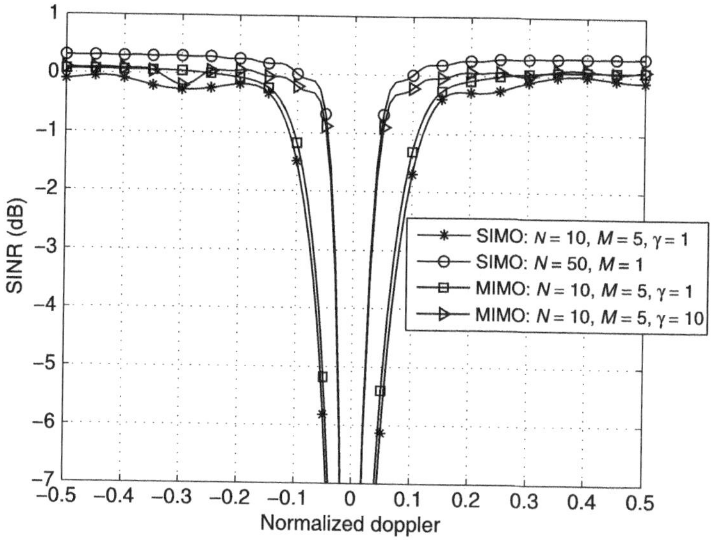

Figure 6.5 The SINR at viewing direction zero as a function of the Doppler frequencies for different SIMO and MIMO systems.

has a much better SINR performance than the case with $\gamma = 1$ and it close to the performance of the SIMO system with NM antennas; this is because the corresponding virtual array of the MIMO system is the same as the SIMO system with NM antenna elements. The small difference in SINR results from the fact that the SIMO system with NM antennas has a better spatial resolution for the jammer signals. This example shows that the choice of γ is very crucial in the MIMO radar. With the choice $\gamma = 10 = N$, the MIMO radar with only 15 antenna elements has about the same performance as the SIMO radar with 51 array elements. This example also shows that the MIMO radar system has a much better spatial resolution for clutter compared to the traditional SIMO system with same number of physical antenna elements.

6.4 CLUTTER SUBSPACE IN MIMO RADAR

In this section, we explore the clutter subspace and its rank in the MIMO radar system. The covariance matrix $\mathbf{R}$ in (6.18) can be expressed as

$$\mathbf{R} = \mathbf{R}_t + \mathbf{R}_c + \mathbf{R}_J + \sigma^2\mathbf{I}$$

where $\mathbf{R}_t$ is the covariance matrix of the target signal, $\mathbf{R}_c$ is the covariance matrix of the clutter, $\mathbf{R}_J$ is the covariance matrix of the jammer, and σ^2 is the variance of the white noise. We define clutter subspace and clutter rank as follows:

$$\text{Clutter subspace} = \text{range space of } \mathbf{R}_c$$
$$\text{Clutter rank} = \text{rank } (\mathbf{R}_c)$$

In the spacetime adaptive processing (STAP) literature, it is well known that the clutter subspace usually has a small rank. It was first pointed out by Klemm [23] that the clutter rank is approximately $N + L$, where N is the number of receiving antennas and L is the number of pulses in a coherent processing interval (CPI). Ward [21] and Brennan and Staudaher [22] had previously proposed a rule for estimating the clutter rank. The estimated rank is approximately

$$N + \beta(L - 1) \tag{6.22}$$

where $\beta = 2vT/d_R$. This is called *Brennan's rule* [21]. Taking advantage of the low-rank property, the STAP can be performed in a lower-dimensional space so that the complexity and the convergence can be significantly improved [31–43]. This result will now be extended to the MIMO radar. These techniques are often called *partially adaptive methods* or *subspace methods*.

6.4.1 Clutter Rank in MIMO Radar: MIMO Extension of Brennan's Rule

We first study the clutter term in (6.15), which is expressed as

$$y_{n,m,l}^{(c)} = \sum_{i=0}^{Nc-1} \rho_i e^{j2\pi f_{s,i}(n+\gamma m+\beta l)}$$

for $n = 0, 1, \ldots, N-1$, $m = 0, 1, \ldots, M-1$, and $l = 0, 1, \ldots, L-1$. Note that $-0.5 < f_{s,i} < 0.5$ because $d_R = \lambda/2$. Define

$$c_{i,n,m,l} = e^{j2\pi f_{s,i}(n+\gamma m+\beta l)} \tag{6.23}$$

and

$$\mathbf{c}_i = (c_{i,0,0,0}, c_{i,1,0,0}, \ldots, c_{i,N-1,M-1,L-1})^T \tag{6.24}$$

By stacking the signals $\{y_{n,m,l}^{(c)}\}$ into a vector, one can obtain

$$\mathbf{y}^{(c)} = \sum_{i=0}^{N_c-1} \rho_i \mathbf{c}_i$$

Assume that ρ_i are zero-mean independent random variables with variance $\sigma_{c,i}^2$. The clutter covariance matrix can be expressed as

$$\mathbf{R}_c = E[\mathbf{y}^{(c)}\mathbf{y}^{(c)\dagger}] = \sum_{i=0}^{N_c-1} \sigma_{c,i}^2 \mathbf{c}_i \mathbf{c}_i^\dagger$$

Therefore, $\text{span}(\mathbf{R}_c) = \text{span}(\mathbf{C})$, where

$$\mathbf{C} \triangleq (\mathbf{c}_0, \mathbf{c}_1, \ldots, \mathbf{c}_{N_c-1})$$

The vector $\mathbf{c}_i$ consists of the samples of $e^{j2\pi f_{s,i}x}$ at points $\{n + \gamma m + \beta l\}$, where γ and β are as defined in (6.16). In general, $\mathbf{c}_i$ is a *nonuniformly* sampled version of the bandlimited sinusoidal waveform $e^{j2\pi f_{s,i}x}$. If γ and β are both integers, the sampled points $\{n + \gamma m + \beta l\}$ can be integers only in

$$\{0, 1, \ldots, N + \gamma(M-1) + \beta(L-1)\}$$

If $N + \gamma(M-1) + \beta(L-1) \le NML$, there will be repetitions in the sample points. In other words, some of the row vectors in $\mathbf{C}$ will be exactly the same and there will be at most $N + \gamma(M-1) + \beta(L-1)$ distinct row vectors in $\mathbf{C}$. Therefore, the rank of $\mathbf{C}$ is less than $N + \gamma(M-1) + \beta(L-1)$. So is the rank of $\mathbf{R}_c$. We summarize this fact in the following theorem.

Theorem 6.1 If γ and β are both integers, then $\text{rank}\,(\mathbf{R}_c) \le \min\,(N + \gamma\,(M-1) + \beta(L-1), N_c, NML)$.

Usually N_c and NML are much larger than $N + \gamma(M-1) + \beta(L-1)$. Therefore $N + \gamma\,(M-1) + \beta(L-1)$ is a good estimation of the clutter rank. This result can be viewed as *the MIMO version of Brennan's rule*, given in (6.22).

Now we focus on the general case where γ and β are real numbers. The vector $\mathbf{c}_i$ in (6.24) can be viewed as a nonuniformly sampled version of the truncated sinusoidal function

$$c(x; f_{s,i}) \triangleq \begin{cases} e^{j2\pi f_{s,i}x}, & 0 \le x \le X \\ 0, & \text{otherwise} \end{cases} \tag{6.25}$$

where $X \triangleq N - 1 + \gamma(M-1) + \beta(L-1)$. Furthermore, $-0.5 \le f_{s,i} \le 0.5$ because d_R is often selected as $\lambda/2$ in (6.14) to avoid aliasing. Therefore, the energy of these signals is confined mostly to a certain time–frequency region. Figure 6.6 shows an example of such a signal. Such signals can be well approximated by linear combinations of $\lceil 2WX + 1 \rceil$ orthogonal functions [24], where W is the one-sided bandwidth and X is the duration of the time-limited functions. In the next section this is discussed in more detail using prolate spheroidal wavefunctions (PSWF). In this case, we have $W = 0.5$ and $2WX + 1 = N + \gamma(M-1) + \beta(L-1)$. The vectors $\mathbf{c}_i$ can be also approximated by a linear combination of the nonuniformly sampled versions of these $\lceil N + \gamma(M-1) + \beta(L-1) \rceil$ orthogonal functions. Thus, in the case where γ and β are nonintegers, we can conclude that only $\lceil N + \gamma(M-1) + \beta(L-1) \rceil$ eigenvalues of the matrix $\mathbf{R}_c$ are significant. In other words

$$\text{rank}(\mathbf{R}_c) \approx \lceil N + \gamma(M-1) + \beta(L-1) \rceil \tag{6.26}$$

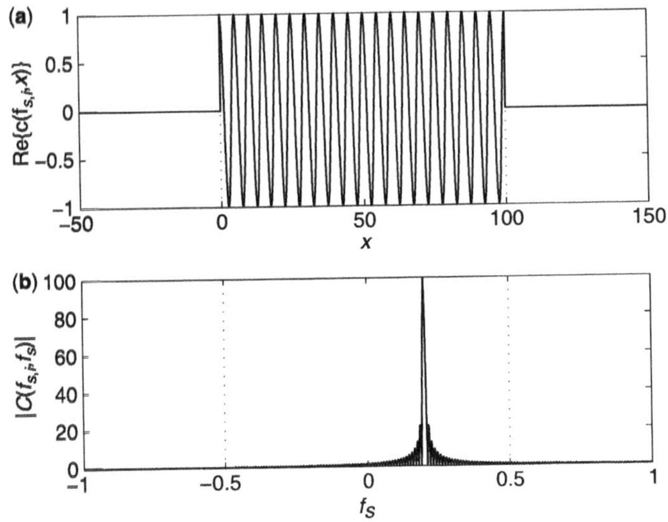

Figure 6.6 Example of the signal $c(x; f_{s,i})$: (a) real part; (b) magnitude response of Fourier transform.

Note that the definition of this approximate rank is actually the number of the dominant eigenvalues. This notation has been widely used in the STAP literature [42,43]. In the SIMO radar case, using Brennan's rule, the ratio of the clutter rank and the total dimension of the spacetime steering vector can be approximated as

$$\frac{N + \beta(L-1)}{NL} = \frac{1}{L} + \frac{\beta(L-1)}{NL}.$$

In the MIMO radar case with $\gamma = N$, the corresponding ratio becomes

$$\frac{N + N(M-1) + \beta(L-1)}{NML} = \frac{1}{L} + \frac{\beta(L-1)}{NML}$$

One can observe that the clutter rank now becomes a smaller portion of the total dimension because of the extra dimension introduced by the MIMO radar. *Thus the MIMO radar receiver can null out the clutter subspace with little effect on the SINR*. Therefore, a better spatial resolution for clutter can be obtained.

The result can be further generalized for the array with arbitrary linear antenna deployment. Let $x_{T,m}$, $m = 0, 1, \ldots, M-1$ be the transmitting antenna locations, $x_{R,n}$, $n = 0, 1, \ldots, N-1$ be the receiving antenna locations, and v be the speed of the radar station. Without loss of generality, we set $x_{T,0} = 0$ and $x_{R,0} = 0$. Then the clutter signals can be expressed as

$$y_{n,m,l}^{(c)} = \sum_{i=0}^{N_c-1} \rho_i e^{j(2\pi/\lambda)\sin\theta_i((x_{R,n}+x_{T,m}+2vTl))}$$

for $n = 0, 1, \ldots, N-1$, $m = 0, 1, \ldots, M-1$, and $l = 0, 1, \ldots, L-1$, where θ_i is the viewing direction of the ith clutter. The term

$$e^{j(2\pi/\lambda)\sin\theta_i(x_{R,n}+x_{T,m}+2vTl)}$$

can also be viewed as a nonuniform sampled version of the function $e^{j(2\pi/\lambda)\sin\theta_i x}$. Using the same argument we have made in the uniform linear array (ULA) case, can obtain

$$\text{rank}(\mathbf{R}_c) \approx \left\lceil 1 + \frac{2}{\lambda}(x_{R,N-1} + x_{T,M-1} + 2vT(L-1)) \right\rceil$$

The quantity $x_{R,N-1} + x_{T,M-1} + 2vT(L-1)$ can be regarded as the total aperture of the spacetime virtual array. One can see that the number of dominant eigenvalues is proportional to the ratio of the total aperture of the spacetime virtual array and the wavelength.

6.4.2 Data-Independent Estimation of the Clutter Subspace with PSWF

The clutter rank can be estimated by using (6.26) and the parameters N, M, L, β, and γ. However, the clutter subspace is often estimated by using data samples instead of using these parameters [31–43]. In this section, we propose a method for estimating the clutter subspace using the geometry of the problem rather than the received signal. The main advantage of this method is that it is data-independent. The clutter subspace obtained by this method can be used to improve the convergence of the STAP. Experiments also show that the estimated subspace is very accurate in the ideal case (i.e., when there are no ICM and array misalignment issues).

6.4.2.1 Prolate Spheroidal Wavefunction From Fig. 6.6, we see that the signal in (6.25) has most of its energy concentrated in the frequency region $[-0.5, 0.5]$. Its duration in the variable x is also limited to the interval $[0, X]$. So, this is analogous to a time-limited signal that is "essentially bandlimited" to $[-0.5, 0.5]$. To approximate the subspace of such functions (the clutter subspace in the MIMO radar context), we propose a set of basis functions

$$\psi_k(x),\ 0 \leq k \leq \infty$$

called the *prolate spheroidal wavefunctions* (PSWFs) [24]. These functions are duration-restricted to

$$0 \leq x \leq X$$

and satisfy the orthonormality property

$$\int_0^X \psi_k(x)\psi_m^*(x)dx = \delta_{km}$$

The reason for choosing the PSWF as the basis for our clutter subspace is the fact that these functions enjoy the following optimality properties:

1. $\psi_0(x)$ has maximum energy in $[-W, W]$ (among all functions with unity energy and support $[0, X]$).
2. For $k > 0$, the function $\psi_k(x)$ has maximum energy in $[-W, W]$, among all functions that are orthogonal to $\psi_0(x), \psi_1(x), \ldots, \psi_{k-1}(x)$ (and support restricted to [0, X]).

With $\{u_k(x)\}$ representing any other orthonormal set with support in $[0, X]$, it can then be shown that, for all $I \geq 0$

$$\sum_{k=0}^{I} \|\text{sinc}(2Wx) * \psi_k(x)\|_X^2 \geq \sum_{k=0}^{I} \|\text{sinc}(2Wx) * u_k(x)\|_X^2 \tag{6.27}$$

where $*$ denotes convolution. Here the notation $\|f(x)\|_X^2$ is defined as the finite integral[1]

$$\|f(x)\|_X^2 = \int_0^X |f(x)|^2 dx$$

Equation (6.27) suggests that, given a finite duration signal $c(x)$ with most of its energy concentrated in $[-W, W]$, we can approximate it more accurately using the truncated prolate set

$$\{\psi_k(x)\}, 0 \leq k \leq I$$

than the truncated set $\{u_k(x)\}$, $0 \leq k \leq I$ for any integer I. This is the motivation for using prolate spheroidal wavefunctions to represent the clutter subspace.

It can be shown that PSWF are solutions to integral equations of the form

$$\int_0^X \psi_k(v)\text{sinc}(2W(x-v))dv = \mu_k\psi_k(x), \quad 0 \leq x \leq X \tag{6.28}$$

Notice that the integral represents a convolution. The meaning of the integral equation is depicted in Fig. 6.7. Thus, if we pass $\psi_k(x)$ through an ideal lowpass filter band-limited to $[-W, W]$ and then truncate the output to the region $[0, X]$, the result is

[1]This notation is necessary because the result of the convolution of $\psi_k(x)$ with the sinc function has infinite duration.

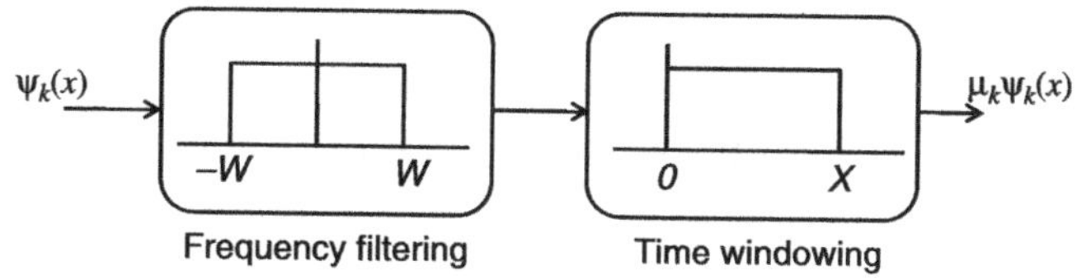

Figure 6.7 Illustration of the integral equation (6.28).

precisely $\psi_k(x)$ scaled by a constant μ_k; that is, $\psi_k(x)$ are eigenfunctions of the linear operator that performs "frequency truncation followed by time truncation." The numbers μ_k can be regarded as eigenvalues of this operator.

It has been shown [24] that the functions $\psi_i(x)$, $i = \lceil 2WX + 1 \rceil, \lceil 2WX + 2 \rceil, \ldots, \infty$ have very small amounts of energy in the frequency region $[-W, W]$. This suggests that we can use the first $\lceil 2WX + 1 \rceil$ basis elements to approximate a finite-duration signal $c(x)$ with most of its energy concentrated in $[-W, W]$

$$c(x) = \sum_{k=0}^{\infty} \alpha_k \psi_k(x) \approx \sum_{k=0}^{\lceil 2WX \rceil} \alpha_k \psi_k(x) \tag{6.29}$$

where $\alpha_k = \int_0^X c(x)\psi_k^*(x)dx$.

6.4.2.2 PSWF Representation of the Clutter Signal The signal $c(x; f_{s,i})$ in (6.25) has support in $[0, X]$, where $X = N - 1 + \gamma(N - 1) + \beta(L - 1)$ and most of its energy is concentrated in the frequency region $[-W, W]$, where $W = 0.5$. Using the result in (6.29), we obtain

$$c(x; f_{s,i}) \approx \sum_{k=0}^{r_c - 1} \alpha_{i,k} \psi_k(x)$$

where $\alpha_{i,k} = \int_0^X c(x; f_{s,i})\psi_k^*(x)\, dx$ and

$$r_c = \lceil N + \gamma(M - 1) + \beta(L - 1) \rceil \tag{6.30}$$

Thus the clutter signals in (6.23), which are the nonuniformly sampled versions of $c(x; f_{s,i})$, can be expressed as

$$\begin{aligned} c_{i,n,m,l} &= c(n + \gamma m + \beta l; f_{s,i}) \\ &= \sum_{k=0}^{r_c - 1} \alpha_{i,k} \psi_k(n + \gamma m + \beta l) \end{aligned}$$

for $n = 0, 1, \ldots, N - 1$, $m = 0, 1, \ldots, M - 1$, and $l = 0, 1, \ldots, L - 1$ where $\alpha_{i,k} = \int_0^X c_i(x; f_{s,i})\psi^*(x)\, dx$. Stacking these elements into $NML \times 1$ vectors, we have

$$\mathbf{c}_i \approx \sum_{k=0}^{r_c - 1} \alpha_{i,k} \mathbf{u}_k$$

where $\mathbf{u}_k$ is a vector that consists of the elements $\psi_k(n + \gamma m + \beta l)$. Finally, we have

$$\text{span}(\mathbf{R}_c) = \text{span}(\mathbf{C}) \approx \text{span}(\mathbf{U}_c) \tag{6.31}$$

where $\mathbf{U}_c \triangleq (\mathbf{u}_0 \quad \mathbf{u}_1 \quad \ldots \quad \mathbf{u}_{r_c-1})$. Note that although the functions $\{\psi_k(x)\}$ are orthogonal, the vectors $\{\mathbf{u}_k\}$ are in general not orthogonal. This is because of the fact that $\{\mathbf{u}_k\}$ are obtained by nonuniform sampling, which destroys orthogonality. In practice, the PSWF $\psi_k(x)$ can be computed offline and stored in memory. When the parameters change, one can obtain the vectors $\mathbf{u}_k$ by resampling the PSWF $\psi_k(x)$ to form the new clutter subspace. In this way, we can obtain the clutter subspace by using only the geometry of the problem.

Example 6.3 Performing the Gram–Schmidt procedure on the basis $\{\mathbf{u}_k\}$, we obtain the orthonormal basis $\{\mathbf{q}_k\}$. The clutter power in each orthonormal basis element can be expressed as $\mathbf{q}_k^{\dagger}\mathbf{R}_c\mathbf{q}_k$. Figure 6.8 shows the clutter power in the orthogonalized basis elements. In this example, $N = 10$, $M = 5$, $L = 16$, $\gamma = 10$, and $\beta = 1.5$. Note that there are totally $NML = 800$ basis elements, but we only show the first 200 on the plot. The clutter covariance matrix $\mathbf{R}_c$ is generated using the model described in Ref. 20. The eigenvalues of $\mathbf{R}_c$ are also shown in Fig. 6.8 for comparison. The estimated clutter rank is $\lceil N + \gamma(M-1) + \beta(L-1) \rceil = 73$. One can see that the subspace obtained by the proposed method captures almost all clutter power. The clutter power decays to less than -200 dB for basis index exceeding 90.

Compared to the eigendecomposition method, the subspace obtained by our method is larger. This is because the clutter spatial bandwidth has been overestimated in this example. More specifically, we have assumed the worst-case situation that the

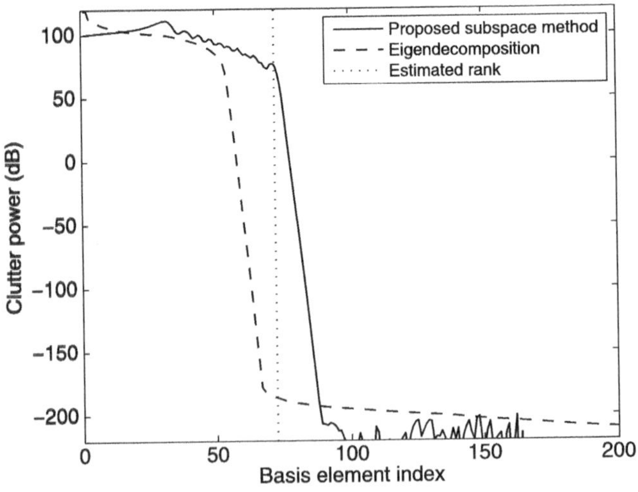

Figure 6.8 Plot of the clutter power distributed in each of the orthogonal basis elements.

clutter spatial frequencies range from -0.5 to 0.5. In fact, however, the range is only from -0.35 to 0.35, because of the specific geometry assumed in this example — the altitude is 9 km, the range of interest is 12.728 km, and a flat ground model is used. Therefore the rank of the subspace is overestimated.

It may seem that our method loses some efficiency compared to the eigendecomposition. However, note that it requires perfect information of the clutter covariance matrix $\mathbf{R}_c$ to achieve the performance indicated by the broken lines in Fig. 6.8, while our method requires no data. In this example, we assume the perfect $\mathbf{R}_c$ is known. In practice, $\mathbf{R}_c$ has to be estimated from the received signals, and it might not be accurate if the number of samples is not large enough. Note that, unlike the eigendecomposition method, the proposed method based on PSWF does not require knowledge of $\mathbf{R}_c$.

6.5 NEW STAP METHOD FOR MIMO RADAR

In this section, we introduce a new STAP method for MIMO radar that uses the clutter subspace estimation method described in the last section. Because the clutter subspace can be obtained by using the parameter information, the performance and complexity can both be improved. Recall that the optimal MVDR beamformer (6.20) requires knowledge of the covariance matrix $\mathbf{R}$. In practice, this has to be estimated from data. For example, it can be estimated as

$$\hat{\mathbf{R}} = \frac{1}{|\mathcal{B}|} \sum_{k \in \mathcal{B}} \mathbf{y}_k \mathbf{y}_k^{\dagger} \tag{6.32}$$

where $\mathbf{y}_k$ is the MIMO-STAP signal vector defined in (6.17) for the kth range bin and $\mathcal{B}$ is a set that contains the neighbor range bin cells of the range bin of interest. However, some nearest cells around the range bin of interest are excluded from $\mathcal{B}$ in order to exclude the target signals [42] as shown in Fig. 6.9. There are two advantages of using the target-free covariance matrix $\mathbf{R}$ in (6.20):

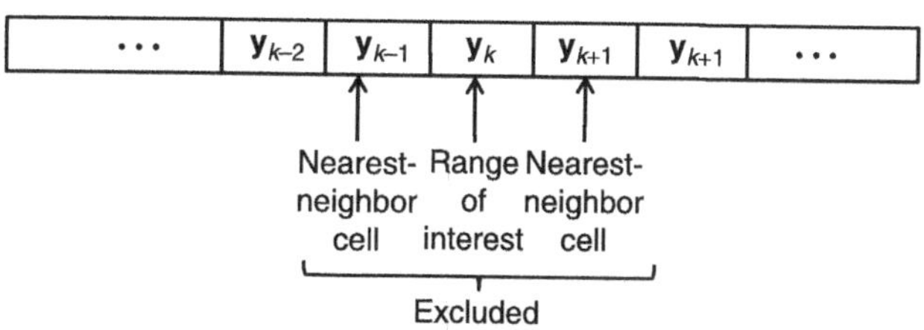

Figure 6.9 Signals of range bins.

1. It is more robust to steering vector mismatch. If there is mismatch in the steering vector $\mathbf{s}(f_s, f_D)$ in (6.18), the target signal is no longer protected by the constraint. Therefore, the target signal is suppressed as interference. This effect, called *self-nulling*, can be prevented by using a target-free covariance matrix. Self-nulling and robust beamforming are discussed in more detail elsewhere (see Refs. 26 and 27 and references cited therein).
2. Using the target-free covariance matrix, the beamformer in (6.20) converges faster than the beamformer using the total covariance matrix. The famous rapid convergence theorem proposed by Reed et al. [30] states that a SINR loss of 3 dB can be obtained by using the number of target-free snapshots equal to twice the size of the covariance matrix. Note that the imprecise physical model which causes steering vector mismatch does not just create the self-nulling problem. It also affects the clutter subspace. Therefore, it affects the accuracy of the clutter subspace estimation in Section 6.4.2.

6.5.1 The Proposed Method

The target-free covariance matrix can be expressed as $\mathbf{R} = \mathbf{R}_J + \mathbf{R}_c + \sigma^2\mathbf{I}$, where $\mathbf{R}_J$ is the covariance matrix of the jammer signals, $\mathbf{R}_c$ is the covariance matrix of the clutter signals, and σ^2 is the variance of the white noise. By (6.31), there exists a $r_c \times r_c$ matrix $\mathbf{A}_c$ so that $\mathbf{R}_c \approx \mathbf{U}_c\mathbf{A}_c\mathbf{U}_c^\dagger$. Thus the covariance matrix can be approximated by

$$\mathbf{R} \approx \underbrace{\mathbf{R}_J + \sigma^2\mathbf{I}}_{\text{call this } \mathbf{R}_v} + \mathbf{U}_c\mathbf{A}_c\mathbf{U}_c^\dagger \tag{6.33}$$

We assume that the jammer signals $y^{(J)}_{n,m,l}$ in (6.15) are statistically independent in different pulses and different orthogonal waveform components [42]. Therefore, they satisfy

$$E\left[y^{(J)}_{n,m,l} \cdot y^{(J)\dagger}_{n',m',l'}\right] = \begin{cases} r_{J,n,n'}, & m = m',\, l = l' \\ 0, & \text{otherwise} \end{cases}$$

for $n, n' = 0, 1, \ldots, N$, $m, m' = 0, 1, \ldots, M$, and $l, l' = 0, 1, \ldots, L$. Applying this fact, the jammer-plus-noise covariance matrix $\mathbf{R}_v$ defined in (6.33) can be expressed as

$$\mathbf{R}_v = \text{diag}(\mathbf{R}_{vs}, \mathbf{R}_{vs}, \ldots, \mathbf{R}_{vs}) \tag{6.34}$$

where $\mathbf{R}_{vs}$ is an $N \times N$ matrix with elements

$$[\mathbf{R}_{vs}]_{n,n'} = r_{J,n,n'} + \sigma^2$$

for $n, n' = 0, 1, \ldots, N$. Therefore the covariance matrix $\mathbf{R}$ in (6.33) consists of a *low-rank* clutter covariance matrix and a *block diagonal* jammer-plus-noise covariance matrix. By using the matrix inversion lemma [48], one can obtain

$$\mathbf{R}^{-1} \approx \mathbf{R}_v^{-1} - \mathbf{R}_v^{-1}\mathbf{U}_c(\mathbf{A}_c^{-1} + \mathbf{U}_c^{\dagger}\mathbf{R}_v^{-1}\mathbf{U}_c)^{-1}\mathbf{U}_c^{\dagger}\mathbf{R}_v^{-1} \tag{6.35}$$

The inverse of the block diagonal matrix $\mathbf{R}_v^{-1}$ is simply

$$\mathbf{R}_v^{-1} = \text{diag}(\mathbf{R}_{vs}^{-1}, \mathbf{R}_{vs}^{-1}, \ldots, \mathbf{R}_{vs}^{-1})$$

and multiplication of the block diagonal matrix with another matrix is simple.

6.5.2 Complexity of the New Method

The complexity of directly inverting the $NML \times NML$ covariance matrix $\mathbf{R}$ is $O(N^3M^3L^3)$. Taking advantage of the block diagonal matrix and the low-rank matrix, in (6.35), the complexity for computing $\boldsymbol{R}_v^{-1}$ is only $O(N^3)$ and the complexity for computing $\mathbf{A}_c^{-1}$ and $(\mathbf{A}_c + \mathbf{U}_c^{\dagger}\mathbf{R}_v^{-1}\mathbf{U}_c)^{-1}$ is only $O(r_c^3)$, where r_c is as defined in (6.30). The overall complexity for computing (6.35) is as thus reduced from $O(N^3M^3L^3)$ to $O(r_cN^2M^2L^2)$. This is the complexity of the multiplication of an $(NML \times r_c)$ matrix with a $(r_c \times NML)$ matrix.

6.5.3 Estimation of the Covariance Matrices

In (6.35), the matrix $\mathbf{U}_c$ can be obtained by the nonuniform sampling of the PSWF as described in the last section. The jammer-plus-noise covariance matrix $\mathbf{R}_v$ and the matrix $\mathbf{A}_c$ both require further estimation from the received signals. Because of the block diagonal structure, one can estimate the covariance matrix $\mathbf{R}_v$ by estimating its submatrix $\mathbf{R}_{vs}$ defined in (6.34). The matrix $\mathbf{R}_{vs}$ can be estimated when there are no clutter and target signals. For this, the radar transmitter *operates in passive mode* so that the receiver can collect the signals with only jammer signals and white noise [43]. The submatrix $\mathbf{R}_{vs}$ can be estimated as

$$\hat{\mathbf{R}}_{vs} = \frac{1}{K_v}\sum_{k=0}^{K_v-1} \mathbf{r}_k\mathbf{r}_k^{\dagger} \tag{6.36}$$

where r_k is an $N \times 1$ vector that represents the target-free and clutter-free signals received by N receiving antennas. By (6.33), one can express $\mathbf{A}_c$ as

$$\mathbf{A}_c = (\mathbf{U}_c^{\dagger}\mathbf{U}_c)^{-1}\mathbf{U}_c^{\dagger}(\mathbf{R} - \mathbf{R}_v)\mathbf{U}_c(\mathbf{U}_c^{\dagger}\mathbf{U}_c)^{-1}$$

Therefore, one can estimate $\mathbf{A}_c$ by using

$$\hat{\mathbf{A}}_c = \frac{1}{K}\sum_{k=0}^{K-1} \mathbf{x}_k\mathbf{x}_k^\dagger - (\mathbf{U}_c^\dagger\mathbf{U}_c)^{-1}\mathbf{U}_c^\dagger\hat{\mathbf{R}}_v\mathbf{U}_c(\mathbf{U}_c^\dagger\mathbf{U}_c)^{-1} \tag{6.37}$$

where $\mathbf{x}_k = (\mathbf{U}_c^\dagger\mathbf{U}_c)^{-1}\mathbf{U}_c^\dagger\mathbf{y}_k$ and $\mathbf{y}_k$ is the $NML \times 1$ MIMO-STAP signal vector defined in (6.17). Substituting (6.36), (6.37), and (6.35) into the MIMO-STAP beamformer in (6.20), we obtain

$$\mathbf{w} \propto \left(\hat{\mathbf{R}}_v^{-1} - \hat{\mathbf{R}}_v^{-1}\mathbf{U}_c(\hat{\mathbf{A}}_c^{-1} + \mathbf{U}_c^\dagger\hat{\mathbf{R}}_v^{-1}\mathbf{U}_c)^{-1}\mathbf{U}_c^\dagger\hat{\mathbf{R}}_v^{-1}\right)\mathbf{s}(f_s, f_d) \tag{6.38}$$

6.5.4 Zero-Forcing Method

Instead of estimating $\mathbf{A}_c$ and computing the MVDR by (6.38), one can directly "null out" the entire clutter subspace as described next. Assuming that the clutter to noise ratio is very large and therefore all of the eigenvalues of $\mathbf{A}_c$ approach infinity, we obtain $\mathbf{A}_c^{-1} \approx 0$. Substituting this into (6.38), one can obtain the MIMO-STAP beamformer as

$$\mathbf{w} \propto (\hat{\mathbf{R}}_v^{-1} - \hat{\mathbf{R}}_v^{-1}\mathbf{U}_c(\mathbf{U}_c^\dagger\hat{\mathbf{R}}_v^{-1}\mathbf{U}_c)^{-1}\mathbf{U}_c^\dagger\hat{\mathbf{R}}_v^{-1})\mathbf{s}(f_s, f_d) \tag{6.39}$$

Thus we obtain a "zero-forcing" beamformer that nulls out the entire clutter subspace. The advantage of this zero-forcing method is that it is no longer necessary to estimate $\mathbf{A}_c$. In this method, we only need to estimate $\mathbf{R}_{vs}$. The method is independent of the range bin. The matrix $\mathbf{R}^{-1}$ computed by this method can be used for all range bins. Because there are many extra dimensions in MIMO radars, dropping the entire clutter subspace will reduce only a small portion of the total dimension. Therefore it will not affect the SINR performance significantly, as we shall demonstrate. Thus this method can be very effective in MIMO radars.

6.5.5 Comparison with Other Methods

In the sample matrix inversion (SMI) method [42], the covariance matrix is estimated to be the quantity $\hat{\mathbf{R}}$ in (33) and $\hat{\mathbf{R}}^{-1}$ is directly used in (6.32), to obtain the MVDR beamformer. However, some important information about the covariance matrix is unused in the SMI method. This information includes the parameters γ and β, the structure of the clutter covariance matrix, and the block diagonal structure of the jammer covariance matrix.

Our method in (6.38) utilizes this information. We first estimate the clutter subspace by using parameters γ and β in (6.31). Because the jammer matrix is block diagonal and the clutter matrix has low rank with known subspace, by using the matrix inversion lemma, we could break the inversion of a large matrix $\mathbf{R}$ into the inversions of some smaller matrices. Therefore the computational complexity was significantly reduced. Moreover, by using the structure, fewer parameters need to be estimated. In our method, only the $r_c \times r_c$ matrix $\mathbf{A}$ and the $N \times N$ matrix $\mathbf{R}_{vs}$ need to be estimated rather than the the $NML \times NML$ matrix $\mathbf{R}$ in the SMI method. Therefore our method also converges much faster.

In subspace methods [32–43], the clutter and the jammer subspace are both estimated simultaneously using the STAP signals rather than from problem geometry. Therefore the parameters γ and β and the block diagonal structure of the jammer covariance matrix are not fully utilized. Klemm [31] also estimated the target-free and clutter-free covariance matrix using (6.36). The jammer and clutter are filtered out in two separate stages. Therefore the block diagonal property of the jammer covariance matrix has been used in that study [31]. However, the clutter subspace structure has not been fully utilized in this method.

6.6 NUMERICAL EXAMPLES

In this section, we compare the SINR performance of our methods and other existing methods. In the example, the parameters are $M = 5$, $N = 10$, $L = 16$, $\beta = 1.5$, and $\gamma = 10$. The altitude is 9 km and the range of interest is 12.728 km. For this altitude and range, the clutter is generated by using the model in Ref. 20. The clutter : noise ratio (CNR) is 40 dB. There are two jammers at 20° and −30°. The jammer : noise ratio (JNR) for each jammer equals 50 dB. The SINR is normalized so that the maximum SINR equals 0 dB. The jammers are modeled as point sources that emit independent white Gaussian signals. The clutter is modeled using discrete points as described in (6.11). The clutter points are equally spaced on the range bin and the RCS (radar cross section) for each clutter is modeled as identical independent Gaussian random variables. In general, the variance of ρ_i will vary along the ground, as we move within one range bin. However, for simplicity we assume that this variance is fixed. The number of clutter points N_c is 10,000. The clutter points for different range bins are also independent. The following methods are compared:

1. *Sample Matrix Inversion (SMI) Method* [42]. This method estimates the covariance matrix $\mathbf{R}$ using (6.32) and directly substitutes it into (6.20).
2. *Loaded Sample Matrix Inversion (LSMI) Method* [28,29]. Before substituting $\hat{\mathbf{R}}$ into (6.20), a diagonal loading $\hat{\mathbf{R}} \leftarrow \hat{\mathbf{R}} + \delta\mathbf{I}$ is performed. In this example, δ is chosen as 10 times the white noise level.
3. *Principal-Component (PC) Method* [42]. This method uses a KLT Karhunen-Loève transform filterbank to extract the jammer-plus-clutter subspace. Then the spacetime beamforming can be performed in this subspace.
4. *Separate Jammer and Clutter Cancellation Method* [31] (*SJCC*). This method also utilizes the jammer-plus-noise covariance matrix $\mathbf{R}_{vs}$, which can be estimated as in (6.36). The covariance matrix can be used to filter out the jammer and form a spatial beam. Then the clutter can be further filtered out

by spacetime filtering [31]. In this example, a diagonal loading is used for the spacetime filtering with a loading factor which equals ten times the white noise level.

5. *The New Zero-Forcing (ZF) Method.* This method directly nulls out the clutter subspace as described in (6.39).
6. *The New Minimum Variance Method.* This method estimates $\hat{\mathbf{R}}_{vs}$ and $\hat{\mathbf{A}}_c$ and uses (6.38). In this example, a diagonal loading is used for $\hat{\mathbf{A}}_c$ with a loading factor that equals 10 times the white noise level.
7. *MVDR with Perfectly Known* **R**. This method is unrealizable because the perfect **R** is always unavailable. It is shown in Fig. 6.10 because it serves as an upper bound on the SINR performance.

Figure 6.10 compares of the SINR for $f_s = 0$ as a function of the Doppler frequencies. The SINR is defined as

$$\mathrm{SINR} \triangleq \frac{|\mathbf{w}^\dagger \mathbf{s}(f_s, f_D)|^2}{\mathbf{w}^\dagger \mathbf{R} \mathbf{w}}$$

where **R** is the target-free covariance matrix. To compare these methods, we fix the number of samples K and the number of jammer-plus-noise samples K_v. In all of the methods except the SMI method, 300 samples and 20 jammer-plus-noise samples are used. We use 2000 samples instead of 300 samples in the SMI method because the estimated covariance matrix in (6.32) with 300 samples

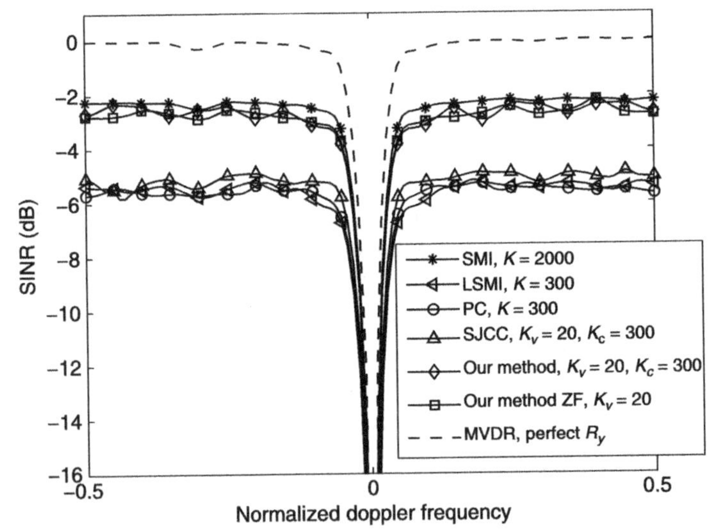

Figure 6.10 The SINR performance of different STAP methods at viewing direction zero as a function of the Doppler frequency.

is not full-rank and therefore cannot be inverted. The spatial beampatterns and spacetime beampatterns for the target at $f_s = 0$ and $f_D = 0.25$ for four of these methods are shown in Figs. 6.11 and 6.12, respectively. The spatial beampattern is defined as

$$\sum_{k=0}^{ML-1} |\mathbf{w}^{\dagger}_{(1:N)+kML}\mathbf{s}(f_s)|^2$$

where $\mathbf{s}(f_s)$ is the spatial steering vector

$$(1 \quad e^{j2\pi f_s} \quad \cdots \quad e^{j2\pi f_s(N-1)})^T$$

and $\mathbf{w}_{(1:N)+kML}$ represents N successive elements of $\mathbf{w}$ starting from $kML + 1$. The spacetime beampattern is defined as

$$|\mathbf{w}^{\dagger}\mathbf{s}(f_s, f_D)|$$

where $\mathbf{s}(f_s, f_D)$ is the spacetime steering vector defined in (6.19). The spatial beampattern represents the jammer and noise rejection, and the spacetime beampattern represents the clutter rejection. In Fig. 6.11, one can see the jammer notches at the corresponding jammer arrival angles $-30°$ and $20°$. In Fig. 6.12, one can also observe the clutter notch in the beampatterns. In Fig. 6.10, lacking use of the covariance matrix structure, the SMI method requires numerous samples to obtain good

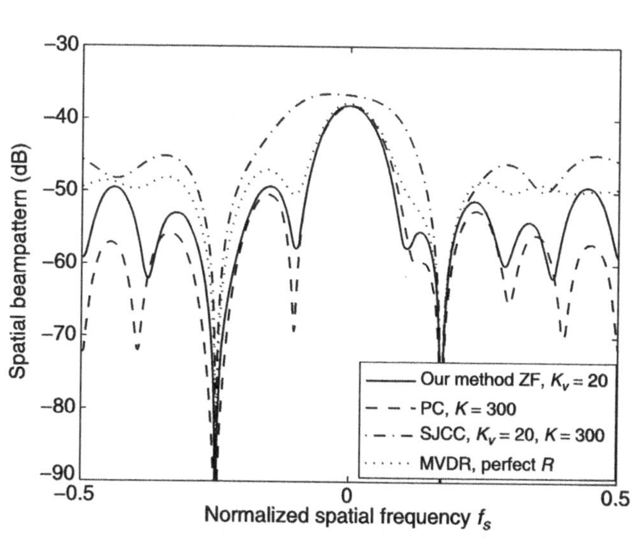

Figure 6.11 Spatial beampatterns for four STAP methods.

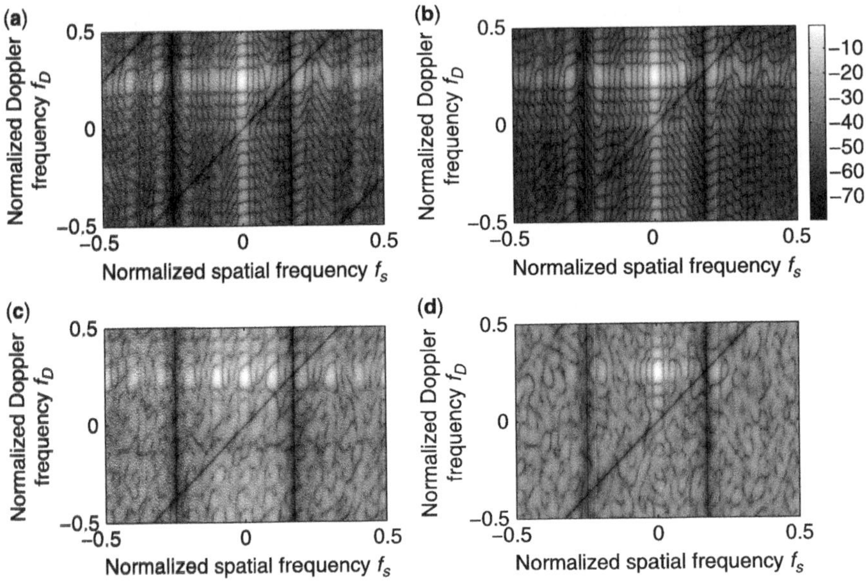

Figure 6.12 Spacetime beampatterns for four methods: (a) proposed zero-forcing method; (b) principal-component (PC) method [42]; (c) separate jammer–clutter cancellation method (SJCC) [31]; (d) sample matrix inversion (SMI) method [42].

performance. It uses 2000 samples, but the proposed minimum-variance method, which has a comparable performance, uses only 300 samples. The PC method and LSMI method utilize the fact that the jammer-plus-clutter covariance matrix has low rank. Therefore they require fewer samples than does the SMI method. The performances of these two are about the same. The SJCC method further utilizes the fact that the jammer covariance matrix is block diagonal and estimates the jammer-plus-noise covariance matrix. Therefore the SINR performance is slightly better than the LSMI and PC methods. Our methods utilize not only the low-rank property and the block diagonal property but also the geometry of the problem. Therefore our methods have better SINR performance than does the SJCC method. The proposed zero-forcing (ZF) method has about the same performance as the minimum variance method. It converges to a satisfactory SINR with very few clutter-free samples. According to (6.26), the clutter rank in this example is approximately

$$\lceil N + \gamma(M-1) + \beta(L-1) \rceil = 73$$

Thanks to the MIMO radar, the dimension of the spacetime steering vector is $MNL = 800$. The clutter rank is just a small portion of the total dimension. This is the reason why the ZF method, which directly nulls out the entire clutter space, works so well.

6.7 SIGNAL DESIGN OF THE STAP RADAR SYSTEM

In Section 6.2, we have demonstrated how to use the orthogonal waveforms $\{\phi_m(t)\}$ to obtain the virtual array that increases the spatial resolution of the radar systems. In this section, we address the issue of designing these waveforms. In fact, the pulses $\{\phi_m(t)\}$ affect not only the spatial resolution but also the range resolution. In the traditional SIMO radar, the ambiguity function ambiguity function [49] is used to characterize the range and Doppler resolution of a radar waveform. More recently, some authors have extended this idea to the MIMO case [37]. In this section, we briefly review the MIMO radar ambiguity function and use the results to formulate the signal design problem in the STAP case. We focus on the case with pulse radar with uniform linear array (ULA).

6.7.1 MIMO Radar Ambiguity Function

In SIMO radars, the ambiguity function of a waveform $u(t)$ is defined as [49]

$$|\chi(\tau,\nu)| \triangleq \left| \int_{-\infty}^{\infty} u(t)u^*(t+\tau)e^{j2\pi\nu t}dt \right| \tag{6.40}$$

This two-dimensional function indicates the matched filter output in the receiver when a delay mismatch τ and a Doppler mismatch ν occur. The value $|\chi(0, 0)|$ represents the matched filter output without any mismatch. Therefore, the sharper the function $|\chi(\tau, \nu)|$ around (0, 0), the better the Doppler and range resolution. Figure 6.13 shows two examples of the ambiguity function. These two ambiguity functions show different Doppler–range tradeoffs. One can see that the LFM pulse has a better range resolution along the cut that Doppler frequency is zero.

In MIMO radar, we have multiple waveforms $\{u_m(t)\}$ emitted from different locations. Since these waveforms are emitted from different locations, the choice of these waveforms also affects the spatial resolution, in addition to Doppler and range resolutions. In this section, we focus only on the case of the ULA shown in Fig. 6.2 in Example 6.1. More general results can be found in Ref. 37. Consider a target at (τ, ν, f_s), where τ is the delay corresponding to the target range, ν is the Doppler frequency of the target, and f_s is the normalized spatial frequency of the target. The demodulated target response in the nth antenna is proportional to

$$y_n^{\tau,\nu,f_s}(t) \approx \sum_{m=0}^{M-1} u_m(t-\tau)e^{j2\pi\nu t}e^{j2\pi f_s(\gamma m+n)}$$

for $n = 0, 1, \ldots, N-1$, where N is the number of receiving antennas, $u_m(t)$ is the radar waveform emitted by the mth antenna, and M is the number of transmitting antennas. If the receiver tries to capture this target signal with a

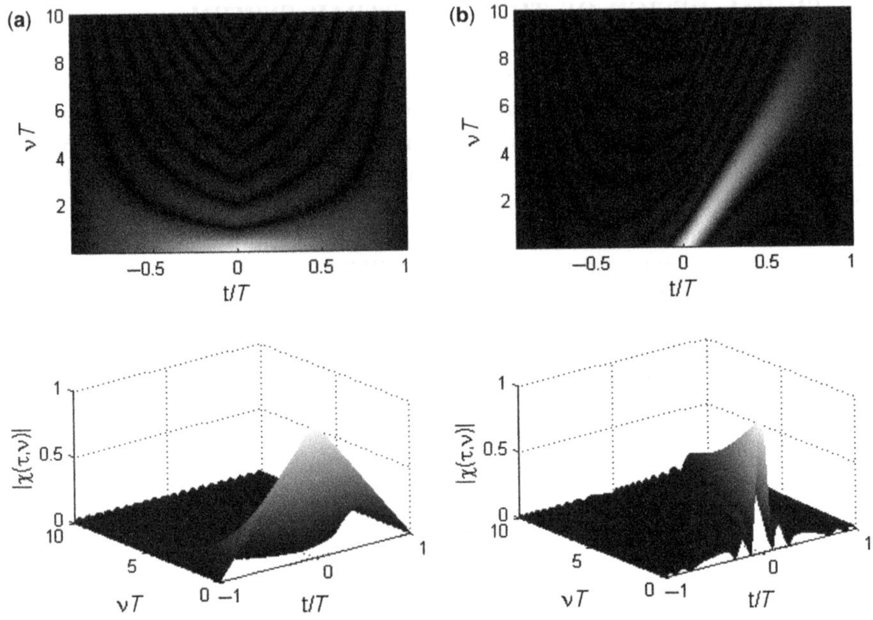

Figure 6.13 Examples of ambiguity functions: (a) rectangular pulse and (b) linear frequency modulation (LFM) pulse with time–bandwidth product 10, where T is the pulse duration.

matched filter with the assumed parameters (τ', ν', f_s') then the matched filter output becomes

$$\sum_{n=0}^{N-1} \int_{-\infty}^{\infty} y_n^{\tau,\nu,f_s}(t) \cdot y_n^{\tau',\nu',f_s'*}(t)dt$$

$$= \left(\sum_{n=0}^{N-1} e^{j2\pi(f_s - f_s')n} \right)$$

$$\left(\sum_{m=0}^{M-1} \sum_{m'=0}^{M-1} \int_{-\infty}^{\infty} u_m(t-\tau)u_{m'}^*(t-\tau')e^{j2\pi(\nu-\nu')t}dt \cdot e^{j2\pi(f_s m - f_s' m)\gamma} \right)$$

The first part in the right side of the equation represents the spatial processing in the receiver, and it is not affected by the waveforms $\{u_m(t)\}$. The second part in the right side of the equation indicates how the waveforms $\{u_m(t)\}$ affect the spatial, Doppler, and range resolutions of the radar system. Therefore, we define the MIMO radar ambiguity function as

$$\chi(\tau, \nu, f_s, f_s') \triangleq \sum_{m=0}^{M-1} \sum_{m'=0}^{M-1} \chi_{m,m'}(\tau, \nu)e^{j2\pi(f_s m - f_s' m')\gamma} \tag{6.41}$$

where

$$\chi_{m,m'}(\tau,\nu) \triangleq \int_{-\infty}^{\infty} u_m(t)u_{m'}^*(t+\tau)e^{j2\pi\nu t}dt \tag{6.42}$$

Note that the MIMO ambiguity function cannot be expressed as a function of the difference of the spatial frequencies, namely, $f_s - f_s'$. Therefore, we need both the target spatial frequency f_s and the assumed spatial frequency f_s' to represent the spatial mismatch. We call the function $\chi_{m,m'}(\tau,\nu)$ the *cross-ambiguity function* because it is similar to the SIMO ambiguity function defined in (6.40) except that it involves two waveforms: $u_m(t)$ and $u_{m'}(t)$. Fixing τ and ν in (6.41), one can view the ambiguity function as a scaled two-dimensional Fourier transform of the cross-ambiguity function $\chi_{m,m'}(\tau,\nu)$ on the parameters m and m'. The value $|\chi(0,0,f_s,f_s')|$ represents the matched filter output without mismatch. Therefore, the sharper the function $|\chi(\tau,\nu,f_s,f_s')|$ around the line $\{(0,0,f_s,f_s)\}$, the better the radar system resolution.

6.7.2 Some Properties of the MIMO Ambiguity Function

We now derive some new properties of the MIMO ambiguity function defined in (6.41). The properties are similar to some of the properties of the SIMO ambiguity functions (e.g., see Ref. 49). We assume that the transmitted waveforms $\{u_m(t)\}$ are orthonormal:

$$\int_{-\infty}^{\infty} u_m(t)u_{m'}^*(t)dt = \delta_{m,m'} \tag{6.43}$$

The following property characterizes the ambiguity function when there exists no mismatch.

Property 6.1

$$\chi(0,0,f_s,f_s) = M, \forall f_s$$

Proof: Using the orthonormality in (6.43), we have

$$\chi_{m,m'}(0,0) = \int_{-\infty}^{\infty} u_m(t)u_{m'}^*(t)dt = \delta_{m,m'}$$

Substituting this equation into (6.41), we obtain

$$\chi(0,0,f_s,f_s) = \sum_{m=0}^{M-1}\sum_{m'=0}^{M-1} \delta_{m,m'}e^{j2\pi\gamma(f_s m - f_s m')} = \sum_{m=0}^{M-1} e^{j0} = M$$

□

This property states that the ambiguity function is a constant along the line $\{(0, 0, f_s, f_s)\}$ that is independent of the waveforms $\{u_m(t)\}$. This means the matched filter output is always a constant when there exists no mismatch, no matter what waveforms are chosen. The following property characterizes the total energy of the cross-ambiguity function.

Property 6.2

$$\int_{-\infty}^{\infty}\int_{-\infty}^{\infty} |\chi_{m,m'}(\tau, \nu)|^2 d\tau\, d\nu = 1$$

Proof: Using Parserval's theorem [47], we have

$$\begin{aligned}
&\int_{-\infty}^{\infty}\int_{-\infty}^{\infty} |\chi_{m,m'}(\tau, \nu)|^2 d\tau\, d\nu \\
&= \int_{-\infty}^{\infty}\int_{-\infty}^{\infty} \left|\int_{-\infty}^{\infty} u_m(t)u_{m'}^*(t+\tau)e^{j2\pi\nu t}dt\right|^2 d\nu\, d\tau \\
&= \int_{-\infty}^{\infty}\int_{-\infty}^{\infty} |u_m(t)u_{m'}^*(t+\tau)|^2 dt\, d\tau
\end{aligned}$$

By changing variables, we obtain

$$\int_{-\infty}^{\infty}\int_{-\infty}^{\infty} |u_m(t)u_{m'}^*(t+\tau)|^2 dt\, d\tau = \int_{-\infty}^{\infty} |u_m(t)|^2 dt \int_{-\infty}^{\infty} |u_{m'}(t)|^2 dt = 1 \qquad \square$$

This property states the fact that the total energy of the cross ambiguity function is a constant that is independent of the waveforms $u_m(t)$ and $u_{m'}(t)$. In the special case of $m = m'$, this property implies that the SIMO radar ambiguity function defined in (6.40) has constant total energy [49]. The following property characterizes the total energy of the MIMO radar ambiguity function.

Property 6.3 If γ is an integer, then

$$\int_0^1\int_0^1\int_{-\infty}^{\infty}\int_{-\infty}^{\infty} |\chi(\tau, \nu, f_s, f_s')|^2 d\tau\, d\nu\, df_s\, df_s' = M^2$$

Proof: By using the definition of MIMO ambiguity function in (6.41) and performing appropriate change of variables, we have

$$\begin{aligned}
&\int_0^1 \int_0^1 \int_{-\infty}^{\infty} \int_{-\infty}^{\infty} |\chi(\tau, \nu, f_s, f_s')|^2 d\tau\, d\nu\, df_s\, df_s' \\
&= \int_{-\infty}^{\infty} \int_{-\infty}^{\infty} \int_0^1 \int_0^1 \left| \sum_{m=0}^{M-1} \sum_{m=0}^{M-1} \chi_{m,m'}(\tau, \nu) e^{j2\pi(f_s m - f_s' m')\gamma} \right|^2 df_s\, df_s'\, d\tau\, d\nu \\
&= \frac{1}{\gamma^2} \int_{-\infty}^{\infty} \int_{-\infty}^{\infty} \int_0^{\gamma} \int_0^{\gamma} \left| \sum_{m=0}^{M-1} \sum_{m=0}^{M-1} \chi_{m,m'}(\tau, \nu) e^{j2\pi(f_s m - f_s' m)} \right|^2 df_s\, df_s'\, d\tau\, d\nu \qquad (6.44)
\end{aligned}$$

Using Parserval's theorem and applying Property 6.2, the integral shown above equals

$$\int_{-\infty}^{\infty} \int_{-\infty}^{\infty} \sum_{m=0}^{M-1} \sum_{m'=0}^{M-1} |\chi_{m,m'}(\tau, \nu)|^2 d\tau\, d\nu = \sum_{m'=0}^{M-1} \sum_{m'=0}^{M-1} 1 = M^2$$

□

This property states that when γ is an integer, the total energy of the MIMO ambiguity function is a constant that is independent of the waveforms $\{u_m(t)\}$. For example, in Section 6.2, where we use $\gamma = 1$ and $\gamma = N$, the total energy of the MIMO ambiguity function is a constant. Recall that Property 6.1 states that the ambiguity function along the line $\{(0, 0, f_s, f_s\}$ is also a constant. This implies that in order to make the ambiguity function sharp around $\{0, 0, f_s, f_s\}$, we can only spread the energy of the ambiguity function evenly on the available time and bandwidth. For the case where γ is not an integer, we cannot directly apply Parserval's theorem. In this case, the volume of the ambiguity function actually depends on the waveforms $\{u_m(t)\}$. However, the following property characterizes the range of the total energy of the MIMO ambiguity function.

Property 6.4

$$\frac{\lfloor \gamma \rfloor^2}{\gamma^2} M^2 \leq \int_0^1 \int_0^1 \int_{-\infty}^{\infty} \int_{-\infty}^{\infty} |\chi(\tau, \nu, f_s, f_s')|^2 d\tau\, d\nu\, df_s\, df_s' \leq \frac{\lceil \gamma \rceil^2}{\gamma^2} M^2$$

where $\lfloor \gamma \rfloor$ is the largest integer smaller than γ, and $\lceil \gamma \rceil$ is the smallest integer larger than γ.

Proof: Using (6.44), we have

$$\int_0^1 \int_0^1 \int_{-\infty}^{\infty} \int_{-\infty}^{\infty} |\chi(\tau, \nu, f_s, f_s')|^2 d\tau\, d\nu\, df_s\, df_s'$$
$$\leq \frac{1}{\gamma^2} \int_{-\infty}^{\infty} \int_{-\infty}^{\infty} \int_0^{\lceil\gamma\rceil} \int_0^{\lceil\gamma\rceil} \left| \sum_{m=0}^{M-1} \sum_{m=0}^{M-1} \chi_{m,m'}(\tau, \nu) e^{j2\pi(f_s m - f_s' m')} \right|^2 df_s\, df_s'\, d\tau\, d\nu$$

Using Parserval's theorem and applying Property 6.2, this value equals

$$\frac{\lceil\gamma\rceil^2}{\gamma^2} \int_{-\infty}^{\infty} \int_{-\infty}^{\infty} \sum_{m=0}^{M-1} \sum_{m'=0}^{M-1} |\chi_{m,m'}(\tau, \nu)|^2 d\tau\, d\nu = \frac{\lceil\gamma\rceil^2}{\gamma^2} M^2$$

Similarly, the lower bound can be proven by the same argument.

□

For the case where γ is not an integer, the total energy of the ambiguity function can actually be affected by the waveforms $\{u_m(t)\}$. However, Property 6.4 implies that the amount of the total energy that can be affected by the waveforms is small. The following property characterizes the symmetry of the cross-ambiguity function.

Property 6.5

$$\chi_{m,m'}(-\tau, -\nu) = \chi_{m',m}^*(\tau, \nu) e^{-j2\pi\nu\tau}$$

Proof: By the definition of the cross ambiguity function (6.42) and changing variables, we have

$$\begin{aligned}\chi_{m,m'}(-\tau, -\nu) &= \int_{-\infty}^{\infty} u_m(t) u_{m'}^*(t - \tau) e^{-j2\pi\nu t} dt \\ &= \int_{-\infty}^{\infty} u_m(t + \tau) u_{m'}^*(t) e^{-j2\pi\nu(t+\tau)} dt \\ &= \chi_{m',m}^*(\tau, \nu) e^{-j2\pi\nu\tau}\end{aligned}$$

□

Using Property 6.5, we can obtain the following property of the MIMO ambiguity function.

Property 6.6

$$\chi(-\tau, -\nu, f_s, f_s') = \chi^*(\tau, \nu, f_s', f_s) e^{-j2\pi\nu\tau}$$

Proof: Using the definition of the MIMO ambiguity function (6.41) and Property 6.5, we have

$$\begin{aligned}
\chi(-\tau, \nu, f_s, f_s') &= \sum_{m=0}^{M-1} \sum_{m'=0}^{M-1} \chi_{m,m'}(-\tau, -\nu) e^{j2\pi\gamma(f_s m - f_s' m')} \\
&= \sum_{m=0}^{M-1} \sum_{m'=0}^{M-1} \chi_{m',m}^{*}(\tau, \nu) e^{-j2\pi\nu\tau} e^{j2\pi\gamma(f_s m - f_s m')} \qquad \text{(Property 6.5)} \\
&= \left(\sum_{m=0}^{M-1} \sum_{m'=0}^{M-1} \chi_{m',m}(\tau, \nu) e^{j2\pi\gamma(f_s' m' - f_s m)} \right)^{*} e^{-j2\pi\nu\tau} \\
&= \chi^{*}(\tau, \nu, f_s', f_s) e^{-j2\pi\nu\tau}
\end{aligned}$$

□

This property implies that given two spatial frequencies, f_s and f_s', it is sufficient to study only $\chi(\tau, \nu, f_s, f_s')$. The function $\chi(\tau, \nu, f_s', f_s)$ can be deduced from the symmetry property. The following property characterizes the cross-ambiguity function of the linear frequency modulation (LFM) signal.

Property 6.7 If $\chi_{m,m'}(\tau, \nu) = \int_{-\infty}^{\infty} u_m(t) u_{m'}^{*}(t + \tau) e^{j2\pi\nu t} dt$, then

$$\begin{aligned}
\chi_{m,m'}^{\text{LFM}}(\tau, \nu) &\triangleq \int_{-\infty}^{\infty} \left(u_m(t) e^{j\pi k\nu t^2} \right) \left(u_{m'}(t + \tau) e^{j\pi k\nu(t+\tau)^2} \right)^{*} e^{j2\pi\nu t} dt \\
&= \chi_{m,m'}(\tau, \nu - k\tau) e^{-j\pi k\tau^2}
\end{aligned}$$

Proof: From direct calculation, we have

$$\begin{aligned}
\chi_{m,m'}^{\text{LFM}}(\tau, \nu) &= \int_{-\infty}^{\infty} u_m(t) u_{m'}^{*}(t + \tau) e^{j\pi k(-2t\tau - \tau^2)} e^{j2\pi\nu t} dt \\
&= \int_{-\infty}^{\infty} u_m(t) u_{m'}^{*}(t + \tau) e^{j2\pi(\nu - k\tau)t} dt \cdot e^{-j\pi k\tau^2} \\
&= \chi_{m,m'}(\tau, \nu - k\tau) e^{-j\pi k\tau^2}
\end{aligned}$$

□

This property states that adding LFM modulation shears off the cross-ambiguity function. We use this property to obtain the following result for the MIMO ambiguity function.

Property 6.8 If $\chi(\tau, \nu, f_s, f_s') = \sum_{m=0}^{M-1} \sum_{m'=0}^{M-1} \chi_{m,m'}(\tau, \nu) e^{j2\pi\gamma(f_s m - f_s' m')}$, then

$$\chi^{\text{LFM}}(\tau, \nu, f_s, f_s') \triangleq \sum_{m=0}^{M-1} \sum_{m'=0}^{M-1} \chi_{m,m'}^{\text{LFM}}(\tau, \nu) e^{j2\pi\gamma(f_s m - f_s' m')}$$
$$= \chi(\tau, \nu - k\tau, f_s, f_s') e^{jk\nu\tau^2}$$

We omit the proof because this property can be easily proved by simply applying Property 6.7. Property 6.8 states that adding LFM modulations shears off the MIMO ambiguity function. This shearing can improve the delay resolution because it compresses the ambiguity function along the direction $(\tau, 0, f_s, f_s)$ [49].

6.7.3 The MIMO Ambiguity Function of Periodic Pulse Radar Signals

In STAP systems, the transmitter often emits periodic pulse as we described in (6.9). The overall waveform $u_m(t)$ consists of the shifted version of a smaller waveform $\phi_m(t)$. In this case, the pulse design problem becomes choosing the waveform $\phi_m(t)$ to obtain a good MIMO ambiguity function $\chi(\tau, \nu, f_s, f_s')$. Therefore, it is important to study the relation between the MIMO ambiguity function and the pulse $\phi_m(t)$. Because modulation and scalar multiplication will not change the shape of the ambiguity function, for convenience, we write the transmitted signals as

$$u_m(t) = \sum_{l=0}^{L-1} \phi_m(t - lT) \tag{6.45}$$

Note that the support of $\phi_m(t)$, namely, T_ϕ, is small enough such that $T_\phi \ll T$. To obtain the relation between $\phi_m(t)$ and the MIMO ambiguity function $\chi(\tau, \nu, f_s, f_s')$, we first derive the cross-ambiguity function. Using (6.42) and (6.45), we can express the cross-ambiguity function as

$$\chi_{m,m'}(\tau, \nu) = \int_{-\infty}^{\infty} u_m(t) u_{m'}^*(t + \tau) e^{j2\pi\nu t} dt$$
$$= \int_{-\infty}^{\infty} \sum_{l=0}^{L-1} \phi_m(t - lT) \sum_{l'=0}^{L-1} \phi_{m'}^*(t - l'T + \tau) e^{j2\pi\nu t} dt$$
$$= \sum_{l'=0}^{L-1} \sum_{l=0}^{L-1} \int_{-\infty}^{\infty} \phi_m(t - lT) \phi_{m'}^*(t - l'T + \tau) e^{j2\pi\nu t} dt$$

By changing variables, this equality becomes

$$\chi_{m,m'}(\tau, \nu) = \sum_{l'=0}^{L-1} \sum_{l=0}^{L-1} \int_{-\infty}^{\infty} \phi_m(t) \phi_{m'}^*(t + lT - l'T + \tau) e^{j2\pi\nu(t+lT)} dt$$
$$= \sum_{l'=0}^{L-1} \sum_{l=0}^{L-1} \chi_{m,m'}^{\phi}(\tau + (l - l')T, \nu) e^{j2\pi\nu lT} \tag{6.46}$$

where $\chi^{\phi}_{m,m'}(\tau, \nu)$ is defined as the cross-ambiguity function of the pulses $\phi_m(t)$ and $\phi_{m'}(t)$:

$$\chi^{\phi}_{m,m'}(\tau, \nu) = \int_0^{T_\phi} \phi_m(t)\phi^*_{m'}(t+\tau)e^{j2\pi\nu t}dt$$

We assume that the Doppler frequency ν and the support of pulse T_ϕ are both small enough such that $T_\phi \nu \approx 0$, so this equation becomes

$$\chi^{\phi}_{m,m'}(\tau, \nu) \approx \int_0^{T_\phi} \phi_m(t)\phi^*_{m'}(t+\tau)dt \triangleq r^{\phi}_{m,m'}(\tau)$$

where $r^{\phi}_{m,m'}(\tau)$ is the cross-correlation between $\phi_m(t)$ and $\phi_{m'}(t)$. Thus, the cross-ambiguity function reduces to the cross-correlation function, and it is no longer a function of Doppler frequency ν. Substituting this result into (6.46), we obtain

$$\chi_{m,m'}(\tau, \nu) \approx \sum_{l'=0}^{L-1}\sum_{l=0}^{L-1} r^{\phi}_{m,m'}(\tau + (l-l')T)e^{j2\pi\nu lT} \tag{6.47}$$

For the delay τ in the range $[0, T - T_\phi]$, the shifted correlation function satisfies

$$r^{\phi}_{m,m'}(\tau + (l-l')T) = \int_0^{T_\phi} \phi_m(\tau)\phi^*_{m'}(t+\tau+(l-l')T)dt = 0$$

when $l - l' \neq 0$. For $\tau > T - T_\phi$, the response in the ambiguity function is created by the second trip echoes. This ambiguity is called "range folding." Such ambiguity can be resolved by using different pulse repetition frequencies (PRFs) from time to time. We do not address this ambiguity in this chapter; rather, we focus on the ambiguity function only when $\tau \in [0, T - T_\phi]$. In this case, we have

$$\chi_{m,m'}(\tau, \nu) \approx r^{\phi}_{m,m'}(\tau)\sum_{l-0}^{L-1} e^{j2\pi\nu lT}$$

for $\tau \in [0, T - T_\phi]$. The Doppler processing is separable from the correlation function, because the duration of the pulses T_ϕ and the Doppler frequency ν are small enough so that $\nu T \approx 0$. This implies the choice of the waveforms $\{\phi_m(t)\}$ does not affect the Doppler resolution. Using the definition of MIMO ambiguity function (6.41), we have

$$\chi(\tau, \nu, f_s, f'_s) = \sum_{m=0}^{M-1}\sum_{m'=0}^{M-1} r^{\phi}_{m,m'}(\tau)e^{j2\pi\gamma(f_s m - f'_s m')} \cdot \sum_{l=0}^{L-1} e^{j2\pi\nu lT}$$

for $\tau \in [0, T - T_\phi]$. The MIMO ambiguity function depends on the cross-correlation functions $r^{\phi}_{m,m'}(\tau)$. Also, the pulses $\{\phi_m(t)\}$ affect only the range and spatial resolution; they do not affect the Doppler resolution. Therefore, to obtain a sharp ambiguity function, we should design the pulses $\{\phi_m(t)\}$ such that the function

$$\Omega(\tau, f_s, f_s') \triangleq \sum_{m=0}^{M-1} \sum_{m'=0}^{M-1} r^{\phi}_{m,m'}(\tau) e^{j2\pi(f_s m - f_s' m)} \tag{6.48}$$

is sharp around the line $\{(\tau, f_s, f_s') | \tau = 0, f_s = f_s'\}$. Note that we have dropped the parameter γ. In MIMO radar, $\gamma > 1$ is usually chosen and the function $e^{j2\pi\gamma(f_s m - f_s' m')}$ is periodic. Therefore, we need to consider this function in only one period. For $M = 1$, the signal design problem reduces to the special case of the SIMO radar. In this case, Eq. (6.48) reduces to the autocorrelation function

$$\Omega(\tau, f_s, f_s') = r^{\phi}_{0,0}(\tau)$$

Thus, in the SIMO radar case, the signal design problem is to generate a pulse with a sharp autocorrelation. The linear frequency modulation (LFM) signal is one of the signals that has sharp autocorrelation [49]. Besides its sharp autocorrelation function, the LFM pulse can be conveniently generated, and it has constant modulus. These reasons make the LFM signal a very good candidate in a pulse repetition radar system. For the MIMO radar case that satisfies $M > 1$, we need to consider not only the autocorrelation functions but also the cross-correlation functions between pulses such that $\Omega(\tau, f_s, f_s')$ can be sharp.

6.7.4 Frequency-Multiplexed LFM Signals

To obtain a good spatial and range resolution in pulse repetition MIMO radar, we should design the pulses $\{\phi_m(t)\}$ such that the function $\Omega(\tau, f_s, f_s')$ in (6.48) can be sharp around the line $\{(0, f_s, f_s)\}$. A natural way to do this is to use a set of LFM signals with different initial frequencies so that their frequency spans do not overlap. In this case, the pulses can be expressed as

$$\phi_m(t) = \begin{cases} e^{j\pi k t^2} e^{j\pi k T_\phi m t}, & t \in [0, T_\phi) \\ 0, & \text{otherwise} \end{cases}$$

for $m = 0, 1, \ldots, M - 1$. For simplicity, we omit the normalization scalar because it does not affect the shape of the ambiguity function. Note that if the initial frequencies between different pulses are not far enough, we may not be able to distinguish the cause of the instant frequency. It may be caused either by some amount of delay or by different pulses. Therefore, we should choose the initial frequency such that the frequency spans do not overlap. If the frequency bands of the pulses do not overlap, the cross-correlation functions will be approximately zero. To compute the function $\Omega(\tau, f_s, f_s')$, we first compute the cross-correlation function $r^{\phi}_{m,m'}(\tau)$, which can be expressed as

$$\begin{aligned} r^{\phi}_{m,m'}(\tau) &= \int_0^{T_\phi} \phi_m(t)\phi^*_{m'}(t+\tau)dt \\ &= \int_0^{T_\phi} e^{j\pi k t^2} e^{j\pi k T_\phi m t} \left(e^{j\pi k (t+\tau)^2} e^{j\pi k T_\phi m'(t+\tau)} \right)^* dt \\ &= \int_0^{T_\phi} e^{-j\pi k(2t\tau+\tau^2)} e^{j\pi k T_\phi (m-m')t} dt \cdot e^{-j\pi k T_\phi m'\tau} \\ &\approx \delta_{m,m'} \int_0^{T_\phi} e^{-j2\pi k t\tau} dt \cdot e^{-j\pi k \tau^2} e^{-j\pi k T_\phi m'\tau} \\ &= T_\phi \delta_{m,m'} \cdot \operatorname{sinc}(kT_\phi\tau) e^{-j\pi k T_\phi \tau} e^{-j\pi k \tau^2} e^{-j\pi k T_\phi m'\tau} \end{aligned}$$

where $\operatorname{sinc}(x) \triangleq (\sin(\pi x)/\pi x$. Substituting the result from above into (6.48), we obtain

$$\begin{aligned} |\Omega(\tau, f_s, f_s')| &\propto \left| \operatorname{sinc}(kT_\phi\tau) \sum_{m=0}^{M-1} \sum_{m'=0}^{M-1} \delta_{m,m'} e^{-j\pi k T_\phi m'\tau} e^{j2\pi(f_s m - f_s' m')} \right| \\ &= \left| \operatorname{sinc}(kT_\phi\tau) \sum_{m=0}^{M-1} e^{j2\pi(f_s - f_s' - (kT_\phi/2)\tau)m} \right| \\ &= \left| \operatorname{sinc}(kT_\phi\tau) \frac{\sin(\pi M(f_s - f_s' - (kT_\phi/2)\tau))}{\sin(\pi(f_s - f_s' - (kT_\phi/2)\tau))} \right|. \end{aligned}$$

One can see that the energy of this function is well concentrated on the line $\{(0, f_s, f_s\}$. For any given f_s, the first null of $|\Omega(\tau, f_s, f_s)|$ occurs at $\tau_0 = 1/kT_\phi$. The pulse compression ratio defined as $T\phi/\tau_0$ is given by

$$\frac{T_\phi}{\tau_0} = kT_\phi^2.$$

The time-bandwidth product of the transmitted pulses $\{\phi_m(t)\}$ is

$$\underbrace{T_\phi}_{\text{time}} \cdot \underbrace{MkT_\phi}_{\text{bandwidth}} = MkT_\phi^2.$$

In the SIMO radar case where $M = 1$, the pulse compression ratio equals the time–bandwidth product [49]. However, in the MIMO radar, the time–bandwidth product is M times the pulse compression ratio if the nonoverlapping LFM signals are used.

This is simply because we use M times the bandwidth to transmit M frequency multiplexed LFM pulses.

6.7.5 Frequency-Hopping Signals

Instead of directly designing the pulses, we can impose some structures on the pulses and design the parameters of the pulses. As an example of this idea, we now consider the pulse generated by frequency-hopping codes. This kind of signal has the advantage of constant modulus. The frequency-hopping pulses can be expressed as

$$\phi_m(t) = \sum_{q=0}^{Q-1} e^{j2\pi c_{m,q}\Delta f t} 1_{(0,\Delta t)}(t - q\Delta t)$$

where

$$1_{(0,\Delta t)}(t) \triangleq \begin{cases} 1, & t \in (0, \Delta t) \\ 0, & \text{otherwise} \end{cases}$$

where $c_{m,q} \in \{0, 1, \ldots, K-1\}$ is the frequency-hopping code and Q is the length of the code. The duration of the pulse is $T_\phi = Q\,\Delta t$, and the bandwidth of the pulses is approximately $K\,\Delta f$. To maintain the orthogonality, the code $\{c_{m,q}\}$ satisfies

$$c_{m,q} \neq c_{m'q}, \quad \text{for} \quad m \neq m', \forall q$$

Now, instead of directly designing the pulses $\phi_m(t)$, we must design the code $c_{m,q}$ for $m = 0, 1, \ldots, M-1$ and $q = 0, 1, \ldots, Q-1$. Recall that our goal is to design the transmitted signals to ensure that the function $\Omega(\tau, f_s, f_s')$ in (6.48) is sharp. So, we should find an expression for the function $\Omega(\tau, f_s, f_s')$ in terms of $\{c_{m,q}\}$ and then find a code $\{c_{m,q}\}$ such that the function $\Omega(\tau, f_s, f_s')$ is sharp. To compute the function $\Omega(\tau, f_s, f_s')$, we first compute the cross-correlation function $r^{\phi}_{m,m}(\tau)$, which can be expressed as

$$\begin{aligned} r^{\phi}_{m,m'}(\tau) &= \int_0^{T_\phi} \phi_m(t)\phi^*_m(t+\tau)dt \\ &= \int_0^{T_\phi} \sum_{q=0}^{Q-1} e^{j2\pi c_{m,q}\Delta f t} 1_{(0,\Delta t)}(t - q\Delta t) \\ &\quad \times \left(\sum_{q'=0}^{Q-1} e^{j2\pi c_{m,q'}\Delta f(t+\tau)} 1_{(0,\Delta t)}(t + \tau - q'\Delta t) \right)^* dt \\ &= \sum_{q=0}^{Q-1} \sum_{q'=0}^{Q-1} \chi^{\text{rect}}(\tau - (q' - q)\Delta t, (c_{m,q} - c_{m',q'})\Delta f) \\ &\quad \times e^{j2\pi\Delta f(c_{m,q} - c_{m',q'})q\Delta t} e^{j2\pi\Delta f c_{m',p'}\tau} \end{aligned} \qquad (6.49)$$

where $\chi^{\text{rect}}(\tau, \nu)$ is the SIMO ambiguity function of the rectangular pulse $1_{(0,\Delta t)}(t)$, given by

$$\begin{aligned}\chi^{\text{rect}}(\tau, \nu) &\triangleq \int_0^{\Delta t} 1_{(0,\Delta t)}(t) 1_{(0,\Delta t)}(t+\tau) e^{j2\pi\nu} dt \\ &= \left(1 - \frac{|\tau|}{\Delta t}\right) \text{sinc}\left(\Delta t\nu\left(1 - \frac{|\tau|}{\Delta t}\right)\right) e^{j\pi\nu(\tau+\Delta t)} \text{ for } |\tau| < \Delta t \end{aligned} \tag{6.50}$$

Substituting (6.49) into (6.48), we obtain

$$\begin{aligned}\Omega(\tau, f_s, f_s') &= \sum_{m=0}^{M-1} \sum_{m'=0}^{M-1} r_{m,m'}^{\phi}(\tau) e^{j2\pi(f_s m - f_s' m')} \\ &= \sum_{m=0}^{M-1} \sum_{m'=0}^{M-1} \sum_{q=0}^{Q-1} \sum_{q'=0}^{Q-1} \chi^{\text{rect}}(\tau - (q' - q)\Delta t, (c_{m,q} - c_{m',q'})\Delta f) \\ &\quad \times e^{j2\pi\Delta f(c_{m,q} - c_{m',q'})q\Delta t} e^{j2\pi\Delta f c_{m',p'}\tau} e^{j2\pi(f_s m - f_s m')} \end{aligned} \tag{6.51}$$

The next step is to choose the frequency-hopping code $\{c_{m,q}\}$ such that the function $\Omega(\tau, f_s, f_s')$ is sharp. To achieve this, we can impose penalty on the sidelobe by using the cost function

$$f_p(\mathbf{C}) \triangleq \int_{-\infty}^{\infty} \int_0^1 \int_0^1 |\Omega(\tau, f_s, f_s')|^p df_s\, df_s'\, d\tau \tag{6.52}$$

where $\mathbf{C}$ is an $M \times Q$ matrix consisting of $\{c_{m,q}\}$ and p is a parameter in the cost function. By minimizing this cost function, we can reduce peaks in the function $\Omega(\tau, f_s, f_s')$. The parameter p affects the relative penalty for large peaks and small peaks. If we choose a larger p, more penalty will be imposed for the large peaks. The number of large peaks can be reduced. However, this might create even more small peaks because their cost becomes a smaller portion in the overall cost function. Thus, there exists a tradeoff between large peaks and small peaks when choosing the parameter p. To obtain the code, we need to solve the optimization problem

$$\begin{aligned} &\min_{\mathbf{C}} f_p(\mathbf{C}) \\ &\text{subject to} \quad c_{m,q} \in \{0, 1, \ldots, K-1\} \\ &\qquad\qquad\quad c_{m,q} \neq c_{m',q}, \text{ for } m \neq m' \end{aligned}$$

This problem is a combinatorial optimization problem. Such problems can be solved by search methods such as the simulated annealing algorithm [40] and genetic algorithm [41].

6.8 CONCLUSIONS

In this chapter, we first studied the clutter subspace and its rank in MIMO radars using the geometry of the system. We derived an extension of Brennan's rule for estimating the dimension of the clutter subspace in MIMO radar systems. This rule is given in (6.26). An algorithm for computing the clutter subspace using nonuniform sampled PSWF was described. Then we proposed a spacetime adaptive processing method for MIMO radars. This method utilizes the knowledge of the geometry of the problem, the structure of the clutter space, and the block diagonal structure of the jammer covariance matrix. Utilizing the fact that the jammer matrix is block diagonal and the clutter matrix has low rank with known subspace, we showed how to break the inversion of a large matrix $\mathbf{R}$ into the inversions of smaller matrices using the matrix inversion lemma. Therefore the new method has much lower computational complexity. Moreover, we can directly null out the entire clutter space for large clutter. In our ZF method, only the $N \times N$ jammer-plus-noise matrix $\mathbf{R}_{vs}$ needs to be estimated instead of the the $NML \times NML$ matrix $\mathbf{R}$ in the SMI method, where N is the number of receiving antennas, M is the number of transmitting antennas, and L is the number of pulses in a coherent processing interval. Therefore, for a given number of data samples, the new method has better performance. In Section 6.6, we provided an example where the number of training samples was reduced by a factor of 100 with no appreciable loss in performance compared to the SMI method. In practice, the clutter subspace might change because of effects such as the internal clutter motion (ICM), velocity misalignment, array manifold mismatch, and channel mismatch [42]. In this chapter, we considered an "ideal model," which does not take these effects into account. When this model is not valid, the performance of the algorithm will degrade. One way to overcome this might be to estimate the clutter subspace by using a combination of both the assumed geometry and the received data. Another way might be to develop a more robust algorithm against the clutter subspace mismatch. These ideas will be explored in the future.

In Section 6.7, we studied the problem of designing the transmitted waveform $\phi_m(t)$. We first reviewed the concept of the MIMO ambiguity function proposed in an earlier paper [37]. Then we derived a number of properties of the MIMO ambiguity function. These properties provide some insights into the MIMO ambiguity function. In STAP, the transmitter emits the periodic pulse signals. Therefore we also derived the MIMO ambiguity for periodic pulse signals. It turned out that the periodic pulse signal affects only the spatial resolution and the range resolution. It does not affect the Doppler resolution. The spatial and range resolutions are determined by the cross-correlations between pulses. By using these results, we derived the ambiguity function for the frequency-multiplexed LFM signals and frequency-hopping signals. An optimization problem for designing the frequency-hopping codes was formulated. The solution to this problem will be explored in the future.

ACKNOWLEDGMENTS

The authors gratefully acknowledge the support from the ONR grant N00014-06-1-0011, which made this research possible.

REFERENCES

1. D. J. Rabideau and P. Parker, Ubiquitous MIMO multifunction digital array radar, *Proc. 37th IEEE Asilomar Conf. Signals, Systems, and Computers*, Nov. 2003, Vol. 1, pp. 1057–1064.
2. D. W. Bliss and K. W. Forsythe, Multiple-input multiple-output (MIMO) radar and imaging: Degrees of freedom and resolution, *Proc. 37th IEEE Asilomar Conf. Signals, Systems, and Computers*, Nov. 2003, Vol. 1, pp. 54–59.
3. E. Fishler, A. Haimovich, R. S. Blum, D. Chizhik, L. J. Cimini, and R. A. Valenzuela, MIMO radar: An idea whose time has come, *Proc. IEEE Radar Conf.*, April 2004, pp. 71–78.
4. E. Fishler, A. Haimovich, R. S. Blum, L. J. Cimini, D. Chizhik, and R. A. Valenzuela, Performance of MIMO radar systems: Advantages of angular diversity, *Proc. 38th IEEE Asilomar Conf. Signals, Systems, and Computers*, Nov. 2004, Vol. 1, pp. 305–309.
5. E. Fishler, A. Haimovich, R. S. Blum, L. J. Cimini, D. Chizhik, and R. A. Valenzuela, Spactial diversity in radars — models and detection performance, *IEEE Trans. Signal Process.* **54**(3):823–837 (March 2006).
6. F. C. Robey, S. Coutts, D. Weikle, J. C. McHarg, and K. Cuomo, MIMO radar theory and experimental results, *Proc. 38th IEEE Asilomar Conf. Signals, Systems, and Computers*, Nov. 2004, Vol. 1, pp. 300–304.
7. K. W. Forsythe, D. W. Bliss, and G. S. Fawcett, Multiple-input multiple-output (MIMO) radar performance issues, *Proc. 38th IEEE Asilomar Conf. Signals, Systems, and Computers*, Nov. 2004, pp. 310–315.
8. G. J. Frazer, Y. I. Abramovich, and B. A. Johnason, Spatially waveform diverse radar: Perspectives for high frequency OTHR, *Proc. IEEE Radar Conf.*, April 2007, pp. 385–390.
9. H. A. Khan, W. Q. Malik, D. J. Edwards, and C. J. Stevens, Ultra wideband multiple-input multiple-output radar, *IEEE Int. Radar Conf.*, May 2005, pp. 900–904.
10. H. A. Khan and D. J. Edwards, Doppler problems in orthogonal MIMO radars, *Proc. IEEE Int. Radar Conf.*, April 2006, pp. 24–27.
11. V. F. Mecca, D. Ramakrishnan, and J. L. Krolik, MIMO radar space-time adaptive processing for multipath clutter mitigation, *Proc. IEEE Workshop on Sensor Array and Multichannel Signal Processing*, July 2006, pp. 249–253.
12. C. Y. Chen and P. P. Vaidyanathan, A subspace method for MIMO radar space-time adaptive processing, *IEEE Int. Conf. Acoustics, Speech, and Signal Processing*, April 15–20, 2007, Vol. 2, pp. 925–928.
13. C. Y. Chen and P. P. Vaidyanathan, MIMO radar space-time adaptive processing using prolate spheroidal wave functions, *IEEE Trans. Signal Process.* (in press).

14. D. R. Fuhrmann and G. S. Antonio, Transmit beamforming for MIMO radar systems using partial signal correlation, *Proc. 38th IEEE Asilomar Conf. Signals, Systems, and Computers*, Nov. 2004, pp. 295–299.
15. G. S. Antonio and D. R. Fuhrmann, Beampattern synthesis for wideband MIMO radar systems, *Proc. 1st IEEE Int. Workshop on Computational Advances in Multi-Sensor Adaptive Processing*, Dec. 2005, pp. 105–108.
16. Y. Yang and R. S. Blum, MIMO radar waveform design based on mutual information and minimum mean-square error estimation, *IEEE Trans. Signal Process.* **43**(1):330–343 (Jan. 2007).
17. L. Xu and J. Li, Iterative generalized-likelihood ratio test for MIMO radar, *IEEE Trans. Signal Process.* **55**(6):2375–2385 (June 2007).
18. K. W. Forsythe and D. W. Bliss, Waveform correlation and optimization issues for MIMO radar, *Proc. 39th IEEE Asilomar Conf. on Signals, Systems, and Computers*, Nov. 2005, pp. 1306–1310.
19. Q. Zhang and W. B. Mikhael, Estimation of the clutter rank in the case of subarraying for space-time adaptive processing, *Electron. Lett.* **33**(5):419–420 (Feb. 1997).
20. N. A. Goodman and J. M. Stiles, On clutter rank observed by arbitrary arrays, *IEEE Trans. Signal Process.* **55**(1):178–186 (Jan. 2007).
21. J. Ward, *Space-Time Adaptive Processing for Airborne Radar*, Technical Report 1015, Lincoln Laboratory, Dec. 1994.
22. L. E. Brennan and F. M. Staudaher, *Subclutter Visibility Demonstration*, Technical Report RL-TR-92-21, Adaptive Sensors Inc., 1992.
23. R. Klemm, Adaptive clutter supression for airborne phased array radars, *IEE Proc. F* **130**(1):125–132 (1983).
24. D. Slepian and H. O. Pollak, Prolate spheroidal wave functions, Fourier analysis and uncertainty-III: The dimension of the space of essentially time-and-band-limited signals, *Bell Syst. Tech. J.* 1295–1336 (July 1962).
25. J. Capon, High-resolution frequency-wavenumber spectrum analysis, *Proc. IEEE* **57**(8):1408–1418 (Aug. 1969).
26. C. Y. Chen and P. P. Vaidyanathan, Quadratically constrained beamforming robust against direction-of-arrival mismatch, *IEEE Trans. Signal Process.* (in press).
27. R. G. Lorenz and S. P. Boyd, Robust minimum variance beamforming, *IEEE Trans. Signal Process.* **53**(5):1684–1696 (May 2005).
28. Y. I. Abramovich, Controlled method for adaptive optimization of filters using the criteriion of maximum SNR, *Radio Eng. Electron. Phy.* **26**:87–95 (March 1981).
29. B. D. Carlson, Covariance matrix estimation errors and diagonal loading in adaptive arrays, *IEEE Trans. Aerospace Electron. Syst.* **24**(4):397–401 (July 1988).
30. J. S. Reed, J. D. Mallett, and L. E. Brennan, Rapid convergence rate in adaptive arrays, *IEEE Trans. Aerospace Electron. Syst.* **AES-10**(6):853–863 (Nov. 1974).
31. R. Klemm, Adaptive air- and spaceborne MTI under jamming conditions, *Proc. IEEE Natl. Radar Conf.*, April 1993, pp. 167–172.
32. J. R. Guerci, J. S. Goldstein, and I. S. Reed, Optimal and adaptive reduced-rank STAP, *IEEE Trans. Aerospace Electron. Systems.* Special section on spacetime adaptive processing, **36**(2):647–663 (April 2000).

33. A. M. Haimovich and M. Berin, Eigenanalysis-based space-time adaptive radar: Performance analysis, *IEEE Trans. Aerospace Electron. Syst.* **33**(4):1170–1179 (Oct. 1997).
34. J. S. Goldstein, I. S. Reed, and L. L. Scharf, A multistage representation of the wiener filter based on orthogonal projections, *IEEE Trans. Inform. Theory* **44**(7):2943–2959 (Nov. 1998).
35. B. Friedlander, A subspace method for space time adaptive processing, *IEEE Trans. Signal Process.* **53**(1):74–82 (Jan. 2005).
36. X. Wen, A. Wang, L. Li, and C. Han, Direct data domain approach to space-time adaptive signal processing, *Proc. ICARCV*, Dec. 2004, Vol. 3, pp. 2070–2075.
37. G. San Antonio, D. R. Fuhrmann, and F. C. Robey, MIMO radar ambiguity functions, *Proc. 40th IEEE Asilomar Conf. Signals, Systems, and Computers*, Nov. 2006, pp. 36–40.
38. R. Hoctor and S. Kassam, The unifying role of the coarray in aperture synthesis for coherent and incoherent imaging, *Proc. IEEE* **78**(4):735–752 (April 1990).
39. R. Kozick and S. Kassam, Application of the singular value decomposition to aperture synthesis with boundary arrays, *Proc. Antennas and Propagation Society Int. Symp.*, June 1991, Vol. 1, pp. 526–529.
40. S. Kirkpatrick, C. D. Gelatt, and M. P. Vecchi, Optimization by simulated annealing, *Science* **220**(4598):671–680 (1983).
41. D. Whitley, A genetic algorithm tutorial, *Stat. Comput.* **4**:65–85 (1994).
42. J. R. Guerci, *Space-Time Adaptive Processing*, Artech House, Norwood, MA, 2003.
43. R. Klemm, *Principles of Space-Time Adaptive Processing*, IEE, 2002.
44. M. A. Richards, *Fundamentals of Radar Signal Processing*, McGraw-Hill, New York, 2005.
45. H. L. Van Tree, *Optimum Array Processing: Part IV of Detection Estimation and Modulation Theory*, Wiley-Interscience, New York, 2002.
46. J. P. Keener, *Principles of Applied Mathematics*, Addison-Wesley, Reading, MA, 1988.
47. A. V. Oppenheim and R. W. Schafer, *Discrete-Time Signal Processing*, Prentice-Hall, Englewood Cliffs, NJ, 1998.
48. R. A. Horn and C. R. Johnson, *Matrix Analysis*, Cambridge, University Press, 1985.
49. N. Levanon and E. Mozeson, *Radar Signals*, Wiley-IEEE Press, Hoboken, NJ, 2004.

7

SLOW-TIME MIMO SPACETIME ADAPTIVE PROCESSING

VITO F. MECCA AND JEFFREY L. KROLIK

Department of Electrical and Computer Engineering, Duke University, Durham, North Carolina

FRANK C. ROBEY

MIT Lincoln Laboratory, Lexington, Massachusetts

DINESH RAMAKRISHNAN

Audio Systems, Qualcomm Inc., San Diego, California

7.1 INTRODUCTION

In this chapter, a multiple-input multiple-output (MIMO) generalization of spacetime adaptive processing (STAP) is presented with the goal of mitigating radar clutter subject to multipath propagation between transmit and receive arrays. Multipath clutter occurs when ground backscatter returns to the receive elements via multiple different paths, each with its own Doppler frequency and wavenumber spreading. Of particular interest here is the problem of multipath clutter mitigation for skywave HF over-the-horizon radar (OTHR) [1]. In this application, multiple ionospheric propagation paths can cause ground returns in transmitter sidelobe directions to return via the receive mainlobe with different Doppler shifts that can mask targets of interest. In such cases, conventional STAP cannot mitigate Doppler spread clutter without also suppressing the target. Similar multipath clutter scenarios can occur in other settings, such as ground moving-target indicator (GMTI) radars operating in complex terrain. Although typically MIMO radar techniques have been proposed using waveforms for which individual pulses are orthogonal (i.e., "fast time"), in

MIMO Radar Signal Processing, edited by Jian Li and Petre Stoica

this chapter, we consider a MIMO STAP approach wherein conventional radar waveforms are used and orthogonality is achieved by phase coding from pulse to pulse. Thus the waveforms are orthogonal over a coherent processing interval (CPI) (i.e., in "slow time"). Slow-time MIMO STAP has the important advantage of being easily implemented using legacy radar hardware without the need for arbitrary waveform generators on each transmit element or digital receivers to facilitate channel separation during pulse compression.

7.1.1 MIMO Radar and Spatial Diversity

An important feature of MIMO radar as studied in previous work is the use of orthogonal transmitted waveforms. If the signals remain orthogonal, and thus separable, at the receiver, MIMO receive processing may be performed to achieve spatial diversity. For example, a MIMO radar method has been presented [2] that achieves a high amount of spatial diversity on a target by using multiple widely spaced transmitting and multiple receiving arrays. With a spatially diverse set of "looks," this method hopes to uncover the target in more than one of the transmit–receive paths to increase direction finding and thus minimize the impact of fading along any particular path. This first implementation of MIMO radar is directly related to the communications-oriented notion of spatial diversity. A second method of achieving spatial diversity exploited by techniques in this chapter focuses on coherently combining orthogonal waveforms sent from the transmit elements to provide processing gain (similar to the methods explored in Refs. 3 and 4). In this class of methods, a weighted sum of the orthogonal transmit channels is formed at the receiver that has the effect of realizing a "virtual" transmitted array pattern. This class of MIMO methods thus has the remarkable ability to alter the effective transmit pattern after the received data have been collected. It is important to distinguish MIMO approaches from traditional phased-array radar systems that perform beamforming at the receiver array, although such single-input multiple-output (SIMO) systems can be realized by MIMO systems with an appropriate choice of parameters in the MIMO formulation.

The work of Rabideau and Parker [5] illustrates how radar system specifications are affected when a MIMO implementation is realized using multiple digital transmitters in a "ubiquitous" (or omnidirectional) manner. For example, using a total of L MIMO transmitting elements, a postbeamforming clutter : noise ratio reduction by a factor of $1/L$ is realized in [5] while the necessary dynamic range of the radar hardware is also decreased by ≈ 30–40 dB over that of the analog phased-array radar. In addition to the relaxation in system hardware requirements, Rabideau and Parker [5] indicate that the variance of the angle estimation error from a MIMO radar is half the error variance of a SIMO phased-array radar. The gains in a MIMO radar do come at a cost, namely, the high price associated with arbitrary waveform generators at the transmit array elements. In the case of HF OTH radar where arrays have hundreds of elements, this cost as well as that of digital receivers can prove to be a significant factor in the implementation of fast-time MIMO systems.

MIMO radar systems that use orthogonal waveforms can also be used to increase the number of degrees of freedom for adaptive processing. Investigations [6] have shown that for a bistatic MIMO radar with L transmitting elements and N receiving elements there are a maximum of LN independent degrees of freedom at a given slant range (or time delay). This is an increase of a factor of L over the N degrees of freedom in the SIMO radar. To achieve LN adaptive degrees of freedom in the SIMO-like case involves passing information to the transmitter that can be used to adapt the transmitter array pattern in real time. As is shown in Section 7.1.3, MIMO radars do not need feedback control from the receiver to the transmitter in order to realize this factor of L additional degrees of freedom.

Forsythe et al. [7] and Robey et al. [8] relate the potential resolution improvements of MIMO radar to an increase in the virtual array size. The virtual array pattern achieved by MIMO processing is calculated by performing a convolution of the real transmitter and receiver element positions. For example, representing five element uniform array weights on transmit and receive as [1 1 1 1 1], the equivalent MIMO virtual array weighting is [1 2 3 4 5 4 3 2 1], which corresponds to a larger virtual aperture size. Note that a sparse transmit–receive array may also be used to produce a virtual array with a large effective aperture, and thus improved resolution over that of the SIMO receive-only sparse array aperture.

As in SIMO systems, waveform design is a critical component that determines MIMO radar performance. Noise levels, clutter manifestations, interference signals, jamming signals, target range spreads, and target Doppler frequency spreads all place limitations on the design process. A single radar can employ different waveforms so performance does not degrade in the presence of different environments or targets [9]. The orthogonality found in MIMO systems is an additional constraint applied to waveform design. More recent work in the area of waveform design has provided some interesting results for MIMO radar processing. In Ref. 10, orthogonal waveforms are designed that optimize the output beamformed radar image, while White and Ray [11] examine waveform design for tracking-based MIMO radar applications. Fuhrmann and San Antonio [12] describe transmit beampatterns in terms of the cross-correlation matrix of the transmitted signals. Their iterative method determines the necessary cross-correlation matrix that is close to a specified transmit beampattern, given a particular cost or penalty function. Much effort has been directed toward waveform optimization for parameter estimation in multiple-target scenarios assuming linear time-invariant models for the combined channel–target response [13–15]. These waveform design techniques can be used to approximate a desired beampattern and minimize the cross-correlation between returns received from targets closely spaced in wavenumber. Stoica et al. [14] provided results illustrating the advantage of MIMO beampattern designs over so-called "rank 1" phased-array designs. Semidefinite quadratic programming (SQP) is utilized to solve the signal design problem for several optimal designs, including maximizing power incident on multiple targets at different aspect angles, minimizing beampattern sidelobe level, and matching to a desired prespecified beampattern. Results [14,15] also show that the Cramér–Rao bound for target angle estimation under MIMO transmit operation is much lower than that of traditional phased-array SIMO transmit operation, especially for situations with multiple targets located at different transmit azimuth angles.

7.1.2 MIMO and Target Fading

In another form of spatial diversity, MIMO techniques inspired by communications applications have been explored to mitigate multipath-related fading of target returns. The MIMO channel model in communications is somewhat similar to that of a MIMO radar channel, which facilitates the use of MIMO communication ideas in radar system development. For diversity, Fishler et al. [2] claim that transmitters will have different views of a target if the transmitting elements are spaced at a distance d_t such that

$$d_t \geq \frac{\lambda R}{D} \tag{7.1}$$

where λ is the radar's operating wavelength, R is the slant range to the target from the transmitter, and D is the target angular spread in radians. As indicated in (7.1), transmitter elements must have a larger spacing as target size decreases to record different aspects of the target.

Fishler et al. [2] provided the derivation of the channel model for the bistatic MIMO radar system. The transmitter and receiver arrays have a total of L and N elements, respectively. The $L \times N$ MIMO channel matrix can be decomposed into three components

$$\mathbf{H} = \mathbf{G}^T \mathbf{\Sigma} \mathbf{K} \tag{7.2}$$

where the columns of $\mathbf{G}$ represent the L directional signals from the transmitting elements that impinge on the target scattering centers, $\mathbf{\Sigma}$ is a diagonal matrix with elements corresponding to the radar cross sections of the scattering centers, and the columns of $\mathbf{K}$ represent the N signals corresponding to the directions from the scattering centers to the receiver elements. According to the criterion in (7.1), the columns of $\mathbf{G}$ are orthogonal. A diversity gain is realized in this case because each column can be processed as if it were an independent source. Via numerical simulations it is shown that as the value of L increases, the resulting signal loss due to target fading decreases rapidly. Other work [16,17] extends the performance characterization of the fading-resistant MIMO radar into the realm of detection theory. A likelihood ratio test is used to compare the performance of the proposed MIMO radar to that of SIMO phased-array radar. Simulations indicated that the MIMO radar outperforms the SIMO phased array radar in the Neyman-Pearson sense for probabilities of detection greater than 0.8 [16].

7.1.3 MIMO and Processing Gain

Beyond improved spatial diversity, MIMO radar can provide significant coherent processing gain and additional degrees of freedom if the diverse, orthogonal channels can be coherently recombined at the receiver. SIMO radars send a single waveform across a phased transmitter array such that the received energy in the hypothesized target direction is high. Direction-finding capabilities are realized at the receive array. The phase path difference of the backscattered waveform across the receive elements is a function of the target's azimuth for a far-field target and one-dimensional (1D) linear array. However, the orthogonal transmitted channels used in MIMO radar offer another dimension over which array processing can be performed if the total phase paths from the transmitter–target–receiver propagation can be used to model the data measured at the receivers [3]. Moreover, when the transmitted waveforms are orthogonal, receive processing can be done on each transmitter "channel" independently, providing an increase in the number of degrees of freedom.

The propagation paths from transmitter to receiver can be identified and separated in a MIMO radar, as illustrated by the number-coded lines in Fig. 7.1. The work of Robey et al. [8,19] consider the case of a ranging radar that sends out only one pulse to detect a nonfluctuating target. In their case, the relative time delays in the transmit–target–receive propagation path can be expressed as

$$\tau_{\ell n} = \tau_{\ell s} + \tau_{sn} \tag{7.3}$$

where $\tau_{\ell s}$ is the time delay from the ℓth transmitter to the scatterer and τ_{sn} is the time delay from the scatterer to the nth receiver element. All of the LN delays can be placed into a steering vector

$$\mathbf{a} = \left[e^{-j(2\pi/\lambda)\tau_{11}}, \ldots, e^{-j(2\pi/\lambda)\tau_{1N}}, e^{-j(2\pi/\lambda)\tau_{21}}, \ldots, e^{-j(2\pi/\lambda)\tau_{LN}}\right] \tag{7.4}$$

Furthermore, the steering vector in (7.4) can be separated into transmit and receive components using a Krönecker product ($\otimes$)

$$\mathbf{a} = \mathbf{a}_{tx} \otimes \mathbf{a}_{rx} \tag{7.5}$$

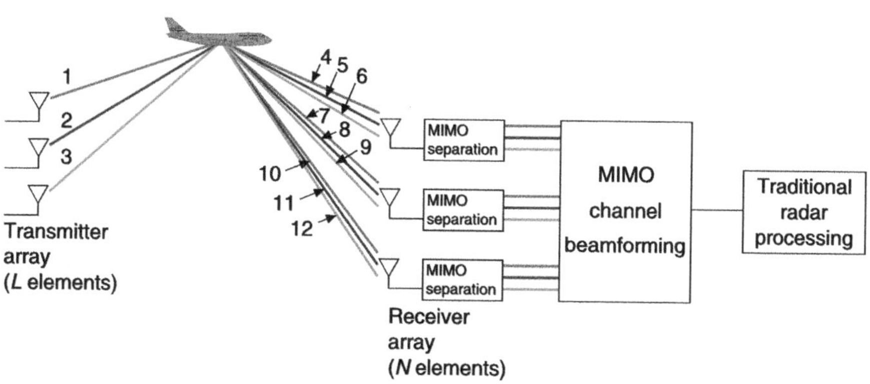

Figure 7.1 Illustration of separable MIMO radar propagation paths (adapted from Ref. 18).

where $\mathbf{a}_{tx}$ and $\mathbf{a}_{rx}$ are defined as

$$\mathbf{a}_{tx} = \left[e^{-j(2\pi/\lambda)\tau_{1s}}, \ldots, e^{-j(2\pi/\lambda)\tau_{Ls}}\right] \tag{7.6}$$

$$\mathbf{a}_{rx} = \left[e^{-j(2\pi/\lambda)\tau_{s1}}, \ldots, e^{-j(2\pi/\lambda)\tau_{sN}}\right] \tag{7.7}$$

When the steering vector in (7.4) is viewed as a steering vector for a SIMO radar, the receiver array has sensor locations that correspond to the spatial convolution of the transmit elements' physical positions with the receive elements' physical positions. This effect results in an increased resolution performance when there are two or more orthogonally transmitted waveforms because the resulting convolution with at least two transmit elements will result in a virtual array that is larger than the physical receive array.

In addition to this improved resolution, significant gains in target scanning times with MIMO radars are also possible. In particular, if a SIMO phased transmitter array with L elements scans the entire space in a total of T seconds, then L beams are formed each with time duration T/L. MIMO systems illuminate the entire surveillance space for the time T, and thus have a factor of L more "looks" at the target. The coherent recombination of received returns from each transmit channel can then be taken into account to provide MIMO systems with faster revisit rate performance with the same probability of detection [17].

For pulse–Doppler radars, Seed et al. [3] extended the idea of MIMO radar to achieve spacetime adaptive processing with an additional adaptive transmit element dimension. Typically, waveforms considered thus far in MIMO radar development [8] are designed to be orthogonal in the "fast time" domain (within each pulse). This requires additional waveform matched filtering at the receiver elements that separates the received data from each different transmit channel. Thus, a MIMO STAP processor will operate on a four-dimensional (4D) data hypercube (see Fig. 7.2) that encompasses the transmit elements L, slow-time Doppler pulses M, receive elements N, and fast-time range samples K. Processing of the data in the 4D hypercube is completely analogous to processing of the 3D hypercube in SIMO STAP [20]. As a consequence, adaptive and partially adaptive implementations of MIMO STAP can be explored as well.

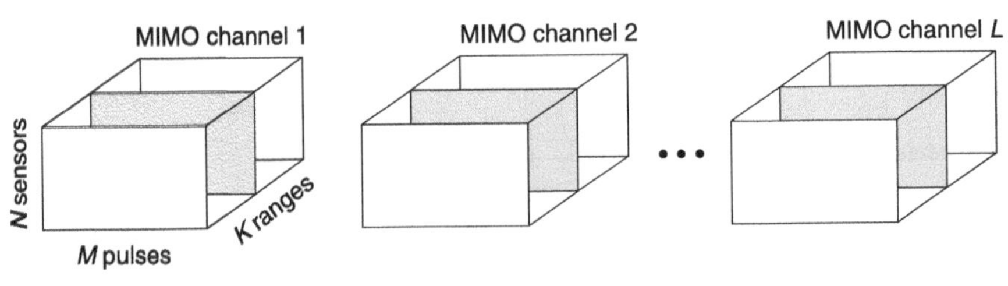

Figure 7.2 Graphical representation of MIMO 4D data hypercube (from Ref. 3).

The MIMO radar implementations reviewed in this section have been implemented in the fast-time, or the within-pulse (intrapulse) domain. As noted above, in order to achieve such strict control over the transmitted waveforms, arbitrary waveform generators and digital receivers are typically required that may preclude implementation in cost-sensitive applications. Orthogonality in fast time applies the condition of orthogonality to only one particular domain of a pulsed radar system. In the remainder of this chapter, a MIMO radar implementation that uses orthogonality in the slow-time, or the pulse-to-pulse, domain is explored. A slow-time MIMO radar is easily implemented with the addition of phase shifters only at the transmitter and with no change to the receive range pulse compression. Slow-time MIMO is thus suited to applications where a legacy SIMO radar is being upgraded to a MIMO design.

7.2 SIMO RADAR MODELING AND PROCESSING

7.2.1 Generalized Transmitted Radar Waveform

Consider a radar system with an L-element 1D transmitter array and an N-element 1D receiver array. Without loss of generality, assume that each array is composed of omnidirectional elements with uniform inter-element spacing d along the x axis. The ℓth transmitter element is excited with a finite-duration pulse waveform $u_p(t;\,\ell)$ repeated in a waveform train. A total of M pulses are transmitted in the slow time (m), with a pulse repetition frequency (PRF) of $f_r = 1/T_r$. The complete time-domain transmitted waveform from the ℓth element is of this form

$$s(t;\,\ell) = w_{tx_\ell} \sum_{m=0}^{M-1} u_p(t - mT_r;\,\ell)\, e^{j2\pi(f(\ell,m)t+\varphi(\ell,m))} \tag{7.8}$$

where $u_p(t)$ is the radar waveform and $f(\ell,\,m)$ and $\varphi(\ell,\,m)$ are the carrier frequency and phase of the ℓth element on the mth pulse, respectively, for $\ell = 0, 1, \ldots, L-1$ and $m = 0,\,1, \ldots, M-1$. The radar waveform is defined such that $u_p(t)$ is zero for $t \leq 0$ and $T_r \leq t$. The transmit array is shaded with an $L \times 1$ complex tapering window $\mathbf{w}_{tx} = [w_{tx_1}, w_{tx_2}, \ldots, w_{tx_L}]^T$ that steers the array to some look (viewing) direction ϕ relative to the array's broadside. It is convenient to think of the tapering window as a vector for the analysis that follows.

After pulse compression at the receiver, the received data from the range bin under consideration are collected into an $N \times M$ data matrix $\mathbf{X}$ consisting of possible target (t), clutter (c), jamming interference (jam), and white noise (v) components:

$$\mathbf{X} = \mathbf{X}_t + \mathbf{X}_c + \mathbf{X}_{jam} + \mathbf{X}_v \tag{7.9}$$

In the following subsections, the terms in this sum are discussed individually.

7.2.2 SIMO Target Model

The majority of the radar array processing literature deals with single-input multiple-output (SIMO) radar. In SIMO radar, an identical waveform is sent from all transmitter elements with an identical starting phase at a common carrier frequency. This is a single input to the system because the L elements are excited identically. The N-element receive array then makes multiple output measurements — hence, single-input multiple-output. A SIMO radar system can be represented in the framework of (7.8) with $u_p(t; \ell) = u_p(t), f(\ell, m) = f_0$, and $\varphi(\ell, m) = 0$. Thus the transmitted radar waveform on the ℓth element is given by

$$s(t; \ell) = w_{tx_\ell} \sum_{m=0}^{M-1} u_p(t - mT_r)\, e^{j2\pi f_0 t} \tag{7.10}$$

Note that the only distinguishing feature between the waveforms across transmit elements is the choice of the complex transmit weight vector $\mathbf{w}_{tx} = [w_{tx_1}, w_{tx_2}, \ldots, w_{tx_L}]^T$.

Consider a point scatterer located at an angle of ϕ_{tx} relative to the transmit array's broadside direction and ϕ_{rx} to the receiver's broadside direction. In addition, the radial velocity of the scatterer relative to both the transmit and receive arrays gives rise to a Doppler frequency shift of f_t. The (nm)th element of $\mathbf{X}_t$ in (7.9) is thus defined as

$$\{\mathbf{X}_t\}_{nm} = \gamma_t\, e^{j(2\pi/\lambda_0)nd \sin\phi_{rx}} e^{j2\pi m f_t T_r} \sum_{\ell=0}^{L-1} w_{tx_\ell}\, e^{-j(2\pi/\lambda_0)\ell d \sin\phi_{tx}} \tag{7.11}$$

where γ_t represents a random complex number that accounts for the target scattering characteristics. The contributions from the transmit array are explicitly retained in (7.11) as the terms to the right of the summation. Physically, the summation develops as a result of the coherent combination of the signals from the transmit array at the target. The energy incident on the scatterer depends on the transmit tapering window $\mathbf{w}_{tx}$ and the scatterer's direction relative to the transmit array ϕ_{tx}. Space-time adaptive processing (STAP) methods [20] are realized by stacking $\mathbf{X}$ column-wise into a vector vec($\mathbf{X}$) such that for the target component (7.11) becomes

$$\text{vec}(\mathbf{X}_t) = \gamma_t \cdot \mathbf{a}_{rx}(\phi_{rx}) \otimes \mathbf{b}_{dop}(f_t T_r) \cdot \mathbf{w}_{tx}^T \mathbf{a}_{tx}(\phi_{tx}) \tag{7.12}$$

where $\mathbf{a}$ and $\mathbf{b}$ are spatial and temporal steering vectors of the scatterer, respectively, defined below in (7.13) through (7.15), and $\otimes$ is the Krönecker product. Subscripts rx and tx on the spatial steering vectors differentiate receive and transmit directions. The subscript dop signifies a slow-time temporal vector. Spatial steering vectors take the form of planar wavefronts for far-field scatterers, and temporal steering vectors take the form of complex sinusoids at the Doppler frequency for objects moving at constant radial velocity:

$$\mathbf{a}_{rx}(\phi) = \left[1,\ e^{j(2\pi/\lambda_0)d\sin\phi}, \ldots, e^{j(2\pi/\lambda_0)(N-1)d\sin\phi}\right]^T \tag{7.13}$$

$$\mathbf{b}_{dop}(f_t T_r) = \left[1,\ e^{j2\pi f_t T_r}, \ldots, e^{j2\pi(M-1)f_t T_r}\right]^T \tag{7.14}$$

$$\mathbf{a}_{tx}(\phi) = \left[1,\ e^{-j(2\pi/\lambda_0)d\sin\phi}, \ldots, e^{-j(2\pi/\lambda_0)(L-1)d\sin\phi}\right]^T \tag{7.15}$$

For $\mathbf{X}_t$ from (7.12), $\text{vec}(\mathbf{X}_t) = \gamma_t \mathbf{v}_t$, where

$$\mathbf{v}_t = \mathbf{a}_{rx}(\phi_{rx}) \otimes \mathbf{b}_{dop}(f_t T_r) \cdot \mathbf{w}_{tx}^T \mathbf{a}_{tx}(\phi_{tx}) \tag{7.16}$$

Note that the steering vector for a scatterer inherently contains illumination components from $\mathbf{w}_{tx}$ in the SIMO case, a term that is usually suppressed in the radar array processing literature.

7.2.3 SIMO Covariance Models

In this section, the second-order statistics of the target and nontarget components are considered. For received data that consist of target, clutter, jamming interference and noise [i.e., $\text{vec}(\mathbf{X}) = \text{vec}(\mathbf{X}_t) + \text{vec}(\mathbf{X}_c) + \text{vec}(\mathbf{X}_{jam}) + \text{vec}(\mathbf{X}_v)$], each term in this sum can be modeled by an associated spacetime covariance matrix defined as

$$\mathbf{R}_a = \text{cov}(\mathbf{X}_a) \tag{7.17}$$

where $\text{cov}(\mathbf{X}_a) = E\{\text{vec}(\mathbf{X}_a)\ \text{vec}(\mathbf{X}_a)^H\}$ and $E\{\cdot\}$ denotes expected value. Thus, the spacetime covariance matrix of the data $\text{vec}(\mathbf{X})$ is

$$\mathbf{R}_x = \mathbf{R}_t + \mathbf{R}_c + \mathbf{R}_{jam} + \mathbf{R}_v \tag{7.18}$$

$$= \mathbf{R}_t + \mathbf{R}_u \tag{7.19}$$

Typically, the strongest target returns are received via the direct path. Direct paths are modeled as those propagation paths where the returned energy is in the same direction of the transmit energy ($\phi_{tx} = \phi_{rx}$). Any direct-path return has a data form identical to that of the target described in (7.16). Thus, a single-target response with received power σ_t^2 has a covariance of

$$\mathbf{R}_t = \sigma_t^2 \cdot \mathbf{a}_{rx}(\phi_t)\mathbf{a}_{rx}^H(\phi_t) \otimes \mathbf{b}_{dop}(f_t T_r)\mathbf{b}_{dop}^H(f_t T_r) \tag{7.20}$$

For multiple targets, (7.18) is simply extended to contain a summation of target terms in the form of (7.20).

Clutter, on the other hand, is often due to distributed scattering, which can arrive via direct or multipath propagation. Namely, the clutter covariance matrix $\mathbf{R}_c$ is the sum of two components: direct path $\mathbf{R}_{cd}$ and multipath $\mathbf{R}_{cm}$. Direct path clutter returns can be modeled as the sum response of point scatterers from N_d multiple directions. Let $\sigma^2_{cd_i}$ define the power received from the ith clutter patch at direction ϕ_i and at Doppler frequency f_i. The direct-path clutter covariance is thus

$$\mathbf{R}_{cd} = \sum_{i=1}^{N_d} \sigma^2_{cd_i} \cdot \mathbf{a}_{rx}(\phi_i)\mathbf{a}^H_{rx}(\phi_i) \otimes \mathbf{b}_{dop}(f_i T_r)\mathbf{b}^H_{dop}(f_i T_r) \tag{7.21}$$

Multipath clutter is attributed to returns that appear at multiple receiver angles, but share a common transmit energy direction. In the SIMO case, transmit direction information from a target response is lost, as evidenced by the inner product in (7.16). Consequently, SIMO multipath clutter returns are treated as having a covariance structure $\mathbf{R}_{cm}$ like the direct-path clutter returns $\mathbf{R}_{cd}$ in (7.21).

In cases where jammers are present, the covariance matrix of the jamming components $\mathbf{R}_{jam}$ represents directional components that appear across all Doppler frequencies, leading to a slow-time covariance matrix as an identity matrix of size $M \times M$. For J sources of jamming occurring at azimuth angles of ϕ_j, $j = 1, \ldots, J$, the covariance matrix is thus

$$\mathbf{R}_{jam} = \sum_{j=1}^{J} \sigma^2_j \cdot \mathbf{a}_{rx}(\phi_j)\mathbf{a}^H_{rx}(\phi_j) \otimes \mathbf{I}_M \tag{7.22}$$

where $\mathbf{I}_M$ is an $M \times M$ identity matrix. The white noise with variance of σ^2_v has a covariance matrix of $\mathbf{R}_v = \sigma^2_v \mathbf{I}_{MN}$.

7.2.4 SIMO Radar Processing

In SIMO STAP, operations are of the form

$$\mathbf{y} = \mathbf{w}^H_{rx,dop}\text{vec}(\mathbf{X}) \tag{7.23}$$

where $\mathbf{y}$ is the processor output and $\mathbf{w}_{rx,dop}$ is a spacetime weight vector. Given the nature of the steering vector in (7.16), it should be clear that SIMO radar can only filter the spatial receive and temporal components; that is, only two dimensions are accessible in the received data. Thus, STAP allows for beampattern control as a function of receive element, pulse number, or some combination of the two. It is important to note that there is no additional transmit beamforming possible in the transmit direction of a SIMO radar, which explains why the transmit array pattern is suppressed in STAP treatments.

The well-known optimal solution to general adaptive beamforming operations that maximizes detection performance takes the form [21]

$$\mathbf{w}^* = \mathbf{R}_u^{-1}\mathbf{v}_t \tag{7.24}$$

where $\mathbf{v}_t$ is a target's hypothesized steering vector in the form of (7.16) and $\mathbf{R}_u$ is the covariance matrix of the nontarget components. In adaptive methods, an estimate of

$\mathbf{R}_u$ obtained from signal-free range bins is used in the calculation of adaptive weights. Although the range dependence is suppressed here to keep the notation simple, implicit in the STAP weight vector is the ability to vary the weights as a function of slant range. For computational efficiency, the STAP weights can be designed in the beamspace and/or Doppler frequency domains instead of receive element and/or pre-Doppler space. Regardless of the implementation, SIMO STAP methods allow for control only of the receive element and Doppler weight design and the fully adaptive SIMO STAP weight vector has NM total degrees of freedom.

7.3 SLOW-TIME MIMO RADAR MODELING

7.3.1 Slow-Time MIMO Target Model

Previous work in MIMO radar systems focused on sending different fast-time waveforms $u_p(t;\ \ell)$ from each of the L transmitters [2,5,8,10,12]. The objective in these systems is to achieve orthogonality in fast time between the transmitted waveforms so that the response from each waveform can be separated within the received data. The remaining parameters in (7.8) [i.e., $f(\ell, m) = f_0$ and $\varphi(\ell, m) = 0$] are typically chosen identical to the SIMO case. A discussion of fast-time MIMO systems may be found in other chapters of this book. The slow-time version of MIMO radar presented here (as well as in Ref. 4) is a special case of (7.8) where $f(\ell, m) = f_0$, a constant carrier frequency, and $\varphi(\ell, m)$ is changed to achieve orthogonality in the Doppler domain. To facilitate C-channel multiple-input operation, consider dividing the full Doppler PRF into C orthogonal sub-PRF channels of width $f_c \leq f_r/C$. Without loss of generality, assume that $f_c = f_r/C$ so that each sub-PRF channel can accommodate M/C Doppler bins. Choosing the ratio of M/C to be an integer ensures that the channels will be distinct and separable in the discrete Doppler domain. Also, f_r must chosen such that the highest Doppler frequency of interest is less than f_c to avoid interchannel aliasing and ambiguous Doppler returns. A full-array slow-time MIMO radar emits a different Doppler-shifted pulse sequence from each transmit element. The number of channels C is equal to the number of transmit elements L.

In particular, consider sending an identical $u_p(t)$ waveform from each of the L transmitter elements, but suppose that the starting phase of each $u_p(t)$ is varied such that the waveform train transmitted from the ℓth transmit element is a function of the slow-time m. This is accomplished by choosing $\varphi(\ell, m) = \alpha_\ell m T_r$, so that the signal transmitted from the ℓth antenna is now

$$s(t;\ \ell) = w_{tx_\ell} \sum_{m=0}^{M-1} u_p(t - mT_r)\, e^{j2\pi(f_0 t + \alpha_\ell m T_r)} \tag{7.25}$$

The instantaneous Doppler frequency of the ℓth channel in (7.25) is given as $T_r \cdot \partial\varphi(\ell, m)/\partial m = \alpha_\ell$. To break up the Doppler domain into L equal-width channels, a simple choice is to use a linear form

$$\alpha_\ell = -\frac{f_c}{2}(L - 1 - 2\ell) \tag{7.26}$$

Operation in this manner establishes $C = L$ orthogonal channels in the Doppler domain, each centered at a unique multiple of f_c. This phase-only adjustment to the transmitted waveform of (7.25) has an additional effect of steering the transmit beampattern, which is explored in Section 7.4.1.

To derive the slow-time MIMO steering vector, consider a far-field direct-path ($\phi_{tx} = \phi_{rx}$) scatterer located at angle ϕ_t relative to the arrays' broadside directions and with Doppler shift f_t. The backscattered target signal component $\mathbf{X}_t$ after carrier demodulation, pulse compression, and range gating at the nth receiver due to the L transmit waveforms at slow-time m is then

$$\{\mathbf{X}_t\}_{nm} = \gamma_t\, e^{j(2\pi/\lambda_0)nd\sin\phi_t} e^{j2\pi m f_t T_r} \sum_{\ell=0}^{L-1} w_{tx_\ell}\, e^{-j(2\pi/\lambda_0)\ell d\sin\phi_t} e^{j2\pi m\alpha_\ell T_r} \tag{7.27}$$

where γ_t represents the complex random amplitude of the target return. When vectorizing $\mathbf{X}_t$, the data vector at each receiver for a single scatterer can be represented as

$$\text{vec}(\mathbf{X}_t) = \gamma_t \mathbf{a}_{rx}(\phi_{rx}) \otimes \sum_{\ell=0}^{L-1} \mathbf{b}_{dop}((\alpha_\ell + f_t)T_r) \cdot w_{tx_\ell} a_{tx_\ell}(\phi_{tx}) \tag{7.28}$$

Note that in contrast to the SIMO case in (7.16), the transmit weights are now "tagged" by a temporal vector $\mathbf{b}_{dop}((\alpha_\ell + f_t)T_r)$ that is orthogonal in the Doppler domain. Thus, the frequency shifts on the different transmitted waveforms in (7.27) allow separation of transmit channels at the receiver simply by Doppler filtering. In particular, the target return is associated with L Doppler shifts $f_t + \alpha_\ell$ corresponding to the L transmit channels $\ell = 0, 1, \ldots, L-1$. This is the key to slow-time MIMO operation. To separate the kth transmitter's response ($k = 0, 1, \ldots, L-1$), a slow-time Doppler demodulation and lowpass filtering can be performed using the α values from (7.26):

$$\begin{aligned} X_{nm,k} &= \{\mathbf{X}_t\}_{nm}\, e^{-j2\pi m\alpha_k T_r} \\ &= \gamma_t\, w_{tx_k}\, e^{j(2\pi/\lambda_0)nd\sin\phi_t} e^{j2\pi m f_t T_r} e^{-j(2\pi/\lambda_0)kd\sin\phi_t} \\ &\quad + \gamma_t\, e^{j(2\pi/\lambda_0)nd\sin\phi_t} e^{j2\pi m f_t T_r} \sum_{\ell \neq k} w_{tx_\ell}\, e^{-j(2\pi/\lambda_0)\ell d\sin\phi_t} e^{j2\pi m(\alpha_\ell - \alpha_k)T_r} \end{aligned} \tag{7.29}$$

These slow-time demodulations shift each L MIMO channel to baseband in the Doppler domain. Applying a lowpass filter with cutoff frequency $L/M \cdot \text{PRF}$ will filter out the other $L-1$ MIMO channels [$\alpha_\ell - \alpha_k$ terms in (7.29)]. Thus, the response from a particular transmit element can be isolated via a Doppler domain lowpass filtering operation with cutoff frequency at $f_c/2$.

Separation of the MIMO channels breaks the M columns into a Doppler–transmit channel matrix of $M/L \times L$ at each receive element. Thus, the full-array MIMO STAP processor operates on a $N \times M/L \times L$ data cube consisting of receive sensor elements, sub-PRF slow-time pulses, and phase-coded transmit channels. In this domain, the target spacetime wavefront is an $MN \times 1$ vector that takes the form

$$\bar{\mathbf{v}}_t = \mathbf{a}_{rx}(\phi_{rx}) \otimes \mathbf{b}_{dop}(f_tT_r) \otimes (\mathbf{w}_{tx} \odot \mathbf{a}_{tx}(\phi_{tx})) \tag{7.30}$$

where $\odot$ represents a Hadamard (element-by-element) vector product; $\mathbf{a}_{rx}$ and $\mathbf{a}_{tx}$ are defined as in (7.13) and (7.15). Because the lowpass filtering in the Doppler domain reduces the number of terms, $\mathbf{b}_{dop}(f_t\, T_r)$ is now

$$\mathbf{b}_{dop}(f_tT_r) = \left[1,\ e^{j2\pi f_tT_r}, \ldots, e^{j2\pi(M/L-1)f_tT_r}\right]^T \tag{7.31}$$

Equation (7.30) is the elemental wavefront in slow-time MIMO processing. Note that the slow-time MIMO elemental wavefront naturally falls into the STAP framework of Ref. 20 with the addition of a dimension corresponding to effective transmit pattern adaptivity. The transmitter–receiver MIMO steering vector appears in work by Robey et al. [8,19], which uses orthogonality of the within-pulse, or fast-time, waveform to achieve MIMO capability. Moreover, this adaptivity can be achieved after the data are collected at the receiver, which permits the effective transmit pattern to be modified at each range bin. This feature is critically important in many clutter mitigation scenarios.

7.3.2 Slow-Time MIMO Covariance Model

Much like the SIMO case, the received data covariance matrix contains contributions from targets (t), direct-path clutter (cd), multipath clutter (cm), jamming (jam), and white noise (v):

$$\overline{\mathbf{R}}_x = \overline{\mathbf{R}}_t + \overline{\mathbf{R}}_{cd} + \overline{\mathbf{R}}_{cm} + \overline{\mathbf{R}}_j + \overline{\mathbf{R}}_v \tag{7.32}$$

$$= \overline{\mathbf{R}}_t + \overline{\mathbf{R}}_u \tag{7.33}$$

Direct paths are modeled as those propagation paths where the returned energy is in the same direction of the transmit energy ($\phi_{tx} = \phi_{rx}$). Any direct-path return has a data form identical to that of the target described in (7.30). Thus, a single-target response has a covariance of

$$\overline{\mathbf{R}}_t = \sigma_t^2 \cdot \mathbf{a}_{rx}(\phi_t)\mathbf{a}_{rx}^H(\phi_t) \otimes \mathbf{b}_{dop}(f_t T_r)\mathbf{b}_{dop}^H(f_t T_r) \otimes [\mathbf{w}_{tx} \odot \mathbf{a}_{tx}(\phi_t)]\,[\mathbf{w}_{tx} \odot \mathbf{a}_{tx}(\phi_t)]^H \tag{7.34}$$

Multiple targets are modeled in (7.32) by including a sum of $\overline{\mathbf{R}}_t$ terms described in (7.34).

The direct-path clutter covariance component is now

$$\overline{\mathbf{R}}_{cd} = \sum_{i=1}^{N_d} \sigma_{cd_i}^2 \cdot \mathbf{a}_{rx}(\phi_i)\mathbf{a}_{rx}^H(\phi_i) \otimes \mathbf{b}_{dop}(f_i T_r)\mathbf{b}_{dop}^H(f_i T_r) \ldots \otimes [\mathbf{w}_{tx} \odot \mathbf{a}_{tx}(\phi_i)]\,[\mathbf{w}_{tx} \odot \mathbf{a}_{tx}(\phi_i)]^H \tag{7.35}$$

The multipath clutter is attributed to returns that appear at multiple receiver angles, but share a common transmit energy direction. Considering multipath returns at a total of N_m angles denoted by ϕ_{rx_i} (where $i = 1, \ldots, N_m$) for a single transmit direction ϕ_{tx}, the multipath clutter covariance is

$$\overline{\mathbf{R}}_{cm} = \sum_{i=1}^{N_m} \sigma_{cm_i}^2 \mathbf{a}_{rx}(\phi_{rx_i})\mathbf{a}_{rx}^H(\phi_{rx_i}) \otimes \overline{\mathbf{R}}_{d_i} \otimes [\mathbf{w}_{tx} \odot \mathbf{a}_{tx}(\phi_{tx})][\mathbf{w}_{tx} \odot \mathbf{a}_{tx}(\phi_{tx})]^H \tag{7.36}$$

$\overline{\mathbf{R}}_{d_i}$ is a general Doppler spread covariance that takes the form

$$\overline{\mathbf{R}}_{d_i} = \sum_{k=1}^{K} \beta_{ki}\mathbf{b}_{dop}(f_k T_r)\mathbf{b}_{dop}^H(f_k T_r) \tag{7.37}$$

and β_{ki} is a power-scaling factor for the kth Doppler on the ith received multipath clutter direction. The model of (7.36) is for multipath on that arises from a single transmit direction. For multiple sources of multipath clutter on different transmit directions, a $\overline{\mathbf{R}}_{cm}$ term should be included in (7.32) for each transmit direction. $\overline{\mathbf{R}}_{jam}$ is the jamming covariance. Since the jamming is independent of the transmitter array, the $\overline{\mathbf{R}}_{tx}$ component is the $L \times L$ identity matrix. For a single source of jamming occurring at an azimuth angle of ϕ_j, the covariance matrix is

$$\overline{\mathbf{R}}_{jam} = \sigma_j^2 \cdot \mathbf{a}_{rx}(\phi_j)\mathbf{a}_{rx}^H(\phi_j) \otimes \mathbf{I}_M \tag{7.38}$$

Finally, the white noise with variance of σ_n^2 has a covariance matrix of $\overline{\mathbf{R}}_v = \sigma_n^2\,\mathbf{I}_{MN}$. Note that in a slow-time MIMO implementation the components in (7.34)–(7.36) have covariance structures that include transmit angles. Slow-time MIMO STAP methods exploit this new dimension, as discussed in the following section.

7.4 SLOW-TIME MIMO RADAR PROCESSING

In terms of the STAP formulation, a MIMO STAP weight vector of the form of (7.24) can be a function of receive element, pulse number, and transmit element. Specifically, $\overline{\mathbf{R}}_u$ obtained from (7.33) and $\overline{\mathbf{v}}_t$ obtained from (7.30) can be substituted into (7.24) to form the optimal slow-time MIMO weight vector:

$$\overline{\mathbf{w}} = \overline{\mathbf{R}}_u^{-1} \overline{\mathbf{v}}_t \tag{7.39}$$

This slow-time weight vector will contain MN complex elements and is in general nonseparable. Essentially, this allows for the same advantages realized in SIMO radar STAP with the additional feature of transmit array pattern shaping. Since $\overline{\mathbf{w}}$ is implicitly a function of range, effective shaping of the transmit beampattern can be achieved at each range.

Observe that for a slow-time MIMO STAP the total adaptive degrees of freedom $[N \cdot (M/L) \cdot L]$ remains the same as that of a SIMO radar STAP processor $(N \cdot M)$. However, MIMO processing offers an additional L degrees of freedom in adapting the transmit beampattern while providing only M/L degrees of freedom to control the Doppler spectrum of the radar return. In OTH multipath scenarios, the tradeoff between reduced maximum unambiguous Doppler and transmit degrees of freedom is discussed in Section 7.4.3.

Although (7.39) is the optimal weight vector, in many cases this fully adaptive solution is computationally impractical. This motivates the use of partially adaptive approximations using a reduced number of degrees of freedom and computational complexity. For example, a MIMO weight vector that approximates SIMO STAP is achieved by constraining the weight vector to $\overline{\mathbf{w}} = \mathbf{w}_{rx,dop} \otimes \mathbf{w}_{tx}$, where $\overline{\mathbf{w}}_{rx,dop}$ is equal to $\mathbf{w}^*$ from (7.24) and $\mathbf{w}_{tx}$ is a nonadaptive transmit weight vector. This SIMO-like weight vector, however, is not capable of mitigating multipath clutter.

To motivate the design of partially adaptive MIMO processing that can mitigate multipath clutter by effectively forming transmit beampattern nulls, note that the transmitter beampattern can be modified by coherently combining the orthogonal slow-time channel data. Specifically, at any particular Doppler frequency and receive azimuth, the target response is the weighted sum of the L $X_{nm,\ell}$ channels [the baseband terms in (7.29)]. Using $\mathbf{w}_c = [w_{c_1}, \ldots w_{c_L}]^T$ to represent the MIMO weighting on the L channels, the combined output is

$$\begin{aligned} y_{nm\ell} &= \sum_{\ell=0}^{L-1} w_{c_\ell}^* \{\mathbf{X}_t\}_{nm,\ell} \\ &= \gamma_t e^{j(2\pi/\lambda_0)nd \sin\phi_t} \cdot e^{j2\pi m f_t T_r} \sum_{\ell=0}^{L-1} w_{c_\ell}^* w_{tx_\ell} e^{-j(2\pi/\lambda_0)\ell d \sin\phi_t}, \end{aligned} \tag{7.40}$$

for $m = 0, 1, \ldots, M/L - 1$ when the PRF is broken into L equal-sized slow-time channels. Again, the summation term corresponds to a transmit beampattern.

The transmit shading pattern has been effectively changed to $\mathbf{w}_{tx} \odot \mathbf{w}_c$, the element-by-element product of the transmitter shading ($\mathbf{w}_{tx}$) and the MIMO channel weighting ($\mathbf{w}_c$). Note that the MIMO weighting is independent of the transmitter element shading, and the choice of MIMO weights affects the effective transmit spectrum after the data are received. Thus, MIMO weights can be used to place nulls in the transmitter directions that are responsible for multipath clutter. On the other hand, the amount of energy incident on the target does not change with application of the MIMO channel weights $\mathbf{w}_c$ — the physically transmitted energy related to the sweeping beampattern of $\mathbf{w}_{tx}$ is effectively reshaped. Beamformer weight design is thus employed to eliminate multipath clutter returns with the use of data-independent or data-dependent MIMO weightings to accomplish this reshaping of transmit energy. From another perspective, consider MIMO transmit beamforming in terms of the elemental wavefront of (7.30). Beamforming only the transmit channels using $\mathbf{w}_c$ results in a $NM/L \times 1$ vector $\mathbf{y}$:

$$\mathbf{y} = \mathbf{a}_{rx}(\phi_{rx}) \otimes \mathbf{b}_{dop}(f_t T_r) \cdot \mathbf{w}_c^H(\mathbf{w}_{tx} \odot \mathbf{a}_{tx}(\phi_{tx})) \tag{7.41}$$

Note that $\mathbf{y}$ in (7.41) is a column vector containing the $y_{nm\ell}$ terms of (7.40). Here it is clear that no additional transmit energy is physically illuminating the target as a result of slow-time MIMO channel combination. To show the effective beamshaping, (7.41) can be rewritten as

$$\mathbf{y} = \mathbf{a}_{rx}(\phi_{rx}) \otimes \mathbf{b}_{dop}(f_t T_r) \cdot (\mathbf{w}_c \odot \mathbf{w}_{tx}^*)^H \mathbf{a}_{tx}(\phi_{tx}) \tag{7.42}$$

In this respect, slow-time MIMO transmit beamforming consists of two stages: (1) Doppler demodulation with filtering to separate these transmit channels and (2) a linear beamforming operation to recombine them at the receiver. Because each channel is associated with only a fraction of the total discrete Doppler bins, decimation in the Doppler domain is a useful operation. Decimation following filtering reduces the amount of data that need to be sent through further processing and removes those out-of-band frequencies in each channel higher than f_c.

Since the MIMO channel weights during this beamforming operation are independent of the receiver elements, the order of the MIMO transmit beamforming and receiver beamforming operations are interchangeable. Nevertheless, the MIMO demodulation should occur before any Doppler domain processing. If traditional Doppler processing were to be performed on the entire PRF, the MIMO channel separability could be compromised. The only restrictions on the structure of the receiver is that the MIMO Doppler demodulation occurs before both the Doppler processing and the MIMO channel combination. From a processing standpoint, it would be wise to perform the Doppler demodulation first to recover the channels and form an $N \times M/L \times L$ data cube. Then the receiver beamforming, Doppler processing, and MIMO channel combination stages are completely interchangeable. Naturally, these stages can be combined to provide fully adaptive beamforming.

Recall that SIMO operation uses identical transmit waveforms but no pulse-to-pulse phase modulation. This mode of operation destroys the ability to separate transmitter responses because the received signals from the transmitters are not orthogonal in any domain. In fact, the responses from each transmitter element are combined in a specific manner predetermined by the transmit beampattern $\mathbf{w}_{tx}$. The ability to reshape the transmit beampattern as a function of range, Doppler frequency, and receive azimuth is a clear advantage of a slow-time MIMO implementation.

A more useful approximation to (7.39) can be obtained by using

$$\tilde{\mathbf{w}} = \mathbf{w}_{rx} \otimes \mathbf{w}_{dop} \otimes \tilde{\mathbf{w}}_{tx} \tag{7.43}$$

where $\mathbf{w}_{rx}$ and $\mathbf{w}_{dop}$ are nonadaptive receive and Doppler weight vectors. The adaptive component $\tilde{\mathbf{w}}_{tx}$ is computed using the target's transmit spatial wavefront $\mathbf{a}_{tx}(\phi_t)$

$$\tilde{\mathbf{w}}_{tx} = \tilde{\mathbf{R}}_{tx}^{-1}\mathbf{a}_{tx}(\phi_t) \tag{7.44}$$

where given a set of N_c transmit directions to null $\{\phi_{tx_i}\}_{i=1}^{N_c}$, $\tilde{\mathbf{R}}_{tx}$ defines a model-based transmit covariance matrix for the transmit dimension

$$\tilde{\mathbf{R}}_{tx} = \sum_{i=1}^{N_c} \mathbf{a}_{tx}(\phi_{tx_i})\mathbf{a}_{tx}^H(\phi_{tx_i}) \tag{7.45}$$

7.4.1 Slow-Time MIMO Beampattern and VSWR

The implementation of slow-time MIMO radar involves generating time-varying transmit weights, which impacts the transmit beampattern. In this section the effect of MIMO beampatterns on the transmitter voltage standing-wave ratio (VSWR) is examined. When using the phasing choice of (7.26), the transmit beampattern of the full-array slow-time MIMO radar system as a function of slow-time pulse m can be expressed as

$$W_{tx}(\phi, m) = \sum_{\ell=0}^{L-1} w_{tx_\ell} e^{-j\pi(f_c/f_r)(L-1)m} e^{-j((2\pi/\lambda_0)d\sin\phi - 2\pi(f_c/f_r)m)\ell} \tag{7.46}$$

This corresponds to a time-varying beampattern with amplitude given by

$$|W_{tx}(\phi, m)| = \left| S_{tx}\left(k_x - \frac{2\pi f_c}{d f_r} m\right)\right| \tag{7.47}$$

where S_{tx} denotes the spatial Fourier transform of $\mathbf{w}_{tx}$ and k_x representing the x-dimension wavenumber. Thus, the physical beampattern on each pulse is simply an angle-shifted version of the transmit beampattern given by $\mathbf{w}_{tx}$. The mainlobe direction of the transmit beampattern given by $\mathbf{w}_{tx}$ periodically sweeps across the wavenumber range $[-(\pi/d), (\pi/d)]$, inclusive, as a function of the slow-time pulse m. An example of this behavior is given in the simulations as Fig. 7.11 (later, in Section 7.6.1). A cycle is completed whenever $(f_c/f_r)m$ is an integer. This occurs on every (M/L)th pulse for the linear phasing described here; however, the ratio f_c/f_r can be modified to adjust the mainlobe's sweep rate. On the other hand, lowering this ratio gives rise to unused sections of the Doppler domain. Because it may not be desirable to send energies in all wavenumber directions $[-(\pi/d), (\pi/d)]$ (especially along the axis of the transmitter array or into nonpropagating directions when the array is oversampled), a variation known as "subarray" slow-time MIMO is presented in the next section as a solution to this problem.

An important consideration for implementation of slow-time MIMO radar is the VSWR exhibited during the course of a CPI. Low VSWR values are critical for efficient radiation or the electric field. Calculation of VSWR relies heavily on a *voltage-scattering parameter* (or S-parameter) characterization of the network [22]. The complex-valued S-parameter matrix $\mathbf{S}$ is populated with values obtained from test measurements of a system with multiple ports and describes the input–output amplitude and phase relationships between all ports. Generally, $\mathbf{S}$ can be a function of frequency or time, but that notation is suppressed here. Given an vector of voltage inputs across vec $\mathbf{a}$, all P ports of a network, the "reflected" voltage value vec $\mathbf{b}$ at all P ports is simply calculated as

$$\text{vec}\,\mathbf{b} = \mathbf{S}\,\text{vec}\,\mathbf{a} \tag{7.48}$$

Note that in the context of VSWR, vec $\mathbf{a}$ and vec $\mathbf{b}$ are voltage input vectors and not spatial or temporal steering vectors. VSWR is then calculated as

$$\text{VSWR} = \frac{|\text{vec}\,\mathbf{a}| + |\text{vec}\,\mathbf{b}|}{|\text{vec}\,\mathbf{a}| - |\text{vec}\,\mathbf{b}|} = \frac{|\text{vec}\,\mathbf{a}| + |\mathbf{S}\,\text{vec}\,\mathbf{a}|}{|\text{vec}\,\mathbf{a}| - |\mathbf{S}\,\text{vec}\,\mathbf{a}|} \tag{7.49}$$

where $|\cdot|$ represents a magnitude operation [23]. VSWR is a nonlinear measure of $\mathbf{S}$ and vec $\mathbf{a}$ that is always greater than or equal to one. Low values of VSWR at a port p indicate that there is a low amount of energy reflected at port p. VSWR is equal to unity only at port p when there is no reflection at that network port, or equivalently, when the pth element of vec $\mathbf{b}$ is zero. High values of VSWR represent cases when great amounts of energy are reflected at an input port p. The maximum value of VSWR is ∞, which is attained when there is perfect reflection at a port, namely, when $|\text{vec}\,\mathbf{a}| = |\text{vec}\,\mathbf{b}|$. Reflection of power at the input to a transmit antenna is not only inefficient but also potentially damaging to the transmitter equipment.

In slow-time MIMO, the values for VSWR are affected by the pulse-to-pulse starting phase variation that is used to establish orthogonal channels in the Doppler domain. In the slow-time MIMO case, the input voltage vector is a function of pulse is denoted as vec $\mathbf{a}(m)$. As a result, the reflected voltage vector and the VSWR become functions of slow-time pulse [vec $\mathbf{b}(m)$] and VSWR(m). For the linear phasing choice of (7.26), vec $\mathbf{a}(m)$ is expressed as

$$\text{vec}\,\mathbf{a}(m) = \begin{bmatrix} e^{j2\pi\alpha_0 T_r(m)} \\ e^{j2\pi\alpha_1 T_r(m)} \\ \vdots \\ e^{j2\pi\alpha_{L-1} T_r(m)} \end{bmatrix} \quad \forall m = 0, 1, \ldots, M-1. \tag{7.50}$$

The VSWR of slow-time MIMO is evaluated for a simulated radar in Section 7.6.1.

7.4.2 Subarray Slow-Time MIMO

Partially adaptive versions of slow-time MIMO may be achieved by limiting the number of channels and grouping transmit elements into subarrays. Given a set of L transmitters, the overall array can be broken into S adjacent subarrays, each consisting of E elements. Thus, $L = SE$ for integer-valued choices of S and E. In this case, the complex-valued transmit shading vector is $\mathbf{w}_{tx} = \mathbf{w}_S \otimes \mathbf{w}_E$, the Krönecker product of the overall array shading is $\mathbf{w}_S$, and the subarray shading, is $\mathbf{w}_E$. The linear choice of α_ℓ is slightly modified because each subarray will be treated as an orthogonal element

$$\alpha_\ell = -\frac{f_c}{2}(S - 1 - 2\ell) \tag{7.51}$$

for $\ell = 0, 1, \ldots, S-1$. Now there are only S unique Doppler shifts, resulting in S distinct slow-time MIMO channels. Clearly, this subarray technique facilitates MIMO processing using a reduced dimensionality in the transmit domain, where there are now S channels instead of L channels. Note that the overall transmit weights in the subarray MIMO case can be represented as the product of the subarray pattern and a sweeping phased-array pattern

$$|W_{tx}(\phi, m)| = |S_E(k_x)| \cdot \left| S_S\left(k_x - \frac{2\pi f_c}{df_r} m\right) \right| \tag{7.52}$$

where S_E and S_S represent spatial Fourier transforms of $\mathbf{w}_E$ and $\mathbf{w}_S$, respectively. Subarray slow-time MIMO methods also handle transmit grating lobes well. The multiple transmit mainlobes are attenuated by the beampattern of $\mathbf{w}_E$, allowing for some transmit energy direction selectivity.

7.4.3 SIMO versus Slow-Time MIMO Design Comparisons

In this section, the operation of full-array and subarray slow-time MIMO radar is compared to that of SIMO radar. To compare the Doppler resolution and maximum unambiguous Doppler of MIMO and SIMO radar, note that for two systems with the same PRFs [i.e. $f_{r_{\text{SIMO}}} = f_{r_{\text{MIMO}}}$] and a CPI consisting of M pulses, both systems have a Doppler bin width proportional to $f_{r_{\text{SIMO}}}/M$. For a slow-time MIMO system, however, the maximum unambiguous Doppler frequency shift is limited because of the sub-PRF width f_c used to establish channels. With an equal sub-PRF f_c for each of the L channels such that $L \cdot f_c \leq f_{r_{\text{MIMO}}}$, the maximum unambiguous Doppler shift is associated with the bin containing the Doppler frequency below the bin containing the channel width

$$\frac{f_{r_{\text{MIMO}}}}{M}\left\lfloor \frac{f_c}{f_{r_{\text{MIMO}}}/M} - \frac{1}{2} \right\rfloor \tag{7.53}$$

where $\lfloor \cdot \rfloor$ represents a "floor" rounding operation. In the case described in Section 7.3.1 with M/L chosen as an integer, and $f_r/f_c = L$, (7.53) reduces to

$$\frac{f_r}{M}\left(\frac{M}{L} - 1\right) \tag{7.54}$$

Thus, with equal PRFs, the MIMO unambiguous Doppler is approximately decreased by a factor equal to the number of channels relative to a comparable SIMO system. The maximum unambiguous range (R_{un}) in a radar system is given as

$$R_{un} = \frac{c}{2f_r} \tag{7.55}$$

The relationship between SIMO and MIMO maximum unambiguous ranges is

$$\frac{R_{un_{\text{SIMO}}}}{R_{un_{\text{MIMO}}}} = \frac{f_{r_{\text{MIMO}}}}{f_{r_{\text{SIMO}}}} \tag{7.56}$$

When comparing SIMO and slow-time MIMO radars with constant Doppler bin widths, the maximum unambiguous range of a slow-time MIMO radar is reduced by a factor equal to the number of slow-time channels assuming the channels completely fill the Doppler space. Alternatively, when it is desirable to have the maximum unambiguous Doppler shift of the MIMO radar equal to that of the SIMO radar, the slow-time MIMO PRF must satisfy $2 \leq f_{r_{\text{MIMO}}}/f_{r_{\text{SIMO}}} \leq L$ to support enough Doppler space in the MIMO implementation. The ratio $\lfloor f_{r_{\text{MIMO}}}/f_{r_{\text{SIMO}}} \rfloor$ represents the number of equal width sub-PRF Doppler channels that can fill the Doppler space. This ratio is also the factor by which the maximum unambiguous range is reduced when the channels do not fill the entire Doppler space.

In both SIMO and slow-time MIMO radar the range resolution is determined solely by the bandwidth of the pulsed waveforms $u_p(t)$. A SIMO and slow-time MIMO system with the same $u_p(t)$ function have identical range resolutions.

7.4.4 MIMO Radar Estimation of Transmit–Receive Directionality Spectrum

In addition to clutter mitigation, a unique feature of MIMO radar is the potential to map out the directionality of the multipath clutter in terms of both transmit and receive angles. By definition, the transmit and receive directions of multipath propagation are different. In this section, a spectrum is presented that can distinguish multipath clutter returns from direct-path returns. Consider the discrete spatially sampled wavefront after the MIMO channels have been separated and brought into the Doppler frequency domain. This resulting data is a $N \times L$ matrix at one of the M/L Doppler frequencies. The response at the nth receiver element from the ℓth channel at one Doppler frequency is denoted $x_{n\ell}$

$$x_{n\ell} = \gamma_t\, e^{j(2\pi/\lambda_0)nd\sin\phi_{rx}}\, w_{tx_\ell}\, e^{-j(2\pi/\lambda_0)\ell d\sin\phi_{tx}} \tag{7.57}$$

where the Doppler-related complex exponential has been suppressed. This portion of the data lends itself to spatial spectral estimation in both the transmit and receive directions — an operation performed across the N and L elements, respectively. The resulting 2D spatial power spectral density will be referred to as a *transmit–receive* spectrum (and is computed in Section 7.6.3, given as Fig. 7.15).

The transmit–receive directionality spectrum can be computed using a minimum variance distortionless response (MVDR) beamformer. This spectral estimation technique requires knowledge of the covariance matrix at any particular Doppler frequency, and is given by

$$S_{\mathrm{MVDR}}(\phi_{rx}, \phi_{tx}) = \frac{1}{\mathbf{v}(\phi_{rx}, \phi_{tx})^H \mathbf{R}^{-1} \mathbf{v}(\phi_{rx}, \phi_{tx})} \tag{7.58}$$

where $\mathbf{v}(\phi_{rx}, \phi_{tx})$ is an $NL \times 1$ steering vector formed

$$\mathbf{v}(\phi_{rx}, \phi_{tx}) = \mathbf{a}_{rx}(\phi_{rx}) \otimes \mathbf{a}_{tx}(\phi_{tx}) \tag{7.59}$$

and $\mathbf{R}$ is the $NL \times NL$ transmit–receive covariance.

When the spatial spectra are represented in this 2D fashion, direct-path propagation returns appear as a line corresponding to $\phi_{tx} = \phi_{rx}$. Multipath propagation returns are manifested as a single transmit direction with multiple receive directions. The transmitter pattern defined by $\mathbf{w}_{tx}$ appears across the transmitter wavenumber direction. If any receiver shading is applied before calculating the transmit–receive spectrum, the receiver beampattern will appear across the receiver wavenumber dimension. The result of the MIMO channel transmit beamforming is the weighted projection of all the transmit wavenumbers at a particular receiver wavenumber as described previously.

Two-dimensional beampatterns can also be represented on the transmit–receive spectrum. This will be useful in future MIMO STAP weight design when trying to mitigate multipath clutter. If the radar system is concerned primarily with direct-path returns, the ideal 2D transmit-receive beampattern is nonzero only when $\phi_{tx} = \phi_{rx}$. Generally, the receiver weighting can be changed as a function of MIMO channel as well as a function of Doppler frequency, allowing some freedom to approach this desirable transmit–receive beampattern. Examples of transmit–receive spectra and beampatterns are presented in Section 7.6.3.

7.5 OTHR PROPAGATION AND CLUTTER MODEL

An important potential application of MIMO radar is in the area of OTH radar. In this section, the propagation modeling which motivates the need for multipath clutter mitigation is presented. In particular, sounding studies have indicated that Doppler frequency spreading of HF electromagnetic radiation is due to scattering off of moving irregularities in the ionospheric plasma [24]. These irregularities create a

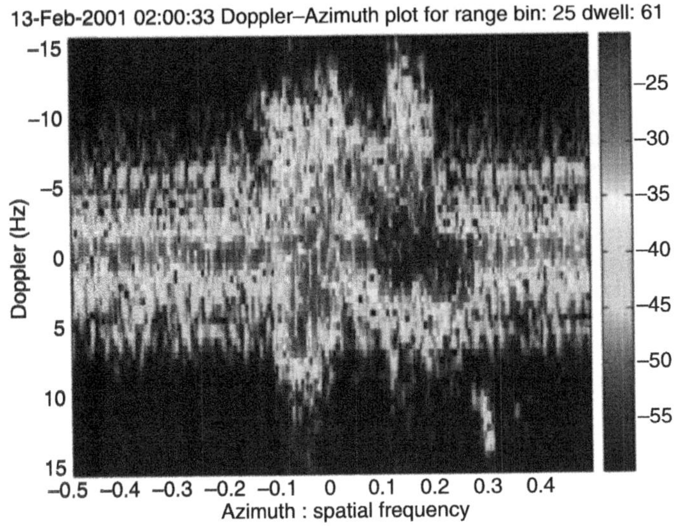

Figure 7.3 Spectrum from measured radar data containing ionospheric clutter.

random fluctuation in the index of refraction in the ionosphere, lengthening or contracting the phase path of the propagating wave. Note that phase path changes from pulse to pulse are manifested as changing Doppler frequencies, or spread, on different ray paths from transmitter to receiver. An example of real data containing ionospheric induced clutter is presented in Fig. 7.3.

Although the exact wave equation solution for the propagation through a complicated medium of this type is difficult to obtain, solutions that are reasonably accurate can be obtained by approximations with phase screens [25]. Phase screens are 2D surfaces that are perpendicular to the wave propagation that impart a single, larger phase change and subsequently scatter the wave into other directions. Phase screens are intended to model sections of the ionosphere where approximate solutions become mathematically tractable. It should be pointed out that alternative approaches to phase screens involving integral equations exist [26]; however, the introduction of ionospheric induced multipath propagation and Doppler frequency spread does not require this alternative approach. One simplifying assumption arises with the fading characteristics of the medium. OTH radar systems generally employ narrowband waveforms with bandwidths that are the inverse of delay spread due to scattering along a ray path. When this is the case, ionospheric effects on the transmitted signal can be modeled as flat fading and the use of phase screens is a valid approximation. In a somewhat simplified view, ionospheric propagation of HF waves can be represented as two rays [27] — a *low ray* that refracts through the most stable part of the ionosphere and a *high ray* that propagates through a more irregular moving phase screen at higher elevation angles. These two propagation paths are depicted as reflections off different layers in Fig. 7.4. Although Fig. 7.4 illustrates in-plane azimuthal scattering, the general scattering model [28] uses a smooth phase screen

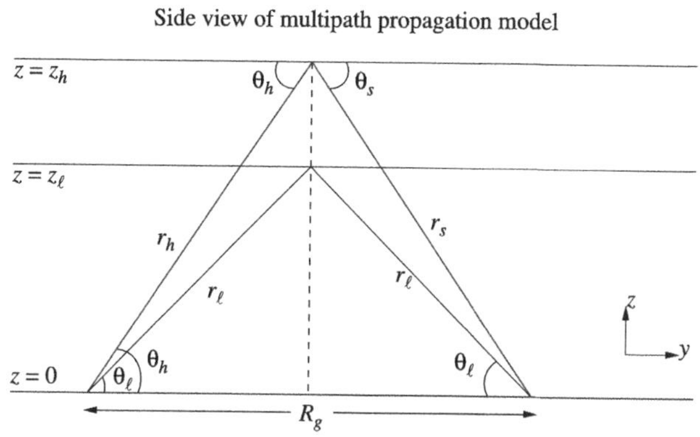

Figure 7.4 Side view of the multipath propagation model with the Tx/Rx arrays at the left and the scatterer of interest on the right.

at the lower ionospheric interface and an irregular moving phase screen at the higher ionospheric interface, which results in out-of-plane azimuthal scattering. Among the paths illustrated in Fig. 7.1, the attenuation is typically greatest on round-trip propagation via the high ray paths exclusively. On the other hand, combination low-ray/high-ray round-trip paths are often strong enough to contribute significantly to Doppler spread clutter. It is thus these clutter paths that MIMO radar is aimed at suppressing while maintaining the exclusively low-ray round trip path that generally contains the target. Analysis of the combination low-ray/high-ray propagation paths yields a Doppler–azimuth relationship that depends on the angles of incidence with respect to the phase screens on the outbound transmit and inbound receive directions.

In the case of skywave HF OTH quasimonostatic radar, the returns from large patches of ground reflect a great deal of energy that can be spread in Doppler due to the random fluctuations of the ionospheric plasma along the high-ray paths. Normally the energy from the ground returns at zero-Doppler frequency, but fluctuations in the ionosphere may cause the ground returns to appear at non-zero-Doppler frequencies, occluding the detection of potential targets. Again, the high-ray path on either the outbound or inbound direction is what imparts the Doppler spread because of the motion of the high phase screen in the model of Ref. 28. In effect, energy from ground returns that is transmitted in an outbound direction can be redirected to a different inbound direction and different Doppler frequency by a moving phase screen that represents the fluctuations in the ionosphere. This scenario is ideally suited to MIMO radar clutter mitigation.

To appreciate the geometry of the multipath return from a ground scatterer, consider a phase screen moving with a velocity v_x parallel to the axis of the array. The elevation angle from the transmitters to the phase screen on the high-ray path is denoted θ_h. ϕ_s represents the azimuthal scattering angle off the high-ray phase screen as in Fig. 7.4. The results that characterize a low-ray propagation path to the scatterers and a high-ray propagation path back to the array (Fig. 7.5) are as follows. Denote the takeoff elevation and azimuth angles from the transmit

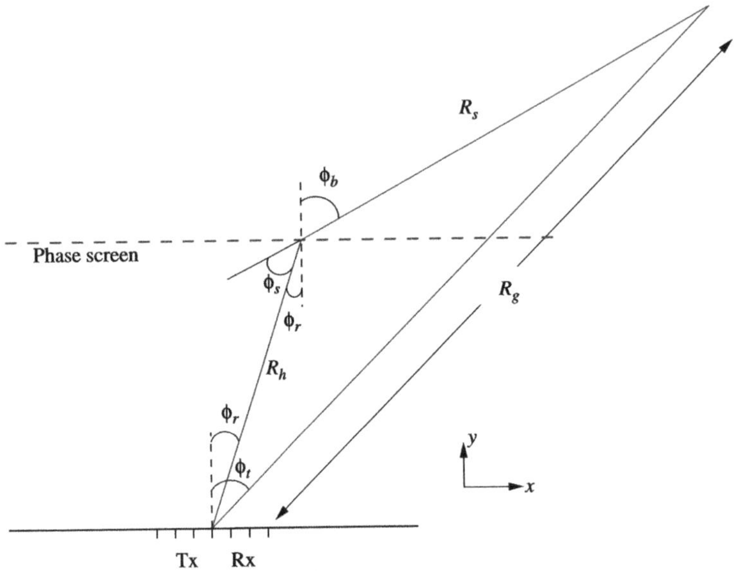

Figure 7.5 Top view of the low-ray/high-ray path in the multipath propagation model.

array to the low, constant phase screen as $\theta_{tx_{\mathrm{low}}}$ and $\phi_{tx_{\mathrm{low}}}$, respectively. The wavenumber of the received returns is given by k_{LH}, and the associated Doppler shift is given as f_{LH}:

$$k_{LH} = -\frac{2\pi}{\lambda_0}\cos(\theta_h)\sin\left(\frac{\phi_{tx_{\mathrm{low}}}}{2} - \frac{\phi_s}{2}\right) \tag{7.60}$$

$$f_{LH} = \frac{4\pi}{\lambda_0} v_x \cos(\theta_h)\sin\left(\frac{\phi_s}{2}\right)\cos\left(\phi_{tx_{\mathrm{low}}}\right) \tag{7.61}$$

For a high-ray propagation path to the scatterers downrange and a low-ray propagation path back to the array (Fig. 7.6), denote the takeoff azimuth angle from the transmit array to the high, variable phase screen as $\phi_{tx_{\mathrm{high}}}$. Denoting the wavenumber of the received returns is given by k_{HL} and the associated doppler shift is given as f_{HL}.

$$k_{HL} = -\frac{2\pi}{\lambda_0}\cos(\theta_h)\sin\left(\frac{\phi_s}{2 - \phi_{rx_{\mathrm{low}}}}\right) \tag{7.62}$$

$$f_{HL} = \frac{4\pi}{\lambda_0} v_x \cos(\theta_h)\sin\left(\frac{\phi_s}{2}\right)\cos\left(\frac{\phi_s}{2 - \phi_{tx_{\mathrm{high}}}}\right) \tag{7.63}$$

Thus, strong ground (or ocean surface) returns can be shifted in wavenumber or Doppler to preclude targets of interest in specific ways. This is directly applicable to the covariance models presented in the previous section for multipath clutter

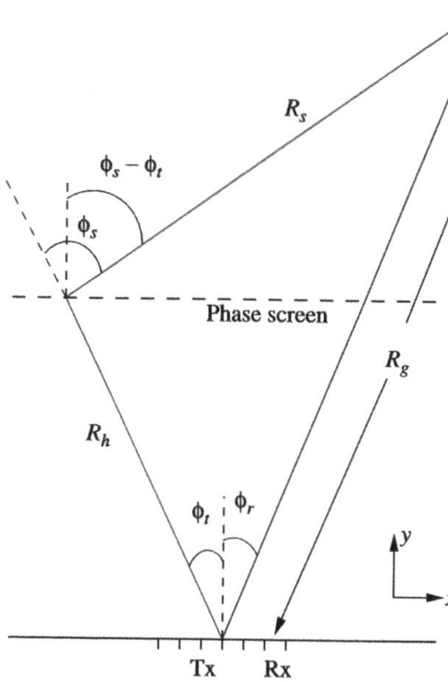

Figure 7.6 Top view of the high-ray/low-ray path in the multipath propagation model.

returns. Traditional SIMO STAP methods break down in situations where the received clutter is in the same receive angle and Doppler frequency as targets of interest. A simple solution to this would be to steer a transmit null in the outbound direction that gives rise to multipath clutter; however, knowledge of these outbound directions is unavailable a priori, and furthermore, changes with range. Thus, in most cases adaptation of the transmit pattern is not feasible. In contrast, the ability of slow-time MIMO radar to modify the effective transmit pattern as a function of range to place nulls in transmit directions that would result in spread clutter make MIMO radar an ideal solution.

7.6 SIMULATIONS EXAMPLES

7.6.1 Postreceive/Transmit Beamforming

To illustrate the potential of slow-time MIMO STAP, consider a simulation with transmitter and receiver arrays with $L = 16$ and $N = 127$ elements placed symmetrically about the origin along the x axis. The operating wavelength was set to $\lambda_0 = 10.7$ m. The elements were spaced at $d = \lambda_0/2$. The choice of $\mathbf{w}_{tx}$ comes from a typical HF OTH radar system. For the slow-time pulses, $M = 528$ and $f_r = 120$ Hz, giving a CPI of 4.4 s. Returns are expected only in the range of Doppler frequencies between ± 3.75 Hz corresponding to a maximum unambiguous speed of $20\,\text{m/s}$.

This choice of operating parameters might be appropriate for detection of surface targets such as ships or ground vehicles.

Consider a far-field direct-path target placed at $\phi_t = 13^\circ$ with an associated Doppler shift of $f_t = -2$-Hz and SNR = 20 dB relative to the background white noise level. Direct path ground clutter appears at all azimuth angles at a Doppler of 0 Hz with a clutter-to-noise ratio (CNR) of 40 dB. A strong multipath clutter patch existed at such that $\phi_{cm,tx}$ ranged between 24° and 26° with a CNR on transmission of 80 dB. On reception, the energy is divided into many directions and spread across Doppler frequencies according to Ramakrishnan's model [28], including a complete masking of the target. No jamming was included. The received data were Taylor windowed in the temporal domain and receiver spatial domain. The resulting Azimuth–Doppler spectrum from traditional SIMO operation over the Doppler frequencies of interest appears in Fig. 7.7. The target at $\sin(\phi) = 0.225$ and $f_t = -2.0$ Hz is masked by the Doppler spread clutter.

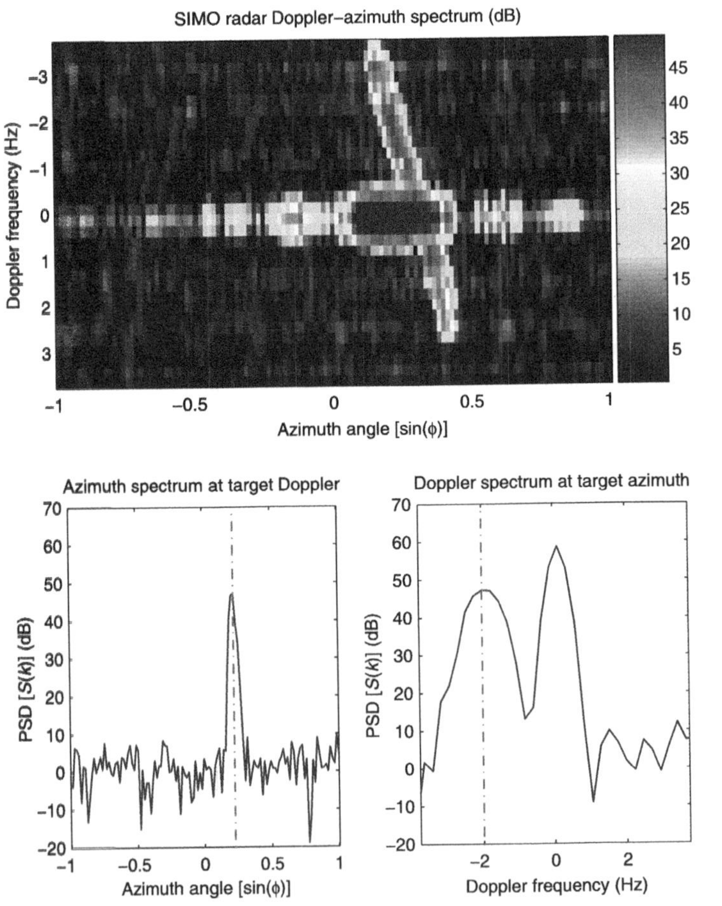

Figure 7.7 Azimuth–Doppler spectrum from SIMO Doppler radar operation.

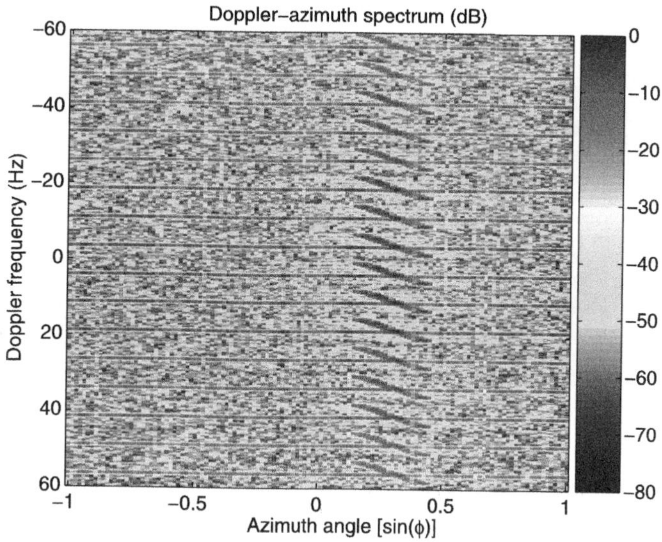

Figure 7.8 Results from conventional Doppler processing on slow-time MIMO implementation for $L = 16$.

For full-array slow-time MIMO operation, the channel sub-PRF was chosen as $f_c = 7.5$ Hz, and the array was phased using the method in (7.26). Figure 7.8 shows the result of conventional SIMO Doppler processing on the slow-time MIMO implementation; $L = 16$ channels are clearly separated in the Doppler domain, each of which is associated with a single transmitter element.

Assuming that the clutter covariance matrix of (7.33) is known, the optimal weight vector of (7.24) may be approximated using (7.43) and (7.45). $\widetilde{\mathbf{R}}_u$ is composed of the multipath clutter component ($\mathbf{R}_{cm,tx}$) and the white noise $(\mathbf{R}_n) = \sigma_n^2\mathbf{I}_L$. Figure 7.9 shows the result of adaptive transmit beamforming that optimizes the CNR on the multipath clutter return. Cuts of the spectrum along the target Doppler and azimuth are provided. The target is clearly unmasked in the Doppler domain of Fig. 7.9, and the multipath clutter is suppressed in both angle and Doppler domains. The improvement in SINR in the target direction and Doppler frequency is approximately 25 dB over conventional SIMO processing. This substantial increase is due to the fact that the multipath clutter is effectively nulled on transmit.

A subarray MIMO technique that uses fewer degrees of freedom is also possible in this transmit array setup. Here, the subarray size is $E = 4$ elements, and there are a total of $S = 4$ subarrays. Again, we use the true transmit dimension covariance matrix to determine the suboptimal $S \times 1$ adaptive weight vector. The results appear in Fig. 7.10. Here, the increase in SINR is closer to 40 dB because of the increased on-target energy caused by the grating lobes in the transmit subarray spacing.

The utility of a subarray technique becomes evident when examining a practical measure of transmitter efficiency — the voltage standing-wave ratio (VSWR).

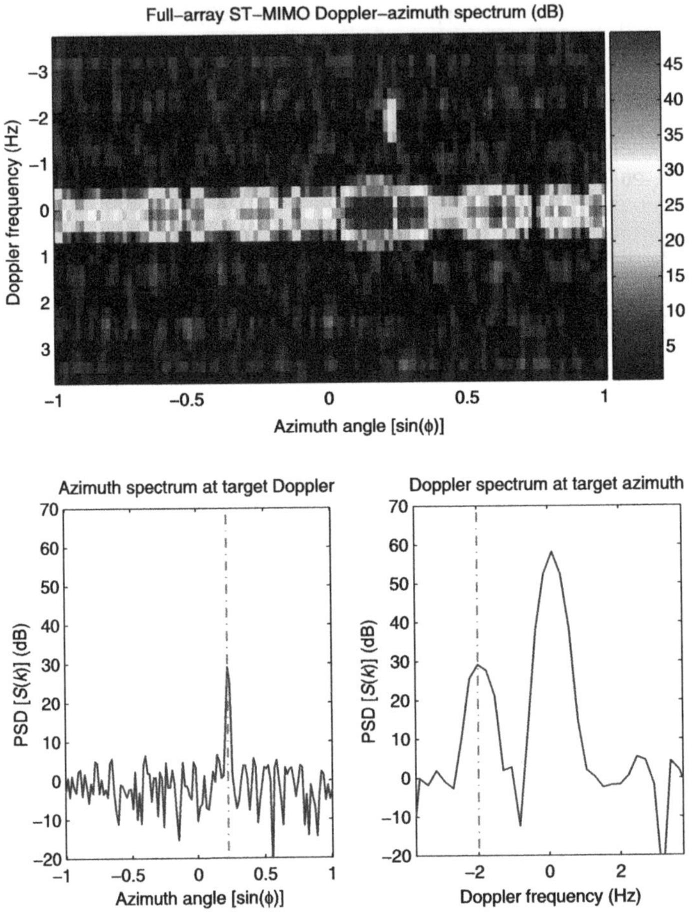

Figure 7.9 Results of 16 adaptively combined full-array slow-time MIMO transmit channels. Cuts of the spectrum are provided at the target azimuth, and Doppler frequency values are provided.

Compared to the full-array in (7.26), which sweeps through all wavenumbers, the subarray slow-time MIMO transmitter allows for modification of the sweeping beampattern of (7.52) via $\mathbf{w}_E$. An example of this is illustrated in Fig. 7.11 for the MIMO scenarios described previously. As noted in (7.47) and (7.52), the physical beampatterns shown in Fig. 7.11 change as a function of slow-time pulse in both cases. Although there are grating lobes in the subarray case, they are never directed along endfire. Thus the VSWR in the transmit elements is reduced. VSWR calculations were performed using the S parameters from a typical OTH HF radar transmit array, and results appear in Fig. 7.12 for SIMO, full-array slow-time MIMO and subarray slow-time MIMO. The SIMO VSWR remains constant over all slow-time pulses because the transmit phasings do not change. In both MIMO cases, the

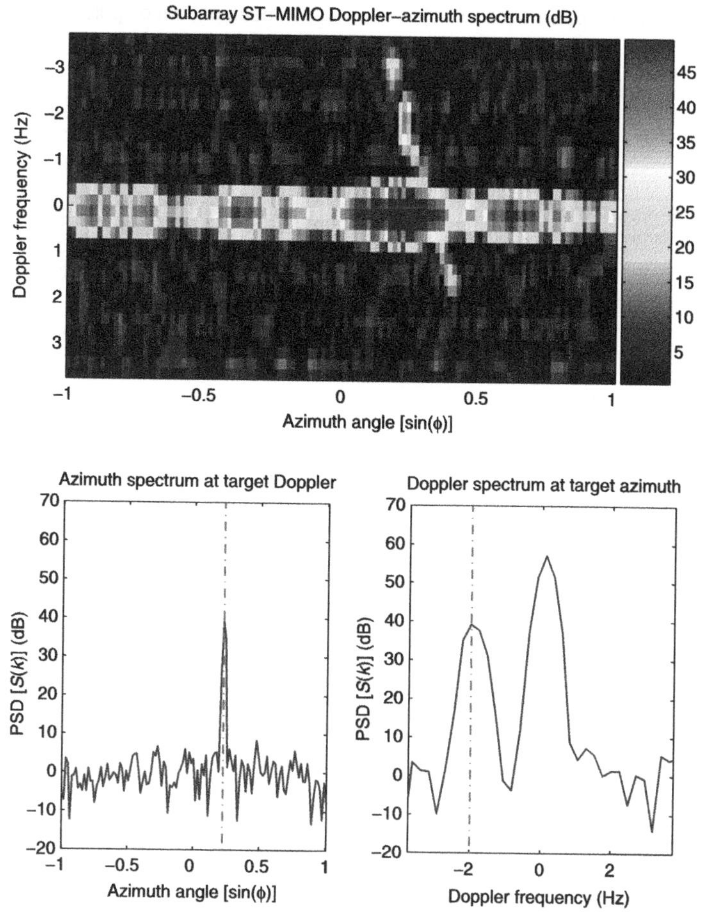

Figure 7.10 Results of eight adaptively combined subarray slow-time MIMO transmit channels. Cuts of the spectrum are provided at the target azimuth, and Doppler frequency values are provided.

VSWR function is periodic; however, the VSWRs in the transmit elements of the subarray MIMO method do not deviate greatly from those of the SIMO case because the transmitted energy is primarily directed away from endfire directions.

7.6.2 SINR Performance

The output signal-to-interference-plus-noise ratio (SINR) of a processor is given as the ratio of the signal power to that of the combined interference and noise powers:

$$\mathrm{SINR} = \frac{\sigma_s^2 |\mathbf{w}^H \mathbf{v}_t|^2}{\mathbf{w}^H \mathbf{R}_u \mathbf{w}}. \tag{7.64}$$

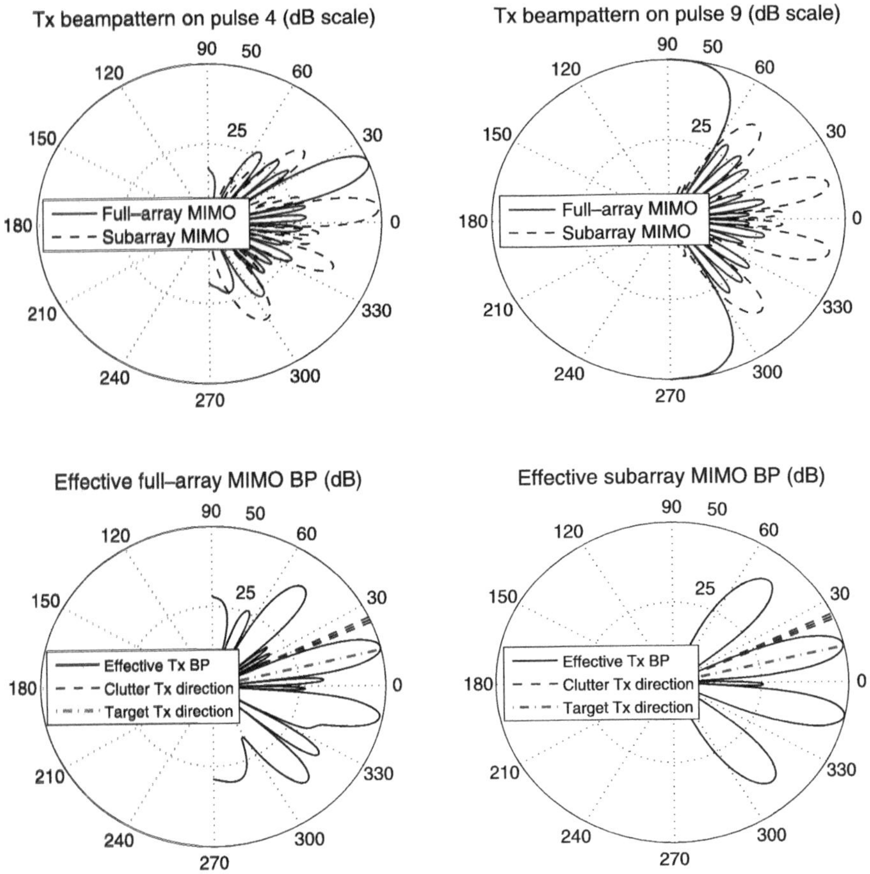

Figure 7.11 Sample physical beampatterns for MIMO operation and effective transmit beampatterns (BPs) after adaptive MIMO channel combination.

SINR loss (L_{SINR}) is a useful measure that compares the processor performance in the presence of interference and noise to the performance of the processor in a noise-only environment. SINR loss falls between the values of 0 and 1, and is defined as

$$L_{\text{SINR}} = \frac{\text{SINR}}{\text{SNR}_o} \tag{7.65}$$

where SNR_o represents the optimum output signal-to-noise ratio in the case where the target is present in white noise only. In (7.65), $L_{\text{SINR}} = 1$ indicates that the target signal is preserved and the interference is completely suppressed, and $L_{\text{SINR}} = 0$ indicates that the target has been completely suppressed along with the interference. To illustrate SINR performance, consider a 30-dB SNR far-field target placed at

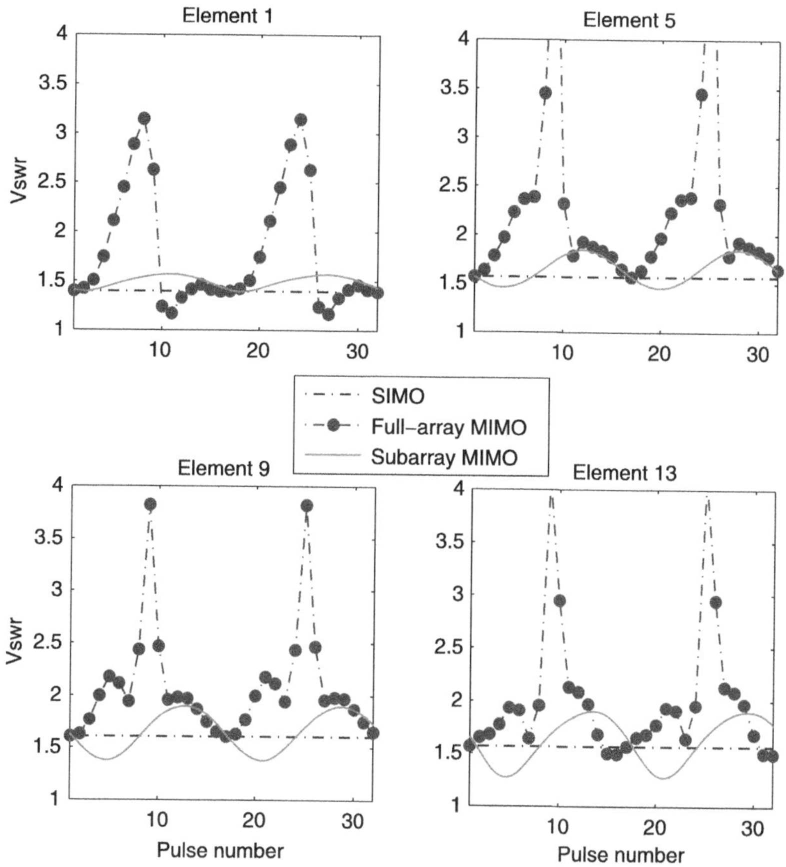

Figure 7.12 Example of VSWR values in four elements as a function of slow-time pulse.

$\phi_t = 0°$ with a Doppler shift of f_t. Multipath clutter with 80-dB CNR is assumed at a single $\phi_{cm,tx}$, with components of $\phi_{cm,rx}$ returning in the target direction $\phi_t = 0°$. The multipath clutter return has a Doppler shift of f_t that completely masks the target in receive azimuth and Doppler domains. No direct-path clutter or jamming is included. Varying $\phi_{cm,tx}$ across the range $[-90, 90]°$, the SINR loss was calculated for the full-array slow-time MIMO case in (7.65) and is plotted in Fig. 7.13. Comparison of L_{SINR} between MIMO and SIMO methods illustrates the advantages of transmit adaptivity.

Note that in Fig. 7.13 the SINR loss comparison of the MIMO transmit adaptive case and the SIMO case assumes the SIMO radar uses M/L pulses in the Doppler domain. In the SIMO radar case, the multipath clutter suppression is accomplished somewhat by the transmit array's beampattern, which explains why the SIMO SINR loss is the inverse of the transmitter array's beampattern and reduces to the MIMO L_{SINR} when ϕ_{tx} for the clutter is in a transmit pattern null. The MIMO

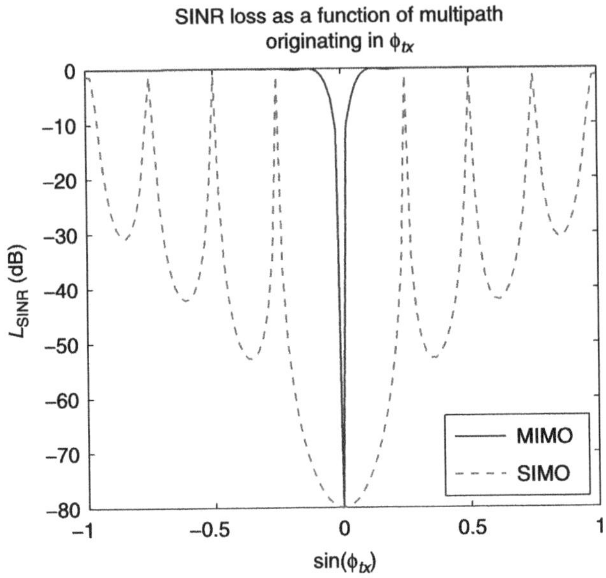

Figure 7.13 SINR loss as a function of multipath clutter source transmit angle (ϕ_{tx}) for the SIMO and MIMO radars.

SINR loss approaches zero when the multipath clutter source is near ϕ_t, signifying suppression of the target along with the clutter. However, the SINR loss remains very close to 1 for $\phi_{tx} \neq 0$, indicating the advantages of combining the MIMO channels as described previously.

SINR loss performance degradation will occur in the case of multipath clutter that is spread in the transmit direction because the MIMO channel weights may not have sufficient degrees of freedom to perform suppression. For example, consider the previous simulation scenario when the multipath clutter on transmit is now spread over a set of transmit azimuth centered at $\phi_{cm,tx}$. The spread is characterized by the parameter Δ, such that the angles ϕ where the multipath clutter exists on transmit satisfies

$$\sin\phi_{cm,tx} - \frac{\Delta}{2} \leq \sin\phi \leq \sin\phi_{cm,tx} + \frac{\Delta}{2} \tag{7.66}$$

With $\phi_{cm,tx} = 0^\circ$ and $\Delta = 1$, for example, the multipath clutter will be located in the transmit directions between -30° and 30°, inclusive.

In Fig. 7.14, transmit azimuth spreads of $\Delta = [0, 0.08, 0.18, 0.39]$ were simulated for all ϕ_{tx} varying between -90° and 90°, and results are shown for the MIMO fully adaptive weights. As the spread Δ increases, the SINR loss performance worsens near the target transmit azimuth. In fact, the SINR loss approaches a value of only 1 beyond $\sin\phi_{tx} = \pm 0.5$, indicating that the multipath clutter can be nulled only between $30^\circ \leq |\phi_{tx}| \leq 90^\circ$ and without suppressing the target response. A greater number of channels are necessary to provide the additional degrees of freedom to improve performance for transmit direction spread multipath clutter.

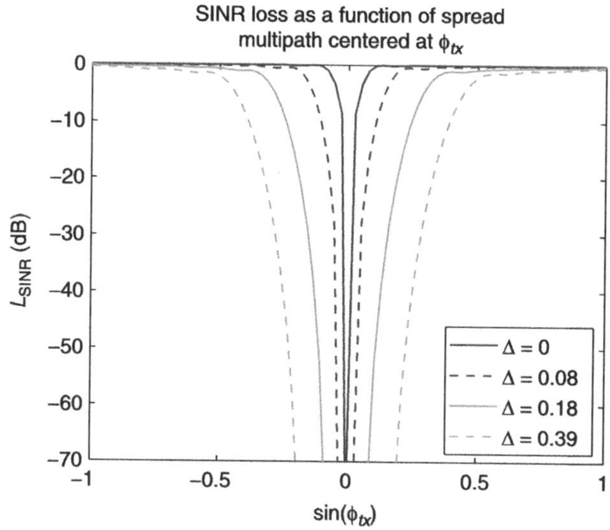

Figure 7.14 Normalized SINR loss as a function of wavenumber spread multipath clutter sources.

7.6.3 Transmit–Receive Spectrum

In order to show the effects of different types of clutter on the transmit–receive directionality spectrum, assume that the target, direct-path clutter, and multipath clutter returns arrive with the same Doppler shift f_t. To illustrate this possibility, consider a simulation using the same MIMO operating parameters from Section 7.6.1 The

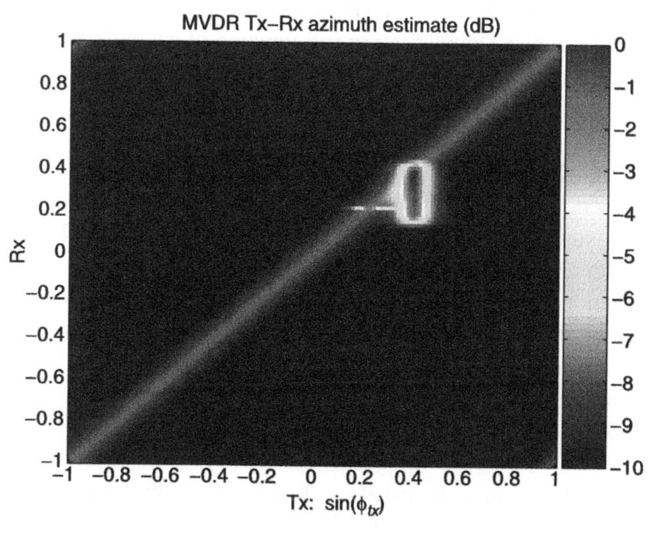

Figure 7.15 MVDR transmit–receive azimuth spectrum for scenario described Section 7.6.3.

resulting transmit–receive directionality spectrum is calculated for this particular Doppler frequency. Without loss of generality, a 20-dB SNR far-field target was placed at $\phi_t = 13°$. Direct-path clutter with CNR = 60 dB was placed at all angles, which masked the target in Doppler as well as azimuth. Multipath clutter with 80-dB CNR with $\sin \phi_{tx} = 0.45$ and various ϕ_{rx} receive angles was simulated as well according to Ramakrishnan's model [28].

The computed MVDR transmit–receive directionality spectrum appears in Fig. 7.15. Again, the direct-path clutter returns and the target return appear along the line $\phi_{tx} = \phi_{rx}$. However, the multipath is clearly visible at $\sin \phi_{tx} \approx 0.45$ and at multiple receive angles. In practice, using MIMO radar and a spectral estimation technique such as that expressed in Eq. (7.58) can thus help determine transmit directions that contribute multipath clutter. In many cases, simply diagnosing the clutter as being due to multipath propagation is the critical first step to successful mitigation.

7.7 CONCLUSION

In MIMO spacetime processing, the slow-time orthogonality exploits previously unaccessible degrees of freedom that may be adapted to improve performance in the presence of multipath clutter. Although establishing slow-time sub-PRFb channels reduces the maximum unambiguous range for this type of MIMO radar, this tradeoff provides the significant advantage of effective transmit beampattern adaptability at each range in OTH applications. Further work is clearly needed to develop methods for covariance estimation with limited training data in both fully adaptive and partially adaptive cases. A robust method for determining transmit directions that result in multipath clutter is also of interest and desirable, and the transmit–receive spectrum presented here is a promising tool in this effort. Finally, although examined here in the fixed OTHR context, MIMO STAP may also be useful in moving platform situations where the clutter Doppler return is a function of azimuth. For this and other situations, a more general framework for slow-time MIMO radar may be desirable.

ACKNOWLEDGMENT

This work was supported in part by ONR code 313, the Air Force Research Laboratory under Air Force Contract FA8721-05-C-0002, and MIT Lincoln Laboratory. Opinions, interpretations, recommendations and conclusions are those of the authors and are not necessarily endorsed by the U.S. Government.

REFERENCES

1. G. J. Frazer, Y. I. Abramovich, and B. A. Johnson, Spatially waveform diverse radar: Perspectives for high frequency OTHR, *Proc. IEEE Radar Conf.*, April 17–20, 2007, pp. 385–390.

2. E. Fishler, A. Haimovich, R. Blum, D. Chizhik, L. Cimini, and R. Valenzuela, MIMO radar: An idea whose time has come, *Proc. IEEE Radar Conf.*, 2004, pp. 71–78.
3. R. G. Seed, A. S. Fletcher, and F. C. Robey, STRAAP: Space-time-radiating array adaptive processing, *Proc. IEEE Intl. Symp. Phased Array Systems and Technology*, 2003, pp. 136–141.
4. V. Mecca, D. Ramakrishnan, and J. Krolik, MIMO radar space-time adaptive processing for multipath clutter mitigation, *Proc. IEEE SAM Workshop*, July 2006, pp. 249–253.
5. D. J. Rabideau and P. Parker, Ubiquitous MIMO multifunction digital array radar, *Proc. 37th Asilomar Conf. Signals, Systems and Computers*, 2003, pp. 1057–1064.
6. D. Bliss and K. Forsythe, Multiple-input multiple-output (MIMO) radar and imaging: Degrees of freedom and resolution, *Proc. 37th Asilomar Conf. Signals, Systems and Computers*, 2003, pp. 54–59.
7. K. Forsythe, D. Bliss, and G. Fawcett, Multiple-input multiple-output (MIMO) radar: Performance issues, *Proc. 38th Asilomar Conf. Signals, Systems and Computers*, 2004, pp. 310–315.
8. F. C. Robey, S. Coutts, D. Weikle, J. C. McHarg, and K. Cuomo, MIMO radar theory and experimental results, *Proc. 38th Asilomar Conf. Signals, Systems and Computers*, 2004, pp. 300–304.
9. M. I. Skolnik, *Introduction to Radar Systems*, 3rd ed., McGraw-Hill, 2001.
10. K. Forsythe and D. Bliss, Waveform correlation and optimization issues for MIMO radar, *Proc. 39th Asilomar Conf. Signals, Systems and Computers*, 2005, pp. 1306–1310.
11. L. B. White and P. S. Ray, Signal design for MIMO diversity systems, *Proc. 38th Asilomar Conf. Signals, Systems and Computers*, 2004, pp. 973–977.
12. D. Fuhrmann and G. San Antonio, Transmit beamforming for MIMO radar systems using partial signal correlation, *Proc. 38th Asilomar Conf. Signals, Systems and Computers*, 2004, Vol.1, pp. 295–299.
13. L. Xu, J. Li, P. Stoica, K.W. Forsythe, and D.W. Bliss, Waveform optimization for MIMO radar: A Cramér-Rao bound based study, *Proc. IEEE Intl. Conf. Acoustics, Speech and Signal Processing*, April 2007.
14. P. Stoica, J. Li, and Y. Xie, On probing signal design for MIMO radar, *IEEE Trans. Signal Process.* **55**(8):4151–4161 (Aug. 2007).
15. J. Li and P. Stoica, MIMO radar with colocated antennas: Review of some recent work, *IEEE Signal Process. Mag.* 106–114 (Sept. 2007).
16. E. Fishler, A. Haimovich, R. Blum, L. Cimini, D. Chizhik, and R. Valenzuela, Performance of MIMO radar systems: Advantages of angular diversity, *Proc. 38th Asilomar Conf. Signals, Systems and Computers*, 2004, pp. 305–309.
17. E. Fishler, A. Haimovich, R. S. Blum, L. J. Cimini, D. Chizhik, and R. A. Valenzuela, Spatial diversity in radars — models and detection performance, *IEEE Trans. Signal Process.* **54**(3):823–838 (2006).
18. F. Robey, S. Coutts, K. Cuomo, and J. Papp, Overview of US/Australian ballistic missile defense adaptive spread-doppler clutter mitigation, presentation for NRL, Feb. 2, 2006.
19. F. Robey, Enhancing radar array performance through space-time coding, *IEEE Trans. Signal Process.* (in press).
20. J. Ward, *Space-Time Adaptive Processing for Airborne Radar*, Lincoln Laboratory, MIT, Technical Report, 1994.

21. L. Brennan and I. Reed, Theory of adaptive radar, *IEEE Trans. Aerospace Electron. Syst.* **9**(2):237–252 (1973).
22. B. Peikari, *Fundamentals of Network Analysis and Synthesis*, Krieger, 1982.
23. C. A. Balanis, *Advanced Engineering Electromagnetics*, Wiley, 1989.
24. R. A. Shepherd and J. B. Lomax, Frequency spread in ionospheric radio propagation, *IEEE Trans. Commun. Technol.* **COM-15**(2):268–275 (April 1967).
25. D. L. Knepp, Multiple phase-screen calculation of the temporal behavior of stochastic waves, *Proc. IEEE* **71**(6):722–737 (June 1983).
26. C. Coleman, A model of HF sky wave radar clutter, *Radio Sci.* **31**(4):869–875 (Aug. 1996).
27. J. M. Headrick and M. I. Skolnik, Over-the-horizon radar in the HF band, *Proc. IEEE* **62**(6):664–673 (June 1974).
28. D. Ramakrishnan, *Adaptive Radar Detection in Non-stationary Doppler Spread Clutter*, Ph.D. dissertation, Duke University, 2006.

8

MIMO AS A DISTRIBUTED RADAR SYSTEM

H. D. GRIFFITHS

DCMT, Shrivenham, Cranfield University, Shrivenham, Swindon, UK

C. J. BAKER AND P. F. SAMMARTINO

Department of Electronic and Electrical Engineering, University College, London, UK

M. RANGASWAMY

Air Force Research Laboratory (AFRL) Sensors Directorate, Hanscom Air Force Base, USA

8.1 INTRODUCTION

This chapter examines MIMO (multiple-input, multiple-output) as a distributed radar concept in an attempt to understand true performance and system utility and to evaluate novel forms of MIMO.

MIMO does not have a strict definition but in essence exploits spatial diversity to obtain performance enhancement. It is possible to conceive of a number of different forms of processing that exploits the spatial diversity offered by a netted or distributed radar system. Therefore the overall aim of this chapter is to compare the performance, under various practical operating conditions, of different ways of implementing a MIMO radar system.

The area of signal processing techniques for radar networks has received increased attention and importance in more recent studies [1–9] that provide a wide variety of system structures and processing methods. Here we define a radar network as a system made up of more than one transmitter and more than one receiver. Transmitters and receivers are distributed over a geographic area such that it is possible to view targets at different aspect angles.

MIMO Radar Signal Processing, edited by Jian Li and Petre Stoica

Other authors [10–12] have been investigating MIMO as a technique based on two arrays of antennas transmitting and receiving different codes from different (sub)elements. Here the distance between the elements is kept relatively small (i.e., in the order of the wavelength), and therefore angular diversity cannot be achieved. However this application, exploiting either orthogonal or partially correlated codes [23–26], has been shown to overcome the standard array of antenna performance and to provide an extra degree of freedom for both beamforming and jammer rejection.

In most general terms the MIMO concept can be applied to radar in a number of ways. For instance, it can be used to isolate scatterers within a monostatic resolution cell; it also offers a means of improving the separation of targets signals from clutter signals. The MIMO concept may additionally be combined with beamforming (e.g., at the subarray level) and employ subarray-level waveform diversity. This provides for a "frequency MIMO" concept within a single radar manifold. However, even more sophisticated waveform diverse beampatterns can potentially be generated offering even further options for performance improvements and versatility. Indeed, the overall degree of coherence from antenna element waveform to antenna element waveform will control the properties of the hybrid beam/signal space that may provide a completely new set of degrees of freedom for the radar designer.

In this chapter we concentrate on showing that MIMO radar systems are able to take advantage of effects such as target scintillation, and we subsequently examine the impact on the signal-to-noise ratio (SNR) in the rest of the radar system. MIMO processing attempts to exploit the same effects to increase performance in a manner similar to that used in telecommunications. Furthermore, the approach potentially offers reduced sensitivity to electronic countermeasures (ECMs) for the same reasons. In this chapter a more fundamental understanding of MIMO as applied to radar is developed. The insight provided is then used to evaluate the most appropriate systems able to exploit this form of processing.

More recent publications have shown that (1) incoherent processing provides the likelihood ratio test (LRT) condition for a Swerling I–II target in a radar network [1–3], (2) there are several different ways to process the same data after matched filtering, and finally (3) diversity (e.g., in space or frequency) can be exploited in radar systems to yield additional improvement that can be used to improve the quality of the information collected.

However, many issues related to the implementation of the network are vital in order to ensure optimal operation and realization of the full performance potential. For example, in coherent MIMO, or in a fully coherent netted radar system, a stringent RF phase resynchronization of all the signals arriving at all the receivers has to be implemented in order to match the returns from the same position in space. This is due to the fact that different delay times will correspond to different paths and hence can lead to complete signal cancellation if they have opposite phase. This is crucial not only for detection but also for tracking and, as will be seen later, has a major impact on performance and hence the eventual choice of system hardware and processing approach. Moreover, when more than one signal is transmitted at different times, all the cross-correlation levels of the waveforms should be ideally minimal [23,24], in order to avoid ghosts and hence an increased false-alarm rate

(FAR). Therefore, it is extremely important to completely understand the characteristics, advantages, disadvantages, and the ability of this class of radar system to exploit the full potential and hence achieve optimal enhanced performance.

Indeed, the same data acquired by a radar network can be processed in a number of different ways, leading to a range of performance levels. In this chapter we describe in detail four different signal processing algorithms. These algorithms vary from processing of fully coherent signals to processing incoherent signals, including incoherent processing with a double threshold. Moreover, the processing approaches can be "centralized" (i.e., the detection decision is made by one processing unit) or "decentralized" (decisions are made at individual receivers in the radar network across all possible mono/bistatic pairs). The statistical properties of the signals resulting from the four differing signal processing approaches lead to substantial differences in their performance in terms of FAR and even more so when subject to either accidental or deliberate interference (jamming). Such an analysis allows a detailed examination of the benefits and drawbacks associated with radar networks and processing schemes. This provides a simple but thorough investigation enabling the potential of radar networks in terms of their false-alarm performance to be assessed. Results for monostatic systems are also reported to provide a well-understood benchmark. The achievable detection performance is computed as a function of processing method, RCS model of the target, and the number of nodes contained in the radar network. Furthermore, the sensitivity of each system is calculated to show the extent of the range coverage that these differing systems could potentially achieve. In order to provide a fair comparison of the processing types, the total effective radiated power (ERP) is maintained constant in transmission regardless of the number of nodes in the network. As a consequence, networks made up of a large number of nodes transmit a reduced amount of power at each transmitter.

8.2 SYSTEMS

In this section the form of the network, the signal model and the system architecture whose performance is to be investigated are described.

A radar network is assumed to consist of a number of transmitters (M) and receivers (N) geographically separated in an arbitrary fashion. Each receiver contains M filters matched to a given transmitted signal. Therefore, a total of MN signals are available for processing. Transmitters and receivers are the devices constituting the nodes of the network of the same specification and are separated in space to provide angular diversity. To provide a deeper insight into the operation and performance of the network, we choose the number of receivers and transmitters to be equal (i.e., $N = M$). In order to simplify the analysis, each pair of these devices is also collocated. In other words, we consider a radar network made up of M radars each transmitting and receiving. Figure 8.1 shows a schematic example radar network with $M = N = 3$. Each can receive both its own transmitted signal and the signals transmitted by the other transmitters. To gain a better understanding of the possible variations in performance for a wide range of parameters, M (and N) are

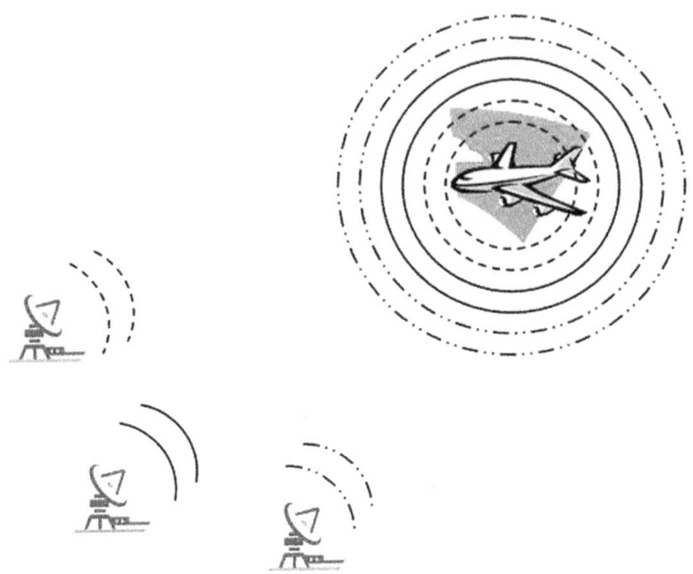

Figure 8.1 MIMO spatial diversity and netted radar configuration.

varied from 1 (the monostatic case) to 5. Transmitter–receiver path pairs are termed "nodes" of the network and the individual transmitter–receiver pair is a "device." As a consequence, the radar networks under investigation are made up of one to five devices and therefore exhibit 1, 4, 9, 16, and 25 nodes, respectively. Hence the number of nodes is the number of signals that it is possible to process.

The network operates with a constant ERP. In other words, as the number of transmitters is increased, the individual emitted power of each transmitter is reduced accordingly. This concept is introduced to allow a fair comparison by keeping the total energy in the system constant. Of course, if each transmitter emits the same power, networks with many nodes will clearly have superior sensitivity. It is also assumed that the transmitted waveforms are orthogonal to one another. However, it should be noted that modulation induced by complex target reflection can cause degradation in orthogonality, and therefore this may not exactly be the case on reception. Each antenna points at the target from a different aspect angle, so that the measurement of radar cross section (RCS) into a particular receiver can differ from the other measurements by several decibels (dB) or more; thus, it is assumed that in this way independent spatial samples of the scattering from the target are obtained.

Although a robust synchronization may have to be performed and an increased quantity of data may have to be jointly processed, the achievable benefits can be worth the effort. Qualitatively, these benefits are:

1. An improved detection capability due to multistatic scintillation of the target that increases the possibility of obtaining one or more high-value echo from a target
2. A joint estimation of target position

3. The capability of resolving multiple targets (as a consequence of benefit 2, above)
4. Increased information in the same bandwidth occupation
5. An increased electronic counter-countermeasures (ECCM)

8.2.1 Signal Model

The systems and a simple model for describing the signals are now introduced. A pure MIMO system is one that operates incoherently, whereas netted radar (NR) systems operate coherently (i.e., exploiting phase) and decentralized radar networks (DRNs) operate incoherently with a two-stage processing. The basic form of the received signal in a MIMO network is now formulated in detail.

In the general case the signal arriving at the kth receiver can be modeled as

$$r_k(t) = \sum_{m=1}^{M} [H_{0/1}\alpha_{m,k}(\sigma)s_m(t - \tau_{m,k}) + c_{m,k}(t - T_{m,k})] + J_k(t) + z_k(t) \tag{8.1}$$

where $H_{0/1}$ is 0 or 1 depending the absence or presence of target, respectively; s_m is the mth transmitted signal; $c_{m,k}$ is the clutter; z_k is the thermal noise; J_k is an external disturbance (such as jamming); $\tau_{m,k}$ and $T_{m,k}$ are the delays occurring during the path between the mth transmitter and the target/clutter, respectively, and the kth receiver; and $\alpha_{m,k}(\sigma)$ is a coefficient that accounts for the parameters of the mono/bistatic radar equation, the phase shift, and the RCS-distribution, specifically

$$\alpha_{k,m}(\sigma) = \sqrt{\frac{P_t}{M}}\sqrt{\frac{G_{tx}G_{rx}\lambda^2\sigma}{(4\pi)^3 R_m^2 R_k^2}} \exp\left\{-j\frac{2\pi R_{m,k}}{\lambda}\right\} \tag{8.2}$$

where G_{tx} and G_{rx} are respectively the gains of the transmitting and receiving antennas; σ is the RCS of the target; P_t is the transmitted power; R_m and R_k are the transmitter–target and target–transmitter distances, respectively; and $R_{m,k}$ is the distance covered by the signal.

At this stage the effects of clutter are not considered, and hence this leads to the following expression for the received signal:

$$r_k(t) = H_{0/1} \sum_{m=1}^{M} \alpha_{m,k}(\sigma)s_m(t - \tau_{m,k}) + J_k(t) + z_k(t) \tag{8.3}$$

The received output may be expressed as the result of the cross-correlation of the received signal with the transmitted waveforms as follows:

$$\begin{aligned}
x_{h,k} &= r_k(t) \otimes s_h(t) \\
&= H_{0/1} \sum_{m=1}^{M} \alpha_{m,k}(\sigma) s_m(t - \tau_{m,k}) \otimes s_h(t) \\
&\quad + [J_k(t) + z_k(t)] \otimes s_h(t) \\
&= H_{0/1} \alpha_{h,k}(\sigma) R_h(t - \tau_{h,k}) + H_{0/1} \sum_{\substack{m=1 \\ m \neq h}}^{M} \alpha_{m,k}(\sigma) R_{m,k}(t - \tau_{m,k}) \\
&\quad + [J_k(t) + z_k(t)] \otimes s_h(t) \\
&= H_{0/1} \alpha_{h,k}(\sigma) R_h(t - \tau_{h,k}) \\
&\quad + H_{0/1} \sum_{\substack{m=1 \\ m \neq h}}^{M} \alpha_{m,k}(\sigma) R_{m,h}(t - \tau_{h,k}) + n_{h,k}(t)
\end{aligned} \tag{8.4}$$

where $R_h(t)$ is the autocorrelation function of s_m and $R_{m,h}$ is the cross-correlation function between s_m and s_h, and $n_{h,k}$ is the component of the overall disturbance incoming in the kth receiver after the hth matched filter. The last two terms in Eq. (8.4) represent the overall noise at the receiver. This equation highlights how the use of low-cross-correlation codes is essential in order to separate the transmitted waveforms from each other and therefore maintain all the benefits of multistatic data collection. This remains an important research challenge in the topic. In the remainder of the chapter, to simplify the analysis and provide a direct insight into the resulting radar performance, it is assumed that waveforms with desired low-cross-correlation properties are employed.

Keeping the geometry of the radar network arbitrary but fixed, four different types of system are analyzed. Indeed, the only difference between the systems is due to the manner in which the signals are processed. It is also assumed that there is adequate synchronization of the delays of the signals due to the different transmitter–target–receiver paths. Consequently, a matrix $\mathbf{M}_X$ signals from the same area may be written as follows:

$$\mathbf{M}_X = \begin{bmatrix} x_{11} & x_{12} & \cdots & x_{1N} \\ x_{21} & x_{22} & \cdots & x_{2N} \\ \vdots & \vdots & \ddots & \vdots \\ x_{M1} & x_{M2} & \cdots & x_{MN} \end{bmatrix}. \tag{8.5}$$

Alternatively, this equation may be rearranged into a vector **X** as follows

$$\mathbf{X} = [x_{11}, \ldots, x_{1M}, x_{21}, \ldots, x_{2M}, \ldots, x_{N1}, \ldots, x_{1N}]^T \tag{8.6}$$

where T is the transpose operator. The differing forms of processing are now considered.

8.2.2 Spatial MIMO System

The MIMO spatial diversity model described below has the conventional form of MIMO that has appeared in the literature [1–3]. This form of MIMO radar system exploits measurements of independent samples of target scattering as the basis for improving the probability of detection.

Here, this processing approach is termed *spatial MIMO* or, simply *MIMO*. Data samples are processed incoherently and in a centralized architecture; that is, there is a central processing unit collecting the receiver outputs from all the nodes and returning a decision about the presence or absence of a target. A decision is determined by

$$\|X\|^2 \underset{H_0}{\overset{H_1}{\gtrless}} \lambda_{\text{MIMO}} \tag{8.7}$$

and arises from the likelihood ratio test (LRT) for this form of processing [3], where λ_{MIMO} is an appropriate threshold. The RCS of the target is assumed to have a Swerling II distribution, which corresponds to a Swerling I distribution when only one pulse is integrated.

8.2.3 Netted Radar Systems

The second and third models that have been developed have the same physical layout as the spatial diversity MIMO, but instead use conventional coherent processing. In this approach the received signals are processed through the bank of matched filters as in Section 8.2.1; after this the coherent summation is performed. As the model of the received signal is the same as represented by Eq. (8.3), the results of processing after the filtering are the same as in Eq. (8.4). Two different forms of netted radar may be formulated: (1) coherent netted radar (NR) and (2) rephased coherent netted radar (RPCNR).

8.2.3.1 Coherent Netted Radar (NR) This processing approach uses the same samples as for the case of the spatial MIMO radar system, but it sums them coherently. This is not the same as the MIMO concept but provides a valuable means of comparison. The phases of the incoming signals are highly correlated as they

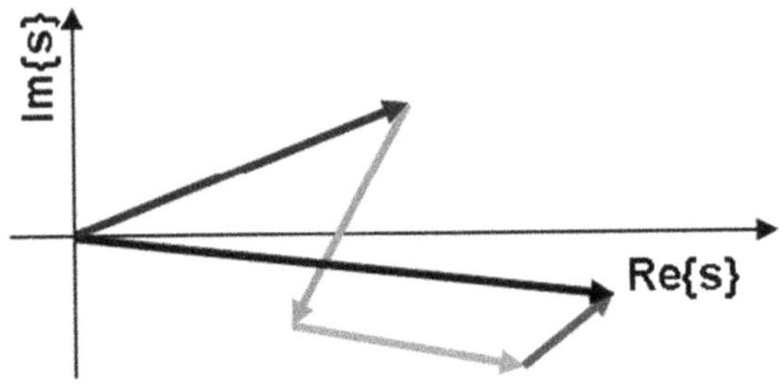

Figure 8.2 The coherent netted radar integration.

depend on the target's position and the geometry of the system. However, it is well known that the phase wraps every half-wavelength, so, given that the position of the target cannot be measured with this degree of accuracy, the signals appear to have uncorrelated phases uniformly distributed between $-\pi$ and π.

Figure 8.2 shows what happens if four signals are coherently summed without preprocessing the phases to align them. In this case the phases will be uniformly distributed and the overall coherent sum will be a signal whose amplitude is much smaller than the sum of the amplitudes of the single elements. In the extreme case, when the amplitude is constant and the sum of the phases is 2π, it is possible to cancel the signal completely. As shown in Section 8.3, in such conditions this processing will provide the lower bound limit for the performance, as its SNR after integration will be statistically the same as for the single-pulse case.

In NR processing the same samples used for the MIMO case are summed coherently and their power compared with the threshold. In this case the detection decision is now given by

$$\left|\sum_{k=1}^{MN} x_k\right|^2 \underset{H_0}{\overset{H_1}{\gtrless}} \lambda_{\mathrm{NR}} \tag{8.8}$$

where λ_{NR} is a suitable threshold.

8.2.3.2 The Rephased Coherent Netted Radar (RPNR) This system again uses the same samples as the previous cases, but performs a rephasing of the vectors according to the exact position of the target in order to maximize the SNR (Fig. 8.3) and subsequently the achieved performance. In other words, the phases of the signals in Fig. 8.2 have been appropriately realigned, so the amplitude of their sum is as great as possible. This approach is extremely challenging and perhaps impossible in practice as it exploits a priori information about the exact location of the target (assumed to be point-like); however, it provides the upper bound limit for the performance as it maximizes the SNR. The decision rule in

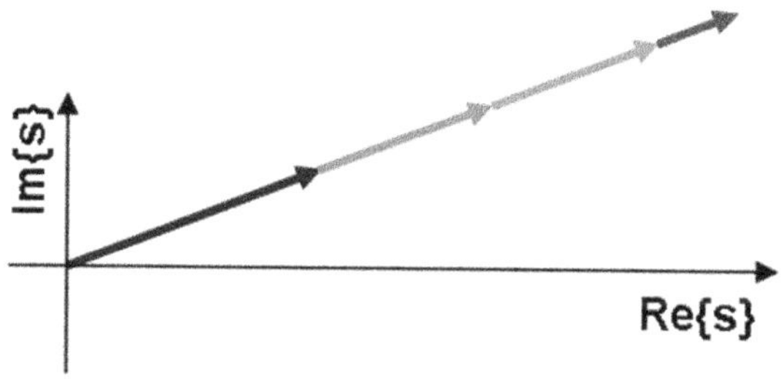

Figure 8.3 The rephased coherent netted radar integration.

this case is therefore

$$\left| \sum_{k=1}^{MN} x_k \exp\{-j\phi_k\} \right|^2 \underset{H_0}{\overset{H_1}{\gtrless}} \lambda_{\mathrm{RPNR}} \tag{8.9}$$

where ϕ_k is the phase of the desired signal only, when present. This system approach is examined as it provides a reference against which the losses of the other systems can be compared and hence may be regarded as an upper benchmark.

8.2.4 Decentralized Radar Network (DRN)

In this section a different suboptimum algorithm is applied to the same distribution of nodes in the radar network. The algorithm is characterized by a double threshold for detection. This is a two-stage approach to detection and is termed a *decentralized radar network* (DRN). The use of this form of processing technique represents an alternative to MIMO processing for radar networks. The radar network operates with the same geometry as the MIMO and NR systems. This is a suboptimum way of processing the incoming signals, where the radar network is assumed to consist of all the possible mono/bistatic radars working separately in the initial stage and where all results are subsequently fused together. The processing therefore consists of two parts: (1) detection is extracted from the signals for each of the mono/bistatic cases (i.e., in decentralized preprocessing), and (2) all the decisions are jointly fused, so the system can provide a final output.

The thresholds at all the nodes can be set to different values in order to guarantee the same FAR. After each mono/bistatic decision has been made, a vector **v**, containing only zeros (i.e., no detections) and ones (i.e., detections), is available for the second stage of processing. At this point the decision rule assumes the target to be present when L elements of the vector are set at 1. The number of L can vary according to the specifications of the system. Thus the decision rule is as follows:

$$\sum_{k=1}^{MN} v[k] \geq L \tag{8.10}$$

This concept is therefore very close to the "moving window" [51] concept often used in monostatic radar systems. However, an important difference between this and the monostatic moving-window algorithm is that samples are gathered in a space-diversity context only and consequently there is no sliding window in time. In other words, at one instant of time all the signals from the available aspect angles are collected and a final detection criterion, as given by Eq. (8.10), is then applied. The overall processing can be considered incoherent, although clearly it is very different from the MIMO case. The number of single detections L that has been shown heuristically to minimize the loss in detection for a relatively wide range of FAR is given by

$$L_{\text{opt}} \approx 1.5\sqrt{MN} \tag{8.11}$$

However, the results obtained when

$$L = L_{50} = \left\lceil \frac{MN}{2} \right\rceil \tag{8.12}$$

will be examined, where $[Q]$ is the largest integer containing Q. In other words, this criterion assumes that the target is present if at least 50% of nodes detect it in a mono/bistatic configuration. This decision rule is applied to provide an increasing jamming rejection of the overall system and is discussed in Section 8.3.3. Given the numbers of collocated transmitters and receivers, it is assumed that a target is detected when $1, 3, 5, 6$, or 8 thresholds of the single nodes are passed for the decision rule in Eq. (8.11) and when $1, 3, 5, 9$, or 13 thresholds of the single nodes are exceeded for the decision rule in Eq. (8.12), with the number of nodes varying between $1, 4, 9, 16$, and 25, respectively. The first three values of L are the same for both decision rules, indicating that the two criteria are equivalent for networks composed of a reduced number of nodes. When the radar network consists of a larger number of nodes, the two decision rules differ considerably, since there is a difference of approximately 20% in the number of nodes contributing to the detection decision. Generally speaking, an important, noteworthy difference between these processing approaches is immediately evident regarding the occupation of bandwidth for transmitting the acquired data from the nodes to the decision unit. NR processing requires the transmission of the amplitudes and the phases of the received signals, MIMO requires their power only, and the DRN requires only the single subsystem decision (i.e., binary information). Therefore, the communication constraints to transmit and receive the radar data vary widely and, of course, should also be accounted for in the design of the complete system.

8.3 PERFORMANCE

8.3.1 False-Alarm Rate (FAR)

Here we considered the case where the probability of false alarm (P_{fa}) is evaluated as a function of threshold value for the case where only white Gaussian noise with zero-mean value and normalized variance is input to the receivers.

In the systems under investigation there is a false alarm when

$$\sum_{k=1}^{MN} |n_k|^2 \geq \lambda_{\mathrm{MIMO}} \tag{8.13}$$

in the MIMO case

$$\begin{aligned} \left|\sum_{k=1}^{MN} n_k\right|^2 &= \sum_{k=1}^{MN} |n_k|^2 + \sum_{k=1}^{MN} \sum_{\substack{h=1 \\ h\neq k}}^{MN} n_k n_h^* \\ &= \sum_{k=1}^{MN} |n_k|^2 + 2\,\mathrm{Re}\left\{\sum_{k=1}^{MN-1} \sum_{h=k+1}^{MN} n_k n_h^*\right\} \geq \lambda_{\mathrm{NR}} \end{aligned} \tag{8.14}$$

in the NR and RPNR cases, and

$$\sum_{k=1}^{MN} v[k] \geq L \tag{8.15}$$

in the DRN case, where

$$v[k] = \begin{cases} 1, & |n_k|^2 \geq \lambda_{\mathrm{DRN}}(k) \\ 0, & \text{otherwise} \end{cases}$$

From a mathematical perspective, it is then clear that, even if the MIMO' and NR' noise samples have the same mean values, an extra variance has to be considered in the probability density function (pdf) of the noise power for the NR. This extra variance leads to poorer performance of the NR systems for a given FAR. The DRN cannot be compared directly since its processing provides a thresholding at the output of each single node only.

Figures 8.4–8.6 show the threshold required to achieve a chosen FAR for a numbers of processed signals. Fixing the FAR to a certain value, say, 10^{-6}, we show that both incoherent systems (Figs. 8.4 and 8.6) have a performance advantage over coherent networks (Fig. 8.5), where a lower threshold is required to achieve the same rate. The advantage achieved by incoherent systems is due to reduction in the variability in the total received signal, hence enabling a lower threshold to be set; in other words, the total noise power contributing to the detection decision is lower in

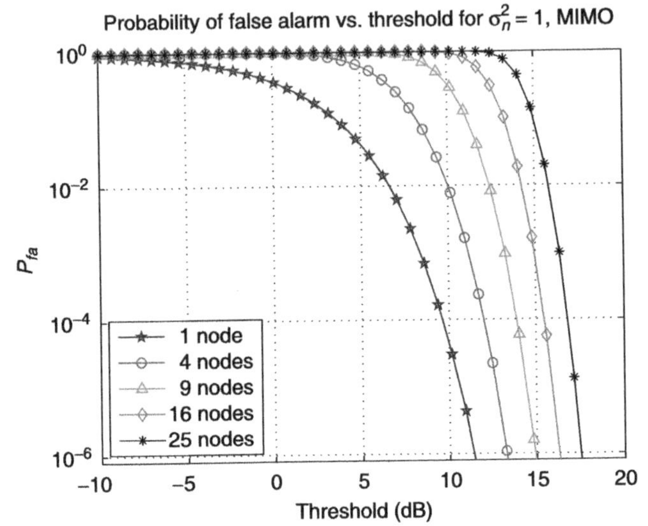

Figure 8.4 MIMO P_{fa} performance.

the incoherent systems than in the coherent ones (e.g., they require a threshold of an additional 3–8 dB to achieve a performance level equivalent to that of MIMO). The reduced threshold set with incoherent processing provides, as seen in the next section, an increased sensitivity when used for detection. It is also evident that for MIMO and NR systems, the higher the number of nodes, the higher the threshold, the opposite

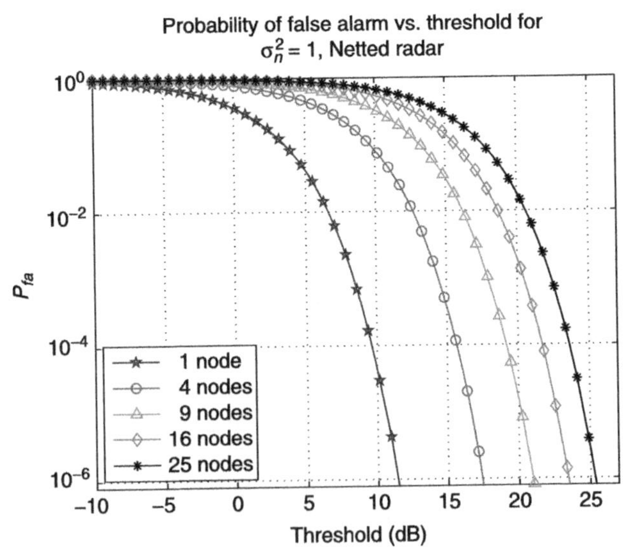

Figure 8.5 NR P_{fa} performance.

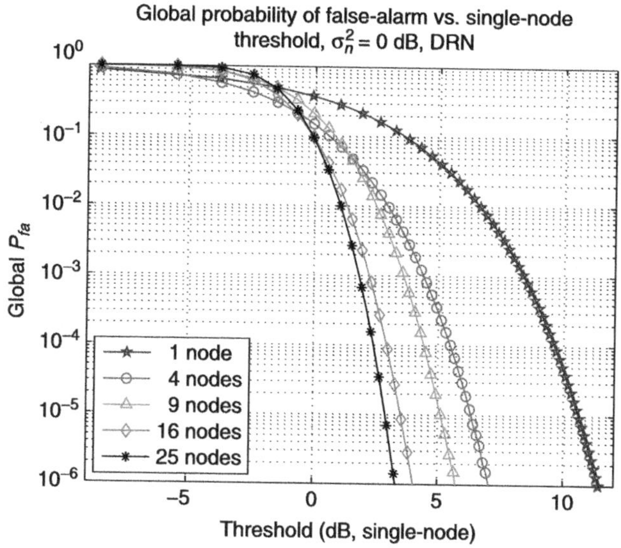

Figure 8.6 DRN P_{fa} performance (50% criterion).

occurs when using a decentralized algorithm. Note that for the decentralized case the threshold is set at the single node, while in the centralized case it is determined for the overall set of received signals.

Figures 8.7 and 8.8 show the results of the overall FAR achievable by the network of radar against the FAR of a single node, respectively, using the two criteria in

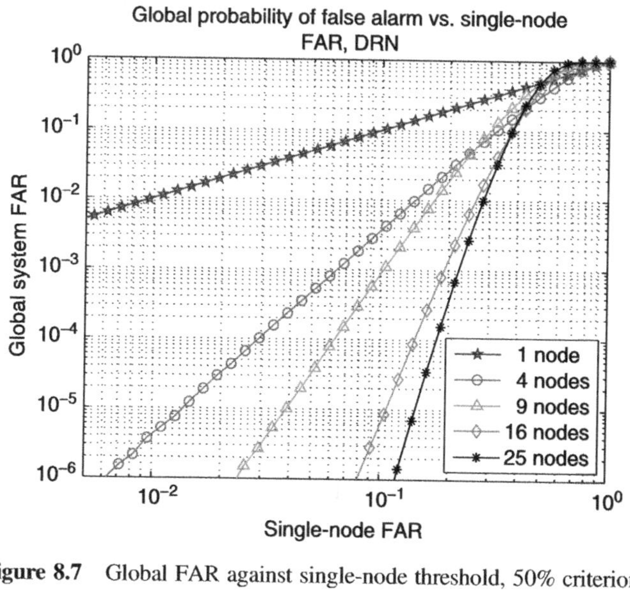

Figure 8.7 Global FAR against single-node threshold, 50% criterion.

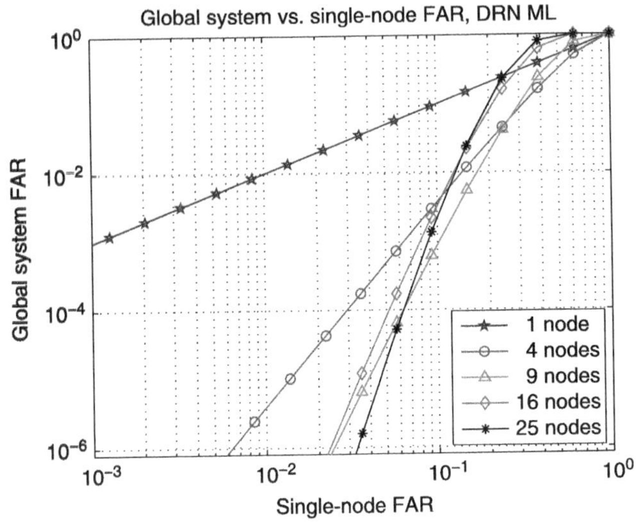

Figure 8.8 Global false-alarm rate (FAR) against single-node threshold L_{opt}.

Eq. (8.12) (i.e., the minimum loss case) and Eq. (8.11) (i.e., the 50% criterion). It is evident that for FAR $\leq 10^{-6}$, the latter case has a more pronounced separation as the number of nodes increases.

Viewing the problem from a statistical perspective, an even deeper understanding can be provided. It is convenient to proceed step by step. First, x and y are termed the I and Q (*in-phase* and *quadrature*) noises, each one independently Gaussian-distributed with mean value 0 and standard deviation σ. Consequently the pdf of n can be expressed as the joint pdf $p(x, y)$ as follows:

$$\begin{aligned} p(n) &= p(x, y) = p(x)p(y) \\ &= \frac{1}{\sqrt{2\pi\sigma^2}} \exp\left\{-\frac{x^2 + y^2}{2\sigma^2}\right\} \end{aligned} \tag{8.16}$$

This leads to a Rayleigh amplitude distribution $\rho = \sqrt{x^2 + y^2}$ with pdf:

$$p(\rho) = \frac{\rho}{\sigma^2} \exp\left\{-\frac{\rho^2}{2\sigma^2}\right\}, \quad \rho \geq 0 \tag{8.17}$$

which in power $w = \rho^2$ becomes an exponential:

$$p(w) = \frac{1}{2\sigma^2} \exp\left\{-\frac{w}{2\sigma^2}\right\}, \quad w \geq 0 \tag{8.18}$$

This expression is simplified to

$$p(t) = \exp\{-t\}u(t) \tag{8.19}$$

In other words, the following transformation is applied

$$t = \frac{w}{2\sigma^2} \tag{8.20}$$

to Eq. (8.18) and the function

$$u(t) = \begin{cases} 0, & t < 0 \\ 1, & t \geq 0 \end{cases}$$

is introduced. This can be interpreted as being equivalent to each noise having unit power.

Under this premise, when the NR processing is applied, the pdf resulting from the coherent summation of *MN* noise variables will be a complex Gaussian with variance *MN* times greater than the original one. Consequently the pdf of the resulting power can be expressed as

$$p(\hat{t}) = \frac{1}{MN}\exp\left\{-\frac{t}{MN}\right\}u(t) \tag{8.21}$$

where $\hat{t} = \left|\sum_{k=1}^{MN} n_k\right|^2$.

This expression for the pdf of the overall noise explains why the curves in Fig. 8.5 have exactly the same shape and the only difference is a shift on the x axis — they are the realization of the same pdf apart from a different variance. Furthermore, the distance, measured on the x axis, between a particular curve and the monostatic case is exactly *MN* converted to dB.

When MIMO processing is applied, the sum of the random variables is in power (i.e., not in amplitude and phase). This means that the pdf of the power of two signals, say, $\tau = |n_1|^2 + |n_2|^2$, can be expressed using Eq. (8.19) as

$$\begin{aligned} p(\tau) &= p(t) * p(t) \\ &= \int_{-\infty}^{+\infty} p(t)p(\tau - t)\,dt \\ &= \int_{-\infty}^{+\infty} \exp\{-t\}u(t)\exp\{-(\tau - t)\}u(\tau - t)\,dt \\ &= \int_0^{\tau} \exp\{-t\}\exp\{t - \tau\}\,dt \\ &= \int_0^{\tau} \exp\{-\tau\}\,dt \\ &= \tau\exp\{-\tau\}u(\tau) \end{aligned} \tag{8.22}$$

where $*$ is the convolution operator. Therefore the distribution resulting from the sum of the power of two-noise distribution is a Rayleigh. Applying this process *MN* times, the pdf of the sum of MN noise powers $\tau = \sum_{k=1}^{MN} |n_k|^2$ is distributed as follows

$$p(\tau) = \frac{\tau^{MN-1}}{(MN-1)!}\exp\{-\tau\}u(\tau) \tag{8.23}$$

that is, it has a Poisson distribution. This distribution of power can be represented as a chi-squared (χ^2) with $2MN$ degrees of freedom:

$$p(\zeta) = \frac{\zeta^{n/2-1}}{2^{n/2}\Gamma(n/2)}\exp\left\{-\frac{\zeta}{2}\right\}u(\zeta) \tag{8.24}$$

with $n = 2MN$. Introducing this in Eq. (8.24), we obtain

$$\begin{aligned} p(\zeta) &= \frac{\zeta^{MN-1}}{2^{MN}\Gamma(MN)}\exp\left\{-\frac{\zeta}{2}\right\}u(\zeta) \\ &= \frac{\zeta^{MN-1}}{2^{MN}(MN-1)!}\exp\left\{-\frac{\zeta}{2}\right\}u(\zeta) \end{aligned} \tag{8.25}$$

Finally, the two expressions in Eqs. (8.23) and (8.24) are exactly the same when the change of variables $\tau = (\zeta/2)$ is applied. This transformation is necessary in order to reduce the mean value of Eq. (8.25) from $2MN$ to MN. This is the expected value, since MN complex noise samples with unit variance (power) are summed. Comparing Eqs. (8.21) and (8.23), it is clear that asymptotically, while the distribution of the power in a MIMO system decays with an exponential coefficient, the distribution of the power in NR systems decays with the same coefficient divided by the number of nodes (i.e., more slowly). Therefore, fixing a value λ for the threshold for both systems, the NR and the MIMO probabilities of false alarm, respectively, can be expressed as

$$\begin{aligned} P_{fa_{\mathrm{NR}}} &= \int_{\lambda}^{+\infty} \frac{1}{MN}\exp\left\{-\frac{t}{MN}\right\} dt \\ &= \exp\left\{-\frac{\lambda}{MN}\right\} \end{aligned} \tag{8.26}$$

and

$$\begin{aligned} P_{fa_{\mathrm{MIMO}}} &= \int_{\lambda}^{+\infty} \frac{t^{MN-1}}{(MN-1)!}\exp\{-t\} \\ &= \exp\{-\lambda\}\sum_{k=0}^{MN-1} \cdot \frac{\lambda^k}{k!} \end{aligned} \tag{8.27}$$

Figures 8.9 and 8.10 show the distributions resulting from Eqs. (8.21) and (8.23) for 4 and 25 nodes. It is possible to discriminate the different slopes of the decay of the power and, consequently, to understand that, from a particular value of the threshold forward, the P_{fa} generated by a MIMO system is smaller than that of NR.

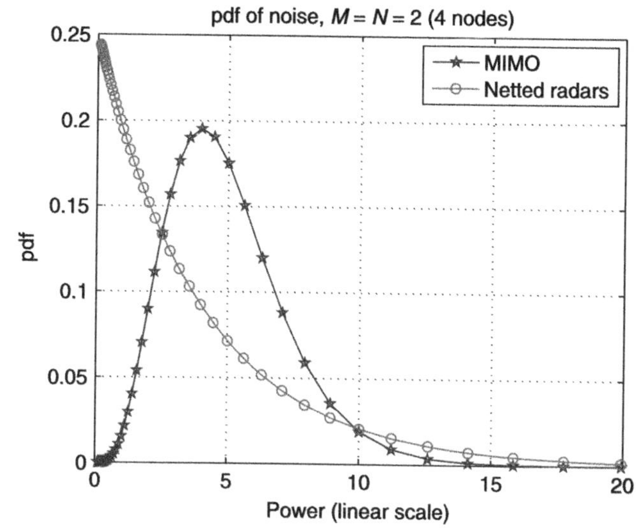

Figure 8.9 Probability density function (pdf) of the noise power in MIMO and NR, four nodes.

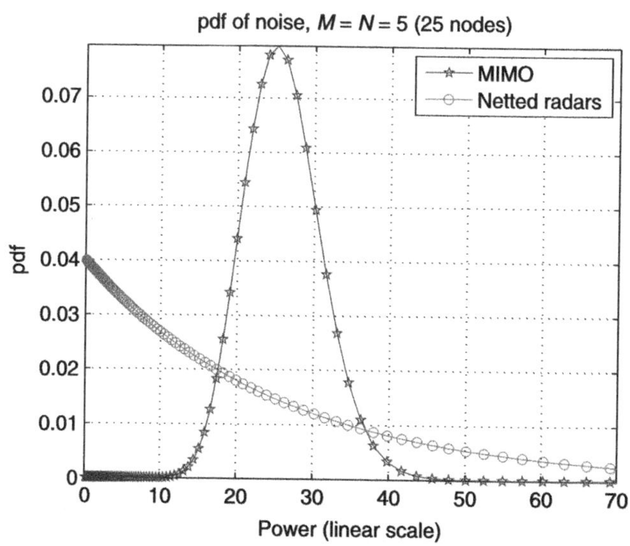

Figure 8.10 Probability density function (pdf) of the noise power in MIMO and NR, 25 nodes.

For a DRN system, from a statistical perspective, if each element of **v** (Section 8.2.4) is independent from the others and each value 1 occurs with a probability P, the overall probability that Eq. (8.10) is verified is given by the following binomial function:

$$\text{Prob}\left(\sum_{k=1}^{MN} v[k] \geq L\right) = \sum_{k=L}^{MN} \binom{MN}{k} P^k (1-P)^{MN-k} \tag{8.28}$$

where

$$\binom{h}{k} = \frac{h!}{k!(h-k)!} \tag{8.29}$$

is the binomial coefficient that factors in all the permutations of the possible positions of 1 in **v**.

In terms of radar, we can hence express the FAR of the overall system as

$$P_{fa_{\text{global}}} = \sum_{k=L}^{MN} \binom{MN}{k} P_{fa}^k (1-P_{fa})^{MN-k} \tag{8.30}$$

and the overall P_d as

$$P_{d_{\text{global}}} = \sum_{k=L}^{MN} \binom{MN}{k} P_d^k (1-P_d)^{MN-k} \tag{8.31}$$

It is worth noting that, when the FAR of each single node is relatively small (i.e., $1 - P_{fa} \approx 1$), we can rewrite Eq. (8.31) as

$$\begin{aligned} P_{fa_{\text{global}}} &= \sum_{k=L}^{MN} \binom{MN}{k} P_{fa}^k (1-P_{fa})^{MN-k} \\ &\approx \sum_{k=L}^{MN} \binom{MN}{k} P_{fa}^k \approx \binom{MN}{L} P_{fa}^L \end{aligned} \tag{8.32}$$

8.3.2 Probability of Detection (P_d)

In this section the performance of MIMO processing in detection is analyzed and compared to those of the other systems. The detection probability is computed for a number of multistatic RCS models so as to obtain a range of performances that might be expected for a variety of scenarios. The target models employed are (1) Swerling II, (2) Swerling IV, (3) Ricean, and (4) a perfectly conducting sphere. The latter model can be considered to be a Swerling 0 model for the case of relatively closely located nodes. When a only single pulse is considered, the two Swerling models become the Swerling I and III cases, respectively.

The model used for the target RCS takes into account the degree of independence in the acquired samples at differing viewing angles, which may be separated by perhaps only a few milliradians [44]. As a consequence, the target RCS at a particular receiver can differ from that at the other nodes by several decibels. This characteristic can be a significant factor in determining the resulting detection performance, which will also depend on the processing approach employed.

The results are presented in two ways: (1) maintaining the RCS model constant and therefore showing the difference in detection between the systems and the number of nodes in the systems (keeping the ERP constant) and (2) maintaining the number of nodes fixed while varying the RCS model as well as the processing approach. In all the results we keep the FAR equal to 10^{-6} for the overall system.

8.3.2.1 P_d as Function of Nodes Here a Swerling II spacetime model for the RCS of the target is assumed. The aim is to explore the effects on detection performance as a function of the number of nodes, when one to five devices are used to transmit and to receive ($M = N = 1 \ldots 5$, $M = N = 1$ corresponds to the monostatic case). The performance for all the systems is examined as a function of the SNR. Each processing approach is now considered in turn.

8.3.2.1.1 NR Processing Figures 8.11 and 8.12 show the detection performance as a function of SNR when rephased and nonrephased integration among all the nodes is performed. In the first case it is clearly evident that the greater the number of nodes, the better the detection performance, even under the bound of

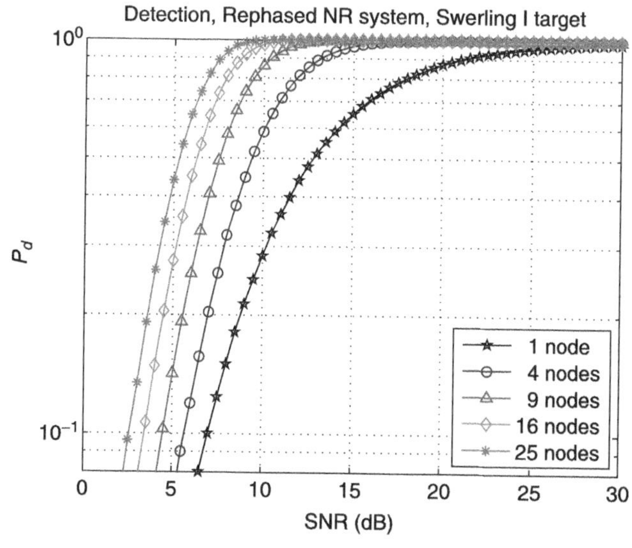

Figure 8.11 RPNR P_d against SNR, FAR = 10^{-6}.

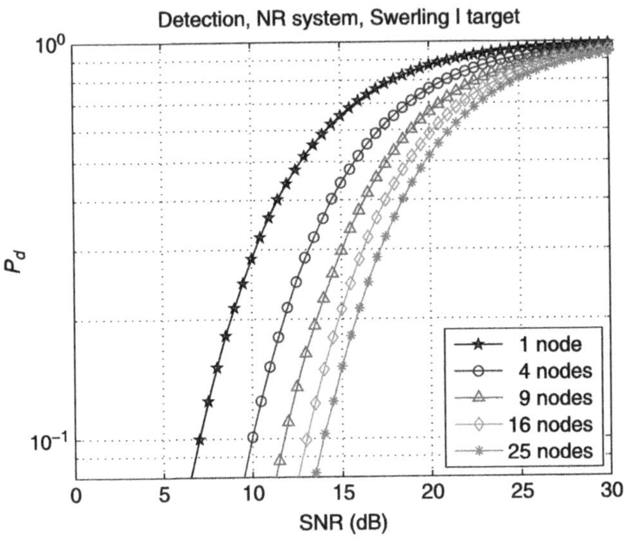

Figure 8.12 NR P_d against SNR, FAR = 10^{-6}.

constant ERP. It is equally clear that the RPNR outperforms the monostatic system. This might be expected, as energy is gathered by the system both bistatically and monostatically. As expected, the RPNR outperforms the simple NR; for example, to achieve a P_d of 80% for $M = N = 5$, a SNR of 7 dB is required in the first case and 25 dB, in the second. However, the RPNR gain tends toward saturation as the number of nodes increases. It is also worth noting that for the simple NR case, the performance decreases as the number of nodes increases. This may seem contrary to expectation, but is explained by the increased randomization of the received signal phases with a higher number of independent looks.

8.3.2.1.2 MIMO Processing Figure 8.13 shows the results for the MIMO processing approach. The MIMO approach has a performance that sits between the two NR cases. Here the losses for incoherent processing, when compared to the rephased NR, lie between approximately 0.5 and 3 dB, when the signals range between 4 and 25, respectively (for P_d = 80%). The RPNR performs better, as it maximizes the SNR; however, MIMO achieves very good results without requiring additional information about the effective position of the target to within a fraction of the wavelength, (i.e., it avoids the need for the stringent rephasing requirement, which implies additional hardware complexity). The good performance of the MIMO processing is due to the fact that it has been possible to set a lower threshold. Here, as in the rephased NR case, the more the nodes, the better the performance. However, the improvement as represented by the separation between the curves of Fig. 8.11 is smaller. Therefore, for this form of processing, it may be more cost-effective to use a reduced number of devices to keep the overall structure as simple and inexpensive as possible.

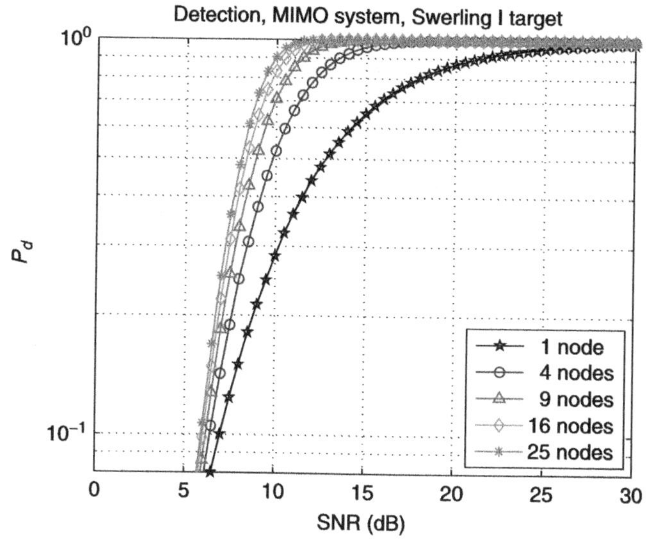

Figure 8.13 MIMO P_d against SNR, FAR = 10^{-6}.

8.3.2.1.3 DRN Processing Figures 8.14 and 8.15 show the detection performance for the DRN processing approach, where detections are extracted independently from the single nodes and fused in a second stage. Two cases are examined, one where a target is declared if 50% of the nodes make a detection and the other where the criterion is set as in Eq. (8.11). These two algorithms have almost identical detection

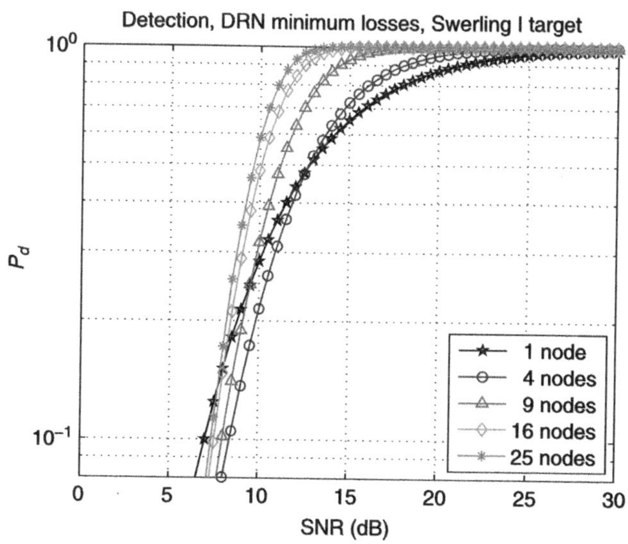

Figure 8.14 Global P_d against SNR, global FAR = 10^{-6}, "50% criterion."

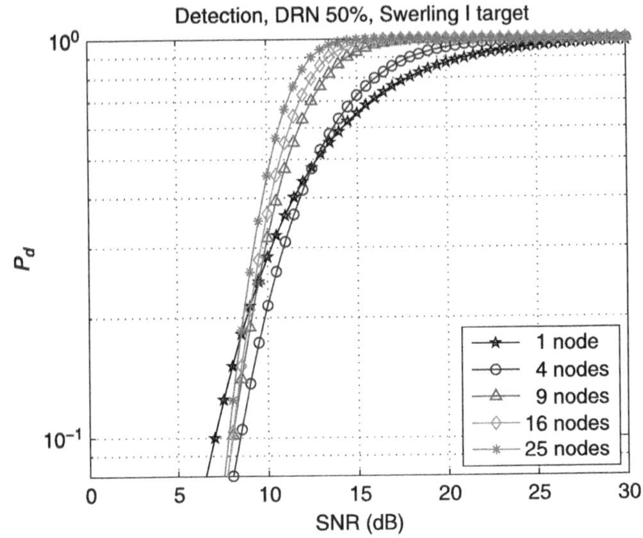

Figure 8.15 Global P_d against SNR, global FAR = 10^{-6}, "ML criterion."

performances. In fact, the "50% criterion" is approximately 1 dB below the "minimum losses," for a detection rate greater than 80%. Both show improved performance as the number of nodes in the network is increased. The slight detection loss, together with an increased tolerance to jamming and interference (as will be seen in Section 8.3.3), also make the 50% criterion an attractive and valid alternative approach. However, although still outperforming the monostatic radar, both these systems suffer greater losses due the double-threshold processing when compared to the centralized algorithms presented above. These losses can be considered as being due to distributed processing and, indeed, they could be expected, for two main reasons:

1. Since the received power is a function of the measured RCS, this system is not able to detect a target when only a few echoes are large and the others are relatively small. This is a drawback of this algorithm that does not affect centralized processing algorithms.
2. Furthermore, for relatively low SNR, while MIMO and NR systems sum up every signal incoming into the receiver, this suboptimum algorithm considers only the signals with a power greater than that of the first threshold. So, while in centralized processing every received signal contributes to the output power, here only a limited number of them are taken into account. This can be considered as a loss of power or sensitivity introduced into the second stage of processing and, consequently, a loss of performance.

8.3.2.2 P_d as a Function of RCS Models In the previous section the number of nodes was varied in order to gain an understanding of whether this improves the detection performance. In this section the performance is computed as a function of the different target models. For brevity, the result for all different numbers of nodes are not presented, but rather for the cases of 4 (i.e., 2 collocated Tx–Rx) and 25 (i.e., 5 collocated Tx–Rx). The first case represents the simplest possible and the second, the maximum number of nodes considered.

8.3.2.2.1 Swerling I–IV Targets The well-known Swerling models correspond to a noise-like RCS. Swerling I and III are models vary from scan to scan, according to exponential and χ^2 distributed RCS values, respectively. However, the RCS is almost constant within a beam dwell. The Swerling II and IV RCS models vary from pulse to pulse, according to the exponential and χ^2 distributions, respectively.

The corresponding pdf of the RCS R can be expressed as

$$p(R) = \frac{1}{\sigma}\exp\left\{-\frac{R}{\sigma}\right\} \tag{8.33}$$

for the Swerling I–II case and

$$p(R) = \frac{4R}{\sigma^2}\exp\left\{-\frac{2R}{\sigma}\right\} \tag{8.34}$$

in the Swerling III–IV cases. σ is the expected RCS values.

The Swerling I model corresponds to a fully noise-like target, while the Swerling III model corresponds to a target consisting of one dominant scatterer plus a number of smaller ones. The resulting pdf can be expressed as χ^2 distributions with 2 and 4 degrees of freedom (DOF).

Figures 8.16 and 8.17 show the probability of detection as a function of SNR for a P_{fa} of 10^{-6}. The MIMO and the DRN systems have performance in between those of the coherent extremes. The RPNR performs best, as it maximizes the SNR, but the incoherent systems achieve good results without requiring additional information about the effective position of the target. The good performance of the incoherent systems is due to the acquisition of independent samples effectively reducing the noise variance, enabling a lower threshold to be set than for the system described in Section 8.3.1. Implementation of these systems is much simpler than the coherent RPNR, due to the very tight tolerance otherwise required for the rephasing the target echo, which may not even exhibit a true single-phase center. Indeed, this may prove impossible for distributed targets. The data communication requirements are also much less severe than for the fully coherent network.

Figures 8.18 and 8.19 show results against a Swerling III target. Here it can be observed that the more spatial samples taken, the better the resulting detection performance. Again, the MIMO and the DRN results fall between the two netted cases. The difference in performance with respect to the netted cases is greater for a Swerling III than for a Swerling I target.

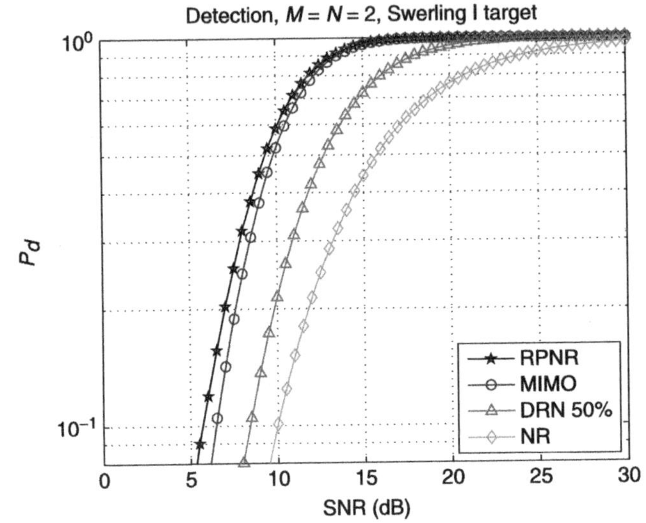

Figure 8.16 Swerling I P_d performances, four processed signals.

8.3.2.2.2 Rician Targets Both the spherical target RCS model and the Swerling I–II models can be considered as a particular realization of the Rician distribution. The Rician distribution arises as the envelope pdf of Gaussian noise with a non-zero-mean level.

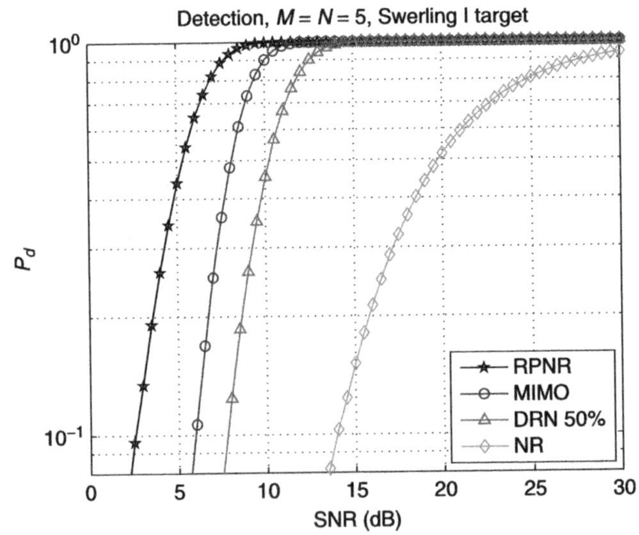

Figure 8.17 Swerling I P_d performances, 25 processed signals.

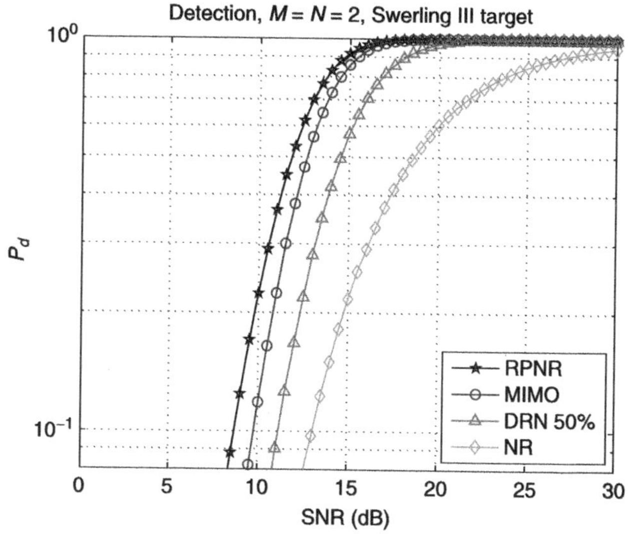

Figure 8.18 Swerling III P_d performances, four processed signals.

Under this assumption, a more realistic statistical model of the RCS of a target (such as a small point-like scatterer in noise) might be described with the pdf expressed as

$$p(R) = \frac{R}{\sigma^2}\exp\left\{-\frac{R^2+q^2}{2\sigma^2}\right\}I_0\left(\frac{Rq}{\sigma^2}\right) \tag{8.35}$$

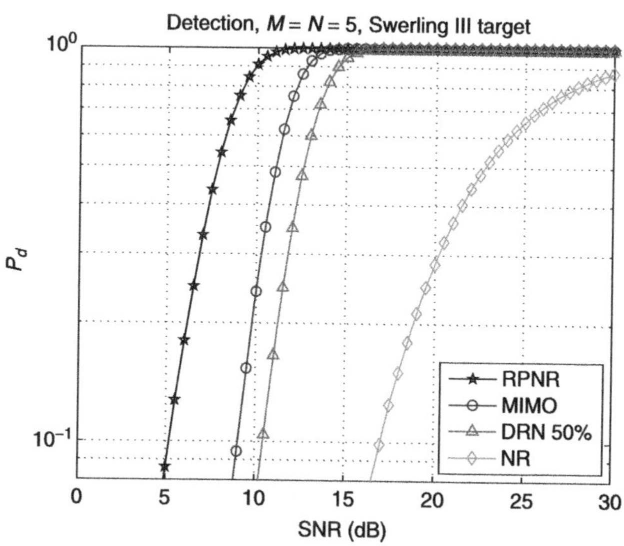

Figure 8.19 Swerling III P_d performances, 25 processed signals.

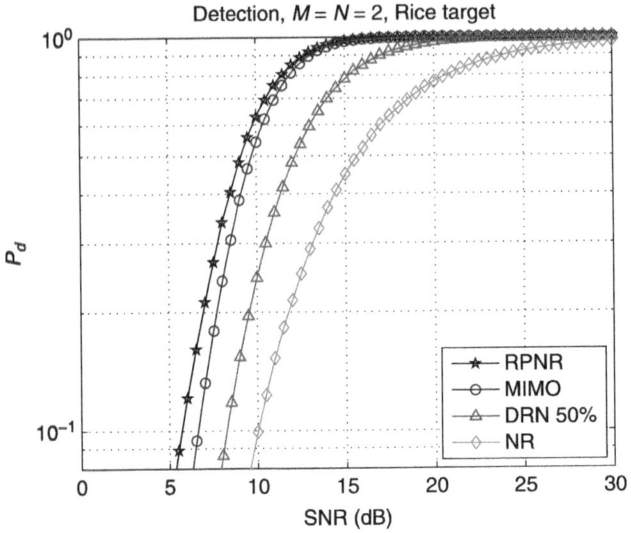

Figure 8.20 Rician P_d performances, 4 processed signals.

where q is the fixed component. Here the noise-like component of the signal is assumed to have, on average, the same power as the fixed component. Varying this ratio alters the pdf of the RCS, as well as the corresponding results, which can approach either a noise-like or a spherical model.

Figures 8.20 and 8.21 show the results for this RCS model. As might be expected, all the systems outperform the simple NR approach. In all cases the centralized

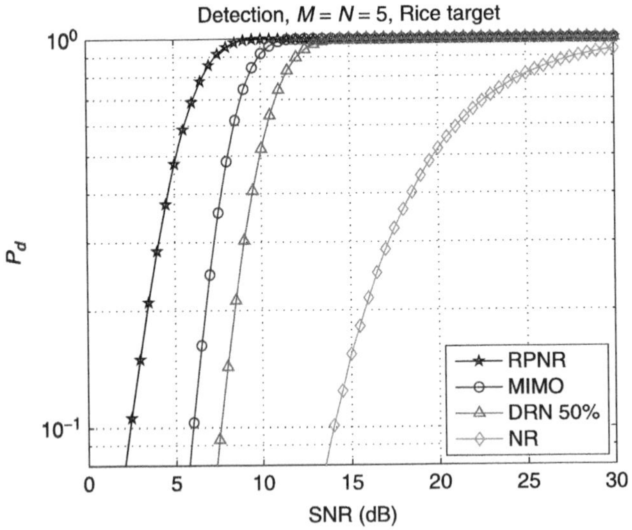

Figure 8.21 Rician P_d performances, 25 processed signals.

processing of MIMO always lies between the centralized coherent processing of the RPNR and the decentralized independent processing of the DRN.

When the network is made up of 2 nodes only, the MIMO concept approaches the upper bound limit of the RPNR. For 25 nodes, a difference between the two systems emerges, due to the higher gain achieved by the coherent (rephased) integration. However, MIMO systems are no more than 4 dB below the theoretical upper bound limit achievable with rephasing. The DRN approach (using a 50% criterion) is approximately 2–4 dB below that of MIMO, but, again, has the advantage of a lower communication bandwidth requirement.

For MIMO, the lower the number of processed signals, the lower the achieved performance. This reinforces the conviction that it is possible to improve the capacity of detection of a radar system by viewing the target from an increasing number of different angles.

However, this assumes that independent samples can always be taken. This may not be the case when considering real targets, and is partly examined by considering a spherical target.

8.3.2.2.3 Spherical Targets This is a special example of a highly correlated RCS and so provides a view of MIMO performance when the target is not noise-like. The detection performance is shown in Figs. 8.22 and 8.23. As in the previous cases, with a limited number of nodes, all the curves are quite close to one another (with the usual exception of the coherent NR). However, with 25 nodes, coherent integration provides the biggest difference between the RPNR performance and the incoherent ones. This is due to the absence of any large backscattered signal that, in MIMO

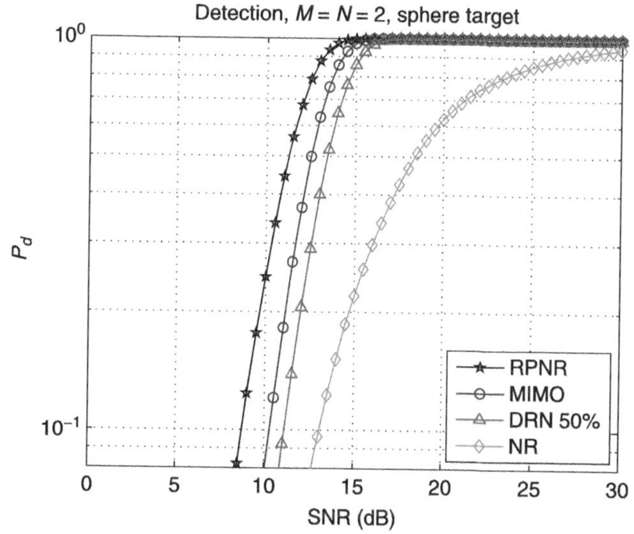

Figure 8.22 P_d performances for spherical target, four processed signals.

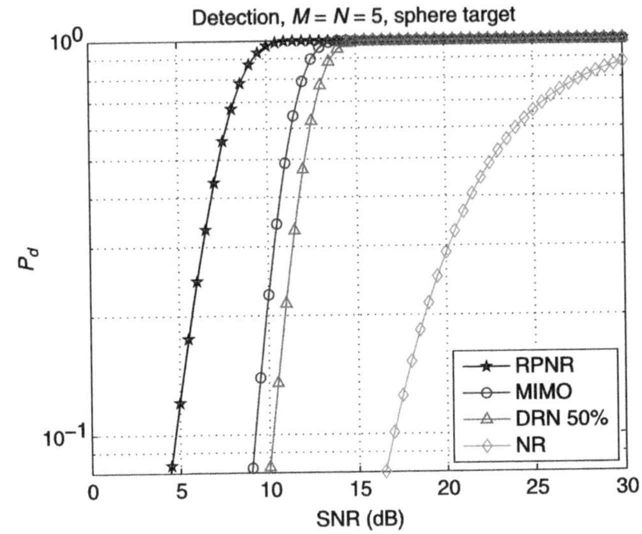

Figure 8.23 P_d performances for spherical target, 25 processed signals.

processing, allows part of the overall detection rate to be recovered. In other words, coherent integration outperforms noncoherent integration, as is well known. The peculiarity of the target model is also the cause of the almost constant distance between the curves of the two incoherent processing schemes (MIMO, DRN) at various SNR levels.

8.3.2.3 Statistics for P_d as a Function of Systems In this section a statistical description of the modification of the signals due to the different processing approach is provided. For simplicity, a multistatic Swerling II target is considered. As is well known, in the monostatic case the distribution of the received signal when a target is present and the resulting probability of detection of a Swerling II target can be expressed respectively as

$$p_{\text{mono}}(\zeta) = \frac{1}{\sigma_n^2 + \sigma_s^2} \exp\left\{ -\frac{\zeta}{\sigma_n^2 + \sigma_s^2} \right\} \tag{8.36}$$

and

$$P_{d_{\text{mono}}}(\lambda) = \int_{\lambda_{\text{mono}}}^{+\infty} \frac{1}{\sigma_n^2 + \sigma_s^2} \exp\left\{ -\frac{\zeta}{\sigma_n^2 + \sigma_s^2} \right\} d\zeta = \exp\left\{ -\frac{\lambda_{\text{mono}}}{\sigma_n^2 + \sigma_s^2} \right\} \tag{8.37}$$

assuming the noise to have variance σ_n^2 and the useful signal σ_s^2.

In the rest of this section we show the different distributions achieved from the processing. In particular, the pdf of the received signal is such that 80% P_d is achieved with thresholds that take into account the differences in processing.

8.3.2.3.1 NR Processing When NR processing is applied, the overall signal is still Gaussian with zero-mean value and variance $MN(\sigma_n^2 + \sigma_s^2)$, and therefore its distribution and probability of detection will be as in Eqs. (8.36) and (8.37) with a factor of only MN modulating the variance and a different threshold for detection.

8.3.2.3.2 RPNR Processing When a target is present, the sum of MN signals is equivalent to the sum of MN Rayleigh distributed variables. Whereas the realignment is made on the phase of the useful signals only, the sum of the noise is still white Gaussian with zero-mean value and variance $MN\sigma_n^2$. Under these assumptions, the signal to process can be expressed as consisting of two parts: (1) the sum of MN Rayleigh signals, say, $\hat{r}$; and (2) a combination of thermal noise, say, $\hat{n}$. Even if there is no closed form for the pdf of the sum of Rayleigh RV, a good approximation can be found using the Nakagami-k distribution as follows:

$$p_{\hat{r}}(\hat{r}) \approx \frac{2}{\Gamma(k)}\left(\frac{k}{\Omega}\right)^k \hat{r}^{2k-1}\exp\left\{-\frac{k}{\Omega}\hat{r}^2\right\} \tag{8.38}$$

Here the following transformation must be applied:

$$\begin{cases} k = MN, \\ \Omega = \dfrac{\pi}{2}\left[\dfrac{\Gamma(MN)}{\Gamma(MN+1/2)}\right]^2 (MN)^3\sigma_s^2 \end{cases} \tag{8.39}$$

and therefore the distribution of $\hat{r}$ can be approximated with

$$\begin{aligned} p_{\hat{r}}(\hat{r}) \approx & \frac{2}{\Gamma(MN)}\left(\frac{2}{\pi\sigma_s^2}\right)^{MN}\left[\frac{\Gamma(MN+1/2)}{MN\Gamma(MN)}\right]^{2MN} \\ & \times \hat{r}^{2MN-1}\exp\left\{-\frac{2}{\pi\sigma_s^2}\left[\frac{\Gamma(MN+1/2)}{MN\Gamma(MN)}\hat{r}\right]^2\right\} \end{aligned} \tag{8.40}$$

Apart from a phase term, it can be assumed that $\hat{r}$ is real and therefore describes the overall received signal as $\hat{r} + \hat{n} = \hat{r} + \hat{n}_R + j\hat{n}_J$, where $\hat{n}_R$ and $\hat{n}_J$ are the I and Q components of the noise. The distribution of the overall signal in power can therefore be expressed as

$$p(|\hat{r}+\hat{n}|^2) = p(\hat{r}^2 + \hat{n}_R^2 + \hat{n}_J^2 + 2\hat{r}\hat{n}_R) \tag{8.41}$$

where $p(\hat{n}_R^2 + \hat{n}_J^2)$ is the exponential pdf derived from the power of the Gaussian noise. Although there is no closed form for the entire Eq. (8.41), single pdf values can be expressed as

$$p(y = \hat{r}^2) \approx \frac{1}{\Gamma k}\left(\frac{k}{\Omega}\right)^k y^{k-1}\exp\left\{-\frac{k}{\Omega}y\right\} \tag{8.42}$$

with k and Ω as in Eq. (8.39) and finally

$$p(w = 2\hat{r}\hat{n}_R) = \frac{1}{\sqrt{2\pi}\sigma_n} \int_0^{+\infty} \frac{1}{\hat{r}} p_{\hat{r}}(\hat{r}) \exp\left\{ -\frac{w^2}{4\sigma_n^2 \hat{r}} \right\} dr \tag{8.43}$$

From Eq. (8.42) it is still possible to distinguish that the power distribution of the sum of k Rayleigh RV approaches a gamma-distributed variable.

8.3.2.3.3 MIMO Processing When MIMO processing is applied, the overall pdf is a gamma RV, and therefore its distribution and probability of detection will be

$$p_{\text{MIMO}}(\zeta) = \frac{1}{\sigma_n^2 + \sigma_s^2} \frac{1}{\Gamma(MN)} \left(\frac{\zeta}{\sigma_n^2 + \sigma_s^2} \right)^{MN-1} \exp\left\{ -\frac{\zeta}{\sigma_n^2 + \sigma_s^2} \right\} \tag{8.44}$$

and

$$\begin{aligned} P_{d_{\text{MIMO}}}(\lambda) &= \int_\lambda^{+\infty} \frac{1}{\sigma_n^2 + \sigma_s^2} \frac{1}{\Gamma(MN)} \left(\frac{\zeta}{\sigma_n^2 + \sigma_s^2} \right)^{MN-1} \exp\left\{ \frac{\zeta}{\sigma_n^2 + \sigma_s^2} \right\} d\zeta \\ &= \exp\left\{ \frac{\lambda_{\text{MIMO}}}{\sigma_n^2 + \sigma_s^2} \right\} \sum_{k=0}^{MN-1} \frac{1}{k!} \left(\frac{\lambda_{\text{MIMO}}}{\sigma_n^2 + \sigma_s^2} \right)^k = \frac{\Gamma(MN, (\lambda_{\text{MIMO}}/(\sigma_n^2 + \sigma_s^2)))}{\Gamma(MN)} \end{aligned} \tag{8.45}$$

8.3.2.3.4 DRN Processing In this case, the single received signal is still Gaussian with zero mean value and variance, with a distribution as in Eq. (8.36), and therefore its probability of detection on a single node will be as in Eq. (8.37). As in all the previous systems, here the threshold must account for the differences in processing.

8.3.2.3.5 Discussion Figures 8.24 and 8.25 show the pdf of the statistics of the power signal for all the systems when the network consists of two and five devices. Each pdf shown is reported for signal level values such that $P_d = 80\%$. As it is possible to see, the MIMO and RPNR have similar shapes even if the gain due to the fully ideal coherent processing is still offering the best scenario. In addition, DRN (on a single node) and the NR systems behave similarly.

8.3.3 Jamming Tolerance

In this section the effects on performance are considered when one of the receivers of the network is jammed with a fully matched transmission, that is, when the received

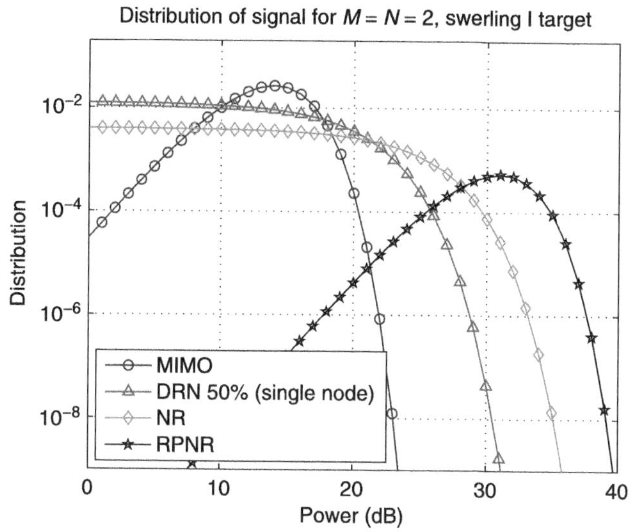

Figure 8.24 Probability density function of the useful signal for the all systems, four processed signals.

signal at the qth receiver is expressed as

$$r_q(t) = H_{0/1} \sum_{m=1}^{M} \alpha_{q,m}(\sigma) s_m\left(t - \frac{R_{m,q}}{c}\right)$$

$$+ n_q(t) + \sum_{m=1}^{M} \beta_{q,m}(\sigma) s_m(t - \tau_m) \tag{8.46}$$

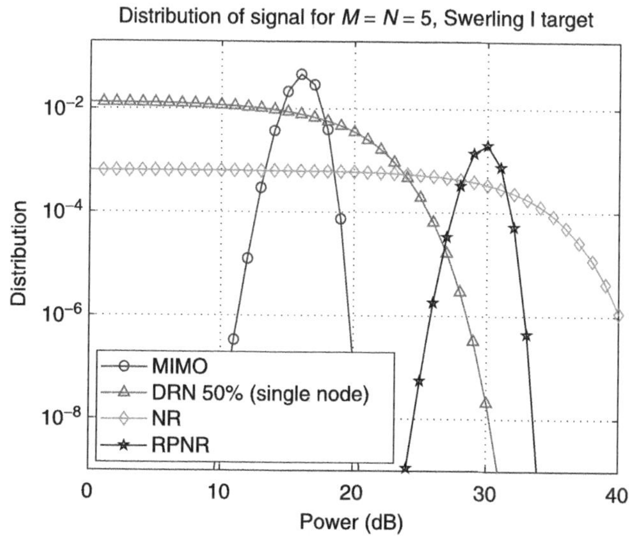

Figure 8.25 Probability density function of useful signal for all different systems, 25 processed signals.

with symbols as in Section 8.2.1 and

$$\begin{cases} |\beta|^2 \gg |\alpha_{q,m}|^2 \\ |\beta|^2 \gg \sigma_n^2 \\ |\beta|^2 \gg \lambda_{q,m} \end{cases} \tag{8.47}$$

and $\lambda_{q,m}$ is the threshold after matched filtering of for the mth waveform at the qth receiver. This represents something of a worst-case scenario but is a useful and illustrative benchmark.

Figures 8.26 and 8.27 show the FAR performance when one of the N receivers is jammed. This means that, due to the collocation of transmitters and receivers, M nodes (from one to five, respectively) are being jammed. The jammer is here assumed to be 40 dB over the noise level. No ECCM has been considered and thresholds are kept fixed at nodes, so applying any sort of ECCM algorithm will provide a higher level of tolerance. These losses are almost of the same magnitude as that of the jammer, and a reduction in performance under jamming is evident for all the cases considered. This drop is particularly severe in all the centralized processing systems, and it is necessary to increase the threshold of the jamming power to ensure the same FAR performance. In each case the detection performance will be severely degraded. However, Fig. 8.28 shows that for DRN using a 50% criterion it is sufficient to increase the (single-node) threshold of just a few decibels to recover the loss in performance. This is due to the decentralized processing where

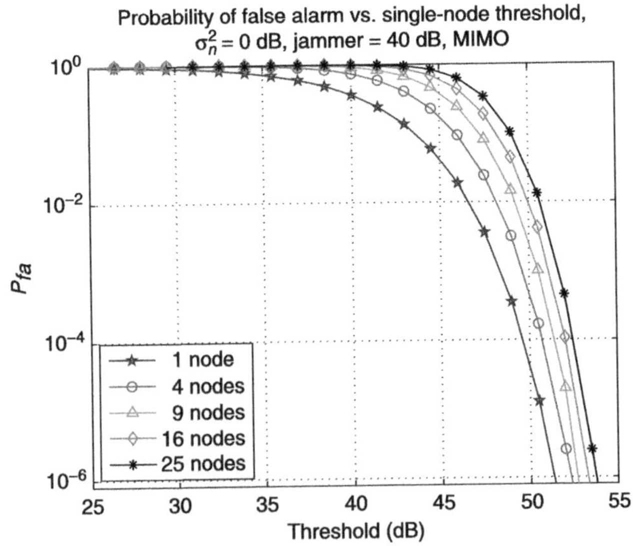

Figure 8.26 MIMO false-alarm rate (FAR), one jammed receiver.

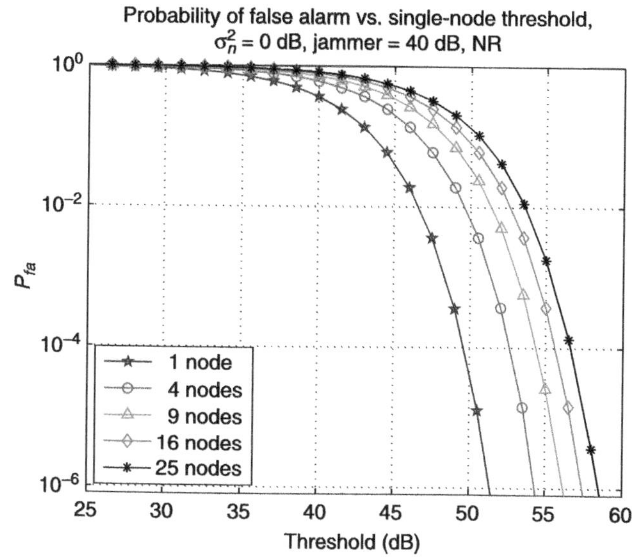

Figure 8.27 NR FAR, one jammed receiver.

the rest of the nodes of the network are in effect used to mitigate the jamming. Finally, Figs. 8.29 and 8.30, as predicted in Section 8.3.2.1, show that the minimum loss criterion performs much worse than the 50% criterion, especially in networks consisting of a large number of nodes.

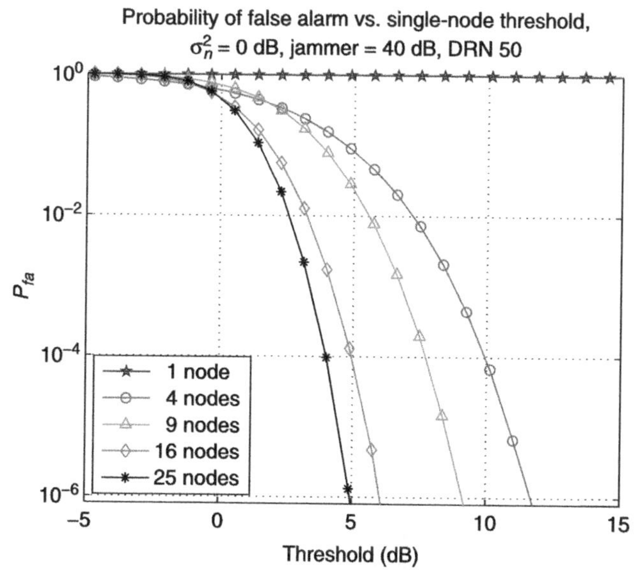

Figure 8.28 DRN FAR, one jammed receiver, "50% criterion."

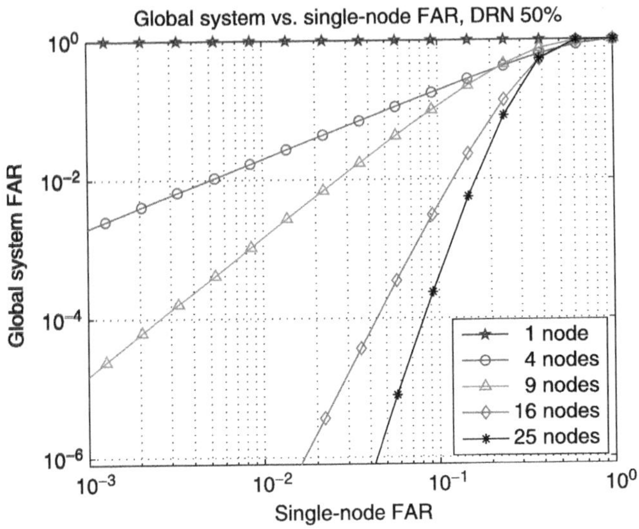

Figure 8.29 DRN global versus single-node FAR, one jammed receiver, "50% criterion."

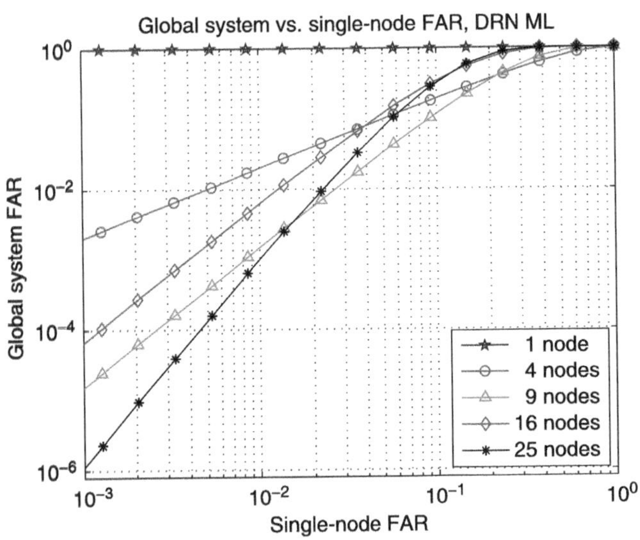

Figure 8.30 DRN global versus single-node FAR, one jammed receiver, "ML criterion."

8.3.4 Coverage

In this section the sensitivity of each processing approach and subsequent coverage is computed as a function of the number of nodes in the radar network. Therefore, the received power and the SNR levels for a target in a particular geometry can be examined. It is recalled that a constant ERP is supplied to the radar network regardless of the number of transmitters. This allows for a straightforward comparison of performance.

8.3.4.1 Sensitivity From the bistatic radar equation the power received $P_r(m, k)$ from a target at a range R_m from the transmitter and R_k from the receiver is given by

$$\mathrm{SNR}(m, k) = \frac{P_t}{M}\frac{G_T(m)G_R(k)\lambda^2\sigma(\theta_{m,k}, \phi_{m,k})}{(4\pi)^3R_m^2R_k^2LKTBF} = \frac{P_r(m, k)}{KTBF} \tag{8.48}$$

where $P_r(m,k)$ is the received power of the useful signal, K is Boltzmann's constant, T is the receiver temperature in kelvins, B is the bandwidth of the system, and F is the noise figure, and the remaining symbols are defined as in Section 8.2. This reduces to the monostatic case when $R_m = R_k$.

All the parameters with no dependence on distance, number of nodes, or RCS are grouped in the symbol $\rho_{m,k}$ (where m refers to the transmitter and k to the receiver). Thus

$$\rho_{m,k} = P_t\frac{G_T(m)G_R(k)\lambda^2}{(4\pi)^3L} \tag{8.49}$$

Therefore the dependence of the power at a single node of the network system on the range and on the number of transmitters is

$$\mathrm{SNR}(m, k) = \frac{1}{M}\frac{\rho_{m,k}\sigma(\theta_{m,k}, \phi_{m,k})}{R_m^2R_k^2}\frac{1}{KTBF} \tag{8.50}$$

In the remainder of this section, for simplicity and to allow an immediate comparison, it is assumed that $P_t = 5\,\mathrm{kW}$, $G_T(m) = G_R(k) = 30\,\mathrm{dB}$, $\lambda = 12.5\,\mathrm{cm}$, $L = 1$, $T = 290\mathrm{K}$, $F = 2$, $B = 10\,\mathrm{MHz}$, and $M = 5$ (i.e., $\rho_{m,k}$ is a constant, say, ρ_0). The length of the transmitted signal is 0.1 μs. In this analysis the peak power is intentionally kept small, compared to common values used in radar systems, so to keep the coverage moderate and therefore allow for estimation, values derived for each case discussed in Section 8.3.4.2, of the distance between the five devices. We still assume the RCS $\sigma(\theta_{m,k}, \phi_{m,k})$ to be Swerling I distributed with 0 mean value and variance $\sigma_0 = 10\,\mathrm{m}^2$.

8.3.4.2 Covered Area The coverage resulting from the differing processing approaches is evaluated as a function of the average SNR for a Swerling I target for a single pulse. A value of SNR is chosen so that the overall detection rate, for each system, is 80% with a FAR equal to 10^{-6}. This means that, given the differences in the underlying statistics of the overall incoming noise, the chosen SNR value for a single pulse at each node is different in every system. This is (approximately) 18 dB in the monostatic case, 7 dB in the RPNR case, 25 dB in the simple NR case, 9.5 dB in the MIMO system, and finally 11.75 dB in the DRN one (Fig. 8.17).

8.3.4.2.1 Monostatic Case In the monostatic case the coverage (in two-dimensional space) is circular and the maximum covered radius can be recovered by inverting Eq. (8.50) and setting $R_m = R_k = R$ and $M = 1$:

$$\max\{R\} = \sqrt[4]{\frac{\rho_0\sigma_0}{KTBF}\frac{1}{\min\{\mathrm{SNR}_{\mathrm{mono}}\}}} \tag{8.51}$$

With $\rho_0\sigma_0/KTBF = 186.9\,\mathrm{dB}\cdot\mathrm{m}^4$ and $\mathrm{SNR}_{\mathrm{mono}} \approx 18\,\mathrm{dB}$, we find that $R_{\max} \approx$ 16.7 km. Figure 8.31 shows the coverage under these assumptions.

8.3.4.2.2 RPNR Here the RPNR is examined. In this case it is convenient not to consider power as these systems operate coherently, but instead to consider the amplitude and phase of all incoming signals. The global received power, after appropriate alignment of the phases from all the signals, can be expressed as

$$P_r = \left|\sum_{m=1}^{M}\sum_{k=1}^{N}\frac{1}{R_m^2R_k^2}\sqrt{\frac{\rho_{m,k}\sigma(\theta_{m,k},\phi_{m,k})}{M}}\exp\{j\phi_0\}\right|^2 \tag{8.52}$$

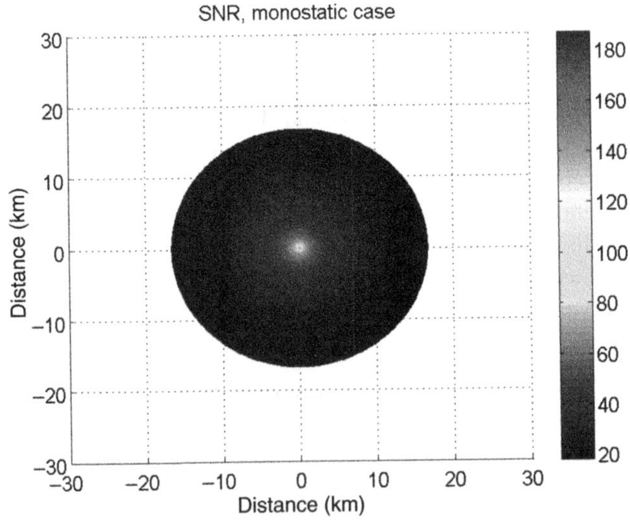

Figure 8.31 SNR and coverage, monostatic case.

where the phases of all the signals have been realigned to ϕ_0. On average and in the far field, where $R_m \approx R_k = R$, the received power can be expressed as

$$P_r \approx \frac{1}{R^4} E\left\{\left|\sum_{m=1}^{M}\sum_{k=1}^{N}\sqrt{\frac{\rho_{m,k}\sigma(\theta_{m,k},\,\phi_{m,k})}{M}}\exp\{j\phi_0\}\right|^2\right\}$$

$$= \frac{\rho_0}{MR^4} E\left\{\left|\sum_{m=1}^{M}\sum_{k=1}^{N}\sqrt{\sigma(\theta_{m,k},\,\phi_{m,k})}\right|^2\right\} \tag{8.53}$$

where $\sqrt{\sigma(\theta_{m,k},\,\phi_{m,k})}$ is Rayleigh-distributed. Therefore Eq. (8.53) can be reduced to

$$P_r \approx \frac{\rho}{MR^4} MN\left[1 + (MN-1)\frac{\pi}{4}\right]\sigma_0 \tag{8.54}$$

Computation of the average SNR after rephased coherent summation of all the signals is therefore the following equation:[1]

$$\mathrm{SNR_{RPNR}} = \frac{(\rho_0/MR^4)MN[1 + (MN-1)(\pi/4)]\sigma_0}{MNKTBF}$$

$$= \frac{\rho_0}{MR^4}\frac{\sigma_0}{MNKTBF}[1 + (MN-1)(\pi/4)] \tag{8.55}$$

where the noise power is increased by a factor MN because of the coherent summation of all the noise samples.

Here, compared to the monostatic case, there is a gain $[1 + (MN-1)(\pi/4)]$ in the global SNR. This can be approximated to $N\pi/4$ as the number of nodes increases. The coverage from the centre of each device is approximately 45 km. The increase in coverage is substantial when compared to the monostatic case, and this is due to the increased global SNR after processing and the reduced minimum SNR

[1] If x_k and y_k are Gaussian RV with 0 mean value and σ^2 and $\xi_k = |x_k + jy_k|$ is their corresponding Rayleigh-distributed envelope, the following applies:

$$E\left\{\left|\sum_{k=1}^{L}\xi_k\right|^2\right\} = E\left\{\sum_{k=1}^{L}\xi_k^2 + \sum_{k=1}^{L}\sum_{\substack{h=1\\h\neq k}}^{L}\xi_h\xi_k\right\} = E\left\{\sum_{k=1}^{L}\xi_k^2\right\} + E\left\{\sum_{k=1}^{L}\sum_{\substack{h=1\\h\neq k}}^{L}\xi_h\xi_k\right\}$$

$$= 2L\sigma^2 + \sum_{k=1}^{L}\sum_{\substack{h=1\\h\neq k}}^{L}E\{\xi_h\xi_k\} = 2L\sigma^2 + \sum_{k=1}^{L}\sum_{\substack{h=1\\h\neq k}}^{L}E\{\xi_h\}E\{\xi_k\}$$

$$= 2L\sigma^2 + \sum_{k=1}^{L}\sum_{\substack{h=1\\h\neq k}}^{L}(\pi/2)\sigma^2 = 2\sigma^2 L[1 + (L-1)(\pi/4)].$$

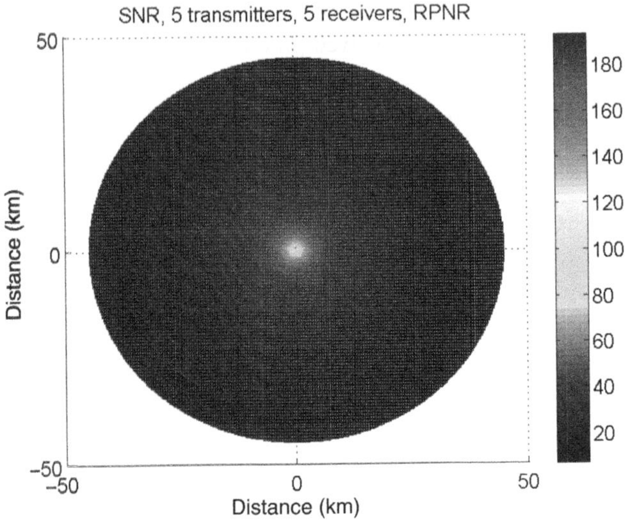

Figure 8.32 SNR and coverage, RPNR case, $d = 500\,\text{m}$.

required at each single node (from 18 to 7 dB). Again, when compared to MIMO and the DRN processing, this gives us the upper bound limit for performance. Figure 8.32 shows the coverage for this kind of radar when devices are spaced 0.5 km away from the original position in the monostatic case when the system is made up of five transmitters and five receivers (collocated devices). This network configuration will be the same for the other processing approaches. As shown, in both cases the coverage is increased and can be further improved by locating the nodes further away from each other (at least until the sensitivity limit is exceeded).

8.3.4.2.3 NR Here we discuss the simple, not rephased, NR system. For the same conditions as used previously, the received power is given by

$$P_r = \frac{1}{R^4}\left|\sum_{m=1}^{M}\sum_{k=1}^{N}\sqrt{\frac{\rho_{m,k}\sigma(\theta_{m,k},\phi_{m,k})}{M}}\exp\{j\phi_{m,k}\}\right|^2$$

$$\approx \frac{1}{M}\frac{\rho_0}{R^4}\left|\sum_{m=1}^{M}\sum_{k=1}^{N}\sqrt{\sigma(\theta_{m,k},\phi_{m,k})}\exp\{j\varphi_{m,k}\}\right|^2 \tag{8.56}$$

where $\varphi_{m,k}$ factors in the phase shift due to the path. On average, this quantity can be expressed as

$$P_r \approx \frac{1}{M}\frac{\rho_0}{R^4}\left|E\left\{\sum_{m=1}^{M}\sum_{k=1}^{N}\sqrt{\sigma(\theta_{m,k},\phi_{m,k})}\exp\{j\varphi_{m,k}\}\right\}\right|^2 \tag{8.57}$$

It is well known that the coherent sum of Gaussian distributed white RV achieves a gain equal to *MN* in power; therefore, the global SNR on the far field can be written as follows:

$$\mathrm{SNR}_{\mathrm{NR}} \approx \frac{1}{M}\frac{\rho_0}{R^4}\frac{MN\sigma_0}{MNKTBF} = \frac{1}{M}\frac{\rho_0}{R^4}\frac{\sigma_0}{KTBF} \tag{8.58}$$

As a consequence, the overall SNR in this system is of a factor *M* smaller than the monostatic case, and the threshold on the single node is, also greater (increasing from 18 to 25 dB). This explains the reduced coverage in Fig. 8.33.

8.3.4.2.4 MIMO When MIMO processing is applied, the received power is computed from the signals as

$$P_r \approx \frac{1}{M}\frac{\rho_0}{R^4} E\left\{ \sum_{m=1}^{M}\sum_{k=1}^{N} \left| \sqrt{\sigma(\theta_{m,k}, \phi_{m,k})}\ \exp\{j\varphi_{m,k}\} \right|^2 \right\} \tag{8.59}$$

On average and in the far field, this can be approximated by

$$\mathrm{SNR}_{\mathrm{MIMO}} \approx \frac{MN(1/M)(\rho_0/R^4)}{MNKTBF} = \frac{1}{M}\frac{\rho_0}{R^4}\frac{1}{KTBF} = \frac{1}{M}\mathrm{SNR}_{\mathrm{MONO}} \tag{8.60}$$

Therefore, in the MIMO case decreased SNR and hence reduced coverage might be expected when compared to the rephased case. Figure 8.34 shows the coverage for MIMO processing. Here the maximum covered distance is approximately 18 km. However, it is worth noting that in this case the coverage is

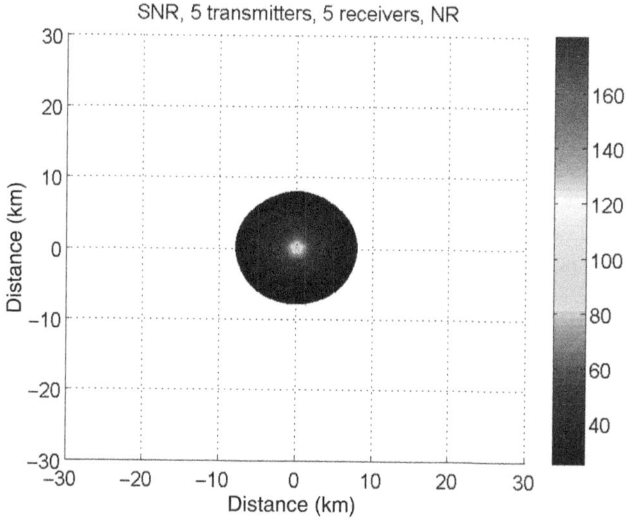

Figure 8.33 SNR and coverage, NR case, $d = 500$ m.

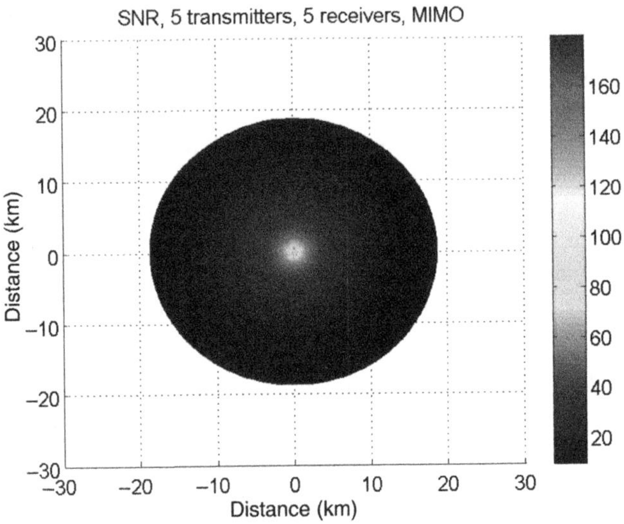

Figure 8.34 SNR and coverage, MIMO case, $d = 500$ m.

still more extensive than in the monostatic case. This reinforces the hypothesis that in real cases, when the RCS exhibits a complex multistatic behavior, the angular diversity provided by MIMO radar systems can exploit scintillation of the target to outperform a monostatic system.

8.3.4.2.5 DRN In DRN processing each node of the system operate as a single mono/bistatic radar system and in a second stage the gathered information is fused together. Under this assumption, when each device works in a monostatic configuration, it can be assumed that the SNR is M times smaller than in the monostatic case (due to the reduced transmitted power); thus the maximum area covered can be obtained from Eq. (8.51). In the far field this assumption is valid, including cases when the receivers work in a bistatic configuration.

Figure 8.35 shows that with this form of processing, the minimum SNR required at 80% of Pd is $\sim$11.75 dB, against 18 dB for the monostatic case (i.e., with a gain of $\sim$6 dB), but has a loss of M (=5, i.e., $\approx$7 dB) due to the bound on the total transmitted power. Therefore, the actual loss of this system in SNR, compared to the monostatic case, is approximately 0.75 dB, and the subsequent loss on the covered distance is $\approx$4.2%. However, this loss in coverage could be fully reversed by spacing the devices in a more optimum fashion. Indeed, in realistic scenarios this will be dictated by the terrain.

Again it is observed that the NR processing approach suffers because of the randomization of the phase, leading to losses. By correcting for this in the rephased case, the best coverage is achieved by far. It is interesting to note that MIMO provides only a small improvement over the monostatic case and that DRN is almost equivalent. Thus the real advantage of MIMO is in more reliable detection rather than in outright sensitivity.

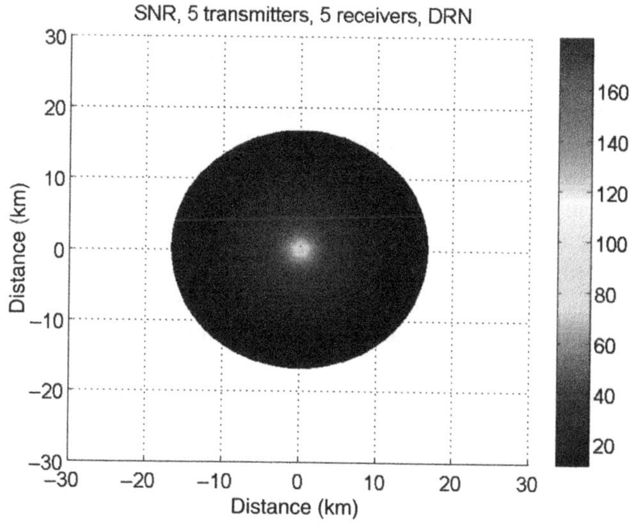

Figure 8.35 SNR and coverage, DRN case, $d = 500$ m.

8.4 CONCLUSIONS

In this chapter the properties and the characteristics of four processing approaches using different algorithms for data processing in a radar network have been examined. The processing approaches exhibit considerable variability in performance with respect to FAR, detection, jamming tolerance, and coverage.

Incoherent systems have an advantage over coherent systems as they reduce the overall power of noise after processing due to the beneficial effect they have on the statistical distribution of the noise model chosen for clutter. The decentralized radar networks show increased tolerance to jamming without any of the countermeasures commonly adopted. These systems have also an advantage over the centralized systems since they require a smaller bandwidth for transmitting the information to the decision unit, and this is an important implementation consideration.

It has been shown that the performance of radar networks can be superior to that of the monostatic case, when compared on a fair basis of constant transmit power for a variety of processing approaches. In addition, these processing approaches have their own varying performance levels. For example, when working incoherently, they can outperform a simple coherent approach. This is because the phases of the incoming signals are totally uncorrelated and hence the coherent sum is a statistically disruptive event. This is true not only for noise-like targets, such as the Swerling models, but also for targets with a constant amplitude and random phase, as in the case of a sphere. In addition, they perform at a level that is a few decibels poorer than the theoretical upper bound limit and therefore are a valid alternative for implementation in real systems. They are also attractive as they have simpler hardware and processing requirements.

In RPNR processing, the extra complexity required for aligning the phases of the signal is potentially extremely demanding, especially when the systems are working over extensive frequency bandwidths and have high carrier frequencies. Indeed, if the target is extended, it may not conform to the concept of the existence of an identifiable single phase center. This is in contrast to both MIMO and DRN which have a relatively simple structure for the detection architecture. The favorable performance of these incoherent systems may also be attributed to the modification of the PDF of the noise in the decision stage of the processing. The PDF is narrowed, allowing a lower threshold to be set and, as a consequence, recovery of part of the losses due to suboptimal processing. In other words, incoherent processing affords a better estimate of the mean noise level, rather in the same way that multiple views in SAR imagery reduce speckle.

It has also shown that RPNR systems have the highest sensitivity and hence largest coverage area. Thus, once again, this processing approach can be considered the upper bound performance limit. However, MIMO and DRN radar systems can cover an area comparable to that of the monostatic case. This area can be further expanded or modified with an appropriate choice of the location of each sensor within the network. The distribution of nodes in the network offers greater flexibility and hence tailored coverage. However, there are various ways to process the same set of signals from a radar network. Each has advantages and disadvantages.

Coherent algorithms perform at the extreme of the possible range. Coherent NR is the easiest to implement, but it underperforms the incoherent systems and even an equivalent monostatic one. The RPNR has the best performance and has the best coverage but requires processing the phase to an accuracy smaller than a wavelength. In addition, NR systems require a considerable communication channel capacity to send data to a centralized processor, as both amplitude and phase have to be transferred.

MIMO processing requires the transmission of only the amplitude (or power) information. It performs close to the upper bound limit but is much easier to implement as there is no rephasing. Its coverage can be considerably larger than that of an equivalent monostatic radar, although its (and the NR's) centralized processing has been shown to be potentially vulnerable to jamming.

Finally, the DRN systems have been shown to have a superior tolerance to directional jamming. Additionally, the requirement that they impose on the communication channel is greatly reduced as only detections have to be passed and they are easiest to implement. In effect, they consist of mono/bistatic independent systems, but have extra losses in detection when compared to the MIMO case.

It is also possible to conceive of systems that jointly process data using all the approaches described and subsequently combine their results to more closely approach optimum performance. This could be attractive where both target and clutter properties are varying in both time and space. Finally, it is worth highlighting that the gain achieved in detection grows rapidly when extra nodes are added but then tends to saturate. Therefore it is possible to obtain the biggest increase in performance simply by adding a relatively small number of nodes.

ACKNOWLEDGMENT

The authors gratefully acknowledge the sponsorship of the Air Force Office of Scientific Research, Air Force Materiel Command, USAF.

REFERENCES

1. E. Fishler, A. Hairmovich, R. Blum, D. Chizhik, L. Cimini, and R. Valenzuela, MIMO radar: An idea whose time has come, *Proc. IEEE Radar Conf. 2004*, Philadelphia, April 2004, pp. 71–78.
2. E. Fishler, A. Hairmovich, R. Blum, L. Cimini, D. Chizhik, and R. Valenzuela, Performance of MIMO radar systems: Advantages of angular diversity, *Conf. Record 38th Asilomar Conf. Signals, Systems and Computers*, Pacific Grove, CA, Nov. 2004, Vol. 1, pp. 305–309.
3. E. Fishler, A. Hairmovich, R. Blum, L. Cimini, D. Chizhik, and R. Valenzuela, Spatial diversity in radars — models and detection performance, *IEEE Trans. Signal Process.*, **54**(3):823–838 (March 2006).
4. C. J. Baker and L. Hume, Netted radar sensing, *IEEE Aerospace Electron. Syst. Mag.* **18**(2):3–6 (Feb. 2003).
5. V. S. Chernyak, Effective simplified decentralized target detection in multisensor systems, *Proc. 3rd Int. Conf. Information Fusion, 2000* (FUSION 2000). July 10–13, 2000, Vol. 2.
6. R. S. Blum, Distributed detection of narrowband signals, *IEEE Trans. Inform. Theory* **41**(2):519–523 (March 1995).
7. R. Viswanathan and P. K. Varshney, Distributed detection with multiple sensors. I: Fundamentals, *IEEE Proc.* **85**(1):54–63 (Jan. 1997).
8. H. D. Griffiths, From a different perspective: Principles, practice and potential of bistatic radar, *Proc. Int. Radar Conf.*, Sept. 3–5, 2003, pp. 1–7.
9. K. Milne, Principles and concepts of multistatic surveillance radars, *Proc. IEE Int. Radar Conf.*, IEE, London, Oct. 1977.
10. J. Li and P. Stoica, MIMO radar — where diversity enables superiority, *Proc. 12th Annual Workshop on Adaptive Sensor Array Processing*, Lexington, MA, June 2006.
11. L. Xu, J. Li, and P. Stoica, Adaptive techniques for MIMO radar, *Proc. 4th IEEE Sensor Array and Multichannel Signal Processing Workshop*, Waltham, MA, July 2006, pp. 258–262.
12. L. Xu, J. Li, and P. Stoica, Radar imaging via adaptive MIMO techniques, *Proc. 14th European Signal Processing Conf. (EUSIPCO)*, Florence, Italy, Sept. 2006.
13. J. Li, P. Stoica, and Y. Xie, On probing signal design for MIMO radar, *Proc. 40th Asilomar Conf. Signals, Systems, and Computers*, Pacific Grove, CA, Oct. 2006.
14. L. Xu, J. Li, P. Stoica, K. W. Forsythe, and D. W. Bliss, Waveform optimization for MIMO radar: A Cramer-Rao bound based study, *Proc. 2007 IEEE Int. Conf. Acoustics, Speech, and Signal Processing*, Honolulu, April 2007.
15. P. Stoica, J. Li, X. Zhu, and B. Guo, Waveform synthesis for diversity-based transmit beampattern design, *Proc. 14th IEEE Workshop on Statistical Signal Processing*, Madison, WI, Aug. 2007.

16. P. Stoica, J. Li, and Y. Xie, On probing signal design for MIMO radar, *IEEE Trans. Signal Process.* (Aug. 2007).
17. J. Li and P. Stoica, MIMO radar with colocated antennas: Review of some recent work, *IEEE Signal Process. Mag.* **24**(5):106–114 (2007).
18. J. Li, P. Stoica, and X. Zheng, MIMO SAR imaging — signal synthesis and receiver design, *Proc. CAMSAP 2007*, US Virgin Islands, Dec. 12–14, 2007.
19. J. Li, P. Stoica, X. Luzhou, and W. Roberts, On parameter identifiability of MIMO radar, *IEEE Signal Process. Lett.* **14**(12):968–971 (Dec. 2007).
20. K. W Forsythe, D. W. Bliss, and G. S. Fawcett, Multiple-input multiple-output (MIMO) radar: Performance issues, *Conf. Record 38th Asilomar Conf. Signals, Systems and Computers*, 2004, Vol. 1, pp. 310–315.
21. G. San Antonio and D. R. Fuhrmann, Beampattern synthesis for wideband MIMO radar systems, *Proc. 1st IEEE Int. Workshop on Computational Advances in Multi-Sensor Adaptive Processing,* Dec. 13–15, 2005, pp. 105–108.
22. F. C. Robey, S. Coutts, D. Weikle, J. C. McHarg, and K. Cuomo, MIMO radar theory and experimental results, *Conf. Record 38th Asilomar Conf. Signals, Systems and Computers*, 2004, Nov. 7–10, 2004, pp. 300–304.
23. H. A. Khan, W. Q. Malik, D. J. Edwards, and C. J. Stevens, Ultra wideband multiple-input multiple-output radar, *Proc. IEE Seminar on Ultra Wideband Communications Technologies and System Design*, 2004, July 8, 2004, pp. 111–118.
24. H. A. Khan and D. J. Edwards, Doppler problems in orthogonal MIMO radars, *Proc. 2006 IEEE Conf. Radar*, April 24–27, 2006.
25. H. A. Khan, Y. Zhang, C. Ji, C. J. Stevens, D. J. Edwards, and D. O'Brien, Optimizing polyphase sequences for orthogonal netted radar, *IEEE Signal Process. Lett.* **13**(10):589–592 (Oct. 2006).
26. D. R. Fuhrmann and G. San Antonio, Transmit beamforming for MIMO radar systems using partial signal correlation, *Conf. Record 38th Asilomar Conf. Signals, Systems and Computers*, 2004, Vol. 1, pp. 295–299.
27. Y. U. Lee, J. Choi, I. Song, and S. R. Lee, Distributed source modelling and direction-of-arrival estimation techniques, *IEEE Trans. Signal Process.* **45**(4):960–969 (April 1997).
28. A. De Maio and M. Lops, Design principles of MIMO radar detectors, *Proc. 2nd Int. Waveform Diversity and Design Conf.* Lihue, HI, Jan. 22–27, 2006.
29. P. Bidigare, MIMO capacity of radar as a communication channel, *Adaptive Sensor and Array Processing Workshop*, March 11–13, 2003.
30. D. J. Rabideau and P. Parker, Ubiquitous MIMO multifunction digital array radar, *Conf. Record 37th Asilomar Conf. Signals, Systems and Computers*, Nov. 9–12, 2003, Vol. 1, pp. 1057–1064.
31. D. W. Bliss and K. W. Forsythe, Multiple–input multiple–output (MIMO) radar and imaging: Degrees of freedom and resolution, *Conf. Record. 38th Asilom Conf. Signals, Systems and Computers*, 2003 Nov. 9–12, 2003, pp. 54–59.
32. K. I. Ziricastro, W. G. Scanlon, and N. E. Evans, Prediction of variation in MIMO channel capacity for the populated indoor environment using a radar cross-section-based pedestrian model, *IEEE Trans. Wireless Commun.* **4**(3):1186–1194 (May 2005).
33. J. S. Jiang and M. A. Ingram, Distributed source model for short range MIMO, *Proc. 58th IEEE Conf. Vehicular Technology*, 2003 (VTC 2003), Oct. 6–9, 2003, Vol. 1, pp. 357–362.

34. R. Venkataramani and Y. Bresler, Multiple–input multiple–output sampling: Necessary density conditions, *IEEE Trans. Inform. Theory.* **50**(8):1754–1768 (Aug. 2004).

35. H. Xu, D. Chizhik, H. Huang, and R. Valenzuela, A generalized space–time multiple–input multiple–output (MIMO) channel model, *IEEE Trans. Wireless Commun.* **3**(3):966–975 (May 2004).

36. G. Latsoudas and N. D. Sidiropoulos, A hybrid probabilistic data association — sphere decoding detector for multiple–input multiple–output systems, *IEEE Signal Process. Lett.* **12**(4):309–312 (April 2005).

37. A. Zanella, M. Chiani, and M. Z. Win, MMSE reception and successive interference cancellation for MIMO systems with high spectral efficiency, *IEEE Trans. Wireless Commun.*, **4**(3):1244–1253 (May 2005).

38. P. Uthansakul and M. E. Bialkowski, Multipath signal effect on the capacity of MIMO, MIMO–OFDM and spread MIMO–OFDM, *Proc. 15th Int. Conf. Microwaves, Radar and Wireless Communications*, 2004 (MIKON 2004), May 17–19, 2004, Vol. 3, pp. 989–992.

39. A. B. Gershman and N. D. Sidiropoulos, *Space–Time Processing for MIMO Communications*, Wiley, 2005.

40. L. B. White and P. S. Ray, Signal design for MIMO diversity systems, *Conf. Record 38th Asilomar Conf. Signals, Systems and Computers*, 2004, Nov. 7–10, 2004, Vol. 1, pp. 973–977.

41. W. Yu and J. M. Cioffi, Sum capacity of a gaussian vector broadcast channel, *IEEE Trans. Inform. Theory*, **50**(9):1875–1892 (Sept. 2004).

42. D. J. MacKay, *Information Theory, Inference and Learning Algorithms*, Cambridge University Press, 2003.

43. N. Levanon and E. Mozeson, *Radar Signals*, Wiley, Hoboken, NJ, 2004.

44. M. I. Skolnik, *Introduction to Radar Systems*, McGraw-Hill, New York, 1981.

45. G. Picardi, *Elaborazione del Segnale Radar*, Franco Angeli, 2000.

46. A. Farina, *Antenna-Based Signal Processing Techniques for Radar Systems*, Artech House, Norwood, MA, Jan. 1992.

47. E. J. Kelly, An adaptive detection algorithm, *IEEE Trans. Aerospace Electron. Syst.* **AES–22** (March 1986).

48. E. J. Kelly, Performance of an adaptive detection algorithm; rejection of unwanted signals, *IEEE Trans. Aerospace Electron. Syst.* **25**(2):122–133 (March 1989).

49. F. C. Robey, D. R. Fuhrmann, E. J. Kelly, and R. Nitzberg, A CFAR adaptive matched filter detector, *IEEE Trans. Aerospace Electron. Syst.* **28**(1):208–216 (Jan. 1992).

50. P. Monticciolo, E. J. Kelly, and J. G. Proakis, A noncoherent adaptive detection technique, *IEEE Trans. Aerospace Electron. Syst.* **28**(1):115–124 (Jan. 1992).

51. M. Schwartz, A coincidence procedure for signal detection, *IEEE Trans. Inform. Theory* (Dec. 1956).

9

CONCEPTS AND APPLICATIONS OF A MIMO RADAR SYSTEM WITH WIDELY SEPARATED ANTENNAS

HANA GODRICH AND ALEXANDER M. HAIMOVICH

Department of Electrical and Computer Engineering, New Jersey Institute of Technology, Newark, New Jersey

RICK S. BLUM

Department of Electrical and Computer Engineering, Lehigh University, Bethlehem, Pennsylvania

The term *multiple-input multiple-output* (MIMO) radar refers to an architecture that employs multiple, spatially distributed transmitters and receivers. In a general sense, while MIMO radar can be viewed as a type of multistatic radar, the separate nomenclature suggests unique features that set MIMO radar apart from the multistatic radar and are closely related to MIMO communications. This chapter reviews some more recent work on MIMO radar with widely separated antennas.

9.1 BACKGROUND

Active arrays, applied to radar systems, have been a topic of intensive research, and are well documented in the literature from the system implementation perspective (see survey of phased-array systems in Ref. 1), as well as processing techniques for target detection and parameter estimation [2]. Target parameters of interest in radar systems include target strength, location, and Doppler characteristics.

MIMO Radar Signal Processing, edited by Jian Li and Petre Stoica

Phased-array radars with digital beamforming at the receiver have the ability to steer multiple, simultaneous beams (see, e.g., Refs. 2–4). Adaptive array radars process the signals received at the array elements in order to optimize some performance figure of merit, such as signal-to-interference ratio [5]. In airborne and other applications, the detection of moving targets and their discrimination against the background clutter are of great interest; this led to the development of array radars with spacetime adaptive processing (STAP) [6–9]. In addition to spatial (or spacetime) beamforming, signals received at an array can be processed to yield high-resolution estimates of angle of arrivals of radar targets, such as MUSIC or maximum likelihood (ML) [2].

Phased arrays with multiple transmit elements are capable of cohering and steering the transmitted energy [4]. Elements of phased-array radars are typically colocated, at both the transmitter and receiver ends. Multiple transmit antennas suitably placed may be configured to operate in multistatic mode. Typically, a multistatic radar is a system that networks multiple, independent radars. Each radar performs a significant amount of local processing. Outcomes of the local processing are delivered to a central processor through a communication link [10]. For example, individual radars of a multistatic system perform local detection decisions, leaving the central processor the task of fusing these decisions.

In the 1990s, a new and interesting concept, employing multiple transmit elements, was introduced — synthetic impulse and aperture radar (SIAR) [11]. The SIAR transmitting array emits orthogonal waveforms. This property ensures that a receiver that processes the aggregate of the transmitted waveforms is capable of separating and processing the waveforms individually. Through synthetic pulse formation, SIAR achieves the advantages of wideband radar (improved range resolution) but with individual antennas transmitting only narrowband waveforms. Unlike conventional beamformers, SIAR features isotropic radiation (an advantage in terms of the probability of intercept of the radar waveform by a third party). This concept involves a higher degree of cooperation among the radar sites than in typical multistatic radar systems. As such, it has parallels with multiple-input multiple-output (MIMO) systems in wireless communications, and subsequent publications have adopted the "MIMO radar" nomenclature for radar systems with cooperating transmitters [12–15].

In this chapter, a MIMO radar is defined broadly as a radar system employing multiple transmit waveforms and capable of jointly processing signals received at multiple receive antennas. Elements of MIMO radar transmit independent waveforms resulting in an omnidirectional beampattern or create diverse beampatterns by controlling correlations among transmitted waveforms [16]. A MIMO radar may be configured with its antennas collocated or widely distributed over an area. MIMO radar is an emerging concept, but it is already apparent that it has the potential to make important contributions to the radar field. Bliss and Forsythe [13] observed that MIMO radar has more degrees of freedom than do systems with a single transmit antenna. These additional degrees of freedom support flexible time–energy management modes [14], lead to improved angular resolution [17,18], and improve parameter identifiability [19]. With widely separated antennas, MIMO radar has the ability to improve radar performance by exploiting radar cross section (RCS) diversity [20], handle slow-moving targets by exploiting Doppler estimates from multiple directions [21], and support high-resolution target location [22].

From a model perspective, widely separated antennas take advantage of the spatial properties of extended targets, while with collocated antennas, the target is modeled as a point with no spatial properties. Each configuration and model has its strengths and challenges. With collocated antennas, MIMO radars can mimic beamformers utilizing low-probability-of-intercept (LPI) waveforms. Rather than focusing energy on a target, the transmitted energy is evenly distributed in space [12]. The loss of processing gain compared to conventional phased-array beamforming due to the uniform illumination is compensated by the gain in time, since a narrow beam does not need to be scanned [15]. MIMO spatial signatures in systems with collocated antennas can be exploited to estimate radar parameters (location, RCS) of multiple targets, but the ability to resolve targets is limited by the Rayleigh resolution limit of the transmit/receive arrays. Widely separated antennas support high-resolution target location, and by focusing on the effect of the channel, create a link between MIMO radar and MIMO communications.

MIMO systems have led to a revolution in wireless communications [23]. More recent publications (e.g., Ref. 20 and references cited therein) indicate that one can exploit similar ideas in radar, suggesting interesting cross-fertilization of ideas between MIMO communications and MIMO radar. For quite some time, it has been understood that radar targets provide a rich scattering environment yielding 5–20-dB target RCS fluctuations, as illustrated in Fig. 9.1. Such targets display essentially independent scattering returns when radiated from sufficiently different directions (see the paper by Fishler et al. [20] for a mathematical illustration of this

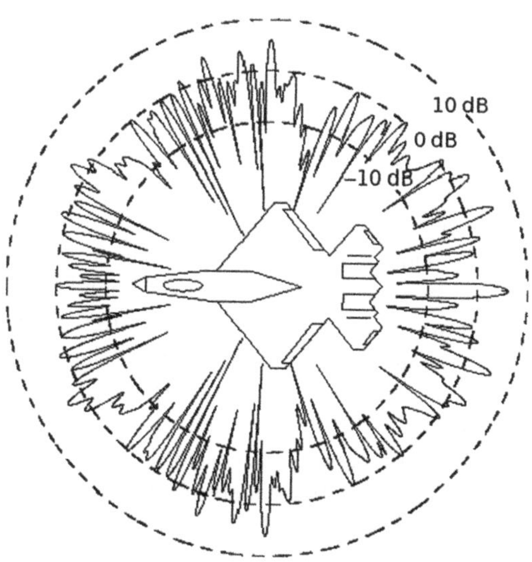

Figure 9.1 Backscatter as function of azimuth.

using a reasonable model). The premise of MIMO radar with widely separated antennas is that angular spread (RCS variations as a function of aspect) can be exploited to improve radar performance in various ways. The parallel to MIMO communication is recognized in the similar roles that the transmission medium (channel) and target play in communication and radar, respectively. In other words, the target serves as the "channel" in the radar problem. For example, combining target returns resulting from independent illuminations yields a *diversity gain* akin to the diversity gain obtained in the communication problem over fading channels when the data are transmitted through independent channels. Diversity gains are well understood in communications [24]. In radar, the idea is that any individual "look" at (view of) the target might have a small-amplitude return with a significant probability, but by increasing the number of looks, the probability that all the looks have small-amplitude returns can be made arbitrarily small. In Ref. 25 and some other publications, the term *statistical* has been used for MIMO radars that exploit a target's spatial diversity.

Diversity gain is only one of two key gains that MIMO communications can provide. The other gain is *spatial multiplexing* [23], which, in MIMO communications, expresses the ability to use the transmit and receive antennas to set up a multidimensional space for signaling. Then, by suitable techniques, it is possible to form uncoupled, parallel channels that enable the rate of communication to grow in direct proportion to the number of such channels. Similarly, in MIMO radar, a multidimensional signal space is created when the returns from multiple scatterers or targets combine to generate a rich backscatter. With proper design, transmit–receive paths can be separated and exploited for improving radar performance.

In communications, the key property that made MIMO such a successful concept has been the ability to substitute the spatial dimension for the bandwidth resource. Narrowband MIMO communication systems perform like wideband systems without MIMO. In radar, bandwidth also plays an important role. Frequency diversity has been applied to decorrelate RCS response of complex targets, and high-resolution location estimation is possible with wideband waveforms [26]. As discussed in Section 9.4, by exploiting the spatial dimension, MIMO radar with widely separated antennas may overcome bandwidth limitations and support high-resolution target localization. At the same time, this type of MIMO radar has the challenge of time and/or phase synchronizing multisite systems, and needs to deal with ambiguities stemming from the large separation between sensors. The lower bound on the attainable accuracy is provided by the Cramér–Rao lower bound (CRLB). Localization estimation accuracy is shown to depend on both the carrier frequency and the geometric positioning of the transmitting and receiving radars versus the inspected target. Two coherent target localization estimators are discussed in this chapter: the maximum-likelihood estimator (MLE) and the best linear unbiased estimator (BLUE). Insight into the relation between sensor location, target location, and localization accuracy is provided by a metric widely used in global positioning systems (GPS) for mapping estimation precision and known as *geometric dilution of precision* (GDOP). GDOP contours map the relative performance accuracy for a given layout of radars over a given geographic area.

The chapter is organized as follows: Section 9.2 introduces the MIMO radar concept and discusses it in the context of other radar architectures. Examples of noncoherent MIMO radar applications are presented in Section 9.3, while Section 9.4 discusses high-resolution target localization. The chapter concludes with a summary.

A comment on notation: vectors are denoted by lowercase bold, while matrices use uppercase bold letters. The superscripts T and $\dagger$ denote the transpose and Hermitian operators, respectively. Locations in the x–y plane are denoted in uppercase $X = (x, y)$.

9.2 MIMO RADAR CONCEPT

This section introduces the MIMO radar signal model, discusses conditions leading to RCS spatial diversity, and contrasts the architecture of MIMO radar with other multiple-element radars.

9.2.1 Signal Model

Assume a distributed target that consists of many, say, Q, independent, isotropic scatterers located in a plane that also contains the transmit and receive antennas. The scatterers are located at coordinates X_q, $q = 1, \ldots, Q$, where $X_q = (x_q, y_q)$. Figure 9.2 provides an illustration of the setup. In the figure, a distributed target is shown to consist of four scatterers. Each scatterer is represented by a circle to emphasize its

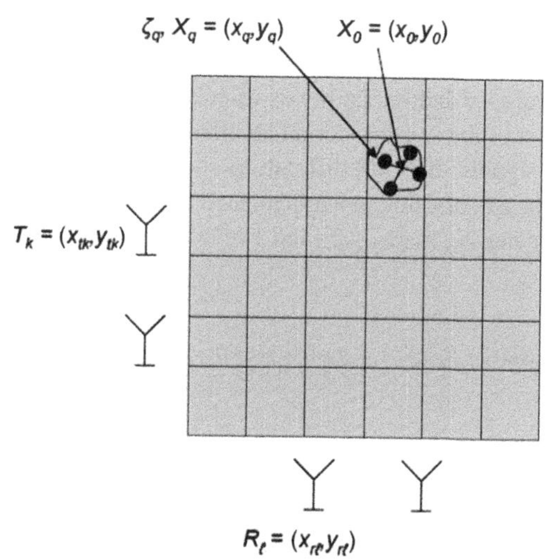

Figure 9.2 MIMO radar with extended target. Target consists of four point scatterers. Scatterers located at points X_q have reflectivity ζ_q. RCS center of gravity located at X_0. Transmit antennas located at T_k receive antennas located at R_ℓ.

isotropic reflectivity. The reflectivity of a scatterer is modeled by a zero-mean, independent and identically distributed (i.i.d.) complex random variable ζ_q with variance $E[|\zeta_q|^2] = 1/Q$. Reflectivity values of the target are organized in a diagonal $Q \times Q$ matrix, $\Sigma = \text{diag}(\zeta_1, \ldots, \zeta_Q)$. Given the reflectivity of an individual scatterer, the target average RCS is $E[\text{tr}(\Sigma\Sigma^\dagger)] = 1$, independent of the number of scatterers in the model. If the RCS fluctuations are fixed during an antenna scan, but vary independently from scan to scan, our target model represents a classical Swerling case I (which represents a target in slow motion) [26]. Now let the target be illuminated by M transmitters arbitrarily located at coordinates $T_k = (x_{tk}, y_{tk}),\ k = 1, \ldots, M$. The signals scattered by the target are collected by N sensors placed at arbitrary coordinates $R_\ell = (x_{r\ell}, y_{r\ell}),\ \ell = 1, \ldots, N$. The set of transmitted waveforms in lowpass equivalent form is $\sqrt{E/M}s_k(t),\ k = 1, \ldots, M,$ where $\int_{\mathcal{T}} |s_k(t)|^2 dt = 1$, E is the total transmitted energy, and $\mathcal{T}$ is the waveforms' duration. Normalization by M makes the total energy independent of the number of transmitters. Let all transmitted waveforms have the same bandwidth W. Then the two-dimensional resolution cells illustrated by the grid in Fig. 9.2 have approximate dimensions $(c/W) \times (c/W)$, where c is the speed of light. Further specification of the waveforms $s_k(t)$ depends on the application. For example, for focused beampatterns, the waveforms $s_k(t)$ may be identical or differ by a phase shift. For target classification or parameter estimation, the statistical model for the observations may be employed for specialized waveform design [27]. In another example, for MIMO radar seeking to exploit target spatial diversity, it might be desired to design orthogonal waveforms, for ease of separation at the receiver. Orthogonality may be imposed in the time domain or frequency domain or in signal space. Throughout the chapter it is assumed that transmitted waveforms are orthogonal and that orthogonality is maintained even for different mutual delays, that is, $\int_{\mathcal{T}} s_k(t)s_m^*(t - \tau)dt = 0$ for all $k \neq m$, and for all time delays of interest.

In the model developed below, path loss effects are neglected; that is, the model accounts for the effect of the sensors/target localizations only through time delays (or phase shifts) of the signals. It is not difficult to show that the lowpass equivalent of the signal observed at sensor ℓ due to a transmission from sensor k and reflection from a scatterer at coordinates $X_q = (x_q, y_q)$ (and excluding noise) is given by

$$z_{\ell k}^{(q)}(t) = \sqrt{\frac{E}{M}}\zeta_q s_k(t - \tau_{tk}(X_q) - \tau_{r\ell}(X_q))\exp(-j2\pi f_c[\tau_{tk}(X_q) + \tau_{r\ell}(X_q)]) \tag{9.1}$$

where $\tau_{tk}(X_q) = d(T_k, X_q)/c$ is the propagation time delay between the kth transmitting sensor located at coordinates T_k and the scatterer at X_q, $d(T_k, X_q) = \sqrt{(x_{tk} - x_q)^2 + (y_{tk} - y_q)^2}$, and f_c is the carrier frequency. The propagation time $\tau_{r\ell}(X_q)$ from the scatterer to the ℓth receiving sensor is defined analogously to $\tau_{tk}(X_q)$. The propagation time $\tau_{r\ell}(X_q)$ from the scatterer to the ℓth receiving sensor is defined analogously to $\tau_{tk}(X_q)$. Note that (9.1) denotes what we call here a near-field signal model, where the phase of the received signal is a function of the transmit and receive element locations as well as the location of the scatterer. In contrast, in a far-field signal model, the target is assumed to be sufficiently far such that it is essentially at the same distance and angle with respect to all the transmit and/or receive elements.

Continuing with the signal model in (9.1), we can interpret the term

$$h_{\ell k}^{(q)} = \zeta_q \exp(-j2\pi f_c[\tau_{tk}(X_q) + \tau_{r\ell}(X_q)]) \tag{9.2}$$

as the equivalent "channel" between transmitter k, scatterer q, and receiver ℓ. A channel element (9.2) consists of three parts: (1) $\exp[-j2\pi f_c \tau_{tk}(X_q)]$, the phase shift due to the propagation from transmitter k to scatterer q, (2) ζ_q, the reflectivity of the scatterer, and (3) $\exp[-j2\pi f_c \tau_{r\ell}(X_q)]$, the phase shift due to the propagation from the scatterer to receiver ℓ. Summing over all the scatterers that make up the target, the model in (9.1) becomes

$$z_{\ell k}(t) = \sqrt{\frac{E}{M}} \sum_{q=1}^{Q} h_{\ell k}^{(q)} s_k(t - \tau_{tk}(X_q) - \tau_{r\ell}(X_q)) \tag{9.3}$$

We assume that the bandwidth of the transmitted waveforms is such that they are not capable of resolving individual scatterers. This is stipulated by the condition that the target has an RCS center of gravity at $X_0 = (x_0, y_0)$, and that $s_k(t - \tau_{tk}(X_q) - \tau_{r\ell}(X_q)) \approx s_k(t - \tau_{tk}(X_0) - \tau_{r\ell}(X_0))$ for all $q = 1, \ldots, Q$. With this condition, (9.3) becomes

$$z_{\ell k}(t) = \sqrt{\frac{E}{M}} h_{\ell k} s_k(t - \tau_{tk}(X_0) - \tau_{r\ell}(X_0)) \tag{9.4}$$

where $h_{\ell k} = \sum_{q=1}^{Q} h_{\ell k}^{(q)}$.

The path gains $h_{\ell k}$ from transmit to receive antennas of the MIMO radar are organized in a $N \times M$ matrix $\mathbf{H}$. The matrix $\mathbf{H}$ can be expressed as [15]

$$\mathbf{H} = \mathbf{K}\Sigma\mathbf{G} \tag{9.5}$$

This equation highlights the contribution of the transmit paths, the target components, and the receive paths; the transmit paths are represented by the $Q \times M$ matrix $\mathbf{G}$, $\mathbf{G} = [\mathbf{g}_1^T; \mathbf{g}_2^T; \ldots; \mathbf{g}_Q^T]$, $\mathbf{g}_q^T = [1, \exp[-j2\pi f_c \tau_{t1}(X_q)], \ldots, \exp[-j2\pi f_c \tau_{tM}(X_q)]]$. Information on the receive paths is embedded in the $N \times Q$ matrix $\mathbf{K}$, $\mathbf{K} = [\mathbf{k}_1, \mathbf{k}_2, \ldots, \mathbf{k}_Q]$, where the semicolon separates rows, and $\mathbf{k}_q^T = [1, \exp[-j2\pi f_c \tau_{r1}(X_q)], \ldots, \exp[-j2\pi f_c \tau_{rN}(X_q)]]$. We recall that the matrix Σ was defined earlier in this section.

From (9.4), letting $\tau_{\ell k}(X_0) = \tau_{tk}(X_0) + \tau_{r\ell}(X_0)$ and accounting for additive noise, it is possible to express the observed waveforms at the receive antenna ℓ as

$$r_\ell(t) = \sqrt{\frac{E}{M}} \sum_{k=1}^{M} h_{\ell k} s_k(t - \tau_{\ell k}(X_0)) + w_\ell(t) \tag{9.6}$$

where $w_\ell(t)$ is circularly symmetric, zero-mean, complex Gaussian noise, spatially and temporally white with autocorrelation function $\sigma_w^2 \delta(\tau)$. We define the vector $\mathbf{r}(t) = [r_1(t), \ldots, r_N(t)]$ for later use. Properties of the elements $h_{\ell k}$ of the channel matrix $\mathbf{H}$ are discussed later in the section.

For a single elemental target located at $X_0 = (x_0, y_0)$, the channel coefficient is given by [from (9.2)] $h_{\ell k} = \zeta \rho_{\ell k}(X_0)$, where ζ is the isotropic reflectivity of the target, $\rho_{\ell k}(X_0) = \exp(-j2\pi f_c \tau_{\ell k}(X_0))$, and $\tau_{\ell k}(X_0) = \tau_{tk}(X_0) + \tau_{r\ell}(X_0)$. The signal waveform received by radar ℓ is then modeled as follows:

$$r_\ell(t) = \sqrt{\frac{E}{M}} \sum_{k=1}^{M} \zeta \rho_{\ell k}(X_0) s_k(t - \tau_{\ell k}(X_0)) + w_\ell(t) \tag{9.7}$$

The noise properties are the same as in the model (9.6). The incoming signal is separated at the receiver end by exploiting the orthogonality between the transmitted waveforms.

9.2.1.1 Linear Perturbation Model In general, the relation between the time delay and the target's coordinates is nonlinear. It is useful to develop a linear model for the time delay relative to a perturbed point. We will use this model in later to obtain insight into the performance of estimates for target location.

We wish to express the time delays $\tau_{\ell k}(X)$ as a linear function of perturbations of the x and y coordinates measured relative to an arbitrary point $X_c = (x_c, y_c)$. For example, the perturbed point X_c can be obtained by an initial, nominal estimation. Using notation defined previously, the time delay between a transmitter–receiver pair is given by

$$\tau_{\ell k}(X) = \frac{1}{c}\left(\sqrt{(x_{tk} - x)^2 + (y_{tk} - y)^2} + \sqrt{(x_{r\ell} - x)^2 + (y_{r\ell} - y)^2}\right) \tag{9.8}$$

Using the Taylor series expansion of $\tau_{\ell k}(X)$ in (9.8) around (x_c, y_c), we can express the time delay as

$$\tau_{\ell k}(X) \approx -\frac{x}{c}\left(\frac{x_{tk} - x_c}{d(T_k, X_c)} + \frac{x_{r\ell} - x_c}{d(R_\ell, X_c)}\right) - \frac{y}{c}\left(\frac{y_{tk} - y_c}{d(T_k, X_c)} + \frac{y_{r\ell} - y_c}{d(R_\ell, X_c)}\right) \tag{9.9}$$

where $d(T_k, X_c) = \sqrt{(x_{tk} - x_c)^2 + (y_{tk} - y_c)^2}$ is the distance from transmitting radar k to the target, $d(R_\ell, X_c) = \sqrt{(x_{r\ell} - x_c)^2 + (y_{r\ell} - y_c)^2}$ is the distance from

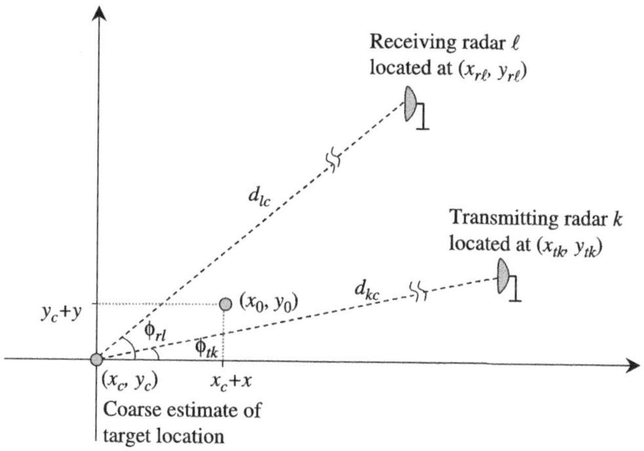

Figure 9.3 System layout: geometric of linear perturbation model.

the target to receiving radar ℓ, $x = x_0 - x_c$, and $y = y_0 - y_c$. This setup is shown in Fig. 9.3.

A convenient way to express (9.9) is to use the angles ϕ_{tk} and $\phi_{r\ell}$ that the transmitting radar k and receiving radar ℓ, respectively, create with the reference axis x_c. This results in the following linear perturbation model for the time delay as a function of the displacements x and y from peturbed point X_c:

$$\tau_{\ell k}(X) \approx -\frac{x}{c}(\cos\phi_{tk} + \cos\phi_{r\ell}) - \frac{y}{c}(\sin\phi_{tk} + \sin\phi_{r\ell}) \tag{9.10}$$

Finally, letting $A_{\ell k} = \cos\phi_{tk} + \cos\phi_{r\ell}$ and $B_{\ell k} = \sin\phi_{tk} + \sin\phi_{r\ell}$, in (9.10), we can express the time delays in the following simplified form:

$$\tau_{\ell k}(X) \approx -\frac{x}{c}A_{\ell k} - \frac{y}{c}B_{\ell k} \tag{9.11}$$

This model is used in the discussion on BLUE target localization in Section 9.4.4.

9.2.2 Spatial Decorrelation

In the previous subsection, we introduced a stochastic model for the channel elements $h_{\ell k}$. Here, we discuss the conditions for statistical decorrelation of these elements. Fishler et al. [20] showed that, if Q is large, all the channel elements $h_{\ell k}$ are jointly Gaussian with zero mean and unit variance. Let two transmit antennas have coordinates (x_{tk}, y_{tk}) and (x_{ti}, y_{ti}), respectively, and let the target dimensions along the x and y axes be D_x and D_y, respectively. Further, let two receive antennas have coordinates $(x_{r\ell}, y_{r\ell})$ and (x_{rj}, y_{rj}), respectively. Recall the definitions of distance $d(T_k, X_0)$ between transmit antenna at T_k and the target at X_0, and of distance $d(R_\ell, X_0)$

between receiver antenna at R_ℓ and the target at X_0. If at least one of the following four conditions are met

$$\begin{aligned}
\frac{x_{tk}}{d(T_k, X_0)} - \frac{x_{ti}}{d(T_i, X_0)} &> \frac{\lambda}{D_x} \\
\frac{y_{tk}}{d(T_k, X_0)} - \frac{y_{ti}}{d(T_i, X_0)} &> \frac{\lambda}{D_y} \\
\frac{x_{r\ell}}{d(R_\ell, X_0)} - \frac{x_{rj}}{d(R_j, X_0)} &> \frac{\lambda}{D_x} \\
\frac{y_{r\ell}}{d(R_\ell, X_0)} - \frac{y_{rj}}{d(R_j, X_0)} &> \frac{\lambda}{D_y}
\end{aligned} \tag{9.12}$$

where λ is the carrier wavelength, then has been shown [20] that the (ℓk)th and (ji)th elements of the channel matrix are *uncorrelated*. Conversely, if

$$\begin{aligned}
\frac{x_{tk}}{d(T_k, X_0)} - \frac{x_{ti}}{d(T_i, X_0)} &\ll \frac{\lambda}{D_x} \\
\frac{y_{tk}}{d(T_k, X_0)} - \frac{y_{ti}}{d(T_i, X_0)} &\ll \frac{\lambda}{D_y} \\
\frac{x_{r\ell}}{d(R_\ell, X_0)} - \frac{x_{rj}}{d(R_j, X_0)} &\ll \frac{\lambda}{D_x} \\
\frac{y_{r\ell}}{d(R_\ell, X_0)} - \frac{y_{rj}}{d(R_j, X_0)} &\ll \frac{\lambda}{D_y}
\end{aligned} \tag{9.13}$$

then the (ℓk)th and (ji)th elements of the channel matrix are *correlated*. Loosely speaking, these results mean that if the distributed target is viewed as an antenna with aperture length D (D could be D_x or D_y from the discussion above), the elements of the channel matrix **H** decorrelate if the target's beamwidth cannot illuminate two sensors simultaneously. This concept is illustrated in Fig. 9.4. For example, for a target at distance $d = 10^4\lambda$ and of dimension $D = 10\lambda$, the separation required between the elements of the MIMO radar is of the order $10^3\lambda$. Conversely, when the distributed target acts as an antenna that has two sensors in the same beamwidth, elements of **H** associated with these sensors are correlated. Since each element of **H** represents the path gain of a transmitter–receiver pair, decorrelation supports a diversity of path gains. We conclude from the foregoing discussion that to obtain path diversity to an extended target, the MIMO radar antennas need to be sufficiently separated. However, even when the MIMO radar antennas are collocated [left side terms in (9.12) are of the order of $\lambda/2$], then decorrelation is still possible for large D. Taking the example we used earlier with a distance to target of $d = 10^4\lambda$, the condition for decorrelation is $D > 2 \times 10^4\lambda$, which is twice the distance to the

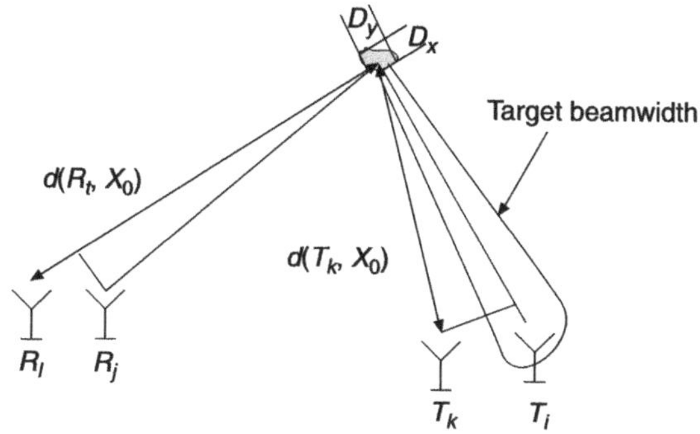

Figure 9.4 Elements of the channel matrix decorrelate when the MIMO radar antennas fall in different beamwidths originating from the target.

"target." In this case, D cannot be the aperture of a single extended target, but rather we could view *two* point targets separated by D as an array with two elements.

For further insight into the conditions for spatial decorrelation, let us consider the role of the target in the spatial decorrelation of the elements of **H**. From the discussion so far, spatial decorrelation of the elements of the channel matrix occurs for complex targets (a large number of scatterers Q) and widely separated sensors. A large number of scatterers results in rich RCS patterns, such as the one in Fig. 9.1. The large number of scatterers, each with its own reflectivity ζ_q and location $X_q = (x_q, y_q)$, combine at the receive sensors to create a fading signal. Widely spaced sensors cause the paths from the transmitters (T_k values in Fig. 9.2) to the scatterers (X_q terms), and from the scatterers to the receivers (R_ℓ values) to be sufficiently diverse such that each receive sensor experiences a *different* fading signal. This is the root of the spatial decorrelation. What happens for simple targets with only a few scatterers? In the extreme of a single spherical scatterer with uniform RCS as a function of angle, there are no multiple signals that combine to create the fading elements of the channel matrix. With the loss of fading, the elements of the channel matrix become correlated.

9.2.3 Other Multiple Antenna Radars

To gain additional insight into the MIMO radar signal model, we discuss here its relation to other multiple antenna radar systems.

9.2.3.1 Conventional Phased-Array Radar In conventional array radar, the antennas are collocated, and the interelement spacing is $\lambda/2$. Typically, the target is in the far field of the antenna array, meaning that all antenna elements are at essentially the same distance and angle with respect to the target. In this case, it is easy to verify that the matrix **G** defined earlier and representing the propagation from the transmit antennas to the target can be expressed as $\mathbf{G} = \mathbf{1}_Q \otimes \mathbf{b}^T(\phi_0')$. Here $\mathbf{1}_Q$ is a column vector of ones, the binary operation is the Krönecker product, and $\mathbf{b}(\phi_0')$ is a $M \times 1$ vector representing the propagation from the transmit antennas to any of the scatterers that make up the target, when the target is at angle ϕ'_0 with respect to the normal to the transmit array [kth element is $b_k(\phi'_0) = \exp[-j2\pi(k-1)\Delta_t \sin(\phi'_0)/\lambda]$]. Similarly, the propagation paths from the target to the receive antennas are represented by the matrix $\mathbf{K} = \mathbf{a}(\phi_0) \otimes \mathbf{1}_Q^T$, where $\mathbf{a}(\phi_0)$ is a $N \times 1$ vector with components $a_\ell(\phi_0) = \exp[-j2\pi(\ell-1)\Delta_r \sin(\phi_0)/\lambda]$. Applying (9.5), it follows that the channel matrix $\mathbf{H}$ equals

$$\mathbf{H} = \mathbf{K\Sigma G} = \mathbf{a}(\phi_0)\mathbf{b}^T(\phi'_0)\sum_q \zeta_q \tag{9.14}$$

Let the random variable $\alpha = \sum_q \zeta_q$. Then, for large Q, by the central-limit theorem, α is a zero-mean, variance 1, complex Gaussian random variable.

In phased arrays, the same waveform $s(t)$ is transmitted by all the antennas. To steer the transmitted beam in direction ϕ' with respect to the transmit array, a steering vector $\mathbf{b}^*(\phi')$ is applied such that $s_k(t) = s(t)b_k^*(\phi')$, where $b_k^*(\phi') = \exp[j2\pi(k-1)\Delta_t \sin(\phi')/\lambda]$. The assumption that the transmit and receive antennas are at the same distance d from the target implies that $s_k(t - \tau_{tk}(X_0) - \tau_{r\ell}(X_0)) = s_k(t-\tau)$ for some τ, and for all k and ℓ. The received signal in (9.6) becomes

$$r_\ell(t) = \sqrt{\frac{E}{M}}\,\alpha a_\ell(\phi_0)\sum_{k=1}^{M} b_k(\phi'_0)b_k^*(\phi')s_k(t-\tau) + w_\ell(t) \tag{9.15}$$

Beamforming at the receiver in the direction ϕ, we obtain at the output of the beamformer $r(t) = \sum_{\ell=1}^{N} a_\ell^*(\phi_0)r_\ell(t)$, which is the model for the received signal in conventional array radars [26].

9.2.3.2 *Adaptive Radar/STAP*

There are two variations of phased-array radar that have been extensively studied in the literature: adaptive radar [5] and spacetime adaptive radar (STAP) [6,7]. In adaptive radar, the output is given by $r(t) = \sum_{\ell=1}^{N} v_\ell^* r_\ell(t)$, where v_ℓ are weights computed based on some optimality criterion, such as maximizing the signal-to-interference ratio. STAP is an extension of adaptive radar that incorporates time domain as well as spatial-domain processing. The STAP filter exploits spatial and time correlations to suppress clutter and interference (including jammers) and to detect moving targets.

9.2.3.3 *Multistatic Radar*

Most of the literature on bistatic/multistatic radars assumes that the individual radars that make up the system operate independently [28]. Detection, target estimation, and other high-level products of local processing are communicated to a central processor. The preprocessing limits the amount of information that needs to be passed on to develop a final detection decision or estimation. The central processor takes advantage of the information delivered by each radar to improve detection/estimation performance [29–31]. While in a general sense MIMO radar can be viewed as a type of multistatic radar, the different designation distinguishes it from the multistatic radar literature by emphasizing the *joint processing* of signals for transmission and reception.

9.3 NONCOHERENT MIMO RADAR APPLICATIONS

In this and the next section, several examples of MIMO radar applications are introduced and studied. This section focuses on applications that do not require phase synchronization between transmit and/or receive antennas. We refer to such applications as *noncoherent*. Applications in which the phase information between antennas is exploited are referred to as *coherent*, and are discussed in the next section. The first example in this section illustrates the application of MIMO radar to obtain a diversity gain to overcome fades in target RCS. The example is concerned with the detection of a stationary target, and it generally follows the observations by Fishler et al. [20]. In the second example, multistatic observations are exploited to improve the detection capability of a moving target [21]. In both cases, the target consists of multiple scatterers, and it is observed against a background of white Gaussian noise. The examples are set up as binary hypothesis testing problems.

9.3.1 Diversity Gain

Diversity processing seeks to exploit multiple paths through which the signals propagate from the transmit antennas to the receive antennas. Such a path starts from a transmitter. The transmitted waveform impinges on the Q scatterers that form the target. The path terminates at a receiver, where the responses from the Q scatterers combine. Since the scatterers are too close to be resolved by the waveform, the Q echoes combine as in a flat fading multipath scenario to form a fading channel coefficient. With M transmit and N receive antennas, there are MN such paths corresponding to the MN elements of the channel matrix $\mathbf{H}$. Under the conditions discussed in Section 9.2.2 (complex targets with large Q and sufficiently spaced antennas), the MN paths are decorrelated. The set of transmitted waveforms has to be designed to facilitate the separation between waveforms arriving simultaneously at the receiver. Simple signal separation is possible when the transmitted waveforms are orthogonal and maintain their orthogonality at the receiver end.

The problem is stated as follows. Given a set of known waveforms $s_k(t - \tau_{\ell k}(X))$ parameterized by the time delays $\tau_{\ell k}$ associated with the location $X = (x, y)$, and given that each receive antenna ℓ observes a mixture of the known waveforms received through random channels $h_{\ell k}$ in white Gaussian noise, where the channel coefficients are modeled as complex, Gaussian random variables, devise a test for the target's presence at X. In practice, the signals at each receive antenna are sampled at time intervals of at least $1/W$ (where W is the bandwidth, common to all transmitted waveforms), and then processed digitally.

For the target detection, we form two hypotheses. Under hypothesis H_1, a target is present at $X = (x, y)$; under the alternative hypothesis, the target is not present. The classic Neyman–Pearson detector is given by the likelihood ratio test

$$\log \frac{p(\mathbf{r}; H_1, X)}{p(\mathbf{r}; H_0, X)} \underset{H_0}{\overset{H_1}{\gtrless}} \gamma \tag{9.16}$$

where $p(\mathbf{r}; H_1, X)$ and $p(\mathbf{r}; H_0, X)$ are the likelihood functions of the observation vectors under the respective hypotheses. The threshold γ is determined by the tolerated level of false alarms. The underlying density functions $p(\mathbf{r}; H_1, X)$, $p(\mathbf{r}; H_0, X)$ are understood in the conventional sense as joint distributions of samples of the real and imaginary parts of $\mathbf{r}(t)$ taken at $1/W$ time intervals (see, e.g., Ref. 32). With this clarification in mind, we will refer to observations of waveforms rather than of samples of waveforms. From the signal model (9.6), the observations $r\ell(t)$ are functions of the channel coefficients $h_{\ell k}$. Concatenating the columns of the channel matrix $\mathbf{H}$ to form the length MN vector $\mathbf{h}$, and recalling the discussion in Section 9.2.2, $\mathbf{h}$ is stochastic with probability density function $p(\mathbf{h}) \propto \exp(-\|\mathbf{h}\|^2)$, where $\|\cdot\|$ denotes the Euclidean norm. This follows since the elements of $\mathbf{h}$ are mutually independent, where each element is complex Gaussian with zero-mean and variance 1. The target detection process involves the simultaneous transmission of orthogonal waveforms and the testing of a decision statistic by a central processor. The central processor is fed with data sampled by the receivers at delays corresponding to resolution cells. By suitable control of sampling instants, this process can test for the presence of a target in each of the resolution cells in Fig. 9.2.

Assuming orthogonality between waveforms and the signal model in (9.6), it is possible to extract a noisy estimate of each element of $\mathbf{H}$ corresponding to a point $X = (x, y)$ in the x–y plane. The estimate for element $h_{\ell k}$ is obtained by applying the matched filter for waveform $s_k(t)$ to the received signal at the ℓth antenna, $r_\ell(t)$, and sampling at a time delay corresponding to location X

$$y_{\ell k}(X) = \int_T r_\ell(t) s_k^*(t - \tau_{\ell k}(X))dt \tag{9.17}$$

where the summation involved in the correlation of sampled waveforms was approximated by an integral, and the waveforms were expressed in continuous time. Notice that by exploiting orthogonality between the transmitted waveforms, $y_{\ell k}(X)$ contains the target response to only $s_k(t)$. Let $\mathbf{y}(X)$ denote the vector formed by $y_{\ell k}(X)$ for $\ell = 1, \ldots, N$ and $k = 1, \ldots, M$. Starting with (9.6) and given the probability density function of $\mathbf{h}$, it is possible to develop an expression for the probability density function of the received vector from a resolution cell containing the target and averaged over the values of $\mathbf{h}$. The detailed steps are omitted here because of space considerations, but following steps similar to those described by Fishler et al. [20], it can be shown that

$$p(\mathbf{r}; H_1, X) = \int_{\mathbf{h}} p_X(\mathbf{r}; H_1, X, \mathbf{h}) p(\mathbf{h}) d\mathbf{h}$$

$$\propto \exp\left[-\frac{1}{\sigma_w^2}\left(\int_T \|\mathbf{r}(t)\|^2 dt\right)\right] \exp\left[\frac{\|\mathbf{y}(X)\|^2}{\sigma_w^2(\sigma_w^2 + (E/M))}\right] \tag{9.18}$$

For hypothesis H_0, we have

$$p(\mathbf{r}|H_0) \propto \exp\left[-\frac{1}{\sigma_w^2}\left(\int_T \|\mathbf{r}(t)\|^2 dt\right)\right] \tag{9.19}$$

It follows that the Neyman–Pearson detector is given by

$$\|\mathbf{y}(X)\|^2 \underset{H_0}{\overset{H_1}{\gtrless}} \gamma_1 \tag{9.20}$$

In order to set the value of the threshold γ_1 for a tolerated level of false alarms, the central processor must have full model knowledge, that is, the values of waveform energy E and noise power σ_w^2. By inspection of (9.20), the optimal processor is a noncoherent combiner of the signals preprocessed by the sensors. With M transmitters and N receivers and under conditions discussed earlier, $\|\mathbf{y}(X)\|^2$ is the sum of the magnitudes of MN uncorrelated noisy estimates of channel coefficients. When, with suitable normalization, a single term $|y_{\ell k}(X)|^2$ has a χ_2^2 [chi-squared (χ^2) with 2-DOF (degrees of freedom)] distribution, the test statistic in (9.20) has the improved statistical properties of a χ_{2MN}^2 random variable.

Closed-form performance expressions are given by Fishler et al. [20] for cases with full and partial model knowledge. The figure reproduced here (Fig. 9.5) illustrates the case in which the signal energy and noise power level are known. Figure 9.5 illustrates the gains provided by MIMO radar (curves labeled "MIMO") over a conventional phased-array radar (labeled "conv"). The curves in the figure represent the miss probability of miss-detection as a function of SNR for a fixed false-alarm probability of 10^{-6} and for various MIMO/phased-array configurations. The slopes of the MIMO curves clearly show diversity gains similar to those obtained in communications (e.g., compare the curves in Ref. 24 to those in Fig. 9.5). For a more extensive set of results, we refer the reader to Ref. 20.

Other applications can benefit from the diversity gains afforded by MIMO radar with widely separated antennas. For example, diversity gains can be exploited for direction finding (DF). In this case, the target spatial diversity is supported by widely separated transmitters, while the receiving sensors are placed at $\lambda/2$ intervals to enable unambiguous DF. Such a system is studied by Alamouti [33].

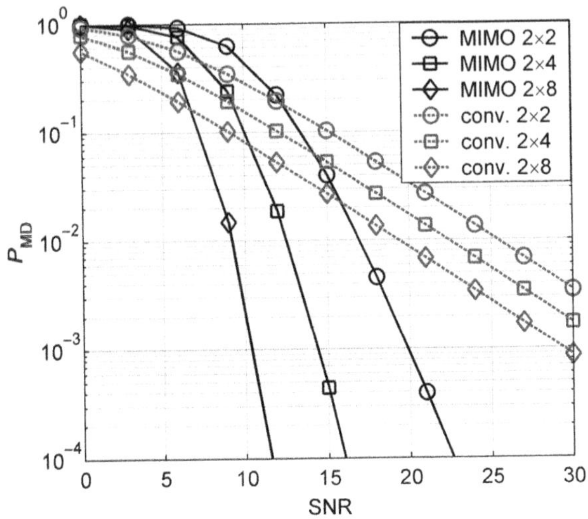

Figure 9.5 Miss probability of MIMO radar compared to conventional (conv.) phased array. Miss probability is plotted versus SNR for a fixed false-alarm probability of 10^{-6}. (Reproduced, with permission, from Ref. 20.)

9.3.2 Moving-Target Detection

Observing targets over wide angular sectors with a MIMO radar offers great benefits for Doppler processing and moving target detection (MTD). MIMO radar is particularly useful in some cases when the targets are difficult to distinguish from the background clutter. The enhanced MTD performance of MIMO radar stems from its distributed nature, which makes it possible to overcome typical problems that plague conventional or array radars such as targets with low radial velocities or blind speeds. Further, the joint processing of the received waveforms results in superior performance compared to multistatic radar, where each sensor typically performs its own Doppler shift estimation.

MTD with MIMO radar was investigated by Lehmann et al. [21]. The scenario considered consists of a single transmitter and $N > 1$ widely dispersed receivers. The transmitter and receivers are stationary. It is assumed that the moving target is sufficiently far from the sensors such that its Doppler shift with respect to the sensors is constant during the observation of K samples. A central processor is fed with test statistics computed at the sensors. The clutter is assumed zero-mean, complex Gaussian with known $K \times K$ temporal covariance matrix $\mathbf{C}$. The clutter is also assumed spatially homogeneous in the sense that it has the same properties for all transmitter–receiver pairs and for all resolution cells. In Section 9.3.1, in discussing target detection, we modeled the channel elements $h_{\ell k}$ as stochastic with known statistics. Here, we take a slightly different approach, and assume that these coefficients are deterministic unknown. The two slightly different approaches follow the respective models employed in the references that are reviewed in this chapter, but should not affect key features of MIMO radar.

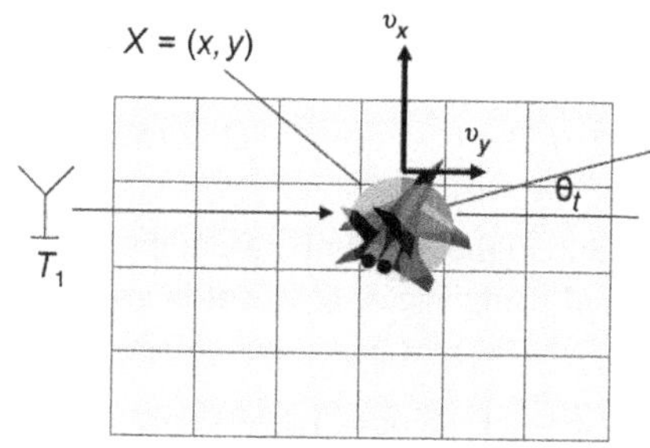

Figure 9.6 Moving-target detection scenario: single transmit antenna at location T_1, multiple receive antennas at location R_l, forming angles θ_l, $l = 1, \ldots, N$. Target at $X = (x, y)$ with velocity components v_x, v_y.

In this subsection, tests for the detection of a single, moving target are discussed. The target, if present, is moving with unknown velocity components (v_x, v_y), where v_x and v_y are deterministic, unknown velocities measured in an arbitrary Cartesian coordinate system (refer to Fig. 9.6 for an illustration of the setup). Given M transmitters and N receivers, there are MN paths resulting in MN different spatial "looks" at the resolution cell under test. The same number of looks can be obtained with a single transmitter and MN receivers; hence, to simplify notation, and without loss of generality, the discussion below assumes $M = 1$. In Fig. 9.6, the locations of the transmitter and the receiving sensors are denoted respectively as T_1 and R_ℓ, $\ell = 1, \ldots, N$. The target and the transmitter are aligned on the x axis, while the receivers form angles ϕ_ℓ with the target–transmitter axis.

To test for the presence of the target in a resolution cell, we develop a generalized likelihood ratio test (GLRT). The GLRT maximizes the likelihood ratio test over the unknown parameters: the target response $h_{\ell k}$ and the velocities v_x, v_y. Under hypothesis H_1, a target is present in the tested resolution cell centered at coordinates $X = (x, y)$; under the alternative hypothesis, the target is not present in the resolution cell. A vector $\mathbf{y}_\ell(X)$ is defined to consist of K time samples of the form $y_{\ell 1}(X)$ defined in (9.17), and taken at the pulse repetition interval T_{PRI}. The GLRT is evaluated by first expressing the likelihood ratio as a function of the unknown parameters $h_{\ell k}$, v_x, and v_y. Subsequently, the likelihood ratio is maximized over the unknown parameters. Skipping the details of the analysis, we list the MTD decision rules obtained for three types of radar: MIMO radar, multistatic radar, and phased-array radar:

The GLRT decision rule for MIMO radar is

$$\max_{v_x, v_y} \sum_{\ell=1}^{N} \frac{|\mathbf{d}^\dagger(f_\ell)\mathbf{C}^{-1}\mathbf{y}_\ell(X)|^2}{\mathbf{d}^\dagger(f_\ell)\mathbf{C}^{-1}\mathbf{d}(f_\ell)} \underset{H_0}{\overset{H_1}{\gtrless}} \gamma_2 \tag{9.21}$$

where $\mathbf{d}(f_\ell)$ is a $K \times 1$ vector representing the matched filter to a target response with Doppler frequency f_ℓ. This vector is composed of K time samples taken at the pulse repetition intervals T_{PRI} and has the form

$$\mathbf{d}(f_\ell) = [1,\ \exp[-j2\pi f_\ell T_{\mathrm{PRI}}], \ldots,\ \exp[-j2\pi f_\ell (K-1)T_{\mathrm{PRI}}]]^T \tag{9.22}$$

where f_ℓ is the Doppler shift due to the motion between the sensor and the target. The Doppler shift can be evaluated using Fig. 9.6, and it is given by

$$f_\ell = \frac{v_x}{\lambda}(1 + \cos\phi_\ell) + \frac{v_y}{\lambda}\sin\phi_\ell \tag{9.23}$$

For example, for a target moving along the x axis with velocity v_x toward the transmitter and receiver, the Doppler shift is composed of the frequency shift due to the motion between transmitter and target (v_x/λ), and the frequency shift due to the target–receiver motion $((v_x/\lambda)\cos\phi_\ell)$. Finally, the threshold γ_2 is determined by the tolerated level of false alarms. Notice that the test statistic in (9.21) is computed for each resolution cell X represented by a square in Fig. 9.6. The location X of the resolution cell determines the values of angle ϕ_ℓ needed in (9.23) and the values of $\mathbf{y}_\ell(X)$ computed using (9.17). A test statistic is computed at each sensor such that the clutter spectrum is whitened. The optimal processor is a noncoherent combiner of the test statistics computed at each sensor. The aggregate statistic is maximized over values of v_x, v_y velocities, and the result is compared to the threshold γ_2 to decide on the target presence.

Moving on to the multistatic radar, each sensor performs its own estimation of the Doppler shift f_ℓ rather than the joint processing carried out by the MIMO radar. This means that rather than finding estimates v_x, v_y and restricting f_ℓ to the form in (9.23), each sensor estimates f_ℓ that maximizes a statistic computed locally. Then, the decision rule is

$$\sum_{\ell=1}^{N} \max_{f_\ell} \frac{|\mathbf{d}^\dagger(f_\ell)\mathbf{C}^{-1}\mathbf{y}_\ell(X)|^2}{\mathbf{d}^\dagger(f_\ell)\mathbf{C}^{-1}\mathbf{d}(f_\ell)} \underset{H_0}{\overset{H_1}{\gtrless}} \gamma_3 \tag{9.24}$$

where $\mathbf{d}(f_\ell)$ is still defined by (9.22), but f_ℓ does not have to obey (9.23).

Finally, the performance of the phased array is limited by its small aperture. Lacking the ability to view the target from multiple aspects, the phased array can observe only a single Doppler shift, $f_\ell = f$. The N transmit elements are beamformed to provide coherent processing gain. The GLRT decision rule is then given by

$$\max_{f} \frac{|\mathbf{d}^\dagger(f)\mathbf{C}^{-1}\mathbf{y}(X)|^2}{\mathbf{d}^\dagger(f)\mathbf{C}^{-1}\mathbf{d}(f)} \underset{H_0}{\overset{H_1}{\gtrless}} \gamma_4 \tag{9.25}$$

where $y(X)$ is the vector of K time samples at the output of the beamformed array corresponding to a target at X.

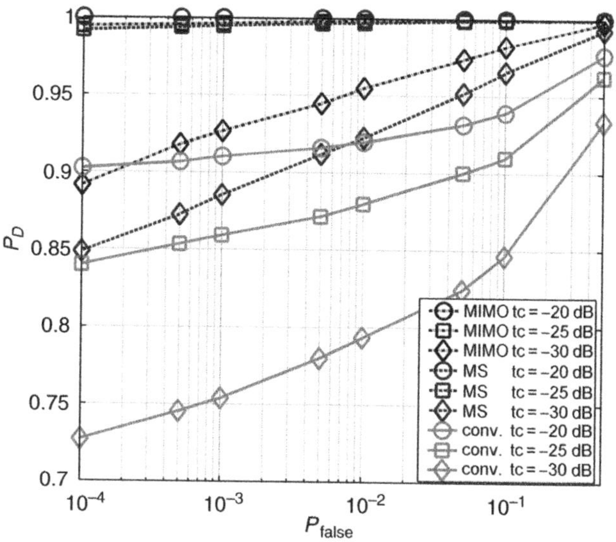

Figure 9.7 ROC for a 300-km/h target with random direction and with RCS fluctuations.

The receiver operating characteristics (ROC) of the three different types of radar systems are as follows: (1) the MIMO radar from (9.21) (labeled "MIMO"), (2) the multistatic radar from (9.24) ("MS"), and (3) the array radar from (9.25) ("conv") (shown in Fig. 9.7). Parameters used in the simulations include $T_{PRI} = 0.5$ ms, $K = 10$ samples, carrier frequency = 1 GHz, clutter : noise ratio = 30 dB, and target angle of arrival 0°. There are $N = 8$ receiving sensors placed at uniformly spaced angles in the sector from $-40°$ to 50° with respect to the target. The target velocity is 300 km/h with random direction. The curves are parameterized by the target : clutter ratio ("tc").

The results demonstrate the superiority of MIMO radar over multistatic or array radars. The advantage over multistatic radar stems from the joint processing that constrains the search to velocities (v_x, v_y) common to all sensors. An additional advantage of MIMO radar is the ability to exploit the spatial diversity of RCS values. The multistatic radar can also benefit from RCS diversity, but since Doppler processing is local, it yields higher false alarms. The increase in false alarms accounts for the gap in performance. The phased-array radar is handicapped by both low Doppler response to low radial velocities and RCS fading.

9.4 COHERENT MIMO RADAR APPLICATIONS

In the previous section we explored the advantages of MIMO radar for applications related to static and moving-target detection. These applications required time synchronization between the transmitting and receiving radars. Thus far, phase synchronization was not a requirement and therefore, these applications were classified as noncoherent. In this section, we seek to study a MIMO radar architecture that exploits knowledge of the phase differences measured at the receive antennas to produce a high-resolution target location estimate. By casting the problem as an estimation problem, we can take advantage of signal processing tools, such as the Cramér–Rao lower bound (CRLB), and known estimators, such as the maximum-likelihood estimator (MLE) and the best linear unbiased estimator (BLUE). These tools enable performance evaluation in closed form and afford insight into what parameters affect performance. In another departure from the previous section, the target is assumed to consist of a single elemental scatterer located at coordinates $X = (x, y)$, expressed relative to a nominal estimate at location X_c. The target has isotropic, unknown reflectivity given by the complex constant ζ. In the previous section, a target with rich scattering supported diversity gain. Here, since we are not interested in diversity gains, the simpler model enables us to focus on the high-resolution properties of MIMO radar. Finally, we will employ the linear perturbation model for the time delay, introduced in Section 9.2.1.1.

The estimation problem is formulated as follows. Given a set of known waveforms $s_k(t - \tau_{\ell k}(X))$ parameterized by the unknown time delays $\tau_{\ell k}$, which in turn are a function of the unknown target location $X = (x, y)$, and given that each receive antenna ℓ observes a mixture of the known waveforms received through random channels $h_{\ell k}$ in white Gaussian noise, estimate the target location X. It is assumed that a nominal estimate X_c of the target location is available and that a linear perturbation model holds for the time delays expressed in terms of the displacement between X and the nominal estimate X_c.

In the following, the ambiguity function, commonly used in radar system for assesing resolution and ambiguity properties, is introduced for the MIMO radar case. A comparative discussion of the coherent and noncoherent ambiguity function is provided. The CRLB on the achievable location accuracy is developed, followed by derivation of two target location estimators: the MLE and the BLUE. In particular, the BLUE provides a close-form estimator, while the MLE can be evaluated only numerically. The main objective of this section is to show that coherent processing of MIMO radars supports a location accuracy that is a function of the carrier frequency f_c, rather than being a function of the signal bandwidth W as in the noncoherent case.

We form the vector of unknown parameters $\boldsymbol{\theta} = [x, y, \zeta_r, \zeta_i]^T$, where $\zeta_r + j\zeta_i = \zeta$. For the signal model (9.7), the joint probability density function (pdf) of the observations (time samples at multiple receive antennas) parameterized by the unknown parameters vector $\boldsymbol{\theta}$ and approximating sums over samples to integrals, is then

$$p(\mathbf{r}; \theta) \exp\left\{-\frac{1}{\sigma_w^2} \sum_{\ell=1}^{N} \int_T \left| r_\ell(t) - \zeta\sqrt{\frac{E}{M}} \sum_{k=1}^{M} \rho_{\ell k}(X) s_k(t - \tau_{\ell k}(X)) \right|^2 dt \right\} \tag{9.26}$$

The phase term $\rho_{\ell k}(X) = \exp(-j2\pi f_c \tau_{\ell k}(X))$ in the waveform statistics provides the additional information that is exploited for attaining better localization. This requires that the phase relationship between all radar sensors be completely known, making coherent processing possible. Next, we discuss the ambiguity function of the coherent MIMO radar and show how the phase data benefits high-resolution target localization.

9.4.1 Ambiguity Function

The ambiguity function concept has long been used in the context of localization [26,37]. It is defined as the absolute value of the envelope of the output of the matched filter. Using the requirement that its values be bounded between 0 and 1, we define the following function

$$A(X) = \frac{1}{M^2N^2}\left|\sum_{k=1}^{M}\sum_{\ell=1}^{N}\rho_{k\ell}^{*}(X)\varphi_{\ell k}(X)\right|^2 \tag{9.27}$$

where the term $\varphi_{\ell k}(X)$ is

$$\varphi_{\ell k}(X) = \int_T s_k(t)s_k^*(t-\tau_{\ell k}(X))dt \tag{9.28}$$

The function $A(X)$ denotes the zero-Doppler cut through the ambiguity function. The ambiguity function represents the response to a point target with reflectivity index $\zeta = 1$ located at X_0, when the processing is adapted to a target located at X. The term $\varphi_{\ell k}(X)$ is the output of sensor ℓ to a signal processed with a filter matched to signal $s_k(t)$. Since $A(X)$ requires phase information, it is referred to as the *coherent* ambiguity function of the MIMO radar. The high-resolution properties of $A(X)$ are determined by the phase terms $\rho_{\ell k}^*(X)$.

The noncoherent ambiguity function (assuming a point target) is given by

$$A_{nc}(X) = \frac{1}{M^2N^2}\left|\sum_{k=1}^{M}\sum_{\ell=1}^{N}\varphi_{\ell k}(X)\right|^2 \tag{9.29}$$

It is clear that the values of the coherent ambiguity function are upper-bounded by the noncoherent ambiguity function [i.e., $A(X) \leq A_{nc}(X)$]. The coherent ambiguity function $A(X)$ is plotted in Fig. 9.8 for a 9×9 MIMO radar system. Each of the nine sensors serves as both transmitter and receiver. The sensors are placed uniformly in the sector between -45° and 45° with respect to the target. The ratio between the bandwidth Δf and the carrier frequency f_c is set to $\Delta f/f_c = 0.1$. The x and y coordinates are expressed in multiples of the carrier wavelength λ.

The noncoherent ambiguity function $A_{nc}(X)$ is shown for comparison in Fig. 9.9. The high-resolution capability is evident in the fact that the width of the mainlobe of the coherent ambiguity function is approximately equal in size to the wavelength λ,

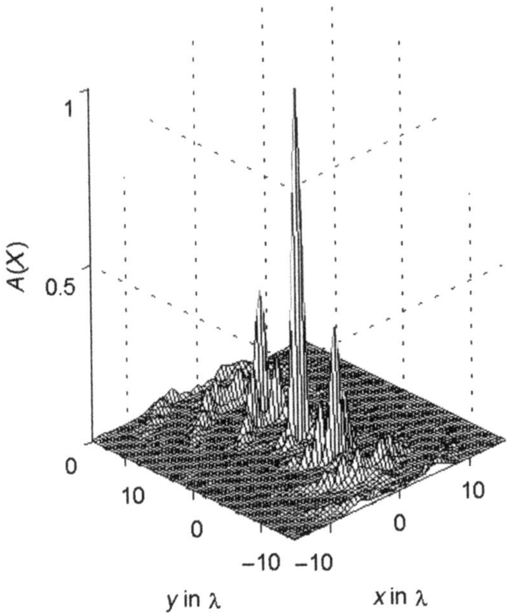

Figure 9.8 Coherent ambiguity function for 9 × 9 MIMO radar.

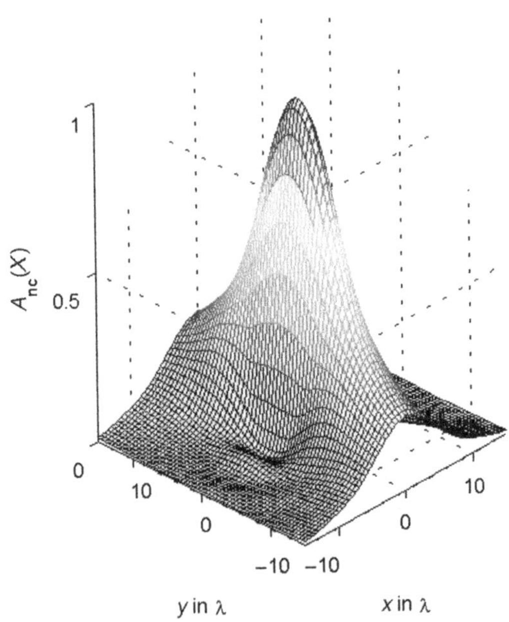

Figure 9.9 Noncoherent ambiguity function for 9 × 9 MIMO radar.

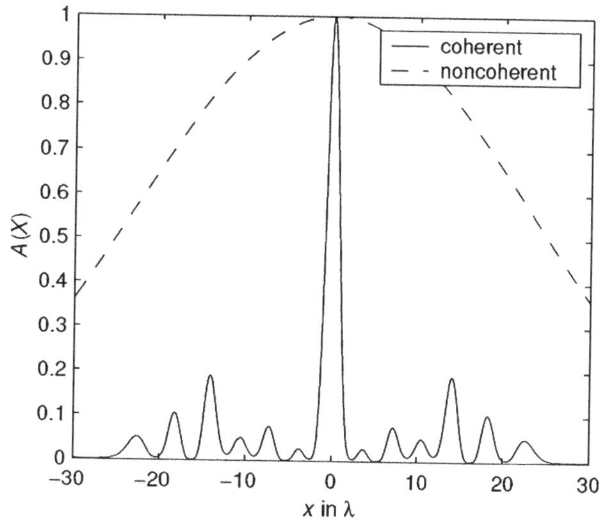

Figure 9.10 Crosscut of coherent and noncoherent ambiguity functions.

while the mainlobe of the noncoherent ambiguity function (determined by the bandwidth) is much wider.

The change in the ambiguity function as a function of the x coordinate is presented in Fig. 9.10. This figure demonstrates how coherent processing enhances the location resolution by narrowing the mainlobe of the ambiguity function. Furthermore, the figure illustrates how the noncoherent ambiguity function serves as an upper bound of the coherent ambiguity function.

The discussion above was concerned with a high-resolution mode of MIMO radar applied to a point target. High-resolution processing can also be applied to multiple targets. Next, we demonstrate the ability of MIMO radar to resolve four targets located at coordinates (0;0), (0;6λ), (6λ;0), and (6λ;6λ), respectively. Figure 9.11 is a three-dimensional representation of the ambiguity function for four targets [22]. To compute the ambiguity function for this case, the first $s_k(t)$ in (9.27) is substituted with $\sum_{k=1}^{4} s_k(t - \tau_{\ell k}(X_k))$; where the X_k terms represent the targets' locations. The four targets are clearly visible.

The high-resolution mode supported by coherent processing of widely distributed sensors merits further discussion. It is well known that given a linear array of length L, fully populated with sensors spaced at $\lambda/2$ intervals, the cross-range resolution (the dimension parallel to the array) at distance d from the array is approximately $\delta = d\lambda/L$. Thinning of the array preserves the resolution, but gives rise to grating lobes. Randomly placing the sensors of the thinned array breaks up the grating lobes at the cost of higher sidelobes than in the fully populated array. Such arrays are referred to as *random arrays*, and their properties have been investigated [36]. Statistical analysis for a single transmitter and N receivers has shown that the mean sidelobe level (MSL) is approximately $1/N$. The resolution of MIMO radar

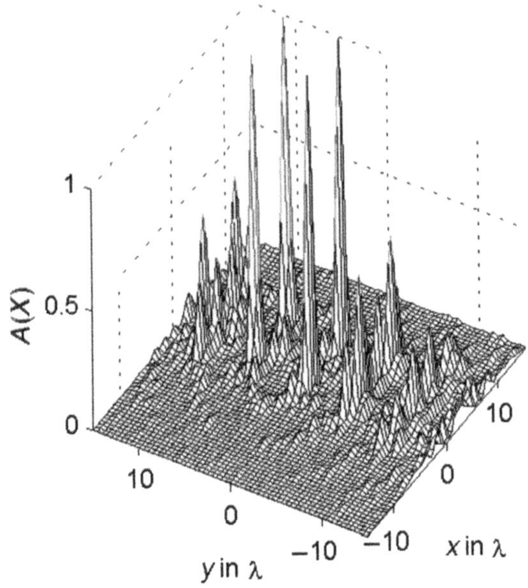

Figure 9.11 Coherent ambiguity function for four scatterers in close proximity and a 9×9 MIMO radar.

configured with widely separated elements scales with the carrier wavelength λ, and thus, can greatly exceed the bandwidth resolution barrier set by the radar-illuminating waveform.

9.4.2 CRLB

The CRLB provides a lower bound for the mean-square error (MSE) of any unbiased estimator for an unknown parameter vector $\boldsymbol{\theta}$ such that [38]

$$E\left\{(\widehat{\boldsymbol{\theta}}_i - \boldsymbol{\theta}_i)(\widehat{\boldsymbol{\theta}}_i - \boldsymbol{\theta}_i)^H\right\} \geq [C_{\mathrm{CRLB}}]_{i,i}; \quad i = 1, \ldots, 4 \tag{9.30}$$

and it is given by the inverse of the Fisher information matrix (FIM) $\mathbf{I}(\theta)$:

$$C_{\mathrm{CRLB}} = \mathbf{I}^{-1}(\theta) \tag{9.31}$$

The FIM for the vector $\boldsymbol{\theta}$ is defined as

$$\mathbf{I}(\theta) = E[\nabla_\theta \log p(\mathbf{r}; \theta)(\nabla_\theta \log p(\mathbf{r}; \theta))^H] \tag{9.32}$$

Detailed derivation of the FIM for $\boldsymbol{\theta} = [x, y, \zeta_r, \zeta_i]^T$, is found in Appendix 9A and results in the following expression:

$$\mathbf{I}(\theta) = \begin{bmatrix} \frac{af_R}{\lambda^2}\sum_{k=1}^{M}\sum_{\ell=1}^{N} A_{\ell k}^2 & \frac{af_R}{\lambda^2}\sum_{k=1}^{M}\sum_{\ell=1}^{N} A_{\ell k}B_{\ell k} & -\frac{b}{\lambda}\zeta_i\sum_{k=1}^{M}\sum_{\ell=1}^{N} A_{\ell k} & \frac{b}{\lambda}\zeta_r\sum_{k=1}^{M}\sum_{\ell=1}^{N} A_{\ell k} \\ \frac{af_R}{\lambda^2}\sum_{k=1}^{M}\sum_{\ell=1}^{N} B_{\ell k}A_{\ell k} & \frac{af_R}{\lambda^2}\sum_{k=1}^{M}\sum_{\ell=1}^{N} B_{\ell k}^2 & -\frac{b}{\lambda}\zeta_i\sum_{k=1}^{M}\sum_{\ell=1}^{N} B_{\ell k} & \frac{b}{\lambda}\zeta_r\sum_{k=1}^{M}\sum_{\ell=1}^{N} B_{\ell k} \\ -\frac{b}{\lambda}\zeta_i\sum_{k=1}^{M}\sum_{\ell=1}^{N} A_{\ell k} & -\frac{b}{\lambda}\zeta_i\sum_{k=1}^{M}\sum_{\ell=1}^{N} B_{\ell k} & \frac{bMN}{2\pi} & 0 \\ \frac{b}{\lambda}\zeta_r\sum_{k=1}^{M}\sum_{\ell=1}^{N} A_{\ell k} & \frac{b}{\lambda}\zeta_r\sum_{k=1}^{M}\sum_{\ell=1}^{N} B_{\ell k} & 0 & \frac{bMN}{2\pi} \end{bmatrix} \tag{9.33}$$

where ζ_r and ζ_i are respectively real and imaginary parts of ζ, the constants a and b are respectively defined as $a = (8\pi^2/\sigma_w^2)\left|\sqrt{E/M}\zeta\right|^2$ and $b = (4\pi/\sigma_w^2)\left|\sqrt{E/M}\right|^2$. The frequency ratio term f_R is defined as $f_R = (1 + (\beta^2/f_c^2))$. λ is the carrier wavelength defined as $\lambda = c/f_c$. The term β is the efficient bandwidth defined as

$$\beta = \frac{\int_W f^2 |S(f)|^2 df}{\int_W |S(f)|^2 df}$$

where the integration is over the bandwidth W. Other quantities were defined previously. The terms $A_{\ell k}$ and $B_{\ell k}$ [defined in connection with (9.11)] reflect the relation between the topology of the radars and the target localization and hence can afford insight into the effect of the topology on target localization estimation. While the CRLB expresses the lower bound on the MSE of the estimate of $\theta = [x, y, \zeta_r, \zeta_i]^T$, we are really interested only in the estimation of x and y. The target reflectivity term ζ_r and ζ_i serve as nuisance parameters. The lower bounds on the variance of x and y are given by the elements of the CRLB submatrix $[\mathbf{C}_{\text{CRLB}}]_{2\times 2} = [\mathbf{I}(\theta)]^{-1}_{2\times 2}$, where from (9.31), we have $\text{var}[x] \geq [\mathbf{I}(\theta)]^{-1}_{1,1}$ and $\text{var}[y] \geq [\mathbf{I}(\theta)]^{-1}_{2,2}$. Deriving the inversion of the FIM in (9.33), needed in order to obtain the CRLB matrix, is an intricate mathematical manipulation. For the covariance matrix of x and y it is sufficient to deriving the submartix $[\mathbf{C}_{\text{CRLB}}]_{2\times 2}$, which include positions $[\mathbf{I}(\theta)]^{-1}_{1,1}$, $[\mathbf{I}(\theta)]^{-1}_{1,2}$, $[\mathbf{I}(\theta)]^{-1}_{2,1}$ and $[\mathbf{I}(\theta)]^{-1}_{2,2}$. Detailed calculation of $[\mathbf{C}_{\text{CRLB}}]_{2\times 2}$ is provided in Appendix 9B, where the following result is obtained:

$$[\mathbf{C}_{\text{CRLB}}]_{2\times 2} = \frac{\lambda^2}{a \cdot f_R} \cdot \frac{1}{u_{\text{CRLB}}} \cdot [\mathbf{G}_{\text{CRLB}}]_{2\times 2} \tag{9.34}$$

The term u_{CRLB} is a summation and multiples of the terms $A_{\ell k}$ and $B_{\ell k}$

$$u_{\mathrm{CRLB}} = \left[\sum_{k=1}^{M}\sum_{\ell=1}^{N} A_{\ell k}^2 - \frac{1}{f_R MN}\left(\sum_{k=1}^{M}\sum_{\ell=1}^{N} A_{\ell k}\right)^2\right] \times \left[\sum_{k=1}^{M}\sum_{\ell=1}^{N} B_{\ell k}^2 - \frac{1}{f_R MN}\left(\sum_{k=1}^{M}\sum_{\ell=1}^{N} B_{\ell k}\right)^2\right] - \left[\sum_{k=1}^{M}\sum_{\ell=1}^{N} A_{\ell k}B_{\ell k} - \frac{1}{f_R MN}\left(\sum_{k=1}^{M}\sum_{\ell=1}^{N} A_{\ell k}\sum_{k=1}^{M}\sum_{\ell=1}^{N} B_{\ell k}\right)\right]^2 \tag{9.35}$$

and the submatrix $[\mathbf{G}_{\mathrm{CRLB}}]_{2\times 2}$ induce the noncommon terms to the covariance of x and y:

$$[\mathbf{G}_{\mathrm{CRLB}}]_{2\times 2} = \begin{bmatrix} \sum_{k=1}^{M}\sum_{\ell=1}^{N} B_{\ell k}^2 - \frac{1}{f_R MN}\left(\sum_{k=1}^{M}\sum_{\ell=1}^{N} B_{\ell k}\right)^2 & \frac{1}{f_R MN}\left(\sum_{k=1}^{M}\sum_{\ell=1}^{N} A_{\ell k}\sum_{k=1}^{M}\sum_{\ell=1}^{N} B_{\ell k}\right) - \sum_{k=1}^{M}\sum_{\ell=1}^{N} A_{\ell k}B_{\ell k} \\ \frac{1}{f_R MN}\left(\sum_{k=1}^{M}\sum_{\ell=1}^{N} A_{\ell k}\sum_{k=1}^{M}\sum_{\ell=1}^{N} B_{\ell k}\right) - \sum_{k=1}^{M}\sum_{\ell=1}^{N} A_{\ell k}B_{\ell k} & \sum_{k=1}^{M}\sum_{\ell=1}^{N} A_{\ell k}^2 - \frac{1}{f_R MN}\left(\sum_{k=1}^{M}\sum_{\ell=1}^{N} A_{\ell k}\right)^2 \end{bmatrix} \tag{9.36}$$

It is apparent that var[x] and var[y] are dependent of the wavelength λ, the reflected energy $\left|\sqrt{(E/M)}\zeta\right|^2$, and the efficient bandwidth : carrier frequency ratio β^2/f_c^2 concealed in the term f_R. The most interesting aspect here is the dependence of the lower bound on the MSE on the carrier frequency. Taken together with our earlier discussion on the ambiguity function, the CRLB expression points to the potential of MIMO radar with widely distributed sensors for supporting high-resolution and high-precision target localization.

9.4.3 MLE Target Localization

Having developed an expression for the lower bound on the MSE for target localization, we turn our attention to target localization techniques and their performance relative to the CRLB. An estimator is *optimal* if it is unbiased and its variance meets the CRLB. In our MIMO radar problem, since the observations are nonlinear functions of the target location unknown parameters x and y, it is unlikely that such an estimator exists. Alternatively, we consider two simpler estimators: the MLE and the BLUE. The MLE is motivated by its asymptotic optimality; the BLUE, by its closed-form expression.

The MLE is a practical estimator in the sense that its application to a problem of observations in white Gaussian noise is relatively straightforward. Moreover, under mild conditions on the probability density function of the observations $p(\mathbf{r}; \theta)$, the MLE of the unknown parameter θ is asymptotically unbiased, and it asymptotically attains the CRLB [38]. Hence the MLE is optimal for large data records.

For the case of MIMO radar, the signal waveform received by radar ℓ is given in (9.7). The MLE of the unknown parameter vector $\boldsymbol{\theta} = [x, y, \zeta_r, \zeta_i]^T$ given the observation vector $\mathbf{r}$ is expressed by [39]:

$$\widehat{\theta}_{ML} = \arg\left\{\max_{\theta}\left[\log p(\mathbf{r}; \theta)\right]\right\} = \arg\left\{\max_{\mathbf{x}}\left\{\max_{\zeta_r,\zeta_i}\left[\log p(\mathbf{r}; \mathbf{x}, \zeta_r, \zeta_i)\right]\right\}\right\} \quad (9.37)$$

where we defined the location vector $\mathbf{x} = [x, y]^T$ and $p(\mathbf{r}; \theta)$ was expressed in (9.26). We are not interested in estimating ζ_r, ζ_i explicitly, but need its estimate for the estimation of x and y. Recall that $\zeta_r + j\zeta_i = \zeta$; hence the MLE of ζ is found from the following condition on the loglikelihood:

$$\frac{\partial}{\partial \zeta} \log p(\mathbf{r}; \mathbf{x}, \zeta)|_{\zeta=\zeta_{ML}} = 0 \quad (9.38)$$

Using (9.26) in (9.38) and the relation

$$\sum_{k=1}^{M}\sum_{\ell=1}^{N}\frac{E}{M}|\zeta|\int |s_k(t-\tau_{\ell k}(X))|^2 dt = \frac{E}{M}MN|\zeta| \quad (9.39)$$

we derive the MLE for the reflectivity index:

$$\widehat{\zeta}_{ML} = \frac{1}{\sqrt{E/M}MN}\sum_{k=1}^{M}\sum_{\ell=1}^{N}\rho_{\ell k}^{*}(\mathbf{x})\int r_\ell(t)s_k^{*}(t-\tau_{\ell k}(\mathbf{x}))dt \quad (9.40)$$

In the last expression, we changed the notation slightly from (9.26), by interpreting the coordinate pair $X = (x, y)$ as a two dimensional vector $\mathbf{x} = [x, y]^T$. This makes the notation consistent with estimation theory literature, where the estimation of the pair of coordinates X amounts to estimation of a two-dimensional unknown vector $\mathbf{x}$.

The MLE of the unknown vector $\mathbf{x}$ is then derived by substituting the reflectivity index MLE $\widehat{\zeta}_{ML}$ (9.40), back into (9.26) and maximizing the loglikelihood function $\log p(\mathbf{r}; \mathbf{x}|\widehat{\zeta}_{ML})$ with respect to $\mathbf{x}$ such that:

$$\widehat{\mathbf{x}}_{ML} = \arg\left\{\max_{\mathbf{x}}\left|\sum_{k=1}^{M}\sum_{\ell=1}^{N}\rho_{\ell k}^{*}(\mathbf{x})\int r_\ell(t)s_k^{*}(t-\tau_{\ell k}(\mathbf{x}))dt\right|\right\} \quad (9.41)$$

Since a closed-form expression cannot be found for the MLE in (9.41), numerical methods need to be applied. A grid search or an iterative maximization of the likelihood function needs to be performed to determine $\widehat{\mathbf{x}}_{ML}$. This might involve a significant computational effort. In Section 9.2.1.1, we introduced the location $X_c = (x_c, y_c)$ as an initial, nominal estimate of the target location. In practice, we can limit the search grid for high-resolution target localization estimation to an area around X_c. When iteration methods are used to determine $\widehat{\mathbf{x}}_{ML}$, the nominal estimate can serve to initialize the search.

To compare the CRLB with the mean-squared error (MSE) of the MLE, we employed a setup similar to the one presented in Section 9.4.1. The MSE is obtained using Monte Carlo simulations with 2000 iterations per SNR value of a 9×9 MIMO system. Figure 9.12 presents results for the CRLB and the MSE of the MLE for estimating x and y, as a function of the SNR, where $\text{SNR} = \left(\sqrt{(E/M)}|\zeta|/\sigma_w\right)^2$. It is observed that the empirical MSEs and the theoretical bounds are in excellent agreement at high SNR values. For low SNR, the CRLB does not provide a tight bound. This is a well-known reflection of the fact that the CRLB is a bound for small errors. As the SNR degrades, the error mechanism changes from noise to ambiguities in the estimate. The high level of errors in the lower-SNR region are a manifestation of the existence of ambiguous estimates consistent with the ambiguities observed in the ambiguity function as discussed earlier. It is also observed in the figure that the x and y estimates have different MSEs. This is indicative of the effect that the geometry of sensors relative to the target may have on accuracy. This topic is further discussed in Section 9.4.5, on GDOP.

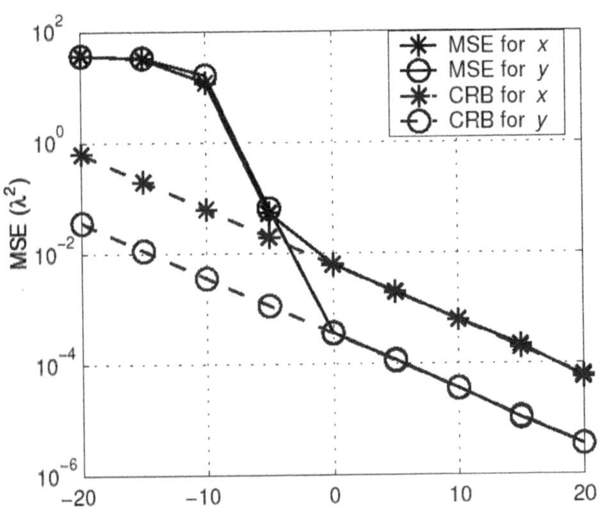

Figure 9.12 MSE of ML estimate and the CRLB for a 9×9 MIMO radar.

9.4.4 BLUE Target Localization

The MLE derived in Section 9.4.3 does not have a closed-form expression, and numerical methods need to be used to solve it. We now exploit the linear perturbations model introduced in Section 9.2.1.1 to develop the minimum variance linear estimate for the unknown vector $\varphi = [x, y, \mu]^T$, where the set (x, y) denotes the target position as defined in Section 9.2.1.1 and μ is an unknown nuisance parameter produced by the time offset between the radars and the target. Beyond its utility as a location estimator, the BLUE provides insights into performance evaluation by facilitating the GDOP analysis undertaken in the next section.

To develop the BLUE, we recall the linear model in (9.11) between time delays and target location. Letting the vector of time delays $\tau = [\tau_{11}, \tau_{12}, \ldots, \tau_{NM}]^T$, where we disregard the dependency on (X) (since in this model, time delays need to be measured/estimated directly), and defining the angle dependent matrix

$$\mathbf{D} = -\frac{1}{c}\begin{bmatrix} A_{11} & B_{11} & 1 \\ A_{12} & B_{12} & 1 \\ \cdots & \cdots & \cdots \\ A_{NM} & B_{NM} & 1 \end{bmatrix}_{NM\times 3} \tag{9.42}$$

we postulate the following model between the observable time delays and the target location

$$\tau = \mathbf{D}\varphi + \epsilon \tag{9.43}$$

where ϵ is the $MN \times 1$ measurement noise vector. We assume that the elements of ϵ are i.i.d. with some variance. Notice that one of the differences between the MLE and BLUE models is that the MLE target localization is carried out utilizing signal observations, while according to (9.43), the BLUE's "observations" are in the form of time delays. So an intermediate step of time-delay estimation is implied, and ϵ are the errors associated with that process. The time-delay estimates used as observations $\tau_{\ell k}$ are derived by the MLE as

$$\tau_{\ell k} = \arg\max_{v}\left[\exp[j2\pi f_c v]\left|\int r_\ell(t)s_k^*(t-v)dt\right|\right] \tag{9.44}$$

where v is a dummy variable for the time delay. This expression differs from (9.17) by the term $\exp[j2\pi f_c v]$, which measures the phase difference between the transmitted waveform $s_k(t)$ and the received waveform $r_\ell(t)$.

In Appendix 9C, it is shown that the maximum-likelihood time-delay estimates are unbiased with error covariance matrix

$$\mathbf{C}_\epsilon = \frac{\sigma_w^2}{8\pi^2\left|\zeta\sqrt{E/M}\right|^2(\beta^2+f_c^2)}\mathbf{I}_{NM\times NM} \tag{9.45}$$

For the linear and Gaussian model in (9.43), the BLUE is computed from the Gauss–Markov theorem [38], which states that the BLUE of the unknown vector φ is given by the expression

$$\widehat{\phi}_{\text{BLUE}} = (\mathbf{D}^T \mathbf{C}_\epsilon^{-1} \mathbf{D})^{-1} \mathbf{D}^T \mathbf{C}_\epsilon^{-1} \tau \tag{9.46}$$

The theorem also determines the error covariance matrix to be

$$\mathbf{C}_{\text{BLUE}} = (\mathbf{D}^T \mathbf{C}_\epsilon^{-1} \mathbf{D})^{-1} \tag{9.47}$$

If the data [time delays in (9.43)] is truly Gaussian, then the BLUE is also the minimum variance estimate, which makes it optimal in the sense of meeting the CRLB [38].

Using the time error covariance matrix $\mathbf{C}_\epsilon$ and the linear transformation matrix $\mathbf{D}$ in (9.46) to get the BLUE for the target localization, we obtain the following estimate for φ:

$$\widehat{\varphi}_{\text{BLUE}} = -c \left[\frac{a \cdot f_R}{\lambda^2} \right] \cdot \mathbf{C}_{\text{BLUE}} \cdot \begin{bmatrix} \sum_{k=1}^{M} \sum_{\ell=1}^{N} A_{\ell k} \tau_{\ell k} \\ \sum_{k=1}^{M} \sum_{\ell=1}^{N} B_{\ell k} \tau_{\ell k} \\ \sum_{k=1}^{M} \sum_{\ell=1}^{N} \tau_{\ell k} \end{bmatrix} \tag{9.48}$$

where $A_{\ell k}$, $B_{\ell k}$, a, and f_R are as defined previously, and $\tau_{\ell k}$ are the observations. The BLUE for the target localization for $\mathbf{x} = [x, y]^T$ is

$$\widehat{\mathbf{x}}_{\text{BLUE}} = [\widehat{\varphi}_{\text{BLUE}}]_{2\times 1} \tag{9.49}$$

Using matrix $\mathbf{C}_\epsilon$ and $\mathbf{D}$ in (9.47), we can obtain an expression of the target localization covariance matrix

$$[\mathbf{C}_{\text{BLUE}}]_{2\times 2} = \frac{\lambda^2}{a \cdot f_R} \cdot \frac{1}{u_{\text{BLUE}}} \cdot \mathbf{G}_{\text{BLUE}} \tag{9.50}$$

where

$$\begin{aligned} u_{\text{BLUE}} = & \left[\sum_{k=1}^{M} \sum_{\ell=1}^{N} A_{\ell k}^2 - \frac{1}{MN} \left(\sum_{k=1}^{M} \sum_{\ell=1}^{N} A_{\ell k} \right)^2 \right] \\ & \times \left[\sum_{k=1}^{M} \sum_{\ell=1}^{N} B_{\ell k}^2 - \frac{1}{MN} \left(\sum_{k=1}^{M} \sum_{\ell=1}^{N} B_{\ell k} \right)^2 \right] \\ & - \left[\sum_{k=1}^{M} \sum_{\ell=1}^{N} A_{\ell k} B_{\ell k} - \frac{1}{MN} \sum_{k=1}^{M} \sum_{\ell=1}^{N} A_{\ell k} \sum_{k=1}^{M} \sum_{\ell=1}^{N} B_{\ell k} \right) \right]^2 \end{aligned} \tag{9.51}$$

and

$$\mathbf{G}_{\text{BLUE}}=\begin{bmatrix} \sum_{k=1}^{M}\sum_{\ell=1}^{N}B_{\ell k}^{2}-\frac{1}{MN}\left(\sum_{k=1}^{M}\sum_{\ell=1}^{N}B_{\ell k}\right)^{2} & \frac{1}{MN}\left(\sum_{k=1}^{M}\sum_{\ell=1}^{N}A_{\ell k}\sum_{k=1}^{M}\sum_{\ell=1}^{N}B_{\ell k}\right) \\ & -\sum_{k=1}^{M}\sum_{\ell=1}^{N}A_{\ell k}B_{\ell k} \\ \frac{1}{MN}\left(\sum_{k=1}^{M}\sum_{\ell=1}^{N}A_{\ell k}\sum_{k=1}^{M}\sum_{\ell=1}^{N}B_{\ell k}\right) & \\ -\sum_{k=1}^{M}\sum_{\ell=1}^{N}A_{\ell k}B_{\ell k} & \sum_{k=1}^{M}\sum_{\ell=1}^{N}A_{\ell k}^{2}-\frac{1}{MN}\left(\sum_{k=1}^{M}\sum_{\ell=1}^{N}A_{\ell k}\right)^{2} \end{bmatrix} \tag{9.52}$$

The terms a and f_R were defined in Section 9.4.2. The matrix in (9.50) provides explicit expressions for the localization estimation error variances, $\sigma_x^2=[\mathbf{C}_{\text{BLUE}}]_{1,1}$ and $\sigma_y^2=[\mathbf{C}_{\text{BLUE}}]_{2,2}$. It can be readily seen that for $\beta^2/f_c^2 \ll 1$, the BLUE covariance matrix for $\mathbf{x}$ is identical to the CRLB in (9.34).

The MSE of the BLUE is dependent on the following parameters:

1. Effective bandwidth β and carrier frequency f_c—for a narrowband signal, $\beta^2/f_c^2 \ll 1$; hence $(1+(\beta^2/f_c^2)) \approx 1$, and the MSE in (9.50) is proportional to the squared carrier signal wavelength λ^2. The proportionality to λ^2 or to $1/f_c^2$ is due to the fact that coherent MIMO systems make use of the phase information across different paths.
2. SNR term a—the MSE is inverse proportional to the SNR term a.
3. Location of the sensing radars—the terms $A_{\ell k}$ and $B_{\ell k}$ relate to the geometric layout of the transmitting and receiving radars. From these above terms one cannot intuitively identify the relation between the radars positions and the obtainable accuracy. For this reason, we need to define a more suitable method to express these relations. This is the focus of the next section.

9.4.5 GDOP

The GDOP metric is commonly used in the global positioning system (GPS) for mapping the attainable localization precision for a given layout of GPS satellites positions [40,41,43]. Here, we seek a metric that expresses the effect of the positions of the transmitting and receiving elements on the relationship between the time-delay estimation errors and the localization errors.

The GDOP metric for the two dimensional case is defined as

$$\text{GDOP} = \sqrt{\frac{\sigma_x^2+\sigma_y^2}{\sigma_\epsilon^2 c^2}} \tag{9.53}$$

where σ_x^2 and σ_y^2 are the variances of localization on the x and y axes, respectively, and σ_ϵ^2 is the time-delay variance in (9.45).

Inherently, this metric provides a normalized value that represents the relative contribution of the radars' location to the overall accuracy. Using GDOP contour maps provides a clear view of high-accuracy areas for a given set of radar locations. GDOP maps could also serve as a tool for choosing favorable radar locations to cover a given target area.

The BLUE enables numerical computation of the GDOP. Substituting $\sigma_x^2 = [\mathbf{C}_{\text{BLUE}}]_{1,1}$ and $\sigma_y^2 = [\mathbf{C}_{\text{BLUE}}]_{2,2}$, where the BLUE covariance matrix $\mathbf{C}_{\mathbf{x}_{\text{BLUE}}}$ is as given in (9.50), we get the following GDOP expression:

$$\text{GDOP} = \sqrt{\frac{\sum_{k=1}^{M}\sum_{\ell=1}^{N}(B_{\ell k}^2 + A_{\ell k}^2) - (1/MN)\left[\left(\sum_{k=1}^{M}\sum_{\ell=1}^{N} B_{\ell k}\right)^2 + \left(\sum_{k=1}^{M}\sum_{\ell=1}^{N} A_{\ell k}\right)^2\right]}{u_{\text{BLUE}}}} \tag{9.54}$$

Sensor locations are embedded in the terms $A_{\ell k}$ and $B_{\ell k}$. The GDOP reduces all the locations to a single metric. Once we get the values mapped, the actual localization error is easily derived by multiplying the GDOP value with $c\sigma_\epsilon$.

In Figs. 9.13 and 9.14, contour plots of the GDOP values are presented for the case of respectively $M = N = 3$ and $M = N = 5$ radars located symmetrically around the origin. The radars are all transmitting orthogonal signals and perform time-delay observations consistent with the BLUE scheme. The first noticeable factor in the comparison of the two plots is the higher accuracy obtained with five radars compared to three radars. For example, the lowest GDOP value for the case of three radars is 0.48, while with five radars, the GDOP is 0.29, corresponding to a 64% improvement. All the targets located inside the virtual N-sided polygon achieve relatively high accuracy, while the most accurate localization is obtained for the target at the center. The increase in GDOP values from the center to the polygon boundaries is slow. Outside this polygon, the GDOP values increase rather rapidly.

In Figs. 9.15 and 9.16 contours of five nonsymmetrically positioned radars are drawn. When the radars are spread around the target, there are still some areas with good measurement accuracy, although the coverage is shrunk compared to the case with symmetric distribution of sensors in Fig. 9.14. When the viewing angle of the target is very restricted, as in the bottom of Fig. 9.16, there is a marked degradation of GDOP values.

These examples have shown that a symmetric deployment of sensors around the target yield the lowest GDOP values. They raise the question as to the lowest GDOP that can be obtained.

9.4.5.1 Lowest GDOP Theoretical analysis of the lowest attainable GDOP for a passive system (such as GPS) was performed by Lee [40] and Leavanon [41]. These studies that for a system with N satellites optimally positioned around the user, where the user is measuring the signals time of arrival to estimate its position, the lowest achievable GDOP value is $2/\sqrt{N}$. In Appendix 9D, we show that for a MIMO radar with M transmitters and N receivers, the result is of the order of $\sqrt{2/MN}$. This result is interpreted as a "MIMO advantage" compared to the single-transmit-antenna case.

As a numerical example, the ratio of the lowest GDOPs in Figs. 9.13 and 9.14 is $0.48/0.29 = 1.66$, whereas the ratio of the number of elements in the systems is given by $5/3 = 1.67$.

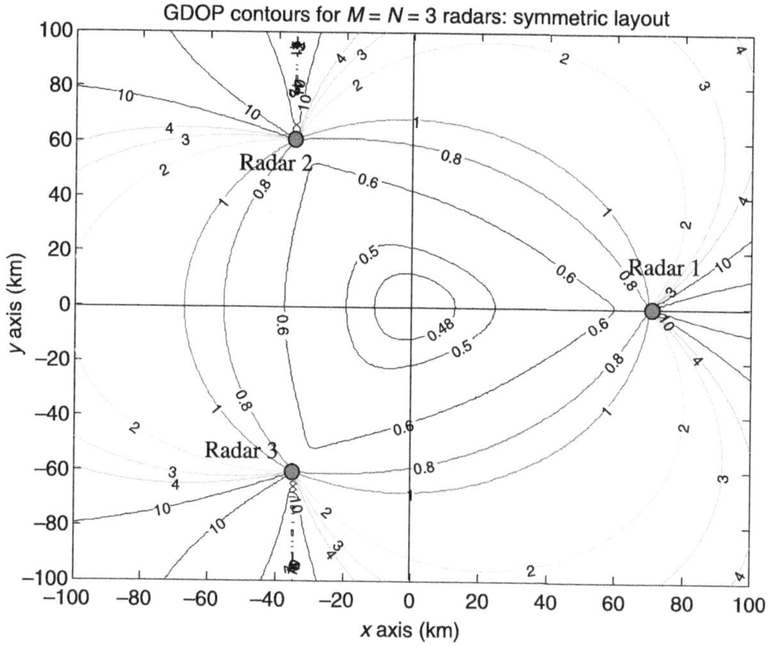

Figure 9.13 GDOP contours for a symmetric positioning of radars around the axis origin: case (a) for $M = 3$ transmitting radars and $N = 3$ receiving radars.

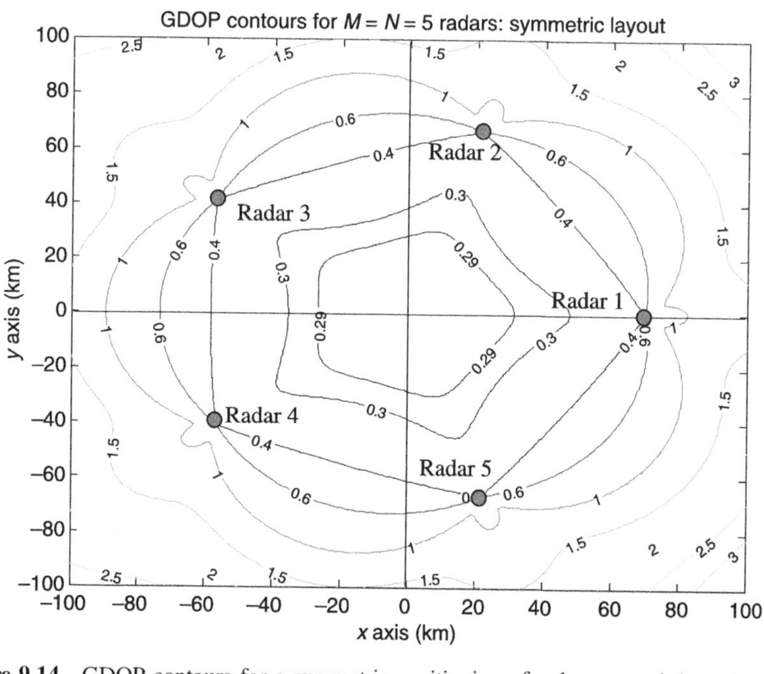

Figure 9.14 GDOP contours for a symmetric positioning of radars around the axis origin: case (b) for $M = N = 5$.

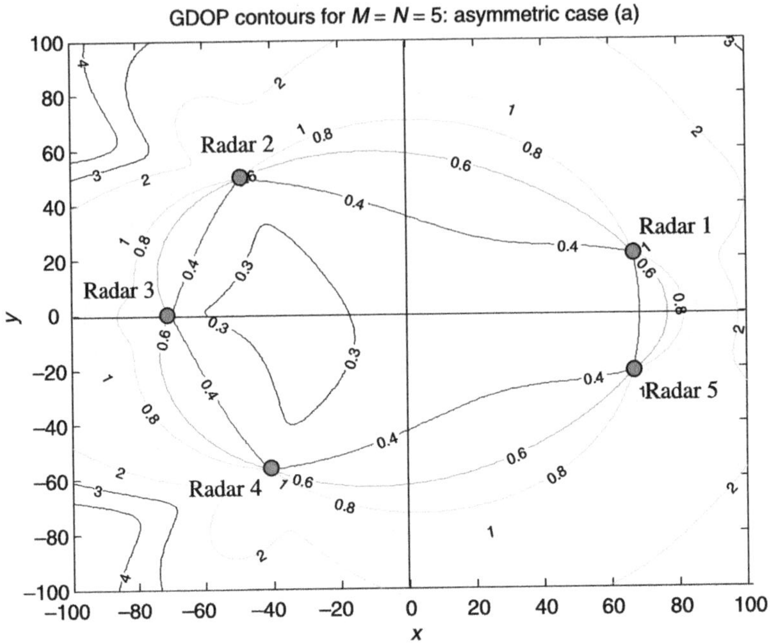

Figure 9.15 GDOP contours for an asymmetric positioning of radars around the axis origin: case (a) for radars surrounding origin in an asymmetric form.

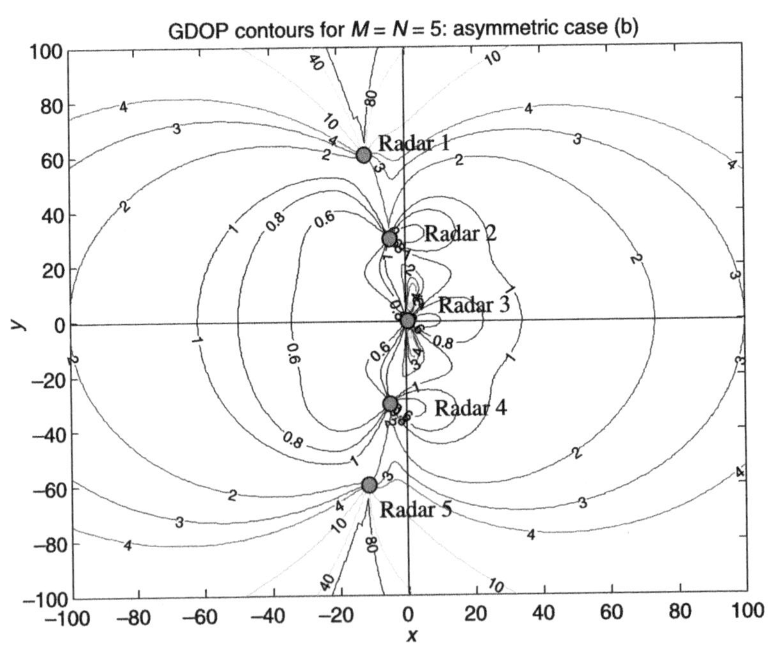

Figure 9.16 GDOP contours for an asymmetric positioning of radars around the axis origin: case (b) for radars are almost aligned.

9.4.6 Discussion

Knowledge of the lower bound on accuracy is very important when its comes to system design. The lower bound on the achievable accuracy, as expressed by the CRLB, provides a tight bound in high SNR, as demonstrated in Fig. 9.12. As for low SNR, the CRLB does not give a rigorous bound. As stated by Weinstein and Weiss [44], the MLE is asymptotically unbiased and its error variance approaches the CRLB arbitrarily close for sufficient long observation time, with the condition that the MLE is not subject to ambiguities. As the ML of the time estimates is based on matched filters at the receiver end, the ambiguity features of the signal waveforms (as presented in Section 9.4.1) arise in low-SNR conditions and predominate the estimation capabilities, causing erroneous time estimates. As the ambiguity problems are usually addressed through the signal waveform design, a more rigid bound needs to be found for the localization error variance in the low-SNR case.

9.5 CHAPTER SUMMARY

In this chapter, we discussed concepts of MIMO radar with noncoherent and coherent processing. The chapter focused on applications with widely dispersed antenna elements. Generally speaking, MIMO radars transmit multiple waveforms, receive signals at multiple antennas, and process them jointly. Processing may be carried out noncoherently (as in Section 9.4) or coherently (as in Section 9.5).

The main topics discussed in the chapter can be summarized as follows:

- Complex targets contain a large number of scatterers that result in diverse RCS patterns as a function of angle. Decorrelation of the elements of the channel matrix occurs for complex targets and widely separated antennas.
- For a target with spatially diverse backscatter, optimal processing of MIMO radar with suitably located sensors leads to diversity gains in the form of improved statistics of the SNR. The processing collects the energy reflected by the scatterers and combines it noncoherently.
- Observations over a wide angular sector can be exploited to detect targets moving in arbitrary directions.
- MIMO radar can locate targets with highresolution and can resolve between closely spaced targets. Estimation errors for the high-resolution mode are dependent on the carrier wavelength. Nevertheless, high-resolution target localization is a challenging proposition due to the ambiguity sidelobes and the need to phase synchronize a multisite radar system.

MIMO radar with widely separated sensors is a very promising concept for future, high-performance radars. Yet many research and engineering challenges need to be addressed to advance MIMO radar from concept to reality. Engineering challenges include centralized coordination of sensor transmissions, synchronized communication with a processing center, and highly precise phase synchronization among sensors (of the order of nanoseconds for resolutions in meters). Research challenges include a better understanding of bistatic and multistatic RCS phenomena, tracking targets with MIMO radar, and MIMO radar on airborne platforms.

APPENDIX 9A DERIVING THE FIM

In this appendix, we develop the FIM for the unknown parameter vector $\boldsymbol{\theta}$ resulting in (9.33).

The FIM is defined as

$$\mathbf{I}(\theta) = E[\nabla_\theta \log p_\theta(\mathbf{r})(\nabla_\theta \log p_\theta(\mathbf{r}))^H] \tag{9A.1}$$

where $\boldsymbol{\theta} = [x, y, \zeta_r, \zeta_i]^T$ and the divergence operator ∇_θ is defined as $[(\partial/\partial x), (\partial/\partial y), (\partial/\partial\zeta_r), (\partial/\partial\zeta_i)]^T$. Since the density function $p_\theta(\mathbf{r})$ in (26) is defined as a function of the time of arrival, $\tau_{\ell k}(X)$, we need to use the alternative form [38] to express the FIM:

$$\mathbf{I}(\theta) = (\nabla_\theta \vartheta^H) E\{\nabla_\vartheta \log p_\theta(\mathbf{r})(\nabla_\vartheta \log p_\theta(\mathbf{r}))^H\}(\nabla_\theta \vartheta^H)^H \tag{9A.2}$$

where $\vartheta = [\tau_{11}(X), \ldots, \tau_{\ell k}(X), \ldots, \tau_{NM}(X), \zeta_r, \zeta_i]^T$ and the time delays $\tau_{\ell k}(X)$ are defined in (9.10).

By using the notation $\mathbf{I}(\vartheta) = E\{\nabla_\vartheta \log p_\theta(\mathbf{r})(\nabla_\vartheta \log p_\theta(\mathbf{r}))^H\}$, we rewrite (9A.2) as follows:

$$\mathbf{I}(\theta) = (\nabla_\theta \vartheta^H)\mathbf{I}(\vartheta)(\nabla_\theta \vartheta^H)^H \tag{9A.3}$$

The value of $\nabla_\theta \vartheta^H$ can then be calculated as

$$\nabla_\theta \vartheta^H = \begin{bmatrix} -\frac{1}{c}A_{11} & \ldots & -\frac{1}{c}A_{NM} & 0 & 0 \\ -\frac{1}{c}B_{11} & \ldots & -\frac{1}{c}B_{NM} & 0 & 0 \\ 0 & \ldots & 0 & 1 & 0 \\ 0 & \ldots & 0 & 0 & 1 \end{bmatrix} \tag{9A.4}$$

where applied the linear model and the notation in Section 9.2.1.1 for $A_{\ell k}$ and $B_{\ell k}$. Now, the expression for $\mathbf{I}(\vartheta) = E\{\nabla_\vartheta \log p_\theta(\mathbf{r})(\nabla_\vartheta \log p_\theta(\mathbf{r}))^H\}$ is derived using

$$\begin{aligned}
[\mathbf{I}(\vartheta)]_{i,i} &= \frac{\partial^2}{\partial \tau_{\ell k}^2(X)} \log p_\theta(\mathbf{r}) && (i = 1 \ldots NM) \\
[\mathbf{I}(\vartheta)]_{i,i} &= \frac{\partial^2}{\partial \zeta_r^2} \log p_\theta(\mathbf{r}) && (i = NM + 1) \\
[\mathbf{I}(\vartheta)]_{i,i} &= \frac{\partial^2}{\partial \zeta_i^2} \log p_\theta(\mathbf{r}) && (i = NM + 2) \\
[\mathbf{J}(\vartheta)]_{MN+1,i} &= \frac{\partial^2}{\partial \tau_{\ell k}(X)\partial\zeta_r} \log p_\theta(\mathbf{r}) && (i = 1 \ldots NM) \\
[\mathbf{J}(\vartheta)]_{MN+2,i} &= \frac{\partial^2}{\partial \tau_{\ell k}(X)\partial\zeta_i} \log p_\theta(\mathbf{r}) && (i = 1 \ldots NM)
\end{aligned} \tag{9A.5}$$

Using the expression for $p_\theta(\mathbf{r})$ given in (9.26), we can calculate each of the terms. Starting with the first term, $[\mathbf{I}(\vartheta)]_{i,i} = (\partial^2/(\partial\tau_{\ell k}(\mathbf{x})\partial\tau_{mn}(\mathbf{x}))\log p_\theta(\mathbf{r})$ for $i = 1\cdots NM$:

$$\begin{aligned}\frac{\partial[\log p_\theta(\mathbf{r})]}{\partial\tau_{\ell k}(X)} = -\frac{1}{\sigma_w^2}\int &\left[\tau_\ell(t) - \sqrt{\frac{E}{M}}\,\zeta\sum_{k=1}^{M}\rho_{\ell k}(X)s_k(t-\tau_{\ell k}(X))\right]\\ &\times\sqrt{\frac{E}{M}}\,\zeta^*\frac{\partial[\rho^*_{\ell k}(X)s^*_k(t-\tau_{\ell k}(X))]}{\partial\tau_{\ell k}(X)}\\ &+\left[r_\ell(t) - \sqrt{\frac{E}{M}}\,\zeta\sum_{k=1}^{M}\rho_{\ell k}(X)s_k(t-\tau_{\ell k}(X))\right]^*\\ &\times\sqrt{\frac{E}{M}}\,\zeta\frac{\partial[\rho_{\ell k}(X)s_k(t-\tau_{\ell k}(X))]}{\partial\tau_{\ell k}(X)}\,dt\end{aligned} \tag{9A.6}$$

Applying the second derivative to (9A.6), we get

$$[\mathbf{I}(\vartheta)]_{i,i} = \frac{8\pi^2\left|\sqrt{(E/M)}\zeta\right|^2}{\sigma_w^2}(f_c^2+\beta^2) \quad (i = 1\cdots NM) \tag{9A.7}$$

As for the nondiagonal elements in the submatrix $[\mathbf{I}(\vartheta)]_{i,j}$ for $i = 1\ldots NM$, $j = 1\ldots NM$, $i \neq j$, it is straightforward to see from (9A.6) that these elements are equal to zero.

For the second term, $[\mathbf{I}(\vartheta)]_{i,i} = (\partial^2/\partial\zeta_r^2)\log p_\theta(\mathbf{r})$ for $(i = NM+1)$, we get the following:

$$\frac{\partial^2}{\partial\zeta_r^2}\log p_\theta(\mathbf{r}) = \frac{1}{\sigma_w^2}\left|\sqrt{\frac{E}{M}}\right|^2\sum_{k=1}^{M}\sum_{\ell=1}^{N}\int|s_k(t-\tau_{\ell k}(X))|^2dt \tag{9A.8}$$

We get a similar results for the third term $(\partial^2/\partial\zeta_i^2)\log p_\theta(\mathbf{r})$, resulting in:

$$[\mathbf{I}(\vartheta)]_{i,i} = \left|\sqrt{\frac{E}{M}}\right|^2\frac{2MN}{\sigma_w^2} \quad (i = NM+1,\, NM+2)$$

It is simple to show that the nondiagonal elements for the submatrix $\mathbf{I}(\vartheta)]_{i,j} = (\partial^2/(\partial\zeta_r\partial\zeta_i))\log p_\theta(\mathbf{r})$ for $i = NM+1,\ NM+2,\ j = NM+1,\ NM+2,\ i \neq j$ are equal to zero.

For the fourth term we calculate the derivative of (9A.6) with respect to the real part of the reflectivity index ζ

$$\frac{\partial^2\log p_\theta(\mathbf{r})}{\partial\tau_{\ell k}(X)\partial\zeta_r} = \frac{1}{\sigma_w^2}\,\mathrm{Re}\left\{\frac{E}{M}\,\zeta\int\frac{\partial[\rho_{\ell k}(X)s_k(t-\tau_{\ell k}(X))]}{\partial\tau_{\ell k}(X)}\rho^*_{\ell k}(X)s^*_k(t-\tau_{\ell k}(X))dt\right\} \tag{9A.9}$$

and after some mathematical manipulation we get

$$[\mathbf{I}(\vartheta)]_{MN+1,i}=[\mathbf{I}(\vartheta)]_{i,MN+1}=-\frac{4\pi f_c\zeta_i}{\sigma_w^2}\left|\sqrt{\frac{E}{M}}\right|^2 \quad (i=1\cdots NM) \tag{9A.10}$$

For the last term, repeating what we did for the real part of the reflectivity index ζ_r, we get

$$\frac{\partial^2 \log p_\theta(\mathbf{r})}{\partial\tau_{\ell k}(X)\partial\zeta_i}=\frac{1}{\sigma_w^2}\mathrm{Im}\left\{\left|\sqrt{\frac{E}{M}}\right|^2\zeta\int\frac{\partial[\rho_{\ell k}(X)s_k(t-\tau_{\ell k}(X))]}{\partial\tau_{\ell k}(X)}\rho_{\ell k}^*(X)s_k^*(t-\tau_{\ell k}(X))dt\right\} \tag{9A.11}$$

and the symmetric results is

$$[\mathbf{I}(\vartheta)]_{MN+2,i}=[\mathbf{I}(\vartheta)]_{i,MN+2}=\frac{4\pi f_c\zeta_r}{\sigma_w^2}\left|\sqrt{\frac{E}{M}}\right|^2 \quad (i=1\cdots NM) \tag{9A.12}$$

Now that we have all the terms for the FIM $[\mathbf{I}(\vartheta)]$, we could write it as

$$\mathbf{I}(\vartheta)=\begin{bmatrix} a(f_c^2+\beta^2) & \cdots & 0 & -bf_c\zeta_i & bf_c\zeta_r \\ \cdots & \cdots & \cdots & \cdots & \cdots \\ 0 & \cdots & a(f_c^2+\beta^2) & -bf_c\zeta_i & bf_c\zeta_r \\ -bf_c\zeta_i & \cdots & -bf_c\zeta_i & \frac{MNb}{2\pi} & 0 \\ bf_c\zeta_r & \cdots & bf_c\zeta_r & 0 & \frac{MNb}{2\pi} \end{bmatrix} \tag{9A.13}$$

where

$$a=\frac{8\pi^2\left|\sqrt{(E/M)}\zeta\right|^2}{\sigma_w^2};\ b=\frac{4\pi\left|\sqrt{(E/M)}\right|^2}{\sigma_w^2};\ f_R=\left(1+\frac{\beta^2}{f_c^2}\right);\ \lambda=\frac{c}{f_c}.$$

In order to determine the value of $\mathbf{I}(\theta)$, we use (9A.4) and (9A.13) in (9A.3) to get

$$\mathbf{I}(\theta)=\begin{bmatrix} \frac{af_R}{\lambda^2}\sum_{k=1}^{M}\sum_{\ell=1}^{N}A_{\ell k}^2 & \frac{af_R}{\lambda^2}\sum_{k=1}^{M}\sum_{\ell=1}^{N}A_{\ell k}B_{\ell k} & -\frac{b}{\lambda}\zeta_i\sum_{k=1}^{M}\sum_{\ell=1}^{N}A_{\ell k} & \frac{b}{\lambda}\zeta_r\sum_{k=1}^{M}\sum_{\ell=1}^{N}A_{\ell k} \\ \frac{af_R}{\lambda^2}\sum_{k=1}^{M}\sum_{\ell=1}^{N}B_{\ell k}A_{\ell k} & \frac{a}{\lambda^2}\sum_{k=1}^{M}\sum_{\ell=1}^{N}B_{\ell k}^2 & -\frac{b}{\lambda}\zeta_i\sum_{k=1}^{M}\sum_{\ell=1}^{N}B_{\ell k} & \frac{b}{\lambda}\zeta_r\sum_{k=1}^{M}\sum_{\ell=1}^{N}B_{\ell k} \\ -\frac{b}{\lambda}\zeta_i\sum_{k=1}^{M}\sum_{\ell=1}^{N}A_{\ell k} & -\frac{b}{\lambda}\zeta_i\sum_{k=1}^{M}\sum_{\ell=1}^{N}B_{\ell k} & \frac{MNb}{2\pi} & 0 \\ \frac{b}{\lambda}\zeta_r\sum_{k=1}^{M}\sum_{\ell=1}^{N}A_{\ell k} & \frac{b}{\lambda}\zeta_r\sum_{k=1}^{M}\sum_{\ell=1}^{N}B_{\ell k} & 0 & \frac{MNb}{2\pi} \end{bmatrix} \tag{9A.14}$$

APPENDIX 9B DERIVING THE CRLB ON THE LOCATION ESTIMATE ERROR

This appendix presents the computation of $[\mathbf{G}_{\mathrm{CRLB}}]_{2\times 2}$ leading to (9.34). The submartix $[\mathbf{C}_{\mathrm{CRLB}}]_{2\times 2}$ is defined by the terms $[\mathbf{I}(\theta)]^{-1}_{1,1}$, $[\mathbf{I}(\theta)]^{-1}_{1,2}$, $[\mathbf{I}(\theta)]^{-1}_{2,1}$, and $[\mathbf{I}(\theta)]^{-1}_{2,2}$.

By definition, the value of $[\mathbf{I}(\theta)]^{-1}_{1,1}$ is obtained by

$$[\mathbf{I}(\theta)]^{-1}_{1,1} = \frac{\left|\tilde{\mathbf{I}}(\theta)_{ex(1,1)}\right|}{|\mathbf{I}(\theta)|} \tag{9B.1}$$

where $|\cdot|$ denotes the determinant, and $\tilde{\mathbf{I}}(\theta)_{ex(1,1)}$ is an $(NM-1)\times(NM-1)$ matrix, obtained by removing the first row and the first column of the $\mathbf{I}(\theta)$ matrix.

We define the following submatrix notations:

$$\Lambda_X = a(f_c^2 + \beta^2)\mathbf{I}_{NM\times NM}$$

$$\Lambda_\zeta = \frac{MNb}{2\pi}\left(\sqrt{\frac{E}{M}}\right)^2 \mathbf{I}_{2\times 2}$$

$$\mathbf{P} = bf_c \begin{bmatrix} -\zeta_i & \zeta_r \\ -\zeta_i & \zeta_r \\ \cdots & \cdots \\ -\zeta_i & \zeta_r \end{bmatrix}_{NM\times 2} \tag{9B.2}$$

$$\mathbf{V} = \begin{bmatrix} -\frac{1}{c}A_{11} & \cdots & -\frac{1}{c}A_{NM} \\ -\frac{1}{c}B_{11} & \cdots & -\frac{1}{c}B_{NM} \end{bmatrix}_{2\times NM}$$

With these notation we could rewrite the FIM in (9A.14) as follows:

$$\mathbf{I}(\theta) = \begin{bmatrix} \mathbf{V} & 0 \\ 0 & \mathbf{I} \end{bmatrix} \begin{bmatrix} \Lambda_X & \mathbf{P} \\ \mathbf{P}^T & \Lambda_\zeta \end{bmatrix} \begin{bmatrix} \mathbf{V}^T & 0 \\ 0 & \mathbf{I} \end{bmatrix} \tag{9B.3a}$$

This results in the following matrix:

$$\mathbf{I}(\theta) = \begin{bmatrix} \mathbf{V}\Lambda_X\mathbf{V}^T & \mathbf{VP} \\ \mathbf{P}^T\mathbf{V}^T & \Lambda_\zeta \end{bmatrix} \tag{9B.4}$$

For the determinant calculation of $\mathbf{I}(\theta)$, we use the property that the determinant of a matrix does not change under a linear operation. We have

$$|\mathbf{I}(\theta)| = \begin{vmatrix} \mathbf{V}\Lambda_X\mathbf{V}^T - \mathbf{VP}\Lambda_\zeta^{-1}\mathbf{P}^T\mathbf{V}^T & 0 \\ \mathbf{P}^T\mathbf{V}^T & \Lambda_\zeta \end{vmatrix} \tag{9B.5}$$

This can be calculated and expressed as

$$|\mathbf{I}(\theta)| = \left|\mathbf{V}\Lambda_X\mathbf{V}^T - \mathbf{V}\mathbf{P}\Lambda_\zeta^{-1}\mathbf{P}^T\mathbf{V}^T\right||\Lambda_\zeta| \tag{9B.6}$$

Repeating the same for the matrix $\tilde{\mathbf{I}}(\theta)_{ex(1,1)}$

$$\tilde{\mathbf{I}}(\theta)_{ex(1,1)} = \begin{bmatrix} \tilde{\mathbf{V}}_{ex(1,1)}\tilde{\Lambda}^T_{X_{ex(1,1)}}\tilde{\mathbf{V}}^T_{ex(1,1)} & \tilde{\mathbf{V}}_{ex(1,1)}\tilde{\mathbf{P}}_{ex(1)} \\ \tilde{\mathbf{P}}^T_{ex(1)}\tilde{\mathbf{V}}^T_{ex(1,1)} & \Lambda_\zeta \end{bmatrix} \tag{9B.7}$$

Using the same matrix manipulation that we applied to extract (9B.6) from (9B.5), we obtain

$$\left|\tilde{\mathbf{I}}(\theta)_{ex(1,1)}\right| = \left|\tilde{\mathbf{V}}_{ex(1,1)}\tilde{\Lambda}^T_{X_{ex(1,1)}}\tilde{\mathbf{V}}^T_{ex(1,1)} - \tilde{\mathbf{V}}_{ex(1,1)}\tilde{\mathbf{P}}_{ex(1)}\Lambda_\zeta^{-1}\tilde{\mathbf{P}}^T_{ex(1)}\tilde{\mathbf{V}}^T_{ex(1,1)}\right||\Lambda_\zeta| \tag{9B.8}$$

and using terms (9B.6) and (9B.8) in (9B.1) yields the term located in the first row and first column of the FIM inverse matrix:

$$[\mathbf{I}(\theta)]^{-1}_{1,1} = \frac{\left|\tilde{\mathbf{V}}_{ex(1,1)}\tilde{\Lambda}^T_{X_{ex(1,1)}}\tilde{\mathbf{V}}^T_{ex(1,1)} - \tilde{\mathbf{V}}_{ex(1,1)}\tilde{\mathbf{P}}_{ex(1)}\Lambda_\zeta^{-1}\tilde{\mathbf{P}}^T_{ex(1)}\tilde{\mathbf{V}}^T_{ex(1,1)}\right|}{\left|\mathbf{V}\Lambda_X\mathbf{V}^T - \mathbf{V}\mathbf{P}\Lambda_\zeta^{-1}\mathbf{P}^T\mathbf{V}^T\right|} \tag{9B.9}$$

This expression is identical to

$$[\mathbf{I}(\theta)]^{-1}_{1,1} = \left[\left(\mathbf{V}\Lambda_X\mathbf{V}^T - \mathbf{V}\mathbf{P}\Lambda_\zeta^{-1}\mathbf{P}^T\mathbf{V}^T\right)^{-1}\right]_{1,1} \tag{9B.10}$$

Repeating the process for the term located at (1,2), (2,1), and (2,2), we see that

$$[\mathbf{C}_{\text{CRLB}}]_{2\times 2} = \left(\mathbf{V}\Lambda_X\mathbf{V}^T - \mathbf{V}\mathbf{P}\Lambda_\zeta^{-1}\mathbf{P}^T\mathbf{V}^T\right)^{-1} \tag{9B.11}$$

Using the matrix values in (9B.2), we can write the CRLB for the location (x, y) as

$$[\mathbf{C}_{\text{CRLB}}]_{2\times 2} = \frac{\lambda^2}{a \cdot f_R} \cdot \frac{1}{u_{\text{CRLB}}} \cdot [\mathbf{G}_{\text{CRLB}}]_{2\times 2} \tag{9B.12}$$

where u_{CRLB} is a summation of multiples of the terms $A_{\ell k}$ and $B_{\ell k}$

$$u_{\mathrm{CRLB}} = \left[\sum_{k=1}^{M}\sum_{\ell=1}^{N} A_{\ell k}^2 - \frac{1}{f_R MN}\left(\sum_{k=1}^{M}\sum_{\ell=1}^{N} A_{\ell k}\right)^2\right] \times \left[\sum_{k=1}^{M}\sum_{\ell=1}^{N} B_{\ell k}^2 - \frac{1}{f_R MN}\left(\sum_{k=1}^{M}\sum_{\ell=1}^{N} B_{\ell k}\right)^2\right] - \left[\sum_{k=1}^{M}\sum_{\ell=1}^{N} A_{\ell k}B_{\ell k} - \frac{1}{f_R MN}\left(\sum_{k=1}^{M}\sum_{\ell=1}^{N} A_{\ell k}\sum_{k=1}^{M}\sum_{\ell=1}^{N} B_{\ell k}\right)\right]^2 \quad (9B.13)$$

and the submatrix $[\mathbf{G}_{\mathrm{CRLB}}]_{2\times 2}$:

$$[\mathbf{G}_{\mathrm{CRLB}}]_{2\times 2} = \begin{bmatrix} \sum_{k=1}^{M}\sum_{\ell=1}^{N} B_{\ell k}^2 - \frac{1}{f_R MN}\left(\sum_{k=1}^{M}\sum_{\ell=1}^{N} B_{\ell k}\right)^2 & \frac{1}{f_R MN}\left(\sum_{k=1}^{M}\sum_{\ell=1}^{N} A_{\ell k}\sum_{k=1}^{M}\sum_{\ell=1}^{N} B_{\ell k}\right) - \sum_{k=1}^{M}\sum_{\ell=1}^{N} A_{\ell k}B_{\ell k} \\ \frac{1}{f_R MN}\left(\sum_{k=1}^{M}\sum_{\ell=1}^{N} A_{\ell k}\sum_{k=1}^{M}\sum_{\ell=1}^{N} B_{\ell k}\right) - \sum_{k=1}^{M}\sum_{\ell=1}^{N} A_{\ell k}B_{\ell k} & \sum_{k=1}^{M}\sum_{\ell=1}^{N} A_{\ell k}^2 - \frac{1}{f_R MN}\left(\sum_{k=1}^{M}\sum_{\ell=1}^{N} A_{\ell k}\right)^2 \end{bmatrix}$$

APPENDIX 9C MLE OF TIME DELAYS — ERROR STATISTICS

In this appendix, we prove that the maximum-likelihood time-delay estimates are unbiased with error covariance matrix given in (9.45). For a set of received waveforms $r_\ell(t)$, $1 \le \ell \le N$, the time-delay estimates $\boldsymbol{\tau} = [\tau_{11}, \tau_{12}, \ldots, \tau_{NM}]^T$ are determined by maximizing the following statistics:

$$\tau_{\ell k} = \arg\max_{v}\left[\exp(j2\pi f_c v)\int_T r_\ell(t)s_k^*(t-v)dt\right] \quad (9C.1)$$

Equivalently

$$\frac{d}{dv}\left[\exp(j2\pi f_c v)\int_T r_\ell(t)s_k^*(t-v)dt\right]_{v=\tau_{\ell k}} = 0 \quad (9C.2)$$

The time-delay estimate (C.2) can be expressed as

$$\tau_{\ell k} = \tau_{0,\ell k} + \epsilon_{\ell k} \quad (9C.3)$$

where $\tau_{0,\ell k}$ is the true time delay associated with the transmitter k receiver ℓ pair. The properties of the noise $\epsilon_{\ell k}$ can be computed from (9.7), (9C.2), and (9C.3). It is not difficult to show that the following relation holds

$$\left.\frac{dg(v)}{dv}\right|_{v=\tau_{lk}} + n_{\ell k} = 0 \tag{9C.4}$$

where

$$g(v) = \sqrt{\frac{E}{M}}\,\zeta \int_T \exp[j2\pi f_c(v - \tau_{0,\ell k})] s_k(t - \tau_{0,\ell k}) s_k^*(t - v)dt \tag{9C.5}$$

and

$$n_{\ell k} = \int_T \frac{d}{dv} w_\ell(t) s_k^*(t - v) \exp(j2\pi f_c v)dt$$

We wish to write (9C.4) in the form of (9C.3). With a few algebraic manipulations, including expanding $g(v)$ in a Taylor series around $\tau_{0,\ell k}$, and neglecting terms $o[(\tau_{\ell k} - \tau_{0,\ell k})^3]$, it can be shown that

$$\tau_{\ell k} = \tau_{0,\ell k} + \frac{n_{\ell k}}{4\pi^2 \zeta \sqrt{(E/M)} (\beta^2 + f_c^2)} \tag{9C.6}$$

Comparing this with (C.3), we have for the error term

$$\epsilon_{\ell k} = \frac{n_{\ell k}}{4\pi^2 \zeta \sqrt{(E/M)} (\beta^2 + f_c^2)} \tag{9C.7}$$

To find the first- and second-order statistics of $\epsilon_{\ell k}$, we need the statitstical characterization of $n_{\ell k}$. As stated previously, we assume the receiver noise $w_\ell(t)$ is a Gaussian random process with zero mean and autocorrelation function $\sigma_w^2 \delta(\tau)$. Since $n_{\ell k}$ is a linear transformation of the process $w_\ell(t)$, and since the mean $w_\ell(t)$ is zero, $E[n_{\ell k}] = 0$. Similary, it can be shown that

$$E[n_{\ell k} n_{nm}] = \left\{ \begin{array}{ll} 0 & \forall \ell k \neq nm \\ 2\pi^2 \sigma_w^2 (f_c^2 + \beta^2) & \forall \ell k = nm \end{array} \right\} \tag{9C.8}$$

Using these results, we finally get

$$\begin{aligned} E[\epsilon_{\ell k} \epsilon_{nm}] &= \frac{E[n_{\ell k} n_{nm}]}{16\pi^4 \left|\zeta \sqrt{(E/M)}\right|^2 (\beta^2 + f_c^2)^2} \\ &= \left\{ \begin{array}{ll} 0 & \forall \ell k \neq nm \\ \dfrac{\sigma_w^2}{8\pi^2 \left|\zeta \sqrt{(E/M)}\right|^2 (\beta^2 + f_c^2)} & \forall \ell k = nm \end{array} \right\} \end{aligned} \tag{9C.9}$$

concluding that the covariance matrix of the terms $\epsilon_{\ell k}$ is given by (9.45).

APPENDIX 9D DERIVING THE LOWEST GDOP FOR SPECIAL CASES

In this appendix, we prove that the minimum GDOP for a MIMO radar is $\sqrt{2/NM}$. Using the expression for the GDOP in (9.54), and the definitions of terms $A_{\ell k}$, $B_{\ell k}$ in (9.11), we expand

$$\sum_{k=1}^{M}\sum_{\ell=1}^{N} A_{\ell k}^2 = \sum_{k=1}^{M}\sum_{\ell=1}^{N} (\cos^2\phi_{tk} + 2\cos\phi_{tk}\cos\phi_{r\ell} + \cos^2\phi_{r\ell})$$

and

$$\sum_{k=1}^{M}\sum_{\ell=1}^{N} B_{\ell k}^2 = \sum_{k=1}^{M}\sum_{\ell=1}^{N} (\sin^2\phi_{tk} + 2\sin\phi_{tk}\sin\phi_{r\ell} + \sin^2\phi_{r\ell})$$

We analyze the following special scenario. Assume that N radars are located symmetrically around the origin. The target is located at the axis origin $X_0 = (0, 0)$. The angles between the radars and the target for this case are $\phi_{tk} = (2\pi(k-1))/N$ and $\phi_{r\ell} = (2\pi(\ell-1))/N$. For this case, the following Fourier summation formulas apply:

$$\sum_{k=1}^{N} \cos^2\frac{2\pi(k-1)}{M} = \sum_{k=1}^{N} \sin^2\frac{2\pi(k-1)}{M} = \frac{N}{2}$$

$$\sum_{k=1}^{N} \cos\frac{2\pi(k-1)}{N}\sin\frac{2\pi(k-1)}{N} = 0$$

9D.1 Special Case: N × N MIMO

Using Fourier summation formulas, we get

$$\sum_{k=1}^{M}\sum_{\ell=1}^{N} A_{\ell k}^2 = \sum_{k=1}^{M}\sum_{\ell=1}^{N} B_{\ell k}^2 = N^2$$

$$\sum_{k=1}^{M}\sum_{\ell=1}^{N} A_{\ell k} = \sum_{k=1}^{M}\sum_{\ell=1}^{N} B_{\ell k} = \sum_{k=1}^{M}\sum_{\ell=1}^{N} A_{\ell k}B_{\ell k} = 0$$

This results in

$$\mathrm{GDOP} = \sqrt{\frac{2}{N^2}} \tag{9D.1}$$

9D.2 Special Case: 1 × N MIMO

Using Fourier summation formulas, we get $\sum_{k=1}^{M}\sum_{\ell=1}^{N} A_{\ell k}^2 = N(0.5 + \cos^2\phi_{tk})$; $\sum_{k=1}^{M}\sum_{\ell=1}^{N} B_{\ell k}^2 = N(0.5 + \sin^2\phi_{tk})$; $\sum_{k=1}^{M}\sum_{\ell=1}^{N} A_{\ell k} = N\cos\phi_{tk}$; $\sum_{k=1}^{M}\sum_{\ell=1}^{N} B_{\ell k} = N\cos\phi_{tk}$; $\sum_{k=1}^{M}\sum_{\ell=1}^{N} A_{\ell k}B_{\ell k} = N\cos\phi_{tk}\sin\phi_{tk}$. This results in

$$\text{GDOP} = \frac{2}{\sqrt{N}} \tag{9D.2}$$

For $M = 1$ the GDOP is reduced to the case similar to the GPS scheme.

9D.3 General Case: M × N MIMO

All N radars are receiving the signal waveforms transmitted by M radars, where $1 < M < N$.

Using the same mechanism as employed previously, and neglecting the sine and cosine terms in the power of ≥ 2, we obtain

$$\text{GDOP}_{\text{BLUE}_{M \neq N}} \approx \sqrt{\frac{2}{NM}} \tag{9D.3}$$

ACKNOWLEDGMENTS

The authors would like to thank Dr. Nikolaus Lehmann for his help in running simulations and generating some of the figures in the chapter.

REFERENCES

1. E. Brookner, Phase arrays around the world — progress and future trends, *Proc. IEEE Int. Symp. Phased Array Systems and Technology,* Oct. 2003, pp.1–8.
2. L. Swindlehurst and P. Stoica, Maximum likelihood methods in radar array signal processing, *Proc. IEEE.* **86**:421–441 (Feb. 1998).
3. S. Haykin, J. Litva, and T. J. Shepherd, *Radar Array Processing*, Springer-Verlag, New York, 1993.
4. A. Farina, *Antenna Based Signal Processing Techniques for Radar Systems*, Artech House, Norwood, MA, 1992, Science Publishers, 1998.
5. L. E. Brennan and I. S. Reed, Theory of adaptive radar, *IEEE Trans. Aerospace Electron. Syst.* **AES-9**:237–252 (March 1973).
6. J. Ward, Space-time adaptive processing for airborne radar, *Proc. IEEE Natl. Radar Conf.*, 1995, pp. 2809–2812.
7. W. L. Melvin, A STAP overview, *IEEE Aerospace Electron. Syst. Mag.* **19**:19–35 (Jan. 2004).

8. R. Klemm, *Principles of Space-Time Adaptive Radar*, IEE Press, 2002.

9. J. R. Guerci, *Space-Time Adaptive Processing for Radar*, Artech House, Norwood, MA, 2003.

10. V. S. Chernyak, *Fundamentals of Multisite Radar Systems*, Gordon & Breach, 1998.

11. J. M. Colin, Phased array radars in France: Present and future, *Proc. IEEE Int. Symp. Phased Array Systems and Technology*, Oct. 1996, pp. 458–462.

12. A. S. Fletcher and F. C. Robey, Performance bounds for adaptive coherence of sparse array radar, *Proc. 11th Conf. Adaptive Sensors Array Processing*, March 2003.

13. D. W. Bliss and K. W. Forsythe, Multiple-input multiple-output (MIMO) radar and imaging: Degrees of freedom and resolution, *Proc. 37th Asilomar Conf. Signals, Systems and Computers*, Nov. 2003, pp. 54–59.

14. D. Rabideau, Ubiquitous MIMO digital array radar, *Proc. 37th Asilomar Conf. Signals, Systems, and Computers*, Nov. 2003, pp. 1057–1064.

15. E. Fishler, A. Haimovich, R. Blum, L. Cimini, D. Chizhik, and R. Valenzuela, MIMO radar: An idea whose time has come, *Proc. 2004 IEEE Int. Conf. Radar*, April 2004, pp. 71–78.

16. D. R. Fuhrmann and G. San Antonio, Transmit beamforming for MIMO radar systems using partial signal correlation, *Proc. 38th Asilomar Conf. Signals, Systems and Computers*, Nov. 2004, pp. 295–299.

17. F. C. Robey, S. Coutts, D. Weikle, J. C. McHarg, and K. Cuomo, MIMO radar theory and experimental results, *Proc. 38th Asilomar Conf. Signals, Systems and Computers*, Nov. 2004, pp. 300–304.

18. I. Bekkerman and J. Tabrikian, Target detection and localization using MIMO radars and sonars, *IEEE Trans. Signal Process* **54**:3873–3883 (Oct. 2006).

19. L. Xu, J. Li, and P. Stoica, Adaptive techniques for MIMO radar, *Proc. 14th IEEE Workshop on Sensor Array and Multi-channel Processing*, Waltham, MA, July 2006.

20. E. Fishler, A. Haimovich, R. Blum, L. Cimini, D. Chizhik, and R. Valenzuela, Spatial diversity in radars — models and detection performance, *IEEE Trans. Signal Process.* **54**:823–838 (March 2006).

21. N. Lehmann, A. M. Haimovich, R. S. Blum, and L. Cimini, MIMO-radar application to moving target detection in homogenous clutter, *Proc. 14th IEEE Workshop on Sensor Array and Multi-channel Processing*, Waltham, MA, July 2006.

22. N. H. Lehmann, A. M Haimovich, R. S. Blum, and L. J. Cimini, High resolution capabilities of MIMO radar, *Proc. 40th Asilomar Conf. Signals, Systems and Computers*, Nov. 2006.

23. G. J. Foschini, Layered space-time architecture for wireless communication in a fading environment when using multiple antennas, *Bell Labs Tech. J.* **1**(2):41–59 (1996).

24. V. Tarokh, N. Seshadri, and A. Calderbank, Space-time codes for high data rate wireless communication: Performance criterion and code construction, *IEEE Trans. Inform. Theory.* **44**:744–765 (March 1998).

25. E. Fishler, A. Haimovich, R. Blum, L. Cimini, D. Chizhik, and R. Valenzuela, Performance of MIMO radar systems: Advantages of angular diversity, *Proc. 38th Asilomar Conf. Signals, Systems and Computers*, Nov. 2004, pp. 305–309.

26. M. Skolnik, *Introduction to Radar Systems*, 3rd ed., McGraw-Hill, New York, 2002.

27. Y. Yang and R. S. Blum, Radar waveform design based on mutual information and mean-square error estimation, *IEEE Trans. Aerospace Electron. Syst.* **43**:330–343 (Jan. 2007).
28. I. Papoutsis, C. J. Baker, and H. D. Griffiths, Fundamental performance limitations of radar networks, *Proc. 1st EMRS DTC Technical Conf.*, Edinburgh, 2004.
29. R. Viswanathan and P. K. Varshney, Distributed detection with multiple sensors I: Fundamentals, *Proc. IEEE.* **85**:54–63 (Jan. 1997).
30. R. S. Blum, S. A. Kassam, and H. V. Poor, Distributed detection with multiple sensors II: Advanced topics, *Proc. IEEE* **85**:64–79 (Jan. 1997).
31. R. S. Blum, Distributed detection for diversity reception of fading signals in noise, *IEEE Trans. Inform. Theory* **48**:158–164 (Jan. 1999).
32. P. M. Woodward, *Probability and Information Theory with Application to Radar*, Artech House, Norwood, MA, 1953.
33. N. Lehmann, E. Fishler, A. M. Haimovich, R. S. Blum, D. Chizhik, L. Cimini, and R. Valenzuela, Evaluation of transmit diversity in MIMO-radar direction finding, *IEEE Trans. Signal Process.* **55**:2215–2225 (May 2007).
34. N. Lehmann, *Some Contributions on MIMO Radar*, Ph.D. dissertation, New Jersey Institute of Technology, 2006.
35. B. D. Steinberg, The peak sidelobe of the phased array having randomly located elements, *IEEE Trans. Anten. Propag.* **AP-20**:129–136 (March 1972).
36. B. D. Steinberg and E. H. Attia, Sidelobe reduction of random arrays by element position and frequency diversity, *IEEE Trans. Anten. Propag.* **AP-31**:922–930 (Nov. 1982).
37. N. Levanon, *Radar Principles*, Wiley, New York, 1988.
38. S. M. Kay, *Fundamentals of Statistical Signal Processing: Estimation Theory*, Prentice-Hall, Englewood Cliffs, NJ, 1993, Vol. 1.
39. H. V. Poor, *An Introduction to Signal Detection and Estimation Theory*, 3rd ed., Springer, New York, 1998.
40. H. B. Lee, A novel procedure for assessing the accuracy of the hyperbolic multilateration systems, *IEEE Trans. Aerospace Electron. Syst.* **11**:2–15 (Jan. 1975).
41. N. Levanon, Lowest GDOP in 2-D scenarios, *IEE Proc.Radar Sonar Navig.* **147**:149–155 (June 2000).
42. Y. Qi, H. Kobayashi, and H. Suda, Analysis of wireless geolocation in non-line-of-sight environment, *IEEE Trans. Wireless Commun.* **5**:672–681 (March 2006).
43. R. Yarlagadda, I. Ali, N. Al-Dhahir, and J. Hershey, GPS GDOP metric, *IEE Proc. Radar Sonar Navig.* **147**:259–264 (Oct. 2000).
44. E. Weinstein and A. J. Weiss, Fundamental limitation in passive time-delay estimation — part II: Wide-band systems, *IEEE Trans. Acoust. Speech Signal Process.* **ASSP-32**(5):1064–1078 (Oct. 1984).
45. A. Papoulis, and S. U. Pillai, *Probability, Random Variables and Stochastic Processes*, 4th ed., McGraw-Hill, New York, 2002.

10

SPACETIME CODING FOR MIMO RADAR

Antonio De Maio

Università degli Studi di Napoli "Federico II," Napoli (Naples), Italy

Marco Lops

Università degli Studi di Cassino, Cassino, Italy

10.1 INTRODUCTION

The concept of multiple-input multiple-output (MIMO) remote sensing systems, equipped with multiple antennas at both the transmitter and the receiver ends, was born as a natural application to the radar of the seminal theoretical results concerning capacity of multiantenna Gaussian channels [1]. In communications, a MIMO structure may be exploited both as a spatial multiplexer — which ultimately allows information transmission at rates that a single-input single-output (SISO) wireless channel would prevent [2] — and as a means to obtain diversity, that is, to overcome the inherent limitations due to fading, once suitable coding is undertaken [3]; tradeoffs between these two major functions are a further degree of freedom [4].

In a radar system, the MIMO concept has been advocated primarily either to achieve *spatial diversity*, so as to cope with target scintillation (an operation also known as *statistical MIMO*), or as a flexible means for beampattern design, even though hybrid applications, compromising between these two different philosophies, have been introduced [5].

Statistical MIMO radars aim at increasing the detector sensitivity by sufficiently spacing the sensors at both the transmitter and the receiver ends, thus generating a number of independent channels, namely, spatial diversity. It is indeed well known

MIMO Radar Signal Processing, edited by Jian Li and Petre Stoica

that target amplitude fluctuations may have a very detrimental effect in the interest region of close-to-one detection rates, while being beneficial in the region of low detection probabilities. As a general trend, if one refers to the "dispersion ratio" (i.e., the ratio between the standard deviation and the average) of the target radar cross section (RCS), the larger such a ratio, the slower the convergence to unity of the detection probability. Diversity is a well-known means to reduce the target RCS fluctuations, that is, to lower the dispersion ratio. For example, the availability of a number, say, D, of equivalent diversity branches, yielding as many independent and identically distributed (i.i.d.) replicas of the backscattered echo, is able to convert, on proper combination at the receiver end, an exponentially distributed fluctuation of the target RCS into a χ^2 with D degrees of freedom, thus reducing the dispersion ratio from 1 to $1/\sqrt{D}$.

In this context, a MIMO radar should, in principle, be able to provide the receiver with as many equivalent diversity branches as possible, which in turn requires stating the conditions under which a "rich scattering environment" can be re-created. Major contributions in this direction have been offered in three studies [6–8], wherein the theoretical background for importing the MIMO concept into radar was laid out, and the interplay between the physical parameters of the target to be detected, those of the transmitted signal, and the transmit/receive architectures were first investigated. The main results can be summarized as follows:

- On a nondispersive channel, and because of the abrupt fluctuations of target RCS with the aspect angle, a MIMO radar is able to grant angle diversity, provided that time differences across the transmit sensors are preserved at the receiver end (waveform-orthogonality-preserving assumption).
- The angle diversity yields a diversity order given by the product of the number of transmit and receive antennas, under additive white Gaussian noise disturbance. In these conditions, waveform orthogonality allows one to adopt a very simple path combination rule.
- In view of the results stated above, MIMO structures are superior to their natural competitors, namely, phased-array, multiple-input single-output (MISO), and single-input multiple-output (SIMO) radars, as far as detection performance is concerned.

The development of a consistent theory for statistical MIMO radar was the focus of three further studies [9–11], where, assuming a general model for the clutter correlation properties, spacetime coding (STC) was recognized as a key ingredient to achieve full diversity, and design criteria for both the transmitter and the receiver were given. In the new framework, the usage of orthogonal waveforms has been shown to be one possible design choice, and only a member of the general class of STC radar waveforms, that can be employed for performance optimization purposes under diverse scenarios. The problem of a global comparison — in terms of detection performance — between MIMO radars, frequency diversity, and coherent netting has been considered [12], while the integration of MIMO strategies with spacetime adaptive processing (STAP) is the focus of another study [13].

Early contributions to the development and implementation of MIMO radars have also appeared in the literature [14–18]. Fletcher and Robey [14] highlighted the fact that the MIMO mode can be conceived of as a means of bootstrapping to obtain larger coherent gain. Rabideau and Parker [15] discuss some practical issues concerning implementation (equipment specifications, dynamic range, phase noise, system stability, isolation, and spurs) of MIMO radars, while waveform design for MIMO beamforming is discussed by Fuhrmann and San Antonio [16]. Finally, experimental results concerning MIMO radars are presented by Robey et al. [17], while MIMO imaging as well as the related resolution issues are addressed in another paper [18]. The interplay between MIMO ambiguity function and resolution requisites are investigated by San Antonio et al. [19], while the impact on the performance of a proper design of the probing signals is studied and the applicability of some known adaptive beamforming techniques is demonstrated for target parameter estimation and interference cancellation purposes by Stoica and others [20] and [21]. The issue of detection and direction estimation has also been considered [22], while Lehman et al. [23] consider the problem of conjugating diversity and estimation accuracy, focusing on the direction finding issue; in both cases, Cramér–Rao bounds for estimation accuracy are evaluated. Waveform design for identification and classification of extended targets is instead the focus of Yang and Blum's study [24]. On the basis of constrained maximization of such measures as the conditional mutual information between scattered signal and target RCS and the minimum mean-square error (MMSE) between estimated and actual target response, some key results concerning the interplay between MMSE and mutual information [25] have indeed been reframed in a MIMO radar context. MIMO radar waveform optimization for parameter estimation of multiple targets in the presence of spatially colored interference and noise has been addressed [26]. Several design criteria, including minimizing the trace, determinant, and the largest eigenvalue of the Cramér–Rao bound matrix are considered and their effectiveness is demonstrated. Finally, a new transmit/receive structure for MIMO radars has been presented [5], aimed at retaining the advantages of both spatial diversity and beamforming; this is achieved by considering transmit/receive sensors made up of several subarrays, with closely spaced elements, but such that the distance between the arrays is sufficient to grant spatial diversity. The present contribution focuses entirely on the statistical MIMO concept, namely, on the use of multiantenna systems as a source of spatial diversity. The scenario considered here is thus one where transmission takes place through a number of transmit antennas whose spacing is sufficient to ensure target scintillation. As a consequence, if the receive antenna consists of several widely spaced sensors, a number of paths — the angle diversity — are generated; needless to say, in a real environment the received signals are inevitably affected by clutter, with negligible spatial correlation, but possibly significant temporal correlation. In this situation a number of problems arise, such as

1. Since the noise is no longer white, transmission of orthogonal waveforms no longer appears to be the wisest choice; rather, the transmitted signal should

be a spacetime-coded (STC) waveform, wherein the code matrix is left as a degree of freedom to be selected on the basis of the clutter statistics and the design constraints.
2. Matched-filter-based receivers are no longer ensured to achieve maximum diversity order.
3. In the presence of unknown time-varying clutter covariance, one should (a) ensure constancy of the false-alarm rate and (b) adjust the transmitted waveforms to the changing scenario. Otherwise stated, the problem arises of designing and assessing *adaptive* transmit/receive structures.

Tasks 1 and 2 cannot be handled separately, in that the path combination rule obviously plays a key role in the achievable diversity order, and strongly depends on the transmitted waveforms. On the contrary, task 3 can be tackled starting with the joint transmitter/receiver design under known clutter statistics, and subsequently borrowing some ideas from adaptive radar processing.

In this context, the balance of the contribution is as follows. In Section 10.2, the signal model for STC MIMO radar under correlated clutter with known statistics is reviewed. This section, which is based on the model outlined by De Maio and Lops [11], introduces the spacetime code — also known as the *code matrix* — as a degree of freedom, thus laying the theoretical foundations for next sections. The model, which is developed with reference to a train of rectangular pulses whose (complex) amplitude is modulated by the spacetime code, can, indeed, be applied to sophisticated waveforms as well. Detection is covered in Section 10.3; the prior uncertainty as to the target parameters is circumvented by using the generalized likelihood ratio test (GLRT), which, not surprisingly, leads to a family of quadratic test statistics (depending on the choice of the code matrix). General formulas for the detection and false-alarm probabilities are also derived. A major property of the proposed receivers, though, is that, under the no-target hypothesis, the test statistics are *ancillary*, that is, their probability density functions are functionally independent of the clutter covariance. The issue of an optimum choice of the code matrix is covered in Section 10.4. Of course, the optimum code matrix turns out to depend on the performance measure to be optimized. Moreover, since the optimization must be necessarily performed under some constraint — such as the overall transmitted power, or the received signal–clutter ratio (SCR) — the nature of the constraint turns out to be very influential on the code choice; these issues are thoroughly discussed in Section 10.4. More specifically, we consider three possible performance measures: (1) the lower Chernoff bound for the detection probability, (2) the mutual information between the set of the received waveforms and the set of the target returns along the diversity paths, and (3) the average received SCR, under the following constraints: (1) fixed received average SCR (applied only to measures 1 and 2 above) and (2) fixed transmitted power.

A thorough study of the interplay between detection performance and code matrix choice is given in Section 10.5 In particular, closed-form asymptotic expressions — holding true in the interest region of close-to-one detection rates — are given for the detection probability, and the interplay between the optimization criterion and the achievable diversity order is discussed in depth.

An extensive analysis of STC MIMO radars under diverse clutter correlation, assumed known, is presented in Section 10.6, also in comparison to uncoded and SISO systems. The issue of adaptive transmit/receive MIMO structures is considered in Section 10.7. In particular, we first deal with the problem of false-alarm rate (FAR) control under unknown clutter covariance. To this end, we first show how a set of secondary (signal-free) data can be collected, and then form a number of possible estimates of the clutter covariance, ranging from the sample covariance matrix based on two consecutive scans to the more refined exponentially weighted mean computed on all of the previous scans. An interesting result is that, under both scenarios, the adaptive form of the GLRT statistics turns out to ensure the constant false-alarm rate (CFAR) property. A simple algorithm for adaptive code selection, so as to track possible variations of the clutter statistics, is also presented; the resulting structure is thus adaptive at both the transmitter and the receiver. A thorough performance assessment under this new scenario is presented, and the cost of the prior uncertainty as to the clutter statistics is assessed by comparing the performances of the nonadaptive and the adaptive systems. Finally, concluding remarks and hints for future research form the object of Section 10.8.

10.2 SYSTEM MODEL

We consider a MIMO radar consisting of $\mathtt{s}$ transmitting and $\mathtt{r}$ receiving elements and assume "widely spaced" antennas at the two ends of the system, which ensures that the returns hitting each transmit/receive pair of sensors are uncorrelated, whether they come from target or from clutter. We denote by $s_i(t)$ the baseband equivalent of the coherent pulsetrain transmitted by the ith antenna

$$s_i(t) = \sum_{j=1}^{N} a_{i,j} p[t - (j-1)T_p], \qquad i = 1, \ldots, \mathtt{s} \tag{10.1}$$

where

- $p(t)$ is the signature of each transmitted pulse, which we assume, without loss of generality, with unit energy and duration τ_p.
- $T_p \geq \tau_p$ is the pulse repetition time (PRT).
- $a_i = [a_{i,1}, \ldots, a_{i,N}]^T$, where $(\cdot)^T$ denotes transpose, is an N-dimensional column vector whose entries are complex numbers that modulate in both amplitude and phase the N pulses of the train. In the following text we refer to a_i as the codeword of the ith antenna.

The baseband equivalent of the signal received at the ith sensor from a target with a two-way time delay[1] τ can be written as

$$r_i(t) = \sum_{l=1}^{\mathtt{s}} \alpha_{i,l} \sum_{j=1}^{N} \alpha_{l,j} p[t - \tau - (j-1)T_p] + n_i(t), \qquad i = 1, \ldots, \mathtt{r} \tag{10.2}$$

[1]For simplicity we assume a zero-Doppler target, but the derivations can be extended to account for a possible known Doppler shift.

where

- $\alpha_{i,l}$, $i = 1, \ldots, r$ and $l = 1, \ldots, s$ are complex numbers accounting for both the target backscattering and the channel propagation effects between the lth transmitter and the ith receiver.
- $n_i(t)$, $i = 1, \ldots, r$ are zero-mean, spatially uncorrelated, complex Gaussian random processes accounting for both the external and the internal disturbance.

We explicitly point out that the validity of the abovementioned model (represented in Fig. 10.1a) requires that the target be in the far-field region and

$$\frac{d^s_{\max} + d^r_{\max}}{c} \ll \frac{1}{B}$$

where B is the bandwidth of the transmitted pulse and $d^s_{\max}$ and $d^r_{\max}$ denote the maximum spacing between two sensors at the transmitter and the receiver end, respectively.

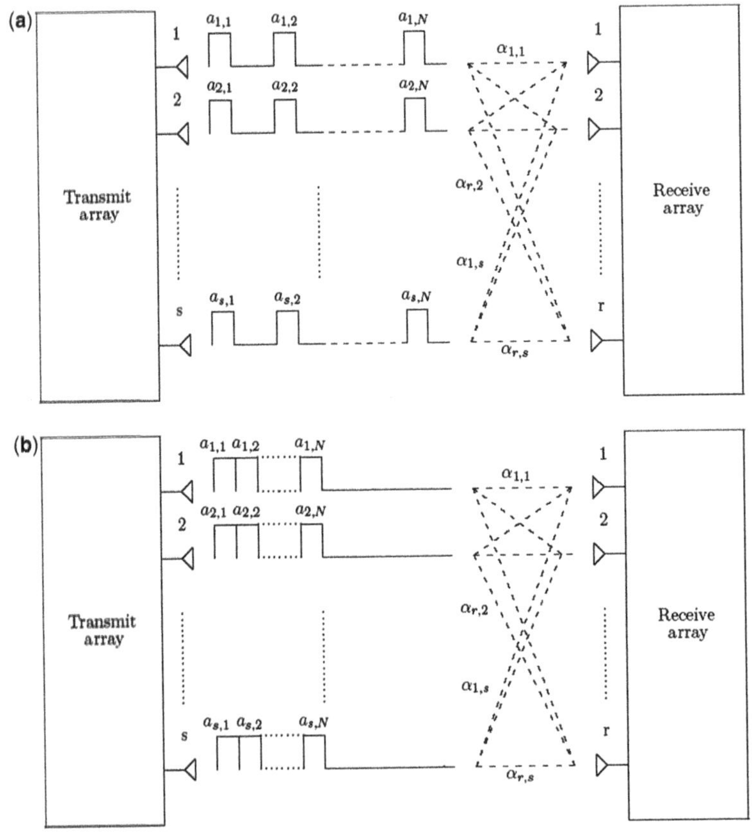

Figure 10.1 Schematic representation of the considered transmit/receive system with a (a) coded pulsetrain; (b) single-coded pulse.

TABLE 10.1 The Most Relevant Variables

Parameter	Symbol
Number of transmitters	s
Number of receivers	r
Number of pulses	N
Code matrix	$\mathbf{A}$
Vectors of received samples	r_i
Disturbance vectors	n_i

The signal $r_i(t)$ is fed to a filter matched to $p(t)$, and the corresponding output is sampled at the time instants $\tau + (k-1)T_p$, $k = 1, \ldots, N$, thus forming the observations

$$r_i(k) = \sum_{l=1}^{s} \alpha_{i,l}\, a_{l,k} + n_i(k), \qquad k = 1, \ldots, N \tag{10.3}$$

where $n_i(k)$ is the filtered disturbance sample.

We may thus define the N-dimensional column vectors $r_i = [r_i(1), \ldots, r_i(N)]^T$ and more compactly rewrite (10.3) as

$$r_i = \mathbf{A}\alpha_i + n_i, \qquad i = 1, \ldots, r \tag{10.4}$$

where $n_i = [n_i(1), \ldots, n_i(N)]^T$, $\alpha_i = [\alpha_{i,1}, \ldots, \alpha_{i,s}]^T$, and the $(N \times s)$-dimensional matrix

$$\mathbf{A} = [a_1, \ldots, a_s] \tag{10.5}$$

has codewords as columns. This last matrix will be referred to as the *code matrix*.

It is worth emphasizing that a model similar to (10.4) can be developed with reference to the case where STC is performed according to Ref. 27, namely, by dividing a single pulse in N subpulses; the code matrix $\mathbf{A}$ thus defines s different codewords of length N, which can be received by a single receive antenna (thus defining the MISO structure of Ref. 27), as well as by a set of r receive antennas, as in the present study (see Fig. 10.1b). The most relevant variables, extensively used in the following text, are summarized in Table 10.1.

10.3 DETECTION IN MIMO RADARS

Target detection under the model (10.4) can be formulated in terms of the following binary hypotheses test

$$\begin{cases} H_0\colon\ r_i = n_i, & i = 1, \ldots, r \\ H_1\colon\ r_i = \mathbf{A}\alpha_i + n_i, & i = 1, \ldots, r \end{cases} \tag{10.6}$$

where n_i, $i = 1, \ldots, r$, are i.i.d. zero-mean complex Gaussian vectors whose covariance matrix

$$E[n_i\, n_i^\dagger] = \mathbf{M}$$

[where $E[\cdot]$ denotes statistical expectation and $(\cdot)^\dagger$ conjugate transpose] is assumed positive definite and known.

The optimum — under the Neyman–Pearson criterion — solution to the hypotheses testing problem (10.6) is the likelihood ratio test (LRT). In the application considered here, however, it cannot be implemented unless priors are assigned to the fluctuation parameters α_i, which would lead to a parametric receiver of little practical interest. A possible way to circumvent this drawback is to resort to the GLRT [28], which relies on replacing the unknown parameters with their maximum-likelihood (ML) estimates under each hypothesis. The GLRT for the detection problem (10.6) results in the decision rule

$$\frac{\max\limits_{\alpha_1,\ldots,\alpha_r} f(r_1,\ldots,r_r|H_1,\mathbf{M},\alpha_1,\ldots,\alpha_r)}{f(r_1,\ldots,r_r|H_0,\mathbf{M})} \underset{H_0}{\overset{H_1}{\gtrless}} T \tag{10.7}$$

where $f(r_1,\ldots,r_r|H_1,\mathbf{M},\alpha_1,\ldots,\alpha_r)$ and $f(r_1,\ldots,r_r|H_0,\mathbf{M})$ denote the probability density functions (pdf's) of the data under H_1 and H_0, respectively. Previous assumptions imply that the aforementioned pdf's can be written as

$$f(r_1,\ldots,r_r|H_0,\mathbf{M}) = \frac{1}{\pi^{Nr}\det^r(\mathbf{M})}\exp\left[-\sum_{i=1}^{r} r_i^\dagger \mathbf{M}^{-1} r_i\right] \tag{10.8}$$

and

$$f(r_1,\ldots,r_r|H_1,\mathbf{M},\alpha_1,\ldots,\alpha_r) = \frac{1}{\pi^{Nr}\det^r(\mathbf{M})}\exp\left[-\sum_{i=1}^{r}(r_i-\mathbf{A}\alpha_i)^\dagger\mathbf{M}^{-1}(r_i-\mathbf{A}\alpha_i)\right] \tag{10.9}$$

where $\det(\cdot)$ denotes the determinant of a square matrix.

Substituting (10.9) and (10.8) in (10.7) and elaborating, we see that the GLRT assumes the form

$$\sum_{i=1}^{r} r_i^\dagger\mathbf{M}^{-1}r_i - \sum_{i=1}^{r}\min_{\alpha_i}(r_i-\mathbf{A}\,\alpha_i)^\dagger\mathbf{M}^{-1}(r_i-\mathbf{A}\,\alpha_i) \underset{H_0}{\overset{H_1}{\gtrless}} T \tag{10.10}$$

where T is a suitable modification of the original threshold. In order to solve the r minimization problems in (10.10), we have to distinguish between two different situations.[2]

[2]The rationale for this distinction will become clear as STC coding is introduced later on.

10.3.1 Full-Rank Code Matrix

First assume that the code matrix $\mathbf{A}$ has full rank. Minimization of (10.10) requires, in turn, considering two different cases:

Case 1: $N \geq s$. In this situation the quadratic form in (10.10) achieves its minimum at

$$\widehat{\alpha}_i = (\mathbf{A}^\dagger \mathbf{M}^{-1}\mathbf{A})^{-1}\mathbf{A}^\dagger \mathbf{M}^{-1} r_i, \qquad i = 1, \ldots, r \tag{10.11}$$

and, as a consequence, the GLRT becomes

$$\sum_{i=1}^{r} r_i^\dagger \mathbf{M}^{-1}\mathbf{A}(\mathbf{A}^\dagger \mathbf{M}^{-1}\mathbf{A})^{-1}\mathbf{A}^\dagger \mathbf{M}^{-1} r_i \underset{H_0}{\overset{H_1}{\gtrless}} T \tag{10.12}$$

Case 2: $N \leq s$. In this situation the minimum of the quadratic form in (10.10) is zero, since each linear system $\mathbf{A}\alpha_i = r_i$ $(i = 1, \ldots, r)$ is underdetermined. As a consequence, the GLRT becomes

$$\sum_{i=1}^{r} r_i^\dagger \mathbf{M}^{-1} r_i \underset{H_0}{\overset{H_1}{\gtrless}} T \tag{10.13}$$

We notice in passing that, for the case $N = s$, the two expressions above coincide. In order to define possible design criteria for the STC, it is useful to establish a direct relationship between the detection performance and the transmitted code, which is thus the main goal of the present subsection.

Under the H_0 hypothesis, the left-hand side (LHS) of the GLRT can be written as

$$\begin{aligned} &\textstyle\sum_{i=1}^{r} z_i^\dagger \boldsymbol{P}_{\mathbf{M}^{-(1/2)}\boldsymbol{A}} z_i && \text{(case 1)} \\ &\textstyle\sum_{i=1}^{r} z_i^\dagger z_i && \text{(case 2)} \end{aligned} \tag{10.14}$$

where $z_i = \mathbf{M}^{-(1/2)} r_i$, $i = 1, \ldots, r$, are i.i.d. zero-mean complex Gaussian vectors with identity covariance matrix and

$$\mathbf{P}_{\mathbf{M}^{-(1/2)}\boldsymbol{A}} = \mathbf{M}^{-(1/2)}\mathbf{A}(\mathbf{A}^\dagger \mathbf{M}^{-1}\mathbf{A})^{-1}\mathbf{A}^\dagger \mathbf{M}^{-(1/2)}$$

is the projector onto the column space of $\mathbf{M}^{-(1/2)}\mathbf{A}$.

The decision statistic is thus a central χ^2 [29] with sr degrees of freedom under case 1, and N_r degrees of freedom under case 2. As a consequence the probability of false alarm (P_{fa}) can be evaluated as

$$P_{f_a} = e^{-T} \sum_{k=0}^{\delta_r - 1} \frac{T^k}{k!} \tag{10.15}$$

where $\delta = \min(s, N)$. This last expression shows that

- The decision statistic is *ancillary*, in the sense that it depends on the actual clutter covariance matrix, but its probability density function is functionally independent of such a matrix.
- The threshold setting is feasible with no prior knowledge as to the clutter power spectrum, namely, the GLRT ensures the CFAR property.

Under H_1, representation (10.14) still holds true and, given α_i, the vectors z_i, $i = 1, \ldots, r$, are statistically independent complex Gaussian vectors with mean value $\mathbf{M}^{-(1/2)}\mathbf{A}\alpha_i$ and identity covariance matrix. It follows that, given α_i, the decision statistic is conditionally noncentral χ^2 distributed, with noncentrality parameter $\sum_{i=1}^{r} \alpha_i^\dagger \mathbf{A}^\dagger \mathbf{M}^{-1} \mathbf{A}\alpha_i$ and degrees of freedom sr in case 1 and Nr in case 2. As a consequence the *conditional* probability of detection (P_d) can be written as[3] [30, p. 2]

$$P_d = Q_{\delta r}(\sqrt{2\beta}, \sqrt{2T}) \tag{10.16}$$

where $\beta = \sum_{i=1}^{r} \alpha_i^\dagger \mathbf{A}^\dagger \mathbf{M}^{-1} \mathbf{A}\alpha_i$ and $Q_k(\cdot, \cdot)$denotes the generalized Marcum Q function of order k. An alternative expression for the conditional P_d, in terms of an infinite series, can be also written as

$$P_d = \sum_{k=0}^{\infty} \frac{e^{-\beta}}{k!} \beta^k [1 - \Gamma_{\text{inc}}(T, k + \delta r)] \tag{10.17}$$

where

$$\Gamma_{\text{inc}}(x, a) = \frac{1}{\Gamma(a)} \int_0^x e^{-t} t^{a-1} dt \tag{10.18}$$

is the incomplete gamma function. Finally, the unconditional P_d can be obtained averaging the last expression over the pdf of α_i, $i = 1, \ldots, r$.

10.3.2 Rank 1 Code Matrix

In this subsection we assume that the code matrix $\mathbf{A}$ is rank 1, specifically, $\mathbf{A} = pq^\dagger$, where p and q are an N-dimensional and an s-dimensional complex vector, respectively. Under this condition, Eq. (10.10) can be written as

$$\sum_{i=1}^{r} r_i^\dagger \mathbf{M}^{-1} r_i - \sum_{i=1}^{r} \min_{\alpha_i} (r_i - c_i p)^\dagger \mathbf{M}^{-1} (r_i - c_i p) \underset{H_0}{\overset{H_1}{\gtrless}} T \tag{10.19}$$

[3]With a slight notational abuse, and to avoid overburdening the chapter with too many symbols, we employ the same notation, P_d, for the conditional and the unconditional detection probability.

where $c_i = q^\dagger \alpha_i$. The quadratic forms in (10.19) are minimum at

$$\widehat{c}_i = q^\dagger \widehat{\alpha}_i = (p^\dagger \mathbf{M}^{-1} p)^{-1} p^\dagger \mathbf{M}^{-1} r_i, \qquad i = 1, \ldots, r, \tag{10.20}$$

leading to the GLRT

$$\sum_{i=1}^{r} \frac{|p^\dagger \mathbf{M}^{-1} r_i|^2}{p^\dagger \mathbf{M}^{-1} p} \mathop{\gtrless}_{H_0}^{H_1} T \tag{10.21}$$

The performance of this receiver can be evaluated observing that, under H_0, the LHS of the GLRT can be written as

$$\sum_{i=1}^{r} z_i^\dagger \mathbf{P}_{\mathbf{M}^{-(1/2)} p} z_i \tag{10.22}$$

where $z_i = \mathbf{M}^{-(1/2)} r_i$, $i = 1, \ldots, r$, are i.i.d. zero-mean complex Gaussian vectors with identity covariance matrix. It follows that the decision statistic is a central χ^2 [29] with r degrees of freedom. As a consequence, P_{fa} is given by

$$P_{fa} = e^{-T} \sum_{k=0}^{r-1} \frac{T^k}{k!} \tag{10.23}$$

As to the detection performance, we observe that, given α_i, the GLRT is noncentral χ^2 distributed, with noncentrality parameter β and r degrees of freedom. This implies that the conditional P_d can be written as [30, p. 2]

$$P_d = Q_r\left(\sqrt{2\beta}, \sqrt{2T}\right) \tag{10.24}$$

Notice that in this specific case the parameter β reduces to

$$\beta = \sum_{i=1}^{r} \alpha_i^\dagger \mathbf{A}^\dagger \mathbf{M}^{-1} \mathbf{A} \alpha_i = p^\dagger \mathbf{M}^{-1} p \sum_{i=1}^{r} |q^\dagger \alpha_i|^2 \tag{10.25}$$

10.4 SPACETIME CODE DESIGN

So far the code structure (i.e., the matrix **A**) has been left as a degree of freedom, which can be exploited for performance optimization purposes. It thus ultimately depends on the performance measure of interest, as will be shortly shown.

Possible figures of merit for the considered detection problem are

1. The unconditional P_d under the assumption that A is full-rank

$$E\left[Q_{\delta r}\left(\sqrt{2\sum_{i=1}^{r}\alpha_i^{\dagger}\mathbf{A}^{\dagger}\mathbf{M}^{-1}\mathbf{A}\alpha_i},\ \sqrt{2T}\right)\right] \tag{10.26}$$

   where the expectation is over α_i, $i = 1, \ldots, r$.
2. The average received SCR per pulse and per diversity path[4]

$$\mathrm{SCR} = \frac{1}{N\delta r}E\left[\sum_{i=1}^{r}\alpha_i^{\dagger}\mathbf{A}^{\dagger}\mathbf{M}^{-1}\mathbf{A}\alpha_i\right] \tag{10.27}$$

   where, again, the expectation is over α_i, $i = 1, \ldots, r$.
3. The mutual information between the random matrix $\mathbf{R} = [r_1, \ldots, r_r]$ — the set of the available observations — and the set of the fluctuation vectors $\alpha = [\alpha_1, \ldots, \alpha_r]$, a random matrix itself [11], specifically

$$I(\alpha, \mathbf{R}) = H(\mathbf{R}) - H(\mathbf{R}|\alpha) \tag{10.28}$$

   where $H(\cdot)$ denotes differential entropy.

Before proceeding in our discussion, it is worth commenting on the figures of merit introduced above.

In principle, the most intuitive criterion to follow in designing the code matrix should be the maximization of the detection probability (10.26), a task that appears as formidable as it appears to be of dubious value. Notice, indeed, that a matrix $\mathbf{A}$ uniformly maximizing P_d is not ensured to exist; for example, it is well known that target fluctuations enhance detection for low values of P_d, while being detrimental in the large signal-to-clutter region. As a consequence, intuition suggests that, for close-to-one detection rates, an optimum code matrix $\mathbf{A}$ should try to generate as many independent and identically distributed replicas of the target as possible, so as to minimize the dispersion ratio. On the other hand, for low signal : clutter ratios the matrix $\mathbf{A}$ should try to make the dispersion ratio as large as possible, so as to grant the maximum gain from target fluctuation; for intermediate values of P_d, finally, it is expected that the optimum matrix $\mathbf{A}$ results from a compromise between the two trends above. On the basis of those arguments, a more reasonable design criterion could focus on the region of close-to-one detection rates; the lower Chernoff bound, first adopted by De Maio and Lops [11] for MIMO radar waveform optimization, is thus a natural design tool.

[4]This definition will become clear later on.

As to the SCR (10.27), notice that it represents the average signal : clutter ratio per pulse and per diversity path, as will be clarified later on. Its maximization does not appear of particular theoretical value, but could be of practical interest for real systems design and implementation. Finally, consider the mutual information (10.28). Its maximization appears particularly meaningful in the GLRT context considered here. Indeed, GLRT amounts to ML-estimating the fading coefficients from the available data, and subsequently plugging these estimates in the likelihood in place of the α_i values; optimizing (10.28) amounts exactly to requiring that the code matrix **A** be designed so as to maximize the mutual information between the available data and the quantities to be estimated.

In subsequent derivations, we assume that α_i, $i = 1, \ldots, r$, are i.i.d. zero-mean complex Gaussian vectors with scalar covariance matrix

$$E[\alpha_i \alpha_i^\dagger] = \sigma_a^2 I$$

where σ_a^2 is a real factor accounting for the backscattered backscattered useful power and I denotes the identity matrix.

10.4.1 Chernoff-Bound-Based (CBB) Code Construction

In this subsection we illustrate the main steps in deriving the code **A**, assumed full-rank, which optimizes the Chernoff bound, deferring to Ref. 11 for further analytical details.

Under hypothesis H_1, representation (10.14), involving the i.i.d. zero-mean complex Gaussian vectors z_i with covariance matrix

$$E[z_i z_i^\dagger] = I + \sigma_a^2 \mathbf{M}^{-(1/2)} \mathbf{A}\,\mathbf{A}^\dagger \mathbf{M}^{-(1/2)}$$

can be recast as

$$\sum_{i=1}^{r} \mathbf{y}_i^\dagger \left(\mathbf{I} + \sigma_a^2 \mathbf{M}^{-(1/2)} \mathbf{A}\,\mathbf{A}^\dagger \mathbf{M}^{-(1/2)}\right)^{1/2} \mathbf{P}_{\mathbf{M}^{-(1/2)}\mathbf{A}} \times \left(\mathbf{I} + \sigma_a^2 \mathbf{M}^{-(1/2)} \mathbf{A}\,\mathbf{A}^\dagger \mathbf{M}^{-(1/2)}\right)^{1/2} \mathbf{y}_i$$

$$\sum_{i=1}^{r} y_i^\dagger \left(\mathbf{I} + \sigma_a^2 \mathbf{M}^{-(1/2)} \mathbf{A}\,\mathbf{A}^\dagger \mathbf{M}^{-(1/2)}\right) \mathbf{y}_i \qquad (10.29)$$

for cases 1 and 2, respectively, where now y_i, $i = 1, \ldots, r$ are i.i.d. zero-mean complex Gaussian vectors with identity covariance matrix.

Denote by $\mathbf{U}^\dagger \boldsymbol{\Lambda} \mathbf{U}$ the eigenvalue decomposition of the matrix $\mathbf{M}^{-(1/2)} \mathbf{A}\,\mathbf{A}^\dagger \mathbf{M}^{-(1/2)}$, where **U** is an $N \times N$ unitary matrix and $\boldsymbol{\Lambda}$ is the diagonal matrix of the corresponding ordered (in decreasing order) eigenvalues:

$$\Lambda = \begin{cases} \mathrm{diag}(\lambda_1, \lambda_2, \ldots, \lambda_s, 0, \ldots, 0) & \text{(case 1)} \\ \mathrm{diag}(\lambda_1, \lambda_2, \ldots, \lambda_N) & \text{(case 2)} \end{cases} \tag{10.30}$$

Let $x_i = \mathbf{U}y_i$ and recast expression (10.29) as

$$\sum_{i=1}^{r}\sum_{j=1}^{\delta}(1+\sigma_a^2\lambda_j)|x_i(j)|^2 \tag{10.31}$$

where $x_i(j)$ denotes the jth entry of x_i, $|\cdot|$ denotes the modulus of a complex number, and the scalars $|x_i(j)|^2$, $i = 1, \ldots, r$ and $j = 1, \ldots, \delta$, are i.i.d. exponentially distributed random variables with unit mean value; that is, their first-order pdf is

$$f_{|x_i(j)|^2}(x) = e^{-x}u(x)$$

where $u(\cdot)$ is the unit step function. Since

$$P_d = P\left(\sum_{i=1}^{r}\sum_{j=1}^{\delta}(1+\sigma_a^2\lambda_j)|x_i(j)|^2 - T > 0\right) = 1 - P(X < 0) \tag{10.32}$$

where $X = \sum_{i=1}^{r}\sum_{j=1}^{\delta}(1+\sigma_a^2\lambda_j)|x_i(j)|^2 - T$, the following Chernoff bound is obtained [31]:

$$P(X < 0) \leq \min_{\gamma \geq 0} \Phi_X(\gamma)$$

where $\Phi_X(\gamma) = E[e^{-\gamma X}]$. As a consequence, the lower Chernoff bound to P_d is

$$P_d \geq 1 - \min_{\gamma\geq 0}\Phi_X(\gamma) = 1 - \min_{\gamma \geq 0}\prod_{i=1}^{r}\prod_{j=1}^{\delta}\Phi_{X_{i,j}}(\gamma) \tag{10.33}$$

with

$$\Phi_{X_{i,j}}(\gamma) = e^{\gamma T/\delta r}\frac{1}{1+\gamma(\lambda_j\sigma_a^2+1)}, \quad i = 1, \ldots, r, \ \ j = 1, \ldots, \delta, \ \ \gamma \geq 0 \tag{10.34}$$

where in (10.33) we have exploited the statistical independence of

$$X_{i,j} = (1+\sigma_a^2\lambda_j)|x_i(j)|^2 - \frac{T}{\delta r} \tag{10.35}$$

and the fact that $X = \sum_{i=1}^{r} \sum_{j=1}^{\delta} X_{i,j}$. Substitution in (10.33) finally yields

$$P_d \geq 1 - \min_{\gamma \geq 0} e^{\gamma T} \prod_{j=1}^{\delta} \left[\frac{1}{1 + \gamma(\lambda_j \sigma_a^2 + 1)} \right]^r, \quad \gamma \geq 0 \tag{10.36}$$

We observe in passing that an upper bound for P_d can be derived with nearly the same procedure. Precisely, exploiting [32, 2.1.172], we get

$$P_d = P(Z > T) \leq \min_{\gamma \in \Xi} \Phi_Z(-\gamma) \tag{10.37}$$

where $Z = \sum_{i=1}^{r} \sum_{j=1}^{\delta} (1 + \sigma_a^2 \lambda_j)|x_i(j)|^2$ and Ξ is the convergence set of $\Phi_Z(-\gamma)$. Moreover, exploiting the statistical independence of the terms in the last summation, we can recast (10.37) as

$$P_d \leq \min_{\gamma \in \Xi} e^{-\gamma T} \prod_{j=1}^{\delta} \left[\frac{1}{1 - \gamma(\lambda_j \sigma_a^2 + 1)} \right]^r \tag{10.38}$$

where $\Xi = [0, \min_{j=1,\ldots,\delta}(1/(\lambda_j \sigma_a^2 + 1))]$.

A possible design criterion is thus maximization of the minimum detection probability, namely of the right-hand side (RHS) of (10.36), under an equality constraint for the average SCR

$$\text{SCR} = \frac{1}{N\delta r} E\left[\sum_{i=1}^{r} \alpha_i^\dagger \mathbf{A}^\dagger \mathbf{M}^{-1} \mathbf{A} \alpha_i \right] = \frac{\sigma_a^2}{N\delta} \text{tr}\left(\mathbf{A}^\dagger \mathbf{M}^{-1} \mathbf{A}\right) = \frac{\sigma_a^2}{N\delta} \sum_{j=1}^{\delta} \lambda_j \tag{10.39}$$

where tr(·) denotes trace, which yields

$$\begin{cases} \min_{\lambda_1,\ldots,\lambda_\delta} \prod_{j=1}^{\delta} \left[\dfrac{1}{1 + \gamma(\lambda_j \sigma_a^2 + 1)} \right]^r \\ \dfrac{\sigma_a^2}{N\delta} \displaystyle\sum_{j=1}^{\delta} \lambda_j = \mu \end{cases} \tag{10.40}$$

or, equivalently

$$\begin{cases} \max_{\lambda_1,\ldots,\lambda_\delta} \sum_{j=1}^{\delta} \log\left[1 + \gamma(\lambda_j \sigma_a^2 + 1)\right] \\ \displaystyle\sum_{j=1}^{\delta} \lambda_j = \dfrac{\mu \delta N}{\sigma_a^2} \end{cases} \tag{10.41}$$

Careful application of the Jensen inequality allows us to derive the following conditions for the optimum code [11]. Indeed, since

$$\sum_{j=1}^{\delta} \log[1 + \gamma(\lambda_j \sigma_a^2 + 1)] \leq \delta \log\left[1 + \gamma\left(\frac{1}{\delta} \sum_{j=1}^{\delta} \lambda_j \sigma_a^2 + 1\right)\right] \tag{10.42}$$

forcing the constraint of (10.41) in the RHS of (10.42) yields

$$\sum_{j=1}^{\delta} \log[1+\gamma(\lambda_j\sigma_a^2+1)] \leq \delta \log[1+\gamma(\mu N+1)] \tag{10.43}$$

whereby equality in (10.43) is achieved if

$$\lambda_k = \frac{\mu N}{\sigma_a^2}, \qquad k=1,\ldots,\delta \tag{10.44}$$

Thus an optimum code must comply with the condition

$$\mathbf{M}^{-(1/2)}\mathbf{A}\,\mathbf{A}^{\dagger}\mathbf{M}^{-(1/2)} = \begin{cases} \dfrac{\mu N}{\sigma_a^2}\mathbf{P}_{\mathbf{M}^{-(1/2)}A} & \text{(case 1)} \\ \dfrac{\mu N}{\sigma_a^2}\mathbf{I} & \text{(case 2)} \end{cases} \tag{10.45}$$

In particular, if the additive disturbance is white (i.e., $\mathbf{M}=\sigma^2\mathbf{I}$), Eq. (10.45) reduces to

$$\mathbf{A}\,\mathbf{A}^{\dagger} = \begin{cases} \dfrac{\mu N\sigma^2}{\sigma_a^2}\mathbf{P}_A & \text{(case 1)} \\ \dfrac{\mu N\sigma^2}{\sigma_a^2}\mathbf{I} & \text{(case 2)} \end{cases} \tag{10.46}$$

Subsumed by (10.46), we find the relevant case of orthogonal space-time codes, as can be readily shown by forcing $N=r=s$, which results into the condition $\mathbf{A}\,\mathbf{A}^{\dagger}=(\mu N\sigma^2/\sigma_a^2)\mathbf{I}$: thus, the code matrix $\mathbf{A}$ should be proportional to any unitary $N\times N$ matrix; namely, any orthonormal basis of $\mathcal{C}^N$ can be exploited to construct an optimum code under case 2 and white noise. In particular, the normalized Alamouti code [33] is an orthonormal basis for $\mathcal{C}^2$, with binary codewords.

10.4.2 SCR-Based Code Construction

The average SCR has been shown to be given by (10.39)

$$\mathrm{SCR} = \frac{\sigma_a^2}{N\delta}\,\mathrm{tr}(\mathbf{M}^{-1}\mathbf{A}\,\mathbf{A}^{\dagger}) = \frac{\sigma_a^2}{N\delta}\,\mathrm{tr}(\mathbf{M}^{-1}\mathbf{Q}) \tag{10.47}$$

with $\mathbf{Q}=\mathbf{A}\mathbf{A}^{\dagger}$. SCR-based methods amount to maximizing (10.47) under a constraint on the transmitted power. In this study we consider just a constraint on the overall average transmitted power $\mathcal{P}$, even though "finer grain" constraints, concerning, for instance, the power transmitted by each antenna, could, in principle, be conceived. The dependence of the SCR on $\mathbf{Q} = \mathbf{A}\mathbf{A}^\dagger$ allows us to state the optimization problem in the following form:

$$\begin{cases} \max_Q \operatorname{tr}(\mathbf{M}^{-1}\mathbf{Q}) \\ \operatorname{rank}(\mathbf{Q}) \leq \delta \\ \operatorname{tr}(\mathbf{Q}) \leq \mathcal{P} \\ \mathbf{Q} \text{ positive semidefinite} \end{cases} \tag{10.48}$$

This problem admits a closed-form solution. To this end we observe that

$$\operatorname{tr}(\mathbf{M}^{-1}\mathbf{Q}) \leq \lambda_{\max}(\mathbf{M}^{-1})\operatorname{tr}(\mathbf{Q}) \leq \frac{\mathcal{P}}{\lambda_{\min}(\mathbf{M})}$$

with $\lambda_{\max}(\cdot)$ and $\lambda_{\min}(\cdot)$ the minimum and the maximum eigenvalue of the argument, respectively. Hence, denoting by v a unit norm eigenvector of $\mathbf{M}$ corresponding to the smallest eigenvalue, we have that equality in the above relationship holds if and only if

$$\mathbf{Q} = \mathcal{P}vv^\dagger \tag{10.49}$$

whereby

$$\mathbf{A} = \sqrt{\mathcal{P}}vu^\dagger \tag{10.50}$$

where u is a unit norm δ-dimensional complex vector.

It is interesting to note that this solution reveals that the optimum code matrix has now rank 1, a fact whose consequences are deeply discussed below. Already at this stage it can be inferred that the optimization criterion leads to exploiting just one of the δ degrees of freedom granted by the STC concept, in order to transmit all of the available energy on the "least interfered" eigenvector.

10.4.3 Mutual-Information-Based (MIB) Code Construction

Mutual information between the random matrices $\mathbf{R}$ and α, defined above, can easily be seen to be

$$\begin{aligned} I(\alpha, \mathbf{R}) &= H(\mathbf{R}) - H(\mathbf{N}) = r\log\left[\det\left(\mathbf{I} + \sigma_a^2\mathbf{M}^{-(1/2)}\mathbf{A}\mathbf{A}^\dagger\mathbf{M}^{-(1/2)}\right)\right] \\ &= r\log\left[\det\left(\mathbf{I} + \sigma_a^2\mathbf{M}^{-(1/2)}\mathbf{Q}\mathbf{M}^{-(1/2)}\right)\right] \end{aligned} \tag{10.51}$$

where $N = [n_1, \ldots, n_r]$ is the random matrix containing the noise vectors and the well-known expressions for differential entropies of $\mathbf{R}$ and $\mathbf{N}$

$$\begin{aligned} H(\mathbf{R}) &= r\log\left[(\pi e)^N \det(\mathbf{M} + \sigma_a^2\mathbf{A}\mathbf{A}^\dagger)\right] \\ H(\mathbf{N}) &= r\log\left[(\pi e)^N \det(\mathbf{M})\right] \end{aligned} \tag{10.52}$$

have been exploited.

MIB methods amount to determining $\mathbf{Q} = \mathbf{A}\mathbf{A}^{\dagger}$ so as to achieve maximization of (10.51) subject to some constraint. In keeping with the previous criteria, we consider both an equality constraint on the received average SCR, and a constraint on the overall transmitted power. The former constraint leads to the following maximization problem

$$\begin{cases} \max_{\lambda_1,\ldots,\lambda_\delta} \sum_{j=1}^{\delta} \log(1 + \sigma_a^2 \lambda_j) \\ \sum_{j=1}^{\delta} \lambda_j = \dfrac{\mu \delta N}{\sigma_a^2}. \end{cases} \tag{10.53}$$

which is equivalent to the formulation (10.41) derived through the Chernoff bound. Thus, MIB optimization and maximization of the lower Chernoff bound yield the same optimum codes if we constrain the average SCR, a result whose complete proof is contained in Ref. 11.

Constraining the overall transmitted power, instead, leads to the optimization problem:

$$\begin{cases} \max_{Q} \log[\det(\mathbf{I} + \sigma_a^2 \mathbf{M}^{-(1/2)} \mathbf{Q} \mathbf{M}^{-(1/2)})] \\ \mathrm{tr}(\mathbf{Q}) \leq \mathcal{P} \\ \mathrm{rank}(\mathbf{Q}) \leq \delta \\ \mathbf{Q} \text{ positive semidefinite} \end{cases} \tag{10.54}$$

This is equivalent to a determinant maximization with a rank constraint. An explicit solution is known when $\delta = N$. This is referred to in information theory as the "water-filling" algorithm [34] and can be found as follows. Considering the eigendecomposition of $\mathbf{M}$, $\mathbf{M} = \mathbf{U_M}\boldsymbol{\Lambda}_\mathbf{M}\mathbf{U}_\mathbf{M}^{\dagger}$ ($\mathbf{U_M}$ unitary and $\boldsymbol{\Lambda}_\mathbf{M} = \mathrm{diag}(\lambda_{1,\mathbf{M}}, \ldots, \lambda_{N,\mathbf{M}})$), and setting $\mathbf{Q}_1 = \mathbf{U}_\mathbf{M}^{\dagger}\mathbf{Q}\mathbf{U_M}$, problem (10.54) can be recast in the following form:

$$\begin{cases} \max_{Q_1} \log[\det(\mathbf{I} + \sigma_a^2 \boldsymbol{\Lambda}_\mathbf{M}^{-(1/2)} \mathbf{Q}_1 \boldsymbol{\Lambda}_\mathbf{M}^{-(1/2)})] \\ \mathrm{tr}(\mathbf{Q}_1) \leq \mathcal{P} \\ \mathbf{Q}_1 \text{ positive semidefinite} \end{cases} \tag{10.55}$$

Then observe that the optimal $\mathbf{Q}_1$ is diagonal since the off-diagonal elements of $\mathbf{Q}_1$ do not appear in the constraint but decrease the objective function (Hadamard's inequality [39, p. 477]). As a consequence, using the Lagrange technique, the optimal solution is $\mathbf{Q}_1(i, i) = \max(0, \nu - (\lambda_{i,\mathbf{M}}/\sigma_a^2))$, $i = 1, \ldots, N$, where the Lagrange multiplier ν can be determined forcing the constraint $\sum_{i=1}^{N} \max(0, \nu - (\lambda_{i,\mathbf{M}}/\sigma_a^2)) = \mathcal{P}$.

Notice that there are two situations of interest:

1. The maximum transmitted power is large enough so as to enable "filling some water" along all of the available directions: in this case the transmitter is required to exploit all N degrees of freedom.

2. The companion situation of a stringent constraint on the transmitted power results in sacrificing some degrees of freedom, with an inevitable (detrimental) effect on the amount of diversity that can be achieved.

10.5 THE INTERPLAY BETWEEN STC AND DETECTION PERFORMANCE

In order to study the impact of the optimization criterion chosen to design the code matrix **A**, we explicitly evaluate the detection probability of the GLRT detector, for the two cases of full-rank and rank 1 code matrix. As already stated, α_i, $i = 1, \ldots, r$, are assumed zero-mean circularly symmetric complex Gaussian with covariance matrix $E[\alpha_i \alpha_i^\dagger] = \sigma_a^2 \mathbf{I}$. Let us consider the CBB codes first. Exploiting representation (10.31) and condition (10.44), we can rewrite the GLRT statistic as

$$(1 + \mu N) \sum_{i=1}^{r} \sum_{j=1}^{\delta} |x_i(j)|^2 = (1 + \text{SCR}N) \sum_{i=1}^{r} \sum_{j=1}^{\delta} |x_i(j)|^2 \tag{10.56}$$

This expression highlights the fact that the decision statistic under the H_1 hypothesis is proportional to a central χ^2 random variable with δr degrees of freedom. As a consequence, the detection performance can be written as

$$P_d = e^{-(T/(1+\text{SCR}N))} \sum_{k=0}^{\delta r - 1} \frac{1}{k!} \left(\frac{T}{1 + \text{SCR}N} \right)^k \tag{10.57}$$

Notice that, for high SCR, this expression admits the following asymptotic development

$$P_d = 1 - e^{-(T/(1+\text{SCR}N))} \sum_{k=\delta r}^{\infty} \frac{1}{k!} \left(\frac{T}{1 + \text{SCR}N} \right)^k \sim 1 - \frac{1}{(\delta r)!} \left(\frac{T}{\text{SCR}N} \right)^{\delta r} \tag{10.58}$$

where $\sim$ denotes asymptotic equivalence. This relationship implies that the CBB-optimal code matrix uses all of the available degrees of freedom to achieve a diversity of order δr, that is, a miss probability that asymptotically decays with the overall SCR (which is N times the SCR per pulse pertaining to each diversity path) according to a polynomial law with exponent δr. This is achieved by pouring along each available channel (i.e., eigenvector) as much energy as necessary to achieve i.i.d. replicas of the reflected echo, namely, to create a set of equivalent diversity channels. It is also interesting to notice the interplay between N, the train length, and s, the number of transmit antennas. Indeed, while the former parameter is the one ruling the amount of coherent energy integration [as revealed by (10.57)], the latter may either be uninfluential (if $s > N$) or determine the diversity order (if $N > s$). On the other hand, the useful signal space coincides with the column span of the matrix **A** [see (10.4)], which is $\min(N, s) = \delta$: if $N > s$, this is an s-dimensional subspace of an N-dimensional space, and we have sr degrees of freedom. If, instead,

$N < s$, the number of transmit antennas, exceeds the number of available dimensions in the signal space: as a consequence, there are at most N linearly independent waveforms, implying that at most Nr distinct diversity paths can be resolved and optimally recombined at the receiver end. The conclusion is that the situation $s > N$ is of no practical interest if the diversity order is the measure of interest, in that the same performance could be achieved by using only N out of the s available sensors, which would yield exactly the same asymptotic behavior of P_d as with s transmitting antennas.

Next, consider the case that the code matrix is chosen so as to achieve maximum SCR (i.e., $\mathbf{A} = \sqrt{\mathcal{P}} v u^\dagger$, where v is the eigenvector corresponding to the smallest eigenvalue of $\mathbf{M}$ and u a unit norm s-dimensional vector). Under this setup the GLRT decision statistic can be recast as

$$\lambda_{\min}(\mathbf{M}) \sum_{i=1}^{r} |v^\dagger \mathbf{M}^{-1} r_i|^2 \tag{10.59}$$

Since $v^\dagger \mathbf{M}^{-1} r_i$, $i = 1, \ldots, r$, are independent zero-mean complex circular Gaussian random variables complying with

$$\begin{aligned} E\left[|v^\dagger \mathbf{M}^{-1} r_i|^2\right] &= v^\dagger \mathbf{M}^{-1} v (1 + \sigma_a^2 \mathcal{P} v^\dagger \mathbf{M}^{-1} v) \\ &= \lambda_{\min}^{-1}(\mathbf{M}) \left[1 + \frac{\sigma_a^2 \mathcal{P}}{\lambda_{\min}(\mathbf{M})}\right] \end{aligned} \tag{10.60}$$

we can write (10.59) as

$$\left[1 + \frac{\sigma_a^2 \mathcal{P}}{\lambda_{\min}(\mathbf{M})}\right] \sum_{i=1}^{r} h_i \tag{10.61}$$

where h_i, $i = 1, \ldots, r$, are independent exponential random variables with unit mean value (i.e., central χ^2 with one degree of freedom). It follows that the GLRT decision statistic is proportional to a central χ^2 random variable with r degrees of freedom, whereby the detection performance is given by

$$P_d = \exp\left(-\left(\frac{T}{1 + (\mathcal{P}\sigma_a^2/\lambda_{\min}(\mathbf{M}))}\right)\right) \sum_{k=0}^{r-1} \frac{1}{k!} \left(T/(1 + (\mathcal{P}\sigma_a^2/\lambda_{\min}(\mathbf{M})))\right)^k \tag{10.62}$$

Moreover, observing that $\mathrm{SCR} = (\mathcal{P}\sigma^2/(N\delta\lambda_{\min}(\mathbf{M})))$, we can rewrite (10.62) as

$$P_d = \exp\left(-\frac{T}{(1 + \mathrm{SCR}N\delta)}\right) \sum_{k=0}^{r-1} \frac{1}{k!} \left(\frac{T}{1 + \mathrm{SCR}N\delta}\right)^k \tag{10.63}$$

Asymptotically, as $\mathrm{SCR} \to \infty$, it reduces to

$$P_d \sim 1 - \frac{1}{r!} \left(\frac{T}{\mathrm{SCR}N\delta}\right)^r \tag{10.64}$$

which shows that the transmitter shoves all of the available energy in the direction wherein the disturbance power is minimum; in doing so, the MIMO radar is equivalent to a SIMO system relying on r receive antennas. This is consistent with the previous comments on the role of N and s; indeed, choosing a rank 1 code matrix amounts to using a single dimension of the $\delta = \min(N, s)$ available in the signal space. As a consequence, the same performance — in terms of both diversity order and SCR — can be obtained by simply not using $s - 1$ transmit antennas, that is, resorting to a SIMO system transmitting a waveform proportional to the eigenvector of $\mathbf{M}$ corresponding to the minimum eigenvalue.

10.6 NUMERICAL RESULTS

The objective of the present section is to illustrate the validity of the proposed encoding and detection schemes under diverse scenarios. In particular, we first assume uncorrelated disturbance, whereby orthogonal STCs are CBB-optimal; in this scenario expressions (10.57) and (10.63) have been compared to the Chernoff bounds derived in Section 10.4 and to the performance achievable through a SISO system. Next, the effect of the disturbance correlation is considered, and the impact of an optimal code choice is studied under different values of transmit/receive antenna numbers. In some cases, the behavior of the mutual information between the observations and the target replicas is also represented, showing that such a measure is itself a useful tool for system design and assessment.

In Figs. 10.2a, b the performance of the GLRT detector under white Gaussian disturbance is shown assuming $P_{fa} = 10^{-4}$ and $N = s = r = 2$. In this situation, the CBB-optimum code is any pair of orthogonal waveforms, and in particular the Alamouti code, corresponding to the matrix

$$\mathbf{A} = \begin{bmatrix} 1 & 1 \\ -1 & 1 \end{bmatrix} \tag{10.65}$$

The GLRT receiver is purely quadratic [see (10.13)], and implements the test

$$\sum_{i=1}^{r} \|r_i\|^2 \underset{H_0}{\overset{H_1}{\gtrless}} T \tag{10.66}$$

where $\|\cdot\|$ denotes the Euclidean norm of a complex vector. An uncoded transmission corresponds, instead, to the code matrix

$$\mathbf{A} = \begin{bmatrix} 1 & 1 \\ 1 & 1 \end{bmatrix} \tag{10.67}$$

and is obviously also SCR-optimum, since the noise is white and no preferred direction exists in the signal space. As to the receiver, we may in principle use the test (10.66). Alternatively, on rewriting (10.67) in the form

$$\begin{bmatrix} 1 & 1 \\ 1 & 1 \end{bmatrix} = \begin{bmatrix} 1 \\ 1 \end{bmatrix} [1 \quad 1] = pp^{\dagger} \tag{10.68}$$

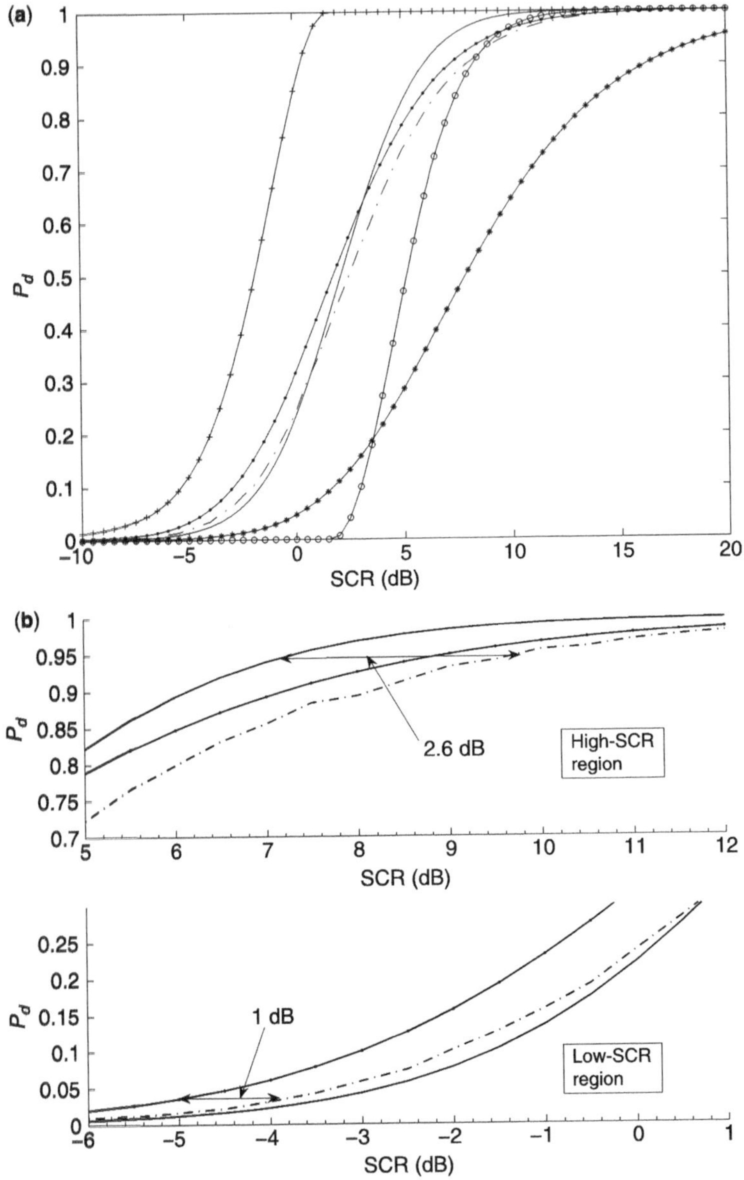

Figure 10.2 P_d versus SCR for $P_{fa} = 10^{-4}$, $N = 2$, $s = 2$, $r = 2$, and white disturbance: (a) CBB-optimal code (solid curve), uncoded with test (10.69) (dotted curve), uncoded with test (10.66) (dashed–dotted curve), upper Chernoff bound (+ curve), lower Chernoff bound (○ curve), SISO system (∗ curve); (b) high/low-SCR region, CBB-optimal code (solid curve), uncoded with test (10.69) (dotted curve), uncoded with test (10.66) (dashed–dotted curve).

and substituting in (10.21), we obtain the following test:

$$\sum_{i=1}^{r} |p^{\dagger} r_i|^2 \underset{H_0}{\overset{H_1}{\gtrless}} T \qquad (10.69)$$

The curves highlight that CBB-optimum waveforms achieve a significant performance gain with respect to both the uncoded and the SISO systems. Precisely, for $P_d = 0.9$, the performance gain over an uncoded transmission format is ~2.6 dB, if test (10.66) is employed, and in the order of 2 dB, if the test (10.69) is used; the advantage over SISO systems is instead in the order of 10 dB. Uncoded systems are instead advantageous for low–medium detection probabilities, as shown in the lower curve of Fig. 10.2b. On the other hand, detection theory predicts that less and less constrained fluctuations are beneficial in the high-SCR region but are detrimental in the low-SCR region; CBB optimization is aimed at reducing the target amplitude fluctuation, and the trends shown in Fig. 10.2b are thus in perfect agreement with theory.

To reinforce the conclusion above, we refer to Fig. 10.3, where the mutual information between the observables and target parameters is represented as a function of SCR. The mutual information corresponding to a random coding, where the code matrix entries are independent realizations of a uniformly distributed random variable within the interval [0, 1], is also plotted in Fig. 10.3. As predicted by theory, such information takes on its maximum — for any SCR — if orthogonal

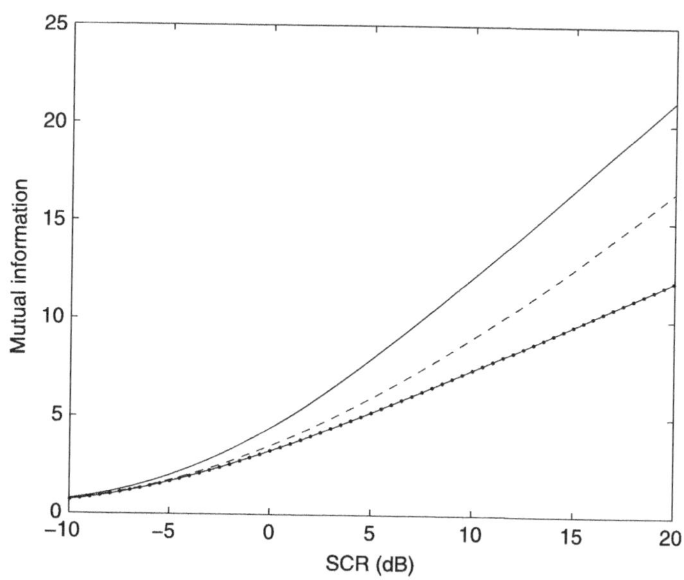

Figure 10.3 Mutual information (in nats) versus SCR for $N = 2$, $s = 2$, $r = 2$, and white disturbance. Optimum CBB code (solid curve), uncoded with test (10.69) (dotted curve), random coding (dashed curve).

waveforms are transmitted; this trend, along with the curves of Fig 10.2, confirms the theory that the mutual information between the observations and the channel vectors is intimately tied to detection performance, at least when GLRT strategies are employed.

The effect of the disturbance correlation is elicited in Fig. 10.4, where the analysis of Fig. 10.2 is reproduced assuming an overall disturbance with exponentially shaped covariance matrix, whose one-lag correlation coefficient ρ is set to 0.95. The remaining parameters of the plots are equal to those of Fig. 10.2. In this case the Alamouti code is no longer optimum. The plots show that the performance gain of the optimum coded systems over both the uncoded and the SISO detector is almost equal to that corresponding to white disturbance. On the other hand, setting $N=r=s=2$ in Eq. (10.45) shows that, under correlated disturbance, the optimum CBB code matrix is proportional to $\mathbf{M}^{1/2}$, specifically, an optimal CBB code tends to restore the "white disturbance condition." This also explains why the conventional Alamouti code follows rather closely the performance of the uncoded system.

The effect of the number r of receive antennas on the performance is analyzed in Fig. 10.5, where P_d is plotted versus SCR for $N=s=8$, exponentially shaped clutter covariance matrix with $\rho=0.95$, and several values of r. The curves highlight that the higher r; namely, the higher the diversity order, the better the performance. Specifically, the performance gap between the case $r=4$ and the case of a MISO

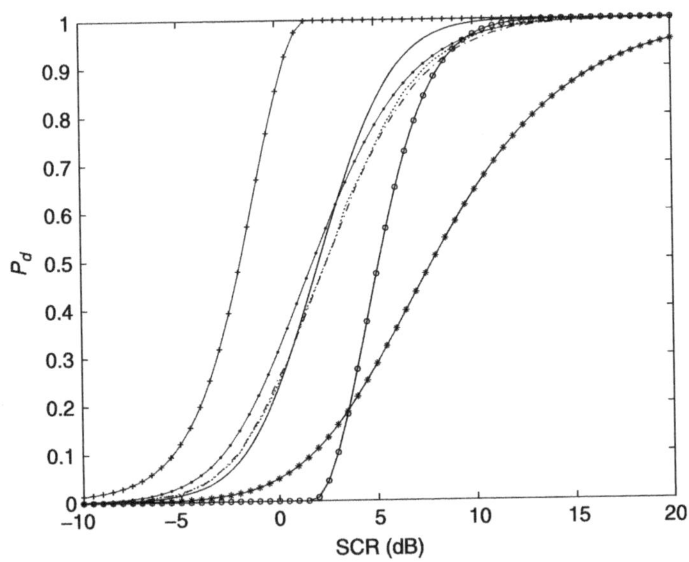

Figure 10.4 P_d versus SCR for $P_{fa}=10^{-4}$, $N=2$, $s=2$, $r=2$, and disturbance with exponentially shaped covariance matrix ($\rho=0.95$); optimum CBB code (solid curve), optimum SCR-based code (dotted curve), uncoded system (dashed–dotted curve), Alamouti code (dashed curve), upper Chernoff bound (+ curve), lower Chernoff bound (○ curve), SISO system (∗ curve).

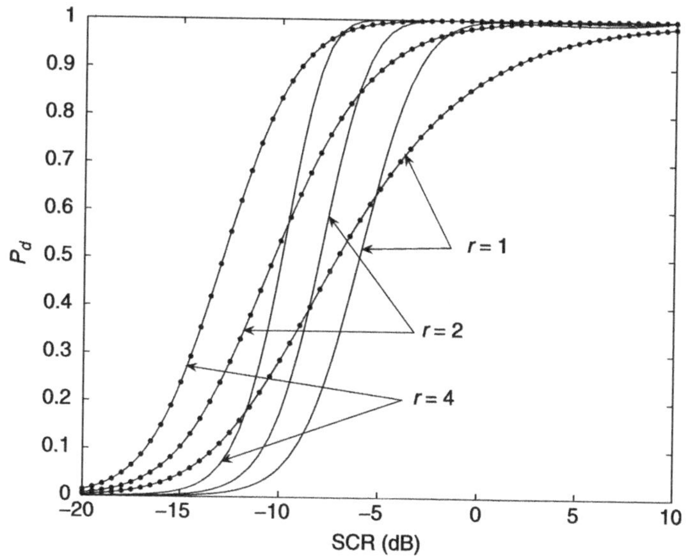

Figure 10.5 P_d of optimum coded systems versus SCR for $P_{fa} = 10^{-4}$, $N = 8$, $s = 8$, disturbance with exponentially shaped covariance matrix ($\rho = 0.95$), and several values of r; optimum CBB code (solid curve), optimum SCR-based code (dotted curve).

radar system (i.e., $r = 1$) is for $P_d = 0.9$ about 5 dB for CBB coded system, and about 10 dB with reference to the SCR-based one. Notice also that the higher the r value the higher the P_d value where the curves of CBB-optimal and SCR-optimal systems cross each other.

In Fig. 10.6 we study the impact of the number s of available transmit antennas on the system performance. To this end, we assume $N = 8$, $r = 2$, exponentially shaped clutter covariance matrix with $\rho = 0.95$, and several values of s. As expected, the plots show that by increasing the number of transmit antennas, namely, increasing the transmit diversity order, increasingly improved detection performances can be achieved. In particular, the gain of a MIMO radar exploiting eight transmit antennas over a SIMO radar, with the same number of receive antennas, is, for $P_d = 0.9$, in the order of 10 dB for the CBB-coded system, and about 9 dB with reference to the SCR-based one.

The monotonic dependence of P_d on the number of transmit antennas s corresponds to a similar behavior of the mutual information. This is shown in Fig. 10.7, where we assume the same simulation setup as in Fig. 10.6 and plot the mutual information versus SCR for several values of s. From the curves we can observe that the higher the s value, the higher the amount of information that the observables carry about the target. Otherwise stated, for increasingly large s, an increasing amount of information on fading affecting prospective targets can be extracted from the received echoes, leading to more significant performance improvements.

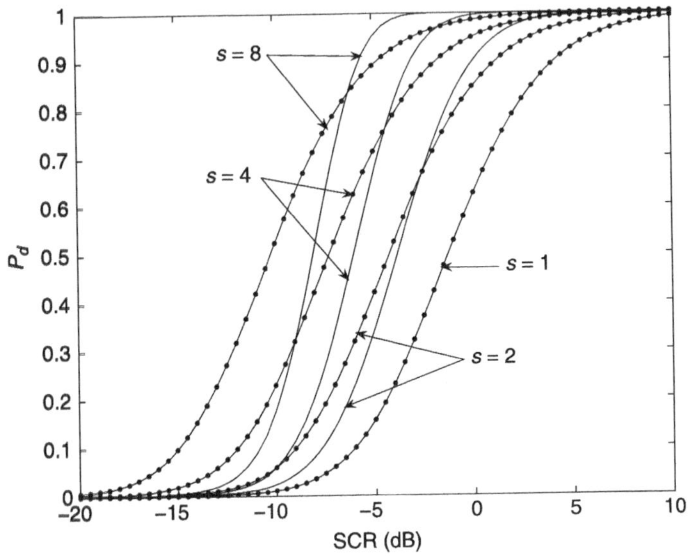

Figure 10.6 P_d of optimum coded systems versus SCR for $P_{fa} = 10^{-4}$, $N = 8$, $r = 2$, disturbance with exponentially shaped covariance matrix ($\rho = 0.95$), and several values of s; optimum CBB code (solid curve), optimum SCR-based code (dotted curve).

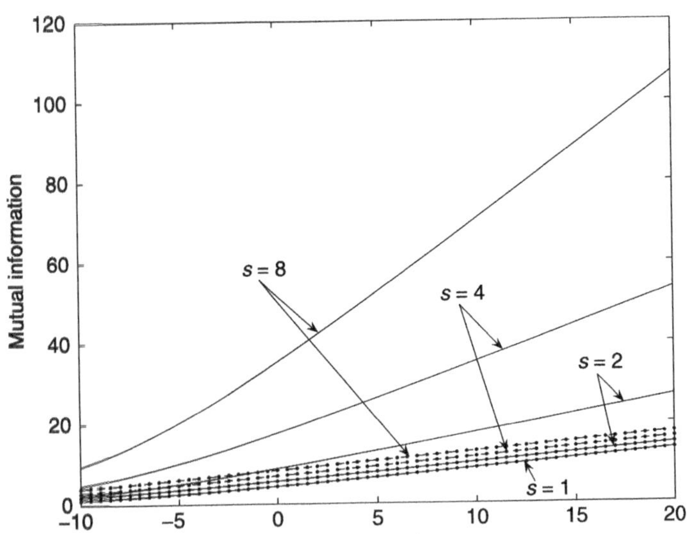

Figure 10.7 Mutual information (in nats) versus SCR for $N = 8$, $r = 2$, disturbance with exponentially shaped covariance matrix ($\rho = 0.95$), and several values of s; optimum CBB code (solid curve), optimum SCR-based code (dotted curve).

10.7 ADAPTIVE IMPLEMENTATION

In this section we suggest an adaptive implementation of the coding–detection procedure based on the assumption that the disturbance spectral properties are constant in two consecutive radar scans (referred to as scans 1 and 2) of the cell under test. We also suppose that each receive sensor collects a secondary dataset of size K, in both scan 1 and scan 2, that the snapshots of the training sets do not contain any useful signal, and exhibit the same covariance matrix as do the data from the cell under test (primary data). Otherwise stated, we suppose that following training data matrices are available:

$$\begin{aligned} \mathbf{Z}_{i,1} &= [r_{i,1,1}, \ldots, r_{i,K,1}] \quad \text{(scan 1)} \\ & \qquad\qquad\qquad\qquad\qquad (i = 1, \ldots, r) \\ \mathbf{Z}_{i,2} &= [r_{i,1,2}, \ldots, r_{i,K,2}] \quad \text{(scan 2)} \end{aligned}$$

Starting from $\mathbf{Z}_{i,1}$ and $\mathbf{Z}_{i,2}$, we construct the sample covariance matrices

$$\mathbf{S}_1 = \frac{1}{rK}\sum_{i=1}^{r} \mathbf{Z}_{i,1}\,\mathbf{Z}_{i,1}^{\dagger}$$

$$\mathbf{S} = \frac{1}{2rK}\sum_{i=1}^{r} (\mathbf{Z}_{i,1}\,\mathbf{Z}_{i,1}^{\dagger} + \mathbf{Z}_{i,2}\,\mathbf{Z}_{i,2}^{\dagger})$$

The former is the sample covariance matrix of the training data from scan 1 and is available before the transmission of the second scan. The latter covariance estimate exploits the data from both scans and is available at the end of the second scan.

The adaptive implementation of the coding–detection procedure is as follows:

- Using the matrix $\mathbf{S}_1$ in place of the true covariance $\mathbf{M}$, determine the STC to be transmitted during scan 2 (either CBB or SCR-based construction).
- Construct the adaptive GLRT (AGLRT) substituting $\mathbf{S}$ in place of $\mathbf{M}$ in the GLRT (10.12) if $N > s$ and CBB coding, in (10.13) if $N \leq s$ and CBB coding, or in (10.21) for SCR-based coding.

$$\sum_{i=1}^{r} r_i^{\dagger}\mathbf{S}^{-1}\mathbf{A}(\mathbf{A}^{\dagger}\mathbf{S}^{-1}\mathbf{A})^{-1}\mathbf{A}^{\dagger}\mathbf{S}^{-1}r_i \underset{H_0}{\overset{H_1}{\gtrless}} T \qquad (N > s \text{ CBB coding}) \tag{10.70}$$

or

$$\sum_{i=1}^{r} r_i^{\dagger}\mathbf{S}^{-1}r_i \underset{H_0}{\overset{H_1}{\gtrless}} T \qquad (N \leq s \text{ CBB coding}) \tag{10.71}$$

or

$$\sum_{i=1}^{r} \frac{|p^{\dagger}\mathbf{S}^{-1}r_i|^2}{p^{\dagger}\mathbf{S}^{-1}p} \underset{H_0}{\overset{H_1}{\gtrless}} T \qquad \text{(SCR-based coding)} \tag{10.72}$$

Notice that the AGLRT ensures the CFAR property with respect to the disturbance covariance matrix. In fact, under the H_0 hypothesis, for $N > s$ and CBB coding, the decision statistic is the sum of subspace adaptive matched filters (AMFs) decision statistics whose distribution, under the H_0 hypothesis, is functionally independent of the actual disturbance covariance matrix [35,36]. If $N \leq s$ and CBB coding is used, the decision statistic can be written as

$$\sum_{i=1}^{r} r_i^{\dagger} \mathbf{S}^{-1} r_i = \operatorname{tr}\left(\mathbf{S}^{-1/2}\mathbf{R}\mathbf{R}^{\dagger}\mathbf{S}^{-1/2}\right)$$

namely, as the sum of the eigenvalues of a complex central F distributed random matrix [37], whose joint probability density function is functionally independent of the actual disturbance covariance matrix $\mathbf{M}$. Finally, with reference to SCR-based coding, the decision statistic coincides with the generalized AMF (GAMF) whose CFAR behaviour is proved by Conte et al. [38].

The performance of the adaptive algorithm is studied in Figs. 10.8 and 10.9, which assume $P_{fa} = 10^{-4}$, $N = s$, and disturbance with exponentially shaped covariance matrix ($\rho = 0.95$). Figure 10.8 refers to $N = 2$, $r = 2$; Fig. 10.9, to $N = 8$, $r = 4$. The plots show that by increasing the number of available secondary data, better performances can be obtained. In particular, for $N = 8$ and $K = 16$, the loss of the adaptive implementations with respect to the nonadaptive counterparts is smaller

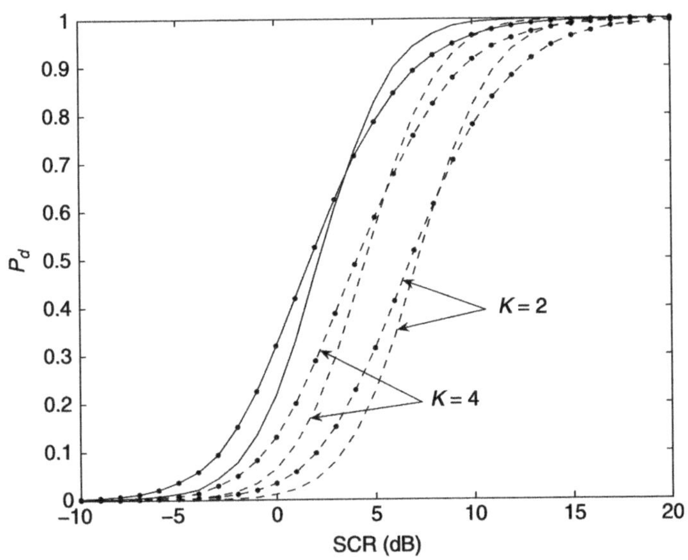

Figure 10.8 P_d versus SCR for $P_{fa} = 10^{-4}$, $N = 2$, $s = 2$, $r = 2$, several values of K, and disturbance with exponentially shaped covariance matrix ($\rho = 0.95$); optimum CBB code (solid curve), adaptive implementation of the optimum CBB system (dashed curve); optimum SCR-based system (dotted curve), adaptive implementation of the optimum SCR-based system (dotted dashed curve).

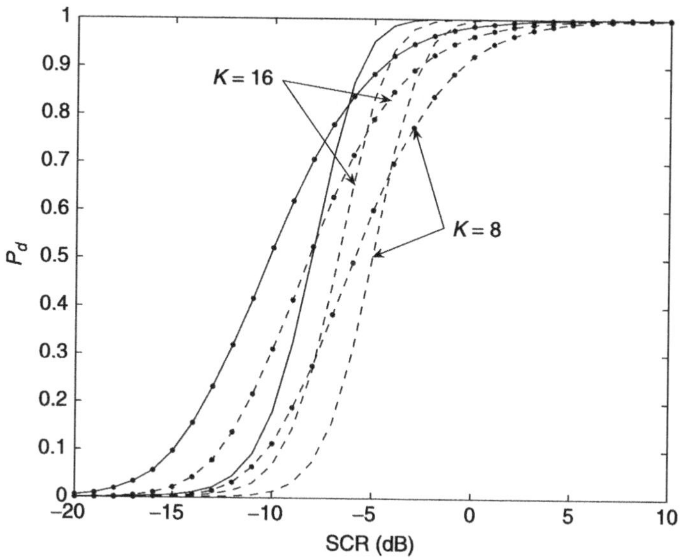

Figure 10.9 P_d versus SCR for $P_{fa} = 10^{-4}$, $N = 8$, $s = 8$, $r = 4$, several values of K, and disturbance with exponentially shaped covariance matrix ($\rho = 0.95$); optimum CBB code (solid curve), adaptive implementation of the optimum CBB system (dashed curve); optimum SCR-based system (dotted curve), adaptive implementation of optimum SCR-based system (dotted dashed curve).

than 2.5 dB for $P_d = 0.9$. As a rule of thumb, $K = 2N$ training data are required in order to achieve a satisfactory performance level (i.e., <3 dB from the benchmark performance). It is also interesting to note that the smaller the number of available training data, the lower the P_d value where crossing occurs between the curves of CBB-optimal and SCR-optimal adaptive procedures.

A possible generalization of the proposed algorithm relies on the use, at scan L, of an exponentially weighted covariance estimate exploiting the secondary data available up to scan L. Otherwise stated, the covariance estimate at the Lth scan is given by

$$\mathbf{S}(L) = \sum_{i=1}^{L} \lambda_F^{L-i} \mathbf{S}_1(i) \tag{10.73}$$

where $\mathbf{S}_1(i)$ is the sample covariance matrix of the secondary data collected during the ith scan and λ_F is the "forgetting" factor, which rules the speed of adaptation. Under this setup the adaptive coding–detection procedure can be formulated as follows:

- Using the matrix $\mathbf{S}(L-1)$ in place of the true covariance $\mathbf{M}$, determine the STC to be transmitted during scan $L(L > 1)$.
- Construct the covariance estimate

$$\mathbf{S}(L) = \lambda_F \mathbf{S}(L-1) + \mathbf{S}_1(L),$$

and evaluate the exponentially weighted AGLRT (EW-AGLRT) as

$$\sum_{i=1}^{r} r_i^\dagger \mathbf{S}^{-1}(L)\mathbf{A}[\mathbf{A}^\dagger \mathbf{S}^{-1}(L)\mathbf{A}]^{-1}\mathbf{A}^\dagger \mathbf{S}^{-1}(L) r_i \underset{H_0}{\overset{H_1}{\gtrless}} T \tag{10.74}$$

if $N > s$ and CBB coding, or

$$\sum_{i=1}^{r} r_i^\dagger \mathbf{S}^{-1}(L) r_i \underset{H_0}{\overset{H_1}{\gtrless}} T \qquad \text{if } (N \leq s \text{ CBB coding}) \tag{10.75}$$

or

$$\sum_{i=1}^{r} \frac{|p^\dagger \mathbf{S}^{-1}(L) r_i|^2}{p^\dagger \mathbf{S}^{-1}(L) p} \underset{H_0}{\overset{H_1}{\gtrless}} T \qquad \text{(SCR-based coding)} \tag{10.76}$$

It is worth pointing out that in operative radar environments, where the spectral properties of the disturbance exhibit temporal variations, the EW-AGLRT no longer ensures the CFAR property with respect to the disturbance covariance matrix. Nevertheless, for the special case where the disturbance covariance remains constant during the training phase of data collection (i.e., L scans), the CFAR property is still maintained.

To prove this claim, we first focus on the case $N > s$ and CBB coding. Hence we write the ith term of the summation in (10.74) as

$$z_i^\dagger \mathbf{S}_w^{-1}(L)\mathbf{M}^{-1/2}\dot{\mathbf{A}}[\mathbf{A}^\dagger \mathbf{M}^{-1/2}\mathbf{S}_w^{-1}(L)\mathbf{M}^{-1/2}\mathbf{A}]^{-1}\mathbf{A}^\dagger \mathbf{M}^{-1/2}\mathbf{S}_w^{-1}(L) z_i \tag{10.77}$$

where, as in Section 10.3, we obtain

$$z_i = \mathbf{M}^{-1/2} r_i$$

and

$$\mathbf{S}_w(L) = \mathbf{M}^{-1/2}\mathbf{S}(L)\mathbf{M}^{-1/2}$$

Moreover, let

$$\mathbf{M}^{-1/2}\mathbf{A} = \mathbf{U}_w \mathbf{\Lambda}_w \mathbf{V}_w^\dagger$$

be the singular value decomposition of $\mathbf{M}^{-1/2}\mathbf{A}$, where $\mathbf{U}_w$ is a $N \times s$ matrix with orthonormal columns, $\mathbf{\Lambda}_w$ is the diagonal matrix of the singular values, and $\mathbf{V}_w$ is a unitary $s \times s$ matrix. Hence we recast (10.77) as

$$z_i^\dagger \mathbf{S}_w^{-1}(L)\mathbf{U}_w[\mathbf{U}_w^\dagger \mathbf{S}_w^{-1}(L)\mathbf{U}_w]^{-1}\mathbf{U}_w^\dagger \mathbf{S}_w^{-1}(L) z_i \tag{10.78}$$

Now, denote by $\mathbf{G}$ a $N \times N$ unitary matrix such that

$$\mathbf{G}\mathbf{U}_w = \begin{bmatrix} \mathbf{I} \\ 0 \end{bmatrix}$$

with 0 an $(N - s) \times s$ matrix of zeros, and write (10.78) as

$$z_{i,G}^{\dagger}\mathbf{S}_{w,G}^{-1}(L)\begin{bmatrix} \mathbf{I} \\ 0 \end{bmatrix}\left(\begin{bmatrix} \mathbf{I} \\ 0 \end{bmatrix}^{\dagger}\mathbf{S}_{w,G}^{-1}(L)\begin{bmatrix} \mathbf{I} \\ 0 \end{bmatrix}\right)^{-1}\begin{bmatrix} \mathbf{I} \\ 0 \end{bmatrix}^{\dagger}\mathbf{S}_{w,G}^{-1}(L)z_{i,G} \qquad (10.79)$$

where

$$z_{i,G} = \mathbf{G}z_i$$

and

$$\mathbf{S}_{w,G}(L) = \mathbf{G}\mathbf{S}_w(L)\mathbf{G}^{\dagger}$$

This expression highlights the fact that the decision statistic depends on the data only through $z_{i,G}$ and $\mathbf{S}_{w,G}$, which are statistically independent and whose distributions, under the H_0 hypothesis, do not functionally depend on the disturbance covariance $\mathbf{M}$. Indeed, under H_0, $z_{i,G}$ is a zero-mean complex circular white Gaussian vector and

$$\mathbf{S}_{w,G} = \sum_{i=1}^{L} \lambda_F^{L-i}\mathbf{G}\mathbf{M}^{-1/2}\mathbf{S}_1(i)\mathbf{M}^{-1/2}\mathbf{G}^{\dagger}$$

is a weighted sum of independent complex Wishart matrices [37], with diagonal matrix parameter functionally independent of $\mathbf{M}$. The CFAR property is thus proved. A similar proof can be developed with reference to the case $N \leq s$ and CBB coding as well as with reference to the SCR-based coding.

10.8 CONCLUSIONS

In this chapter we have addressed the synthesis and the analysis of MIMO radar detectors exploiting STC. To this end, after a short description of the MIMO radar signal model, we have devised the GLRT detector under the assumption of additive Gaussian disturbance.

Remarkably, the decision statistic is *ancillary*, and, as a consequence, CFAR property is ensured; specifically, the detection threshold can be set independent of the disturbance spectral properties. We have also assessed the performance of the GLRT detector providing closed-form expressions for both P_{fa} and P_d. Lacking a manageable expression for P_d under arbitrary target fluctuation models, we restricted our attention to the case of Rayleigh-distributed amplitude fluctuation; code optimization has been shown to be feasible under different criteria, the Chernoff bound, SCR optimization, and the maximization of the mutual information between received signals and fading coefficients, which reassuringly lead to the same optimality condition as the Chernoff bound.

The performance assessment, which has been undertaken under several instances of number of receive and transmit antennas, and of clutter covariance, has confirmed that MIMO radars with a suitable STC achieve significant performance gains over SIMO, MISO, or conventional SISO systems.

Finally, we have developed an adaptive implementation of the proposed coding–detection algorithms, which ensure the CFAR property with respect to the disturbance covariance matrix, exhibit a satisfactory performance level, and, for increasingly large size of the training set, converge to their nonadaptive counterparts (i.e., those assuming perfect knowledge of the clutter spectral properties).

Future research might concern the exploitation of the transmitted pulses shape as a further degree of freedom for performance optimization, for example, in the framework of joint target detection and parameter estimation. On a parallel track, of primary importance appears to be the issue of joint transmit/receive structures design under the very common situation of non-Gaussian radar clutter as well as wide distributed apertures, where the farfield target hypothesis can be no longer met [40].

ACKNOWLEDGMENT

The authors gratefully acknowledge Dr. E. Biglieri, Dr. G. Foglia, Dr. E. Grossi and Dr. L. Venturino for careful reading of the manuscript and useful suggestions.

REFERENCES

1. I. E. Telatar, Capacity of multi-antenna Gaussian channels, *Eur. Trans. Commun.* **10**:585–595 (Nov. 1999).
2. G. J. Foschini, Layered space-time architecture for wireless communication in a fading environment using multi-element antennas, *Bell Labs. Tech. J.* **1**:41–59 (1996).
3. V. Tarokh, N. Seshadri, and A. R. Calderbank, Space-time codes for high data-rate wireless communication: Performance criterion and code construction, *IEEE Trans. Inform. Theory* **44**:744–765 (March 1998).
4. L. Zheng and D. N. C. Tse, Diversity and multiplexing: a fundamental tradeoff in multiple-antenna channels, *IEEE Trans. Inform. Theory* **49**:1073–1096 (March 2003).
5. L. Xu and J. Li, Iterative generalized-likelihood ratio test for MIMO radar, *IEEE Trans. Signal Process.* **55**:2375–2385 (June 2007).
6. E. Fishler, A. Haimovich, R. Blum, L. Cimini, D. Chizhik, and R. Valenzuela, MIMO radar: An idea whose time has come, *Proc. 2004 IEEE Radar Conf.*, April 2004, pp. 71–78.

7. E. Fishler, A. Haimovich, R. Blum, L. Cimini, D. Chizhik, and R. Valenzuela, Performance of MIMO radar systems: Advantages of angular diversity, *Proc. 38th Asilomar Conf. Signals, Systems and Computers*, Pacific Grove, CA, Nov. 2004, pp. 305–309.

8. E. Fishler, A. Haimovich, R. Blum, L. Cimini, D. Chizhik, and R. Valenzuela, Spatial diversity in radars—models and detection performance, *IEEE Trans. Signal Process.* **54**:823–838 (March 2006).

9. A. De Maio and M. Lops, Design principles of MIMO radar detectors, *Proc. Waveform Diversity and Design Conf.*, Lihue, HI, Jan. 2006.

10. A. De Maio and M. Lops, Achieving full diversity in MIMO radar: Code construction and performance bounds, *Proc. 7th Int. Radar Symp. (IRS)*, May 2006, pp. 153–156.

11. A. De Maio and M. Lops, Design principles of MIMO radar detectors, *IEEE Trans. Aerospace Electron. Syst.* (July 2007).

12. P. F. Sammartino, C. J. Baker, and H. D. Griffiths, Target model effects on MIMO radar performance, *Proc. 2006 IEEE Int. Conf. Acoustic, Speech and Signal Processing*, Toulouse, France, 2006.

13. C. Y. Chen and P. P. Vaidyanathan, Beamforming issues in modern MIMO radars with Doppler, *Proc. 40th Asilomar Conf. Signals, Systems and Computers*, Pacific Grove, CA, Oct.–Nov. 2006, pp. 41–45.

14. A. S. Fletcher and F. C. Robey, Performance bounds for adaptive coherence of sparse array radar, *Proc. 11th Workshop on Adaptive Sensor Array Processing (ASAP)*, MIT Lincoln Laboratory, Lexington, MA, March 2003.

15. D. J. Rabideau and P. Parker, Ubiquitous MIMO multifunction digital array radar, *Proc. 37th Asilomar Conf. Signals, Systems and Computers*, Pacific Grove, CA, Nov. 2003, pp. 1057–1064.

16. D. R. Fuhrmann and G. San Antonio, Transmit beamforming for MIMO radar systems using partial signal correlation, *Proc. 38th Asilomar Conf. Signals, Systems and Computers*, Pacific Grove, CA, Nov. 2004, 295–299.

17. F. C. Robey, S. Coutts, D. Weikle, J. C. McHarg, and K. Cuomo, MIMO radar theory and experimental results, *Proc. 38th Asilomar Conf. Signals, Systems and Computers*, Pacific Grove, CA, Nov. 2004, pp. 300–304.

18. D. W. Bliss and K. W. Forsythe, Multiple-input multiple-output (MIMO) radar and imaging: Degrees of freedom and resolution, *Proc. 37th Asilomar Conf. Signals, Systems and Computers*, Pacific Grove, CA, Nov. 2003, pp. 54–59.

19. G. San Antonio, D. R. Fuhrmann, and F. C. Robey, MIMO radar ambiguity functions, *IEEE J. Select. Topics Signal Process.* **1**:167–177 (June 2007).

20. J. Li, P. Stoica, and Y. Xie, On probing signal design in MIMO radar, *Proc. 40th Asilomar Conf. Signals, Systems and Computers*, Pacific Grove, CA, Oct.–Nov. 2006, pp. 31–35.

21. L. Xu, J. Li, and P. Stoica, Adaptive techniques for MIMO radar, *Proc. 4th IEEE Workshop on Sensor Array and Multichannel Signal Processing*, Waltham, MA, July 2006, pp. 258–262.

22. I. Bekkerman and J. Tabrikian, Target detection and localization using MIMO radars and sonars, *IEEE Trans. Signal Process.* **54**:3873–3883 (Oct. 2006).

23. N. Lehman, E. Fishler, A. Haimovich, R. Blum, D. Chizhik, L. Cimini, and R. Valenzuela, Evaluation of transmit diversity in MIMO-radar direction finding, *IEEE Trans. Signal Process.* **55**:2215–2225 (May 2007).

24. Y. Yang and R. Blum, MIMO radar waveform design based on mutual information and minimum-mean square error estimation, *IEEE Trans. Aerospace Electron. Syst.* **43**:330–343 (Jan. 2007).

25. D. Guo, S. Shamai, and S. Verdu, Mutual information and minimum mean-square error in Gaussian channels, *IEEE Trans. Inform. Theory* **51**:1261–1282 (April 2005).

26. J. Li, L. Xu, P. Stoica, D. Bliss, and K. Forsythe, Range compression and waveform optimization for MIMO radar: A Cramer-Rao bound based study, *IEEE Trans. Signal Process.* **56**(1):218–232 (Jan. 2008).

27. P. Calvary and D. Janer, Spatio-temporal coding for radar array processing, *Proc. 1998 IEEE Int. Conf. Acoustics, Speech, and Signal Processing*, May 1998, pp. 2509–2512.

28. H. L. Van Trees, *Detection, Estimation, and Modulation Theory*, Part 1, Wiley, New York, 1968.

29. E. J. Kelly and K. Forsythe, *Adaptive Detection and Parameter Estimation for Multidimensional Signal Models*, Technical Report 848, MIT Lincoln Laboratories, April 19, 1989.

30. J. Omura and T. Kailath, *Some Useful Probability Distributions*, Technical Report 7050-6, Stanford Electronics Laboratories, Stanford University, Stanford, CA, 1965.

31. S. Benedetto and E. Biglieri, *Principle of Digital Transmission with Wireless Applications*, Kluwer Academic/Plenum Publishers, New York, 1999.

32. J. G. Proakis, *Digital Communications*, 3rd ed., McGraw-Hill, New York, 1995.

33. S. M. Alamouti, A simple transmit diversity technique for wireless communications, *IEEE J. Select. Areas Commun.* **16**:1451–1458 (Oct. 1998).

34. T. M. Cover and J. A. Thomas, *Elements of Information Theory*, Wiley Series in Telecommunications, 1991.

35. R. S. Raghavan, N. Pulsone, and D. J. McLaughlin, Performance of the GLRT for adaptive vector subspace detection, *IEEE Trans. Aerospace Electron. Syst.* **32**:1473–1487 (Oct. 1996).

36. S. Kraut, L. L. Scharf, and L. T. McWhorter, Adaptive subspace detectors, *IEEE Trans. Signal Process.* **49**:1–16 (Jan. 2001).

37. A. T. James, Distribution of matrix variates and latent roots derived from normal samples, *Ann. Math. Stat.* **35**:475–501 (June 1964).

38. E. Conte, A. De Maio, and G. Ricci, GLRT-based adaptive detection algorithms for range-spread targets, *IEEE Trans. Signal Process.* **49**:1336–1348 (July 2001).

39. R. A. Horn and C. R. Johnson, *Matrix Analysis*, Cambridge University Press, 1985.

40. R. Adve and L. Landi, Time-orthogonal-waveform-space-time adaptive processing for distributed aperture radars, *Proc. Waveform Diversity and Design Conf.*, Pisa, Italy, June 2007.

INDEX

MIMO Radar Signal Processing, edited by Jian Li and Petre Stoica

Printed and bound by CPI Group (UK) Ltd, Croydon, CR0 4YY
18/09/2026
14964970-0002